国家自然科学基金项目（42176080，42076073）
山东省泰山学者攀登计划项目（tspd20181216）
中国科学院南海海洋研究所自主部署项目（SCSIO2023PT10, SCSIO202201）
中国 IODP 办公室（同济大学中央高校基本科研业务费）（22120230087）
资助出版

南极罗斯海的新生代放射虫

张兰兰　张　强　陈木宏　著

科学出版社

北　京

内 容 简 介

本书是南极罗斯海新生代放射虫化石群的综合性研究最新成果。首次建立了南极罗斯海较为完整的新生代放射虫生物地层年代框架，为地层界线的厘定提供了化石种的初现面与末现面以及组合变化的证据；与南大洋已有的相关资料进行对比，分析了放射虫发育在南极海区与南大洋海区新生代地质历史中的阶段性差异特征。书中还较为全面地分析了新生代以来罗斯海放射虫的种类组成与系统分类，共发现放射虫 37 科 138 属 502 种，建立 4 个新属和 126 个新种，均作了详细的古生物学描述，列出同物异名表及讨论，并提供全部种类的标本照相图版共 154 幅及其说明，所有种类照片均按照分类系统的顺序在图版中进行编排，以方便读者参考使用。本书填补了南极海区放射虫分类系统及生物地层的研究空白，提供新认识与新证据，对深入了解该海区新生代放射虫的演变过程及其对全球气候变化和构造运动的响应具有重要的科学意义与应用价值。

本书可供海洋地质学、海洋生物学、古生物学和极地研究等相关领域工作者以及高等院校师生参考。

图书在版编目（CIP）数据

南极罗斯海的新生代放射虫/张兰兰，张强，陈木宏著. —北京：科学出版社，2023.11

ISBN 978-7-03-077067-7

Ⅰ. ①南… Ⅱ. ①张… ②张… ③陈… Ⅲ. ①罗斯海-放射虫目-动物化石-研究 Ⅳ. ①Q915.2

中国国家版本馆 CIP 数据核字（2023）第 220937 号

责任编辑：孟美岑/责任校对：何艳萍

责任印制：赵 博/封面设计：北京图阅盛世

科学出版社 出版

北京东黄城根北街 16 号

邮政编码：100717

http://www.sciencep.com

北京建宏印刷有限公司印刷

科学出版社发行 各地新华书店经销

*

2023 年 11 月第 一 版 开本：787×1092 1/16

2024 年 8 月第二次印刷 印张：18 1/2 插页：78

字数：670 000

定价：375.00 元

（如有印装质量问题，我社负责调换）

Cenozoic Radiolarians from the Ross Sea, Antarctic

Zhang Lanlan, Zhang Qiang and Chen Muhong

Science Press
Beijing

前　言

南极海区是指在南极圈内（66°33′39″S 以南）的海洋区域，代表了地球上最为寒冷的气候环境，具有典型的生物生态条件、生物群落组合与生物地理特征；同时，也是放射虫系统分类研究最为欠缺的区域之一。

罗斯海位于南极圈内，是辽阔太平洋南端的最高纬度的边缘海，完全深入南极地区，其南部海区被陆缘冰架所覆盖，具有典型的极地环境特征；同时又与南太平洋紧密相连，海洋环境和水团发育在地质历史中相互影响与作用，是探索南太平洋古海洋及极地气候历史演变的关键区域。放射虫作为微体古生物的一个重要门类，是开展高纬度海区生物地层研究的绝佳材料，也是重建过去水团活动和古生态环境变化的有效工具。目前，人们对南极罗斯海新生代放射虫的系统分类、种类组成、群落演变、生物地理，及其生物地层特征的了解仍较欠缺，制约了放射虫化石在海洋地质、生物地理和全球变化中的应用程度。开展南极罗斯海放射虫的种类组成和系统分类研究是一项关键的基础性工作，有助于深入了解南极圈内海区的放射虫多样性与历史变化规律，对进一步探讨太平洋-南极的海洋环境以及生物演变的动力特征与过程将可发挥特殊的作用，具有重要的科学意义。

此项研究得到国际大洋钻探计划组织（Gulf Coast Repository of International Ocean Discovery Program）的支持并提供了库存样品；感谢美国得克萨斯农工大学菲尔·拉姆福德（Phil Rumford）博士和日本高知大学高知岩心中心的拉利亚纳·古普塔（Lallan Gupta）博士帮助采集所需岩心样品；特别感谢中国大洋钻探办公室和自然资源部第一海洋研究所石学法教授对此项工作的关心与支持；感谢国家自然科学基金项目（42176080，42076073）、山东省泰山学者攀登计划项目（tspd20181216）、中国科学院南海海洋研究所自主部署项目（SCSIO2023PT10, SCSIO202201）、中国 IODP 办公室（同济大学中央高校基本科研业务费）（22120230087）的资助。

目　　录

Contents

第一章 研究背景

以往国际上放射虫生物地层的研究工作主要集中在各大洋的中-低纬度海区，已建立了较为成熟的中-低纬度新生代放射虫化石带序列（Sanfilippo and Nigrini, 1998）；然而对高纬度尤其是极地海区的放射虫生物地层学研究仍不完善，未能形成整个区域性的系统剖面。尽管国际深海钻探计划—国际大洋发现计划（DSDP–IODP）的早期系列钻探计划已在南极邻近海区开展了许多工作，但受到已有井位的区域及沉积条件等因素影响，尚未能系统了解南极海区的生物地层特征，近年来的相关研究进展缓慢。由于历史条件的局限，罗斯海（Ross Sea）的相关研究至今仍较为欠缺，仅 Chen（1974）描述了 15 个放射虫新种（部分属同物异名），以及 Chen（1975）进行了初步的鉴定统计，并讨论了放射虫的地层特征与区域分布，但未能开展较为完整的种类鉴定与统计和详细的系统分类工作，也存在一些遗漏和误解。此后，尚未有人继而开展深入的研究工作，尤其缺乏对该海区放射虫群落演变与气候变化或大洋环流的关系讨论。我们曾分析了白令海的放射虫生物地层和西北太平洋的放射虫生物地理特征等，发现北太平洋的环流系统与边界流活动对大区域范围或不同纬度带的浮游性放射虫分布起到重要的驱动与影响作用，而且与全球气候变化密切相关（Chen et al., 2014；Zhang et al., 2014, 2016, 2018；Liu et al., 2016；陈木宏等，2017）。然而，辽阔太平洋在南、北半球海区的洋流与水团对生态环境的影响，以及两者是否存在一定程度的关联仍不清楚。利用系统性的放射虫分类学与沉积学研究，将可进一步揭示地球大洋与气候系统各自独立又相互关联的一些神秘自然现象与规律。

我们已经对白令海等北太平洋高纬度亚极区沉积物中的放射虫进行过较为详细的种类鉴定和系统分类（陈木宏等，2017），初步发现在已描述的上新世—全新世白令海 397 个放射虫种类中，总共 46 种（约占 12%的种类）在南极邻近海区中也曾有记载，说明现在或历史中南、北两极存在着一定程度上相近的地理气候环境。另外，大洋环流与水团交换也可以给一些海洋浮游动物的迁移提供自然条件，造成了这些种类在不同区域出现的时间差特征。迄今为止，国际上针对这些问题的相关研究仍然较为欠缺，进一步分析南极邻近海区，尤其是罗斯海的放射虫动物群演变特征，可为了解新生代以来的全球气候变化及环境演变历史提供有力的科学证据。

第一节 南极放射虫的生物地层问题

国际上最早的南极放射虫研究工作开始于 19 世纪，Ehrenberg（1844）首先对取自詹姆斯·克拉克·罗斯爵士航次（1841–1843）样品中的放射虫标本做了描述，接着 Haeckel（1887）描述了“挑战者”号（H.M.S. Challenger, 1873–1876）全球大洋调查中的 25 个南极放射虫种类；进入 20 世纪，Popofsky（1908, 1912）和 Schröder（1913）利用德国南极探险（Deutsche Südporlar-Expedition, 1901–1903）的样品，对活体放射虫的一些种类进行了初步的描述与记录。直到 20 世纪中期，Riedel（1958）研究了南极科学考察（B.A.N.Z. Antarctic Research Expedition, 1929–1931）沉积物样品中的放射虫类群；Hays（1965）利用取自南大洋约 40ºS–62ºS 区域范围的 95 个站位柱状样进行南极邻近海区的放射虫与新生代历史分析，讨论与南极锋相关的三个现代分布区域特征，共描述 31 个种类，其中包含 9 个暖水种类；而 Petrushevskaya（1967）则利用苏联的南极探险（The Soviet Antarctic Expedition, 1955–1958）所获取的南极样品，对南极海区的现代放射虫做了更为详细的报道。上述这些工作主要侧重分类、生态及其沉积分布的研究，为人类认识南极海区的放射虫类群以至微体古生物特征描绘了初步的基础资料。

进入 20 世纪 60 年代之后，人类才开始利用重力岩心取样对南极邻近海区的放射虫化石及其生物地层学进行初步研究（Hays, 1965；Hays and Opdyke, 1967）。70 年代开始，借助国际 DSDP 和 ODP 在南极周围一些海区的相继实施，钻探取心才获得了较老地层的海底连续沉积样品，如 Chen（1975）对 DSDP Leg 28 航次的南极周边不同区域多个井位进行了初步分析，其中 Sites 270–273 较集中位于罗斯海区域内（仅 Site 273 的放射虫化石保存较好），虽然描述了该海区井位岩心中的一些放射虫种类的分布状况，但遗憾的是未能清楚地划分出罗斯海的放射虫化石带特征。在南大洋的其他区域，Petrushevskaya（1975）利用 Leg 29 的 Sites 278, 280, 281 的岩心样品，首先在西南太平洋的亚南极海域建立了较高纬度的南半球新生代放射虫化石带，并发现一些事件记录；涉及南极邻近海域的其他放射虫地层工作主要有：Abelmann（1990, 1992a）和 Lazarus（1992）分别利用 Leg 113 和 Leg 120 的航次样品探讨南极的渐新世—中中新世放射虫地层特征，以及 Vigour 和 Lazarus（2002）利用 Leg 183 的 Site 1138 岩心分析晚中新世—早上新世的放射虫地层分布。然而，以往这些大洋钻探的研究区域覆盖了辽阔南极周边的一些有限范围，尤其是对具有最高纬度地理位置的罗斯海的研究程度较低，使人们对南极放射虫生物地层的系统认识与了解尚存明显的欠缺。本书旨在开展该海域新生代放射虫特征事件与地层年代的系统分析，将其结果与南极周边或南大洋其他区域进行对比，进一步完善南极与南大洋放射虫生物地层的系统信息，了解不同地理位置的区域特征，建立较为全面的南极放射虫生物地层及其年龄框架。

南极陆地的地质剖面中，较早的中生代沉积地层出现于东部半岛上，在亚历山大岛（Alexander Island）呈现许多构造褶皱与断裂的沉积层中，放射虫的 *Archaeodictyomitra apiara* 组合与 *Protostichocapsa* cf. *stocki* 组合的地质年代分别属于晚侏罗世和早白垩世（Holdsworth and Nell, 1992）。在南极周边海域中，年代最老的化石记录报道于 Leg 113 航次在亚历山大岛东侧的威德尔海（Weddell Sea）钻探的 Hole 689B 井位中，丰度较高的放射虫组合 *Cromyodruppa concentrica*、*Dictyomitra multicostata* 和 *Protostichocapsa stocki* 归属白垩纪的坎潘期—马斯特里赫特期（Campanian–Maastrichtian）地层（Ling and Lazarus, 1990），其中的马斯特里赫特期放射虫组合也被发现于纬度稍低的南大西洋亚极区的 Leg 114 航次 Holes 698A 和 700B 井位中（Ling, 1991），初步揭示了南极及其邻近海区地层发育的纬度变化与关联。然而，这几个位置的钻孔岩心中却缺失新生代的放射虫化石记录。

第二节 南极邻近海区的放射虫生物地层及种类组成

南极邻近海区（含亚极区的南大洋）的新生代放射虫生物地层研究已有一些成果，如 Petrushevskaya（1975）研究了 Leg 29 Site 278 的新生代较连续沉积岩心样品，结合位于 45ºS–55ºS 的 Sites 280 和 281 的分布特征，讨论始新世、渐新世和中新世的放射虫地层对比与划分，初步建立南极邻近海区的较老地层放射虫化石带序列；较为详细的始新世和渐新世放射虫地层出现在普里兹湾以北的亚南极区（Leg 119 Sites 738, 744）井位中，并可清楚划分始新世-渐新世的界线，还发现 27 个放射虫地层事件及其与古地磁对比的年龄（35–24Ma）（Caulet, 1991），而晚中新世以来的较完整放射虫地层事件及其年龄却仅见于更加趋于东北向区域的 Sites 745 和 746 井位中（Caulet, 1991）；在南印度洋凯尔盖朗海台（Kerguelen Plateau）的 Sites 748 和 749 则主要发育了中始新世至早中新世（40–23Ma）的 3 个放射虫化石带：*Eucyrtidium spinosum* 带、*Axoprunum* (?) *irregularis* 带和 *Lychnocanoma conica* 带，其特征与中-低纬度完全不同（Takemura, 1992），但似乎成为南半球较高纬度海区的最老放射虫沉积记录（该处出现中生代的钙质超微化石）。在威德尔海东南部的毛德隆起（Maud Rise）区，接近南极圈（Antarctic Circle）的 Sites 689 和 690 井位岩心中保存了渐新世—中中新世的 10 个放射虫地层事件，但其中存在着 6 处地层缺失，尤其在晚渐新世与早中新世之间的地层缺失时间跨度达 6Ma（Abelmann, 1990）。进而，Abelmann（1992a）还对凯尔盖朗海台的 Sites 747, 748, 751 岩心样品做了进一步研究，详细描述早-中中新世的 13 个放射虫地层事件和 7 个化石带特征，对早中新世的化石带作了修订，并结合古地磁地层进行年龄校正。在该海台中央的近极峰处，Site 1138A 岩心保存了晚中新世—早上新世的放射虫地层，但在 6.1–4.6Ma 仍有缺失（Vigour and Lazarus, 2002）。而该缺失时段在西南大西洋亚极区的福克兰海台（Falkland Plateau）却发现了放射虫的低纬度暖水组合（Weaver, 1983）。福克兰海台的综合研究共描述了晚中新世—上新世的 19 个放射虫地层事件（Weaver, 1983）。这些工作结果表明南极邻近各海区的不同位置钻井中的放射虫地层不仅连续性普遍较差，存在着不同的层位缺失，而且沉积过程的环境差异较大，除了地形因素外，还受不同的水团与海流的影响和控制。

在南太平洋别林斯高晋海的深海平原（Bellingshausen abyssal plain）及其南部陆隆区，4 个 DSDP 井位 Sites 322, 323, 324, 325 的岩心样品中也仅有少部分的样品保存了中新世到更新世的不连续放射虫地层化石，认为是该区域的高沉积速率、极大水深和成岩作用变化等因素所造成（Weaver, 1976）。威德尔海的 ODP Leg 113 航次钻井中保存了中中新世到现代的较为连续放射虫地层，共有 8 个化石带，并获得较清楚的化石带界线及年龄（Lazarus, 1990）。

至今为止，南极邻近的极区或亚极区海区的放射虫地层研究已在几个不同区域展开，获得中生代与新生代的一些基本地层分布特征，尤其是凯尔盖朗海台和威德尔海的中新世—全新世放射虫地层较为详细，而以凯尔盖朗海台的中新世—全新世放射虫地层更加完整，自 22.4Ma 以来共定义有 17 个较高纬度的放射虫化石带（Lazarus, 1992）。然而，凯尔盖朗海台并未进入南极圈范围，其东南-西北向展布的区域跨越15个纬度以上，均在65ºS以北，属亚极区范围。在南极圈内海区，人们至今尚未发现一个完整记录整个新生代放射虫的地层剖面，在前人工作基础上的进一步寻找与探索，有助于全面了解和探讨南半球高纬度的极区-亚极区放射虫地层分布特征，尤其有助于认识地质历史的全球变化过程中世界各大洋对南极海区的不同反应与综合作用。具最高纬度的罗斯海极端生态与沉积环境较为特殊，受极地冰盖环境和环南极流的综合影响较为明显，同时与南太平洋环流也可能存在着历史关联，应隐藏着许多新的放射虫地层信息，具有典型的区域特征。因此，开展南极罗斯海的放射虫地层学工作将填补整个南极海区地层发育在具有代表性区域的研究空白或不足，对系统掌握高纬度区域的生物地层演变及其与其他不同纬度区域的关联具有十分重要的科学意义。

将南极海区与亚极区南大洋已建立的放射虫化石带和地层事件与低纬度的热带海区和北半球高纬度的极区-亚极区进行初步对比，发现新生代仅有一个化石带RN15或Zone *Stylatractus universus* 在这些不同区域近乎同时出现，均以 *Stylatractus universus* 的末现面年龄0.40–0.43Ma为界线（不同区域的时间面存在一定差异），而其他年代的化石带则呈现出南极-热带-北极基本不同的特征。它们之间无法进行地层对比，这可能反映了地球两极与赤道热带的明显生物地理与生物地层的差异，也可能是因为更加详细的放射虫地层学信息尚未完全被挖掘。尽管化石带的特征各有不同，但它们之间的年代信息是否关联仍不清楚。通过更进一步的分类学研究与种类鉴定，也许地球海洋及其生物发育的系统性特征将被揭示。

有关南极邻近海区新生代放射虫的种类组成及其分类学早期研究开展较多，已经形成主要的基础性资料，但仍缺少系统性的梳理和总结。在南大洋的约40ºS–66ºS区域范围，各类大洋钻探计划的航次与航次后研究已经对各个不同海区的放射虫系统分类分别做了一些描述报道。较为详细的工作包括：在南太平洋与南印度洋的交汇区，Petrushevskaya（1975）记录了181种放射虫，可惜的是他对这些种的系统分类较为混乱，归并了许多属，但无详细说明，修订了一些“科”和“属”，但在有些“科”中缺少“属”的记述与说明，或仅记述属而无种一级的说明，所建立的新种也缺少必要的标本测量，其他一些描述种类也无图版说明，因此需要对该作者所建立的新属、新种及其分类原则等的合理性与实用性重新进行审核；在南印度洋的凯尔盖朗海台，Caulet（1991）记录了184个种，但仅对新建立的2个新属和17个新种做了详细描述，其余种类只按字母顺序列出种名清单，显然缺乏具体的系统分类工作。其他的还有：在威德尔海东北部有113种（Abelmann, 1990, 1992a）；在南太平洋东部有61种，在南大西洋的福克兰海台有113种（Weaver, 1976, 1983），在南太平洋西部有101种等。在这些文献中，作者们对南极邻近海区或南大洋放射虫的分类方法不尽一致，甚至有的仅列出种名清单或按字母排列，尚未形成一个合理而统一的分类方案，必然存在重复定种或同物异名现象。据初步的资料分析，在整个南大洋中已经报道的新生代放射虫种类总共有近500种（含80个以上的未定种），去除一些同物异名种，估计南大洋的放射虫多样性与北半球相似纬度区接近或略少，但放射虫群的构成和特征种组成存在着明显的南、北半球生物地理差异，有待于进一步的定量统计和对比分析。

第三节　南极区域的放射虫分类学概要

至今为止，人们早期已有对南极圈内（66°33′39″S 以南）海域放射虫分类系统的研究程度较低，主要是对国际大洋钻探计划早期航次的 3 个井位进行了初步的描述与报道（在南极圈内发现放射虫化石保存较好的仅有这 3 个井位），分别是 Chen（1974, 1975）描述了罗斯海西北部 Site 274 岩心样品中放射虫的 60 个种类（其中包含 15 新种和 11 未定种）、Weaver（1976）报道在别林斯高晋海（Bellingshausen Sea）Site 324 岩心中的 27 个种，以及 Lazarus（1990）、Ling 和 Lazarus（1990）对威德尔海东北部 Site 693 的放射虫分别描述了 19 种和 34 种，在已有的这些不同作者发表的种类中存在部分相互重叠。根据初步的统计分析，南极圈内海区的已知放射虫累计有 113 种，其中包括 34 个未定种（占总种数的近 1/3）以及部分同物异名种。由于上述作者仅有初步报道，未能对这些种类开展详细的分类系统讨论，如 Lazarus（1990）、Ling 和 Lazarus（1990）、Weaver（1983）均只将各个种类按照字母顺序排列，缺乏系统的分类学研究。实际上，在南极海区应该还有很大一部分种类尚未被鉴定、记述和报道，较为完整的系统分类工作更是有待于进一步详细开展和深入探讨，努力填补此基础性研究的空缺，以期更加全面地认识南极放射虫类群的基本面貌及其反映的地理特征、生物地层、地质历史和环境变化。

本书揭示了地质历史中南极圈内海区范围的放射虫种类多样性及其生物分类学特征，探讨和建立了南极海区的新生代放射虫生物地层及其年代框架，填补了国际上在南极海区的放射虫研究空白，完善了人们对放射虫的全球性生物地理特征、高纬度海区生物地层特征，以及生物多样性与生物分类系统的认识。我们利用位于罗斯海西北部的南极圈内放射虫化石保存较为丰富和完整的 Site 274（68°59.81′S，173°25.64′E，水深 3305m）岩心样品（于 1973 年 2 月由 DSDP Leg 28 航次钻取），在细致而充分的样品处理、标本鉴定和资料阅读基础上，开展了较为详细的新生代南极放射虫的系统分类工作，共发现、鉴定和描述放射虫 37 科 138 属 502 种，其中泡沫虫类 286 种、罩笼虫类 216 种，建立了放射虫的 4 个新属和 126 个新种，尚存未定种 68 个。在全部种类中，仅限于南极新生代的放射虫有 217 种，占 43.2%，曾报道分布于南大洋的有 166 种，占 33.1%，记录于北太平洋的有 119 种，占 23.7%，反映出明显的地理特征，较大程度地拓展了人们对南极放射虫的认识，并提供了翔实的基本信息和分析资料，包括全部种类的标本照相图版。

第二章　南极罗斯海的环境概况

第一节　南极大陆的形成历史

在地质历史上，南极大陆是由冈瓦纳古陆分离解体而来。根据阿尔弗雷德·韦格纳提出的大陆漂移学说，地球上的各个分散大陆在早古生代（约 4 亿年前）曾经在赤道附近相互拼接，共同成为一个“联合古陆”或“超级大陆”，之后超级大陆逐渐裂解为北部的劳亚大陆和南部的冈瓦纳大陆，大约 1.7 亿年前的早侏罗世末期，冈瓦纳大陆开始进一步分解，逐渐形成东面的南极洲、印度、新西兰和澳大利亚联合古陆以及西面的南美洲和非洲联合古陆，之后它们又分别开始进一步离散，大约在 0.53 亿年前的早始新世，澳大利亚与南极洲开始分离，到晚始新世，即 0.39 亿年前，南极洲完全脱离澳大利亚，并且南极半岛与南美洲也在同时期或稍后分离开，形成现在的德雷克海峡（Drake Passage），至此南极大陆在地理上完全独立，并在逐渐南移过程中形成了南半球具有宽阔海域的西风带以及现今地球上的基本地理格局。因此，随着南极大陆在漫长的地质历史过程中从地球的低-中纬度区域逐渐漂移到高纬度区域，其邻近海域也必然经历了不同地质时期由地质构造运动引起的不同纬度气候带环境变化，产生了各个时期的生物群类型，分别代表不同的生存环境特征，尤其是在南极海区发现大量新生代的新种类更加显示出它们的独特属性，因此详细解译南极海区放射虫种类的多样性与分类系统及其在地质历史中的变化过程，为深入了解南极环境变化提供了重要的技术手段和基础信息。

第二节　南极罗斯海的地理环境特征

罗斯海有着宽广的大陆架和大陆坡（图 1），位于太平洋的最南端，近似喇叭状嵌入南极大陆，顶部接近南纬 85°S，在纬度 80°S 以南区域为大陆架之上的罗斯冰架，冰架北侧主要为南极圈内的海洋环境，是一个典型的南极大陆边缘海，其东侧在西经 158°W，西侧在东经 170°E，之间跨越东半球与西半球的分界线（180° meridian）。罗斯海的南-北跨度近 18 个纬度，东-西跨度达 32 个经度，是南极大陆周边仅次于威德尔海的第二大边缘海（图 2–图 3）。

罗斯海的海底地形、地貌与地质特征较为复杂，总体上冰架之外的海域可以分为东部与西部两种类型，平均水深是东部海域大于西部海域，在中部海域发育有南-北向延伸的艾斯林海滩（Iselin Bank），自陆架到陆坡深海呈长条形状分布。在其东侧的近陆架区有一条希拉里峡谷（Hillary Canyon），东侧其他区域海底地貌单元简单，深海地形较为平缓，是罗斯海的主要沉积盆地，而且该东部区域无地球物理的重力异常现象；而在西侧却相继发育有中央海盆（Central Basin）与阿代尔海盆（Adare Basin），它们被南-北向

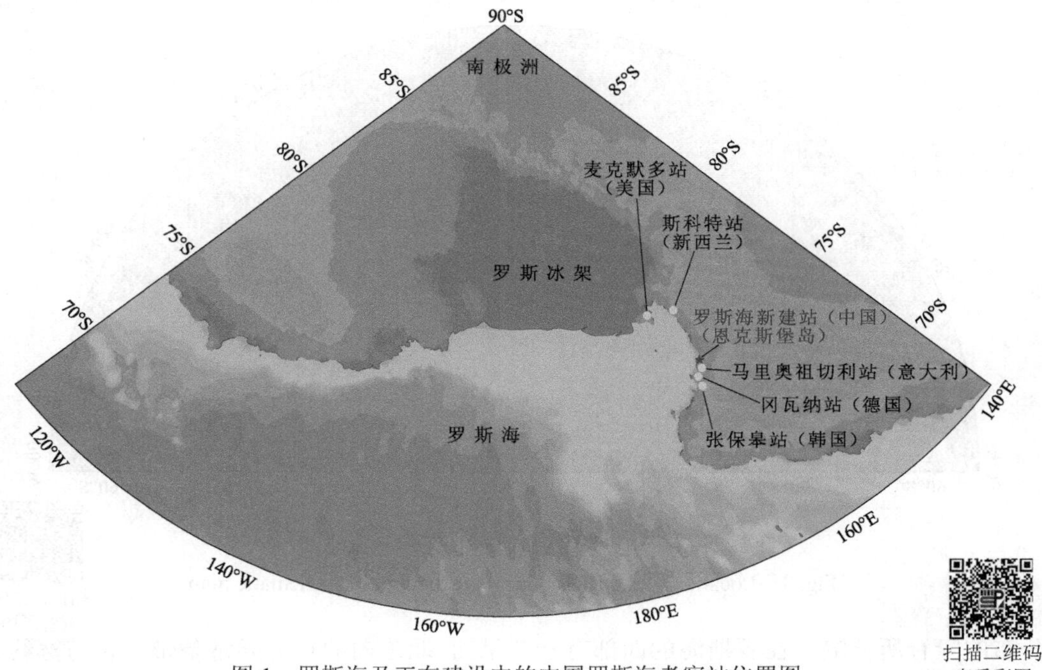

扫描二维码
查看彩图

图 1　罗斯海及正在建设中的中国罗斯海考察站位置图

Fig. 1　Geological location of the Ross Sea and Chinese expedition site in Ross Sea under construction

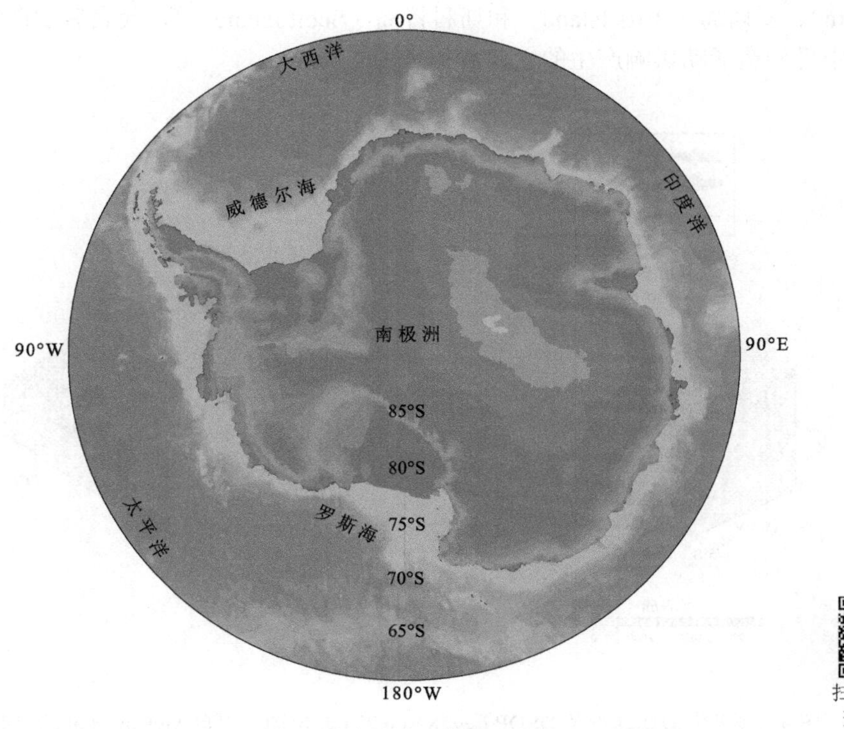

扫描二维码
查看彩图

图 2　南极洲与罗斯海地图

Fig. 2　Geological location of the Antarctica and the Ross Sea

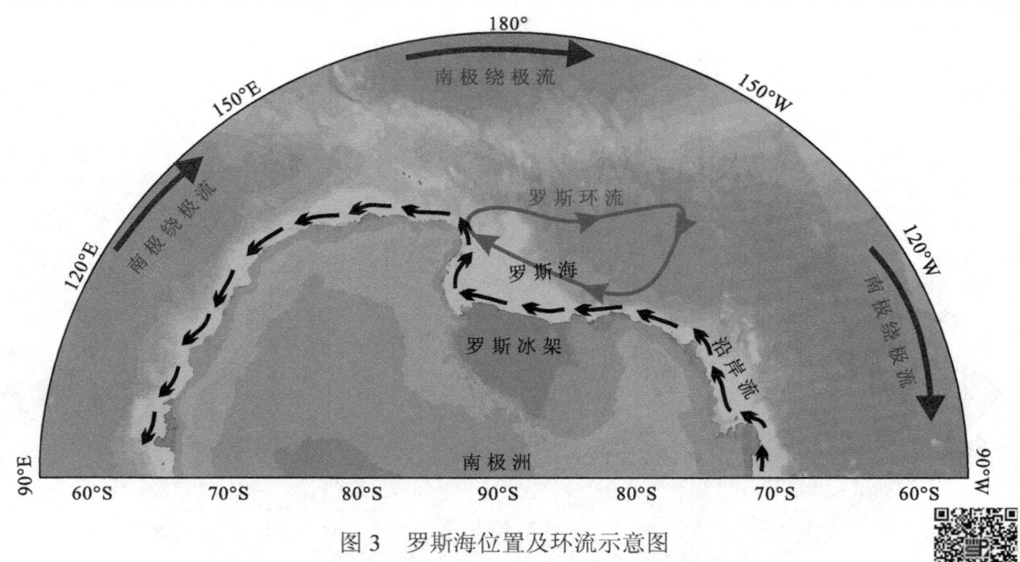

图 3　罗斯海位置及环流示意图

Fig. 3　Location of the Ross Sea and its circulation schematic map

扫描二维码查看彩图

分布的隆脊所分隔。在罗斯海的西部有一条呈南-北走向的重力异常条带，长与宽约为700km 和 50–100km，被推测为板块运动中脱离澳大利亚与新西兰的一条古裂缝（Hayes and Davey, 1975），该区域还发育有一些海底火山与隆起[如埃里伯斯火山（Erebus Mountain）、罗斯岛（Ross Island）和斯科特岛（Scott Island）]，代表着罗斯海在地质历史过程中受构造活动影响产生的主要特征（图 4）。

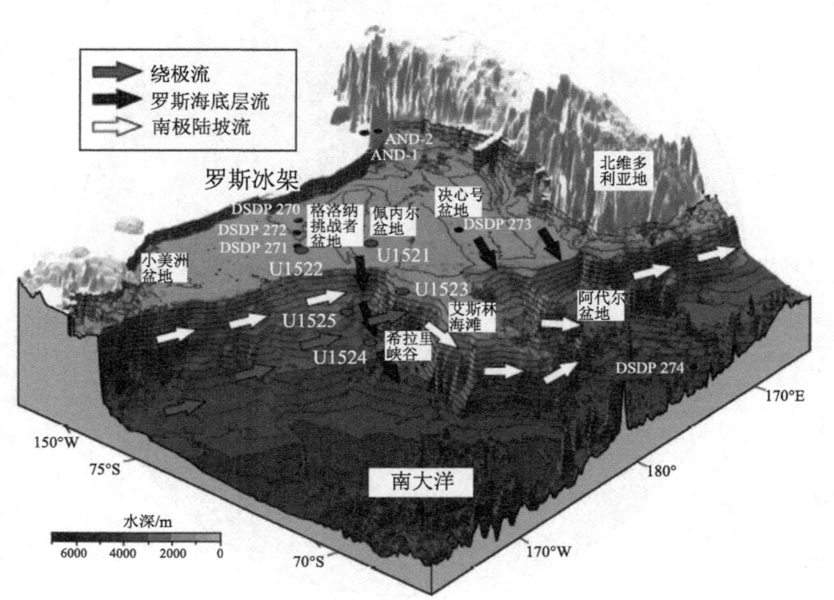

扫描二维码查看彩图

图 4　罗斯海海底地形及 DSDP Leg28 Site 274 站位图（引自 McKay et al., 2018）

Fig. 4　Seafloor topography of the Ross Sea and the location of DSDP Leg 28 Site 274

第三节 南极罗斯海及其邻近海区的气候环境与海洋学特征

罗斯海属于极地区域，冬季的太阳始终在地平线以下（极夜），夏季的太阳却始终在地平线之上（极昼）。冬季的空气温度可降至–40℃，形成大量的陆缘冰，而到了夏季气温常回升至冰点以上，陆缘冰迅速消融，这种罗斯海海冰的季节性变化是世界上最为激烈的海洋物理事件之一，也是驱动该海区生态系统形成与转变的重要因素。在冬季，风速高达 200km/h 的南极风经过罗斯冰架进入罗斯海。

在罗斯海，东南风盛行，驱动沿岸流自东向西流动，并与西风漂流汇合继续往西，而罗斯海的底层水却是自南往北，从陆架到深海往大洋迁移。

南极的威德尔海与罗斯海产生了大洋中大部分的深海高密度冷水团，因此罗斯海的温度或溶解水变化控制着全球子午线（经向的）翻转流的循环（meridional overturning circulation, MOC）（Jacobs et al., 2002；Orsi et al., 2002；Purkey and Johnson, 2010）。近几十年来，罗斯海产生的南极底层水淡化结果增加了其东边的阿蒙森海（Amundsen Sea）和别林斯高晋海的冰水输入影响（Jacobs et al., 2002）。

南极外围的主要海流是宽阔的南极绕极环流（Antarctic circumpolar current, ACC）。除南极沿岸一小股流速很弱的东风漂流外，其主流是自西向东运动的西风漂流，为宽阔、深厚而强劲的风生漂流，南北跨距大概在南纬 35°–65°，与西风带平均范围一致。在罗斯海以东的阿蒙森海和别林斯高晋海，明显受南极绕极环流影响，穿过陆架时诱发稍暖的环极深水（circumpolar deep water, CDW）在内陆架的出现。而在罗斯海则不同，有罗斯旋涡沿着陆坡带来更冷的环极深水（Whitworth et al., 1998），形成强劲的向西流动的南极陆坡流（Antarctic slope current, ASC）。

现今，罗斯海以北的广阔海域是太平洋南部的南大洋，南大洋环绕着整个南极洲，连接罗斯海与太平洋的过渡海区作为南大洋的组成部分，在强西风带的作用下流急浪高，形成一个自西往东的洋流（南极绕极环流），与沿岸环南极流的方向相反。因此，一定程度隔离或限制了太平洋与罗斯海的上层水交换，形成各自的海洋环境与生态条件，产生不同海区生物群落的差异特征。在过渡海区，海洋环境气压梯度大，风向稳定，风力强劲，成为威胁航行的“咆哮西风带”。盛行的西风在罗斯海与南太平洋之间形成了一堵“风壁”，阻挡着低–中纬区暖空气的入侵，维持着罗斯海及南极大陆的气候稳定。

第三章　南极罗斯海的新生代放射虫生物地层划分

至今为止，DSDP Leg 28 Site 274 是南极圈内区域范围的唯一一个较为完整保留新生代沉积岩心记录的钻探井位，为初步探讨与建立南极海区的新生代放射虫生物地层提供了分析依据，也将为进一步的高纬度综合地层年代讨论与区域性对比奠定基础条件。

前人在南极海区做了大量的放射虫研究工作，但由于样品条件限制等，在该海区的生物地层分析较少，仅涉及部分的晚新生代地层（Hays, 1965；Chen, 1974, 1975；Abelmann, 1992a；Lazarus, 1992），而对早新生代的地层讨论近乎空缺。最近，Hollis 等（2020）分析了西南太平洋与东南印度洋的中-高纬度早新生代放射虫生物地层，建立了地层年代框架，为南大洋或邻近海区的放射虫生物地层对比与划分提供了有价值的化石带及其年龄资料。然而，由于生态环境或生物地理特征的差异，西南太平洋与东南印度洋的新生代放射虫组合面貌与演变特征与南极罗斯海存在着明显的不同，反映出不同的放射虫动物群类型，因此难以将两者进行整体的化石带与地层对比。尽管如此，我们还是发现了少量或个别的共同特征种类，可应用于罗斯海放射虫生物地层划分的讨论与借鉴。

本书主要利用 Site 274 岩心样品中放射虫特征种的分布与演变特征（图 5），讨论罗斯海新生代放射虫生物地层界线的划分、厘定与年龄，分析各个年代地层中的特征化石组合以及放射虫事件，辅以典型冷水种在地层中的出现特征说明南极冰盖的形成过程，及其年代与全球气候在渐新世变冷的对比结果，初步建立真正属于南极海区的新生代放射虫生物地层年代框架（图 6），进一步揭示沉积序列反映的环境演变过程。

第一节　第四纪地层界线

1. 全新世/更新世

界线年龄 12ka，位于样品 274-1-5，井深 6.75m。

该年代的时间跨度较短，难以用放射虫的种类演化特征进行判断或地层对比，但群落的突然改变表明生态环境发生了明显的转换。在 Site 274 岩心剖面中，发现一些放射虫种类的末现面（LAD）位于该界线处，它们是：*Solenosphaera bitubula* sp. nov.、*Zygocircus productus* (Hertwig)、*Antarctissa clausa* (Popofsky)和 *Cycladophora conica* Lombari and Lazarus，代表了南极罗斯海区域对更新世与全新世环境更替的响应，但尚无其他分析资料可供参照。新种 *Solenosphaera bitubula* sp. nov.仅分布在罗斯海的晚上新世—晚更新世地层中，消失于全新世。典型南极种 *Antarctissa denticulata* (Ehrenberg) group 自中新世开始在罗斯海出现，一直延续至全新世，也是近代该海区中的优势种之一。

在晚更新世中南极罗斯海放射虫化石动物群的变化不明显。

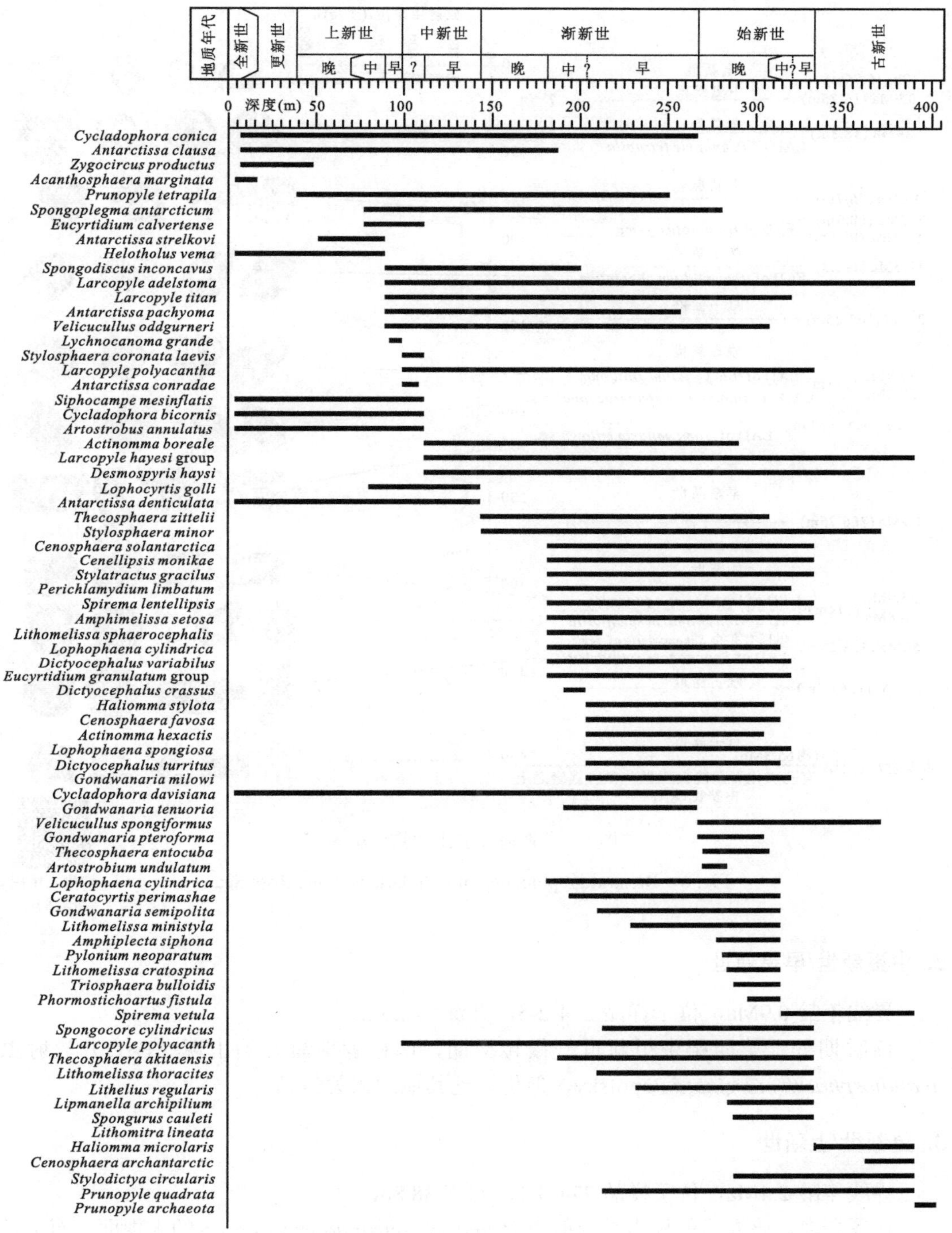

图 5 罗斯海 Site 274 井位主要放射虫特征种地层分布图

Fig. 5 Stratigraphic distribution of major characteristic species of radiolarians from Site 274 in the Ross Sea

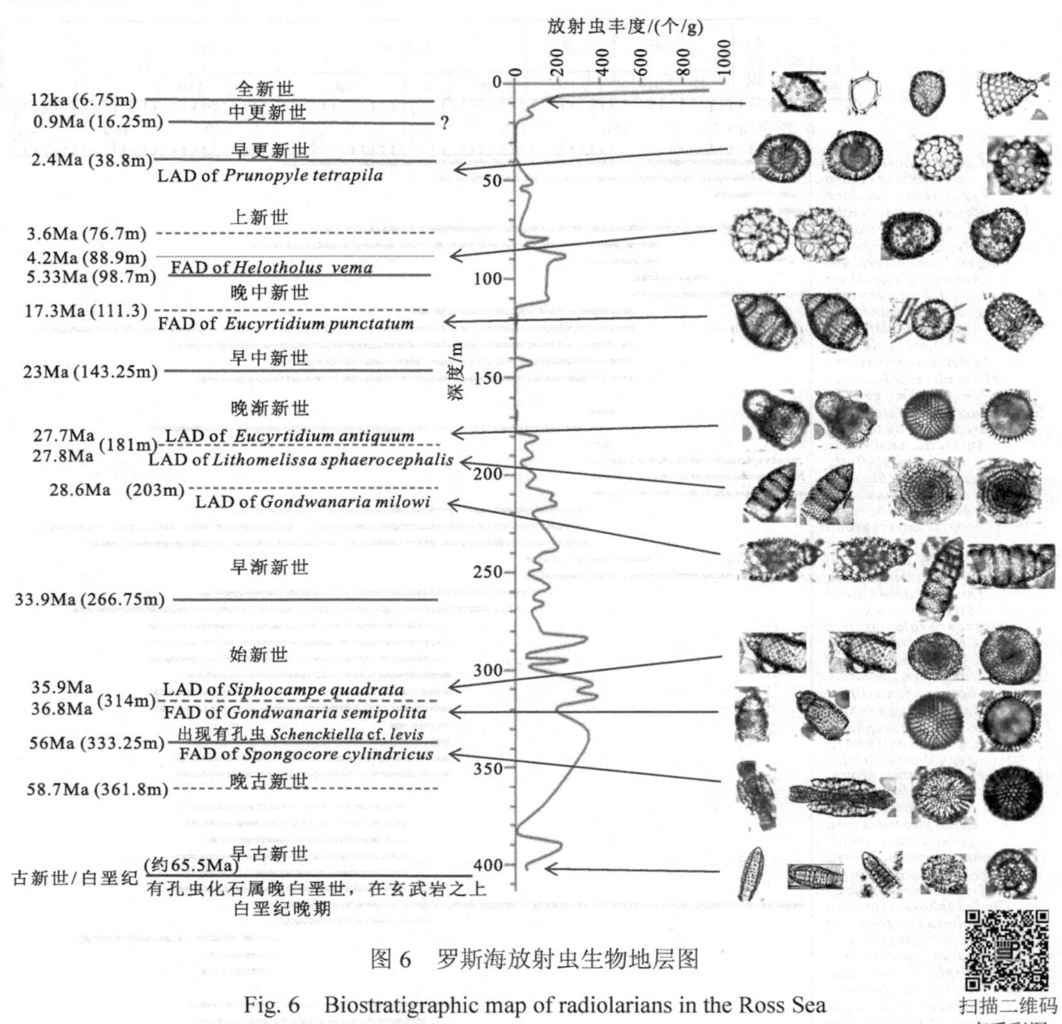

图 6　罗斯海放射虫生物地层图

Fig. 6　Biostratigraphic map of radiolarians in the Ross Sea

2. 中更新世/早更新世

界线年龄 0.9Ma，位于样品 274-2-5，井深 16.25m。

该时期处于第四纪中更新世气候转型期，南极罗斯海开始出现（FAD）放射虫 *Acanthosphaera marginata* Popofsky，但仍缺乏其他的相关信息。

3. 更新世/上新世

界线年龄 2.4Ma，位于样品 274-5-1，井深 38.8m。

在罗斯海，该界线的标志是放射虫 *Prunopyle tetrapila* Hays group 的末现面，而不是 *Helotholus vema* Hays，前者在罗斯海自渐新世开始出现，分布上限为上新世晚期，进入第四纪即消失，而后者则自中上新世开始出现，一直延续至全新世。这两个种在罗斯海的地层分布与南大洋或其他区域存在一定差别。

第二节　新近纪地层界线

1. 晚上新世/中上新世

界线年龄 3.6Ma，位于样品 274-9-1，井深 76.7m。

该界线是放射虫两个种类 *Spongoplegma antarcticum* Haeckel 和 *Eucyrtidium calvertense* Martin 的末现面（LAD）。

2. 中上新世/早上新世

界线年龄 4.2Ma，位于样品 274-10-3，井深 88.9m。

在该界线，放射虫化石组合发生了较为明显的变化，种类 *Antarctissa strelkovi* Petrushevskaya 和 *Helotholus vema* Hays 开始出现（FAD），其中 *Helotholus vema* Hays 在东南印度洋的初现年龄是 4.2Ma（Lazarus, 1992）。同时，该界线也是一些种类的末现面（LAD），如：*Spongodiscus inconcavus* sp. nov.、*Larcopyle adelstoma* (Kozlova and Gobovets)、*Larcopyle titan* (Campbell and Clark) group、*Antarctissa pachyoma* sp. nov.、*Velicucullus oddgurneri* Bjørklund 和 *Pseudodictyophimus galeatus* Caulet，包括两个新种，推测该时期气候环境迅速转冷。

3. 上新世/中新世

界线年龄 5.33Ma，位于样品 274-11-3，井深 98.7m。

在罗斯海，该界线的标志是开始出现（FAD）放射虫 *Helotholus vema* Hays 和 *Lychnocanoma grande* (Campbell and Clark) group，同时，*Stylosphaera coronata laevis* Ehrenberg、*Larcopyle polyacantha* (Campbell and Clark)与 *Antarctissa conradae* Chen 种呈现为最后的分布界线（LAD）。此界线的年龄有待于其他的地层学年龄手段提供进一步的证据，或修订确认。

4. 晚中新世/中中新世

地层中，未能发现放射虫化石初现面或末现面的特征事件，因此罗斯海的该地层放射虫生物地层界线未能提供生物演化的证据。

然而，在南极冰川变化的历史记录中，中中新世期间（约 17–15Ma）为气候最适宜期（middle Miocene climatic optimum, MMCO），冰盖后退，大量融化的冰水入海，紧随其后的是一个短暂的中中新世气候过渡期（middle Miocene climate transition, MMCT, 14.2–13.8Ma），其间南极冰盖增长，气候转冷（Kennett, 1977；Flower and Kennett, 1994；Zachos et al., 2001；Shevenell and Kennett, 2004；Cramer et al., 2009）。因此，在中中新世，南极冰盖发育与全球气候变化经历了较为动荡的时期，罗斯海的海洋生态与沉积环境受到的影响较大，可能存在一些沉积地层的缺失（或取样间隔原因），未能记录到该冷事件。总体上，在中-晚中新世罗斯海出现的一些放射虫暖水种类 *Tetrapyle* spp. 表明气候以温暖期为主，有部分的中−低纬度暖水团进入罗斯海，形成冷−暖水团的混合生态环境。我们推测，晚中新世/中中新世的地层界线应该存在于样品 274-11-3 与 274-12-5 之间的层位中，由于暂未获得相关样品进行分析，有待于进一步补充取样验证。

5. 中中新世/早中新世

界线年龄 17.3Ma，位于样品 274-12-5，井深 111.3m。

开始出现（FAD）*Stylosphaera coronata laevis* Ehrenberg、*Desmospyris spongiosa* Hays、*Botryopera cylindrita* sp. nov.、*Cycladophora bicornis* (Popofsky) group、*Artostrobus annulatus* (Bailey)、*Eucyrtidium calvertense* Martin、*Eucyrtidium punctatum* (Ehrenberg) group 和 *Siphocampe mesinflatis* sp. nov.等种类；而早期出现的种类 *Actinomma boreale* Cleve 和 *Larcopyle hayesi* (Chen) group 在该界线之上却基本消失（LAD）。其中的 *Eucyrtidium punctatum* (Ehrenberg) group 种的初现年龄为 17.3Ma（Abelmann, 1992a; Lazarus, 1992），是此界线年龄初步判断的主要依据。

6. 中新世/渐新世

界线年龄 23Ma，位于样品 274-16-1，井深 143.25m。

在该界线，首次并开始大量出现（FAD）南极特有种放射虫 *Antarctissa denticulata* (Ehrenberg) group，该种的特征有一定变化范围，又难以相互区分，故被归为一个种群；同时，末次出现（LAD）的种有 *Thecosphaera zittelii* Dreyer 和 *Stylosphaera minor* Clark and Campbell。依据这些明显的放射虫特征事件，结合上下层位的地层关系，推测此处为中新世/渐新世的地层界线。在渐新世时期，放射虫动物群的演化中开始出现一些与南大洋同步发生的现象，尽管两地的总体组合面貌不同，各自的生态环境也存在明显差异。

第三节 古近纪地层界线

1. 晚渐新世/中渐新世

界线年龄 27.7–27.8Ma，位于样品 274-20-1，井深 181m。

该界线是一些放射虫特征种的末现面（LAD），包括早期出现的放射虫新种类，它们是：*Cenosphaera solantarctica* sp. nov.、*Cenellipsis monikae* (Petrushevskaya)、*Stylatractus gracilus* sp. nov.、*Perichlamydium limbatum* Ehrenberg、*Spirema lentellipsis* Haeckel、*Amphimelissa setosa* (Cleve)、*Lithomelissa sphaerocephalis* Chen、*Lophophaena cylindrica* (Cleve)、*Dictyocephalus variabilus* sp. nov.和 *Eucyrtidium granulatum* (Petrushevskaya) group（该种与 *Eucyrtidium antiquum* Caulet 为同物异名种）。这一界线年龄的确定主要依据 Site 274 剖面的层位关系和放射虫组合的变化特征，并参照了南大洋的研究结果（Hollis et al., 2020）。

Lithomelissa sphaerocephalis Chen 在南大洋的末现年龄为 27.8Ma（Hollis et al., 2020），*Eucyrtidium granulatum* (Petrushevskaya) group（与 Hollis 等鉴定的 *Eucyrtidium antiquum* 同物异名）在南大洋的末现年龄也是 27.8Ma（Hollis et al., 2020），该放射虫事件的年龄与罗斯海地层的上下关系较为吻合，应用于罗斯海的放射虫生物地层较为合理。

2. 中渐新世/早渐新世

界线年龄 28.6Ma，位于样品 274-22-3，井深 203m。

在该界线，开始出现（FAD）放射虫种 *Dictyocephalus crassus* Carnevale，并且是一些放射虫种类的末现面（LAD），如：*Cenosphaera favosa* Haeckel、*Haliomma stylota* sp. nov.、*Actinomma hexactis* Stöhr、*Lophophaena spongiosa* (Petrushevskaya)、*Dictyocephalus turritus* sp. nov.、*Gondwanaria milowi* (Riedel and Sanfilippo) group。Hollis 等（2020）将 *Gondwanaria milowi* (Riedel and Sanfilippo) group 定为 *Lophocyrtis* (*Paralampterium*) *longiventer* 种，并认为其在南大洋的最后出现时间为 28.6Ma，分类分析的结果显示两者实属同物异名，为同一个种类。在罗斯海是否该种的末现面及中/早渐新世的界线在更早时期还有待于进一步的验证。在该界线的略下井深样品（274-23-3）中，开始出现典型冷水种 *Spongotrochus glacialis* Popofsky，可能指示南极冰盖及其环境在早渐新世的晚期基本形成（约 30Ma）。另一个典型冷水种 *Cycladophora davisiana* Ehrenberg 最早发现于样品（274-29-1）中，其地层年龄为 33.9Ma，可能该时期南极冰盖开始出现。因此，南极冰盖从开始出现到基本形成大约经历 4Ma 的过程。

3. 渐新世/始新世

界线年龄 33.9Ma，位于样品 274-29-1，井深 266.75m。

在该样品中，发现一些放射虫特征种的初现（FAD）：*Cycladophora conica* Lombari and Lazarus、*Cycladophora davisiana* Ehrenberg group 和 *Gondwanaria tenuoria* sp. nov.，以及另外一些放射虫种类的末现（LAD）：*Velicucullus spongiformus* sp. nov.和 *Gondwanaria pteroforma* sp. nov.。尤其是在早渐新世开始出现 *Cycladophora davisiana* Ehrenberg group，该种几乎遍布全世界，是南极冰盖开始出现的一个标志性指标。

尤其值得关注的是，其他较早期出现的许多种类却在该界线处（274-29-1 样品中）突然消失，如：*Thecosphaera entocuba* Chen et al.、*Artostrobium undulatum* (Popofsky)、*Solenosphaera monotubulosa* (Hilmers) group、*Amphiplecta siphona* sp. nov.、*Lithostrobus* cf. *longus* Grigorjeva、*Pylonium neoparatum* sp. nov.、*Actinomma holtedahli* Bjørklund、*Lithomelissa cratospina* sp. nov.、*Lipmanella archipilium* (Petrushevskaya)、*Spongurus cauleti* Goll and Bjørklund、*Stylodictya circularis* (Clark and Campbell)、*Triosphaera bulloidis* sp. nov.、*Sethocorys odysseus* Haeckel、*Artostrobus multiartus* sp. nov.，反映了生态环境的重大变化，进入渐新世气候开始明显变冷，罗斯海的放射虫动物群发生了较大的响应，成为该地层界线的重要标志。

4. 晚始新世/中始新世

界线年龄 35.9–36.8Ma，位于样品 274-34-1，井深 314m。

在该界线处，首次出现（FAD）的放射虫特征种有：*Cenosphaera favosa* Haeckel、

Actinomma impolita sp. nov.、*Pylonium neoparatum* sp. nov.、*Triosphaera bulloidis* sp. nov.、*Lithomelissa cratospina* sp. nov.、*Lithomelissa ministyla* sp. nov.、*Lophophaena cylindrica* (Cleve)、*Amphiplecta siphona* sp. nov.、*Ceratocyrtis irregularis* sp. nov.、*Ceratocyrtis perimashae* sp. nov.、*Gondwanaria semipolita* (Clark and Campbell) 和 *Phormostichoartus fistula* Nigrini；末次出现（LAD）的特征种有：*Spirema vetula* sp. nov.和 *Siphocampe quadrata* (Petrushevskaya and Kozlova)。Hollis 等（2020）报道 *Siphocampe quadrata* 的末现年龄为35.9Ma，在这些种类的组合中，许多是本次研究罗斯海时发现的新种类，形成一个罗斯海在该时期出现的特别的放射虫组合特征，与南大洋等其他海区在该时期的放射虫组合有着明显的区别。

此外，经过系统分类学分析，上述的 *Gondwanaria semipolita* (Clark and Campbell)与Hollis 等（2020）鉴定的 *Aphetocyrtis rossi* 是同物异名种，Hollis 等（2020）认为该种的初现年龄为 36.8Ma。因此，Hollis 等（2020）分析的 *Aphetocyrtis rossi* 与 *Siphocampe quadrata* 这两个特征种的事件年龄与罗斯海较为吻合，提供了地层划分的借鉴依据。

5. 中始新世/早始新世

界线年龄 48.6Ma，位于样品 274-34-5，井深 320m。

由于岩心取样间隔关系，放射虫化石分布的特征事件等信息也较为单调或缺乏，因此该地层界线在 Site 274 井位中尚无法真正确定。初步分析显示 *Eucyrtidium granulatum* (Petrushevskaya)开始出现（FAD）于样品（274-34-5）层位中，在晚始新世/中始新世的界线（样品 274-34-1）之下，可能是中始新世开始或早始新世结束的标志。我们未能在样品 274-34-5 与 274-36-1 之间取到任何样品，而两者相距超过 13m，年龄跨度近 19Ma，该种的初现层位很大可能是在更深更老的岩心样品中。此外，放射虫种 *Eucyrtidium granulatum* (Petrushevskaya)与 Hollis 等（2020）鉴定的 *Eucyrtidium antiquum* 是同物异名，应属同一种，Hollis 等（2020）认为该种在澳大利亚或南大洋的初现年龄为 34.70Ma，显然晚于在罗斯海的初现年代。

Site 274 剖面的分析与对比结果表明，罗斯海与南大洋两个不同区域的放射虫特征种类的演变趋势并不完全一致（同步），有些种类的出现时间在罗斯海明显早于南大洋。

6. 始新世/古新世

界线年龄 56Ma，位于样品 274-36-1 之下，井深>333.25m。

在 274-36-1 样品中（即该界线之上），许多放射虫种类开始出现（FAD），包括若干新种，如：*Cenosphaera solantarctica* sp. nov.、*Carposphaera globosa* Clark and Campbell、*Thecosphaera akitaensis* Nakaseko、*Thecosphaera reticularis* sp. nov.、*Solenosphaera monotubulosa* (Hilmers) group、*Actinomma plasticum* Goll and Bjørklund、*Cenellipsis monikae* (Petrushevskaya)、*Stylatractus gracilus* sp. nov.、*Spongurus cauleti* Goll and Bjørklund、*Spongocore cylindricus* Haeckel、*Stylodictya heliospira* Haeckel、*Larcopyle frakesi* (Chen) group、*Staurodictya medusa* Haeckel、*Larcopyle polyacantha* (Campbell and Clark)、*Spirema lentellipsis* Haeckel、*Lithelius rotalarius* sp. nov.、*Amphimelissa setosa* (Cleve)、

Lithomelissa thoracites Haeckel、*Lipmanella archipilium* (Petrushevskaya)、*Pterocanium polypylum* Popofsky、*Gondwanaria robusta* (Abelmann)、*Lithomitra lineata* (Ehrenberg) group 和 *Lithocampe subligata* Stöhr group。在该界线，也出现早始新世的有孔虫 *Schenckiella* cf. *levis*（Hayes et al., 1974），与放射虫动物群的新组合特征出现较为一致，但有孔虫种类单调，放射虫更加适应当时的罗斯海生态环境。

Hollis 等（2020）认为 *Larcopyle frakesi* (Chen) group 在澳大利亚或南大洋的初现年龄为 35.4Ma，比该种在罗斯海的出现晚了约 20.6Ma。

7. 晚古新世/早古新世

界线年龄 58.7Ma，位于样品 274-39-1，井深 361.8m。

初现 *Hexalonche anaximensis* Haeckel、*Actinomma henningsmoeni* Goll and Bjørklund、*Astrophacus perplexus* (Clark and Campbell)、*Larcopyle ovata* (Kozlova and Gorbovetz)和 *Desmospyris haysi* (Chen)。罗斯海古新世地层进一步划分缺乏可供对比或借鉴的参考资料，只能依据放射虫化石种的特征事件及其组合进行分析，参考事件发生的所在层位，做出初步界定。

8. 古新世/白垩纪

界线年龄 65.5Ma，位于样品 274-43-3–274-42-2，井深 402.77–390.27m。

在该界线之上（274-42-2）开始出现大量早新生代的放射虫新种或特有种，如首次出现（FAD）的 *Cenosphaera archantarctica* sp. nov.、*Haliomma microlaris* sp. nov.、*Stylodictya circularis* (Clark and Campbell)、*Larcopyle adelstoma* (Kozlova and Gobovets)、*Larcopyle hayesi* (Chen) group、*Pylozonium saxitalum* sp. nov.、*Prunopyle quadrata* sp. nov.、*Spirema vetula* sp. nov.、*Streblacantha circumyexta* (Jørgensen) 和 *Sethopyramis quadrata* Haeckel；同时消失（LAD）的有 *Prunopyle archaeota* sp. nov. 和 *Pylozonium saxitalum* sp. nov.。作为新生代最老地层，其放射虫种类的标本特征是多数个体较小、结构简单、发育不足、略显模糊，普遍呈现为较原始的种征状态。

岩心底部样品 274-43-3 的井深是 402.77m，直接位于基底玄武岩之上，是最早的沉积记录。据 Hayes 等（1974, p. 375）分析，发现该最早的沉积物中出现有孔虫化石 *Globotruncana* 或 *Rugoglobigerina*，属晚白垩世晚期地层，这是唯一的年代依据（该井位研究及报告中缺乏对磁性地层和渐新世以下生物地层的分析）。我们在该样品中发现放射虫的新种 *Prunopyle archaeota* sp. nov.和 *Pylozonium saxitalum* sp. nov.，提供了新的化石信息。因此，我们认为该岩心剖面的古新世/白垩纪界线位于样品 274-43-3 与 274-43-2 之间，274-43-3 样品的沉积年龄>65.5Ma。

第四节　地层划分、化石组合与全球变化

上述地层划分与分析的结果表明，南极罗斯海 Site 274 所在海区在新生代地质时期中有着独特的生态与沉积环境，其放射虫动物群或组合现象与邻近的南大洋海区之间虽

然曾有过紧密交流的历史，但更多的是各自独立，存在着较大的生物地理与地层差异特征。在早古新世，罗斯海的放射虫组合完全不同于南大洋，罗斯海区出现大量的放射虫新种。在晚古新世，罗斯海继续有一些新种出现，也发现个别与南大洋相同的放射虫化石种类，但出现的地质年代却明显早于南大洋。始新世是罗斯海与南大洋交融的过渡期，早始新世的罗斯海放射虫继续保持着独特的组合特征，同一化石种的出现仍然要比南大洋早很多。从中-晚始新世开始，罗斯海的放射虫动物群不仅增加许多新的种类，还开始出现一些与南大洋同时发生的事件种类，此现象一直延续到渐新世、中新世和早上新世，一些化石种类可用于不同区域间的地层年代对比。自中-晚上新世至第四纪以来，罗斯海的放射虫动物群又显示出典型的区域特征，与南大洋有较大的差别，因而难有相互联系的地层种供对比分析。

南极罗斯海放射虫生物地层划分的结果与全球气候及冰盖变化的步调与节奏基本一致（图 7）。早古新世，全球气候较暖并不断升温，到晚古新世与早始新世达到最暖期，此阶段南极放射虫动物群的丰度逐渐增加，还出现许多小个体或原始状态的暖水种，如

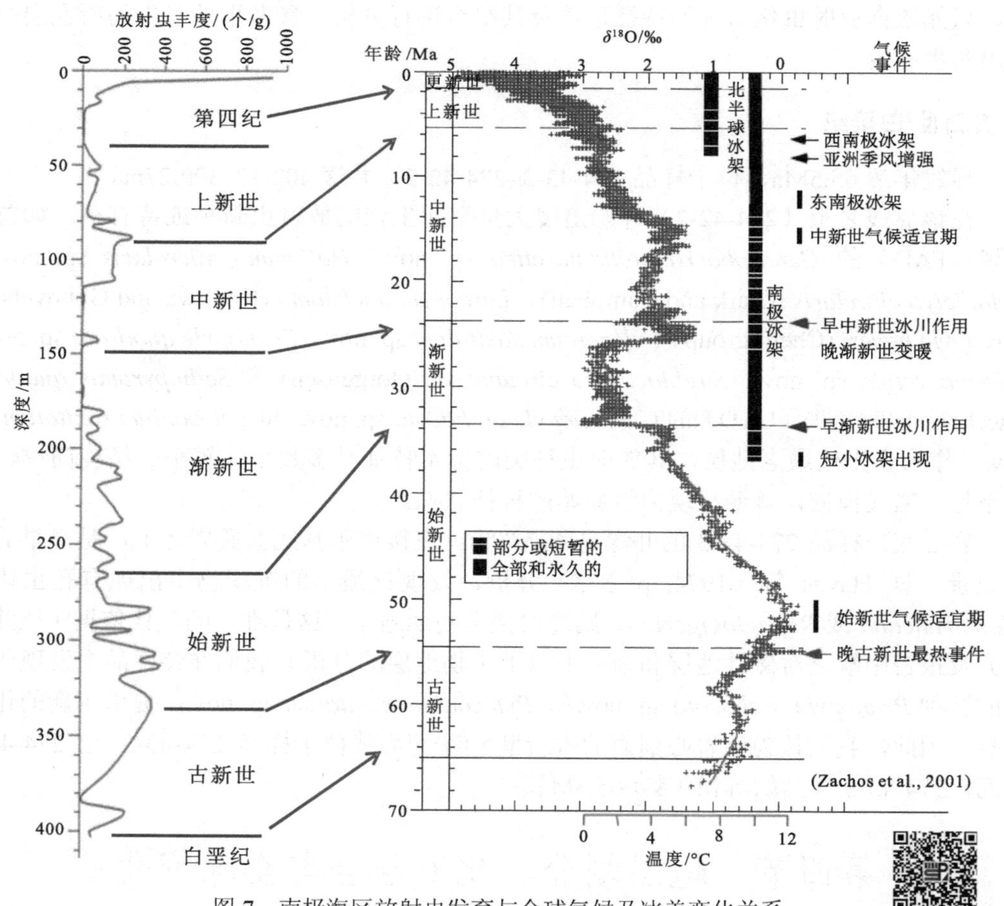

图 7　南极海区放射虫发育与全球气候及冰盖变化关系

Fig. 7　Relationship between the radiolarian development in the Antarctic Sea Region and the changes of global climate and ice sheet

四门孔虫属（*Tetrapyle*）的种类，在始新世/古新世转换过程中发生组合类型的变化，提供了地层界线划分的依据。之后，中始新世开始气候迅速变冷，晚始新世出现短暂的小冰川（Ice-sheet），罗斯海放射虫动物群的丰度较低，暖水种类逐渐消失，至渐新世冷水种 *Cycladophora davisiana* Ehrenberg 在该海区开始出现，此时期南极冰盖开始形成，全球气候进入较为寒冷的时期，罗斯海放射虫动物群的总体丰度也随之逐渐减少，在早渐新世的中-晚期还开始出现另一个典型的冷水种 *Spongotrochus glacialis* Popofsky，该种与 *Cycladophora davisiana* Ehrenberg 一起自此延续至今。第四纪是全球冰川形成的时代，罗斯海的放射虫动物群整体发育呈现为萎靡状态，但从全新世开始又迅速繁盛起来，暖水种与冷水种共同生存。根据这些现象特征，推测南极罗斯海放射虫生物的地层形成和群落演变除了受全球气候的主要影响之外，还与地质板块运动产生的地理迁移结果及其形成的海流与水团交换密切相关（图 8），这些因素导致了南极罗斯海与南大洋的阶段性差异特征。

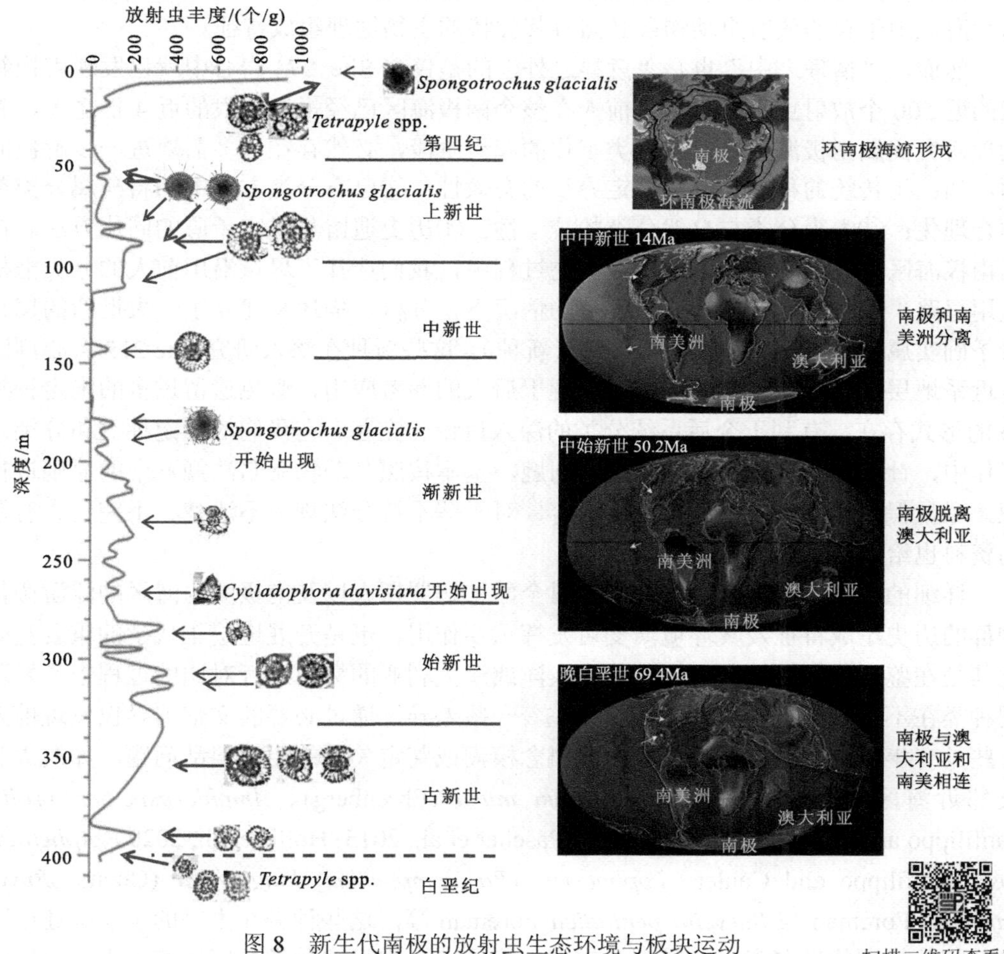

图 8　新生代南极的放射虫生态环境与板块运动

Fig. 8　The ecological environment of radiolarians in the Antarctic and the plate movement during the Cenozoic

新生代板块运动产生的地理格局控制了南极放射虫的生态环境

第四章 南极罗斯海新生代放射虫的系统分类

第一节 南极海区放射虫的种类组成与分类问题

现代严格定义的南极海区应该是南极圈以南的有限海域，围绕着南极大陆，与巨厚的冰盖相邻，接受着大量来自冰盖边缘脱落的冰山，同时与南大洋（太平洋、大西洋和印度洋）之间还被宽阔的西风带与环南极流海域所分隔，因此，南极海区具有“独一无二”和接近“与世隔绝”的特殊地理环境，存在着地球上典型的极地恶劣生态条件。在这片海域中生存的放射虫动物群必然有其独特的生物地理组成特征。

然而，“情理之中”也是“意料之外”的是仅在 Site 274 岩心中就已发现南极新生代的近 500 个放射虫种类，达到前人在整个南极海区已经报道种数的近 4 倍之多，表明此项研究尚属南极海区放射虫分类工作的起步阶段，必然存在许多需要进一步解决的问题，如：①传统的种类定义与鉴定界线的有效性；②已有分类方案中的科与属分类单元的合理化；③人为分类与自然分类的统一性；④历史遗留与现实矛盾的同化方法。在处理南极海区大量出现的放射虫新面孔的过程中，我们选择了尽量沿用前人的分类框架与适用思路，在标本特征比较清晰完整的情况下，分析、描述和建立了一大批新的属种，给予确实属于新类群的种以新的名称（新种）。这些新种在深入研究 Site 275 岩心剖面的古近系地层中发挥了重要的作用，也便于后人的参考应用，避免遗留过多的未定种类以 sp.的形式存在，有利于今后系统分类的深入讨论和完善。在我们的鉴定种类和分类系统工作中，针对各种历史文献中存在的问题，基本按照《动物命名法规》中的原则进行处理，尽量保留前人成果中的合理性内容，对一些不符合法规或不清楚、不完善、有矛盾的资料也给予适当的讨论与处理。

详细的放射虫种类系统分类描述对全面了解罗斯海以及南极圈内海区的浮游微体动物群的历史组成特征及其环境演变可发挥重要作用，也是开展地层年代学的重要基础。尤其是在鉴定地层特征种及将其与南大洋或澳大利亚的标本进行对比的过程中，发现相同种类在不同海区的个体形态特征存在着一些差异，通过必要的文献追寻也发现前人对一些具有年龄标记的主要种类存在着概念模糊或鉴定欠妥的历史混乱问题，如澳大利亚及邻近海区古近纪种类：*Zealithapium mitra* (Ehrenberg)、*Amphicraspedum prolixum* Sanfilippo and Riedel（O’Connor, 1999; Pascher et al., 2015; Hollis et al., 2020）、*Aphetocyrtis rossi* Sanfilippo and Caulet、*Lophocyrtis* (*Paralampterium*) *longiventer* (Chen)、*Buryella tetradica* Foreman 与 *Buryella pentadica* Foreman 等，这些种类在本书的系统描述中可见于各相关的同物异名表中，以利于辨别和讨论，也许相关种类在罗斯海中也具有同样的年代记录，或存在着一定区域年龄差异。除此之外，某些“新属”的建立也存在着与“已有属”类同或难以辨别的问题，应用中必然也会引起一些混乱。如果所有种类都能够合理地安排在同一个分类系统之中，相互比较和取舍，遵循生物或古生物命名法规，而不是各顾各的，缺乏较全面的系统认识，上述问题可能就会更加清晰。

分类描述中仍有未定种问题，针对一些尚未确定种类的标本，我们均已详细查阅了相关资料，并确定属于非已知种，虽然已在属级分类单元中找到了位置，但由于这些标本的数量限制或不够完整等，未能将之确定为新种并冠予名称，因此暂记为未定种（sp.，附有编号），有待于发现新的标本补充和完善。

关于放射虫分类系统的历史遗留问题，我们已在《南海中、北部沉积物的放射虫》（陈木宏和谭智源，1996）、《中国近海的放射虫》（谭智源和陈木宏，1999）和《西北太平洋及其边缘海沉积物中的放射虫》（陈木宏等，2017）各书中从不同角度分别做了一些说明。由于种类较多，放射虫分类系统的建立是一项长期的任务，需要不断地补充、修改和完善，最后才能得到真正的确立。

第二节　分类名录

泡沫虫目 Order SPUMELLARIA Ehrenberg, 1875

球虫亚目 Suborder SPHAEROIDEA Haeckel, 1887

光滑球虫科 Family Liosphaeridae Haeckel, 1887, emend.

空球虫属 Genus *Cenosphaera* Ehrenberg, 1854

1. 原始南极空球虫（新种）*Cenosphaera archantarctica* sp. nov.
2. 坚空球虫 *Cenosphaera compacta* Haeckel
3. 冠空球虫 *Cenosphaera cristata* Haeckel
4. 福地空球虫 *Cenosphaera elysia* Haeckel
5. 巢空球虫 *Cenosphaera favosa* Haeckel
6. 果空球虫 *Cenosphaera mellifica* Haeckel
7. 南极小空球虫（新种）*Cenosphaera miniantarctica* sp. nov.
8. 娜莱空球虫 *Cenosphaera nagatai* Nakaseko
9. 假空球虫（新种）*Cenosphaera pseudocoela* sp. nov.
10. 里德空球虫 *Cenosphaera riedeli* Blueford
11. 太阳南极空球虫（新种）*Cenosphaera solantarctica* sp. nov.
12. 海绵空球虫（新种）*Cenosphaera spongiformis* sp. nov.
13. 女神空球虫 *Cenosphaera veneris* Clark and Campbell
14. 蜂空球虫 *Cenosphaera vesparia* Haeckel
15. 剑空球虫（新种）*Cenosphaera xiphacantha* sp. nov.
16. 空球虫（未定种 1）*Cenosphaera* sp. 1
17. 空球虫（未定种 2）*Cenosphaera* sp. 2
18. 空球虫（未定种 3）*Cenosphaera* sp. 3
19. 空球虫（未定种 4）*Cenosphaera* sp. 4

果球虫属 Genus *Carposphaera* Haeckel, 1881

20. 角果球虫 *Carposphaera angulata* (Haeckel)
21. 异果球虫（新种）*Carposphaera anomala* sp. nov.
22. 圆果球虫 *Carposphaera globosa* Clark and Campbell
23. 大孔果球虫 *Carposphaera magnaporulosa* Clark and Campbell
24. 结实果球虫（新种）*Carposphaera sterrmona* sp. nov.
25. 果球虫（未定种 1）*Carposphaera* sp. 1
26. 果球虫（未定种 2）*Carposphaera* sp. 2

偏球虫属（新属）Genus *Eccentrisphaera* gen. nov.

27. 双皮偏球虫（新种）*Eccentrisphaera biderma* sp. nov.
28. 双髓壳偏球虫（新种）*Eccentrisphaera bimedullaris* sp. nov.
29. 小偏球虫 *Eccentrisphaera minima* (Clark and Campbell)
30. 孔偏球虫（新种）*Eccentrisphaera porolaris* sp. nov.
31. 三髓壳偏球虫（新种）*Eccentrisphaera trimedullaris* sp. nov.

光滑球虫属 Genus *Liosphaera* Haeckel, 1881

32. 果光滑球虫（新种）*Liosphaera carpolaria* sp. nov.
33. 光滑球虫（未定种 1）*Liosphaera* sp. 1

荚球虫属 Genus *Thecosphaera* Haeckel, 1881

34. 尖纹荚球虫 *Thecosphaera akitaensis* Nakaseko
35. 内方荚球虫 *Thecosphaera entocuba* Chen et al.
36. 土豆荚球虫 *Thecosphaera glebulenta* (Sanfilippo and Riedel)
37. 日本荚球虫 *Thecosphaera japonica* Nakaseko
38. 细孔荚球虫（新种）*Thecosphaera minutapora* sp. nov.
39. 显赫荚球虫 *Thecosphaera nobile* (Ehrenberg)
40. 卵荚球虫（新种）*Thecosphaera ovata* sp. nov.
41. 厚壁荚球虫（新种）*Thecosphaera pachycortica* sp. nov.
42. 帕维阿荚球虫 *Thecosphaera parviakitaensis* (Kamikuri)
43. 网荚球虫（新种）*Thecosphaera reticularis* sp. nov.
44. 桑氏荚球虫 *Thecosphaera sanfilippoae* Blueford
45. 东芝荚球虫 *Thecosphaera tochigiensis* Nakaseko
46. 兹特荚球虫 *Thecosphaera zittelii* Dreyer
47. 荚球虫（未定种 1）*Thecosphaera* sp. 1

洋葱球虫属 Genus *Cromyosphaera* Haeckel, 1881

48. 四壳洋葱球虫 *Cromyosphaera quadruplex* Haeckel
49. 粗糙洋葱球虫（新种）*Cromyosphaera asperata* sp. nov.

编枝球虫属 Genus *Plegmosphaera* Haeckel, 1882

50. 糙编枝球虫（新种）*Plegmosphaera asperula* sp. nov.
51. 球编枝球虫（新种）*Plegmosphaera globula* sp. nov.
52. 陪氏编枝球虫（新种）*Plegmosphaera petrushevia* sp. nov.

53. 针编枝球虫（新种）*Plegmosphaera spiculata* sp. nov.

柳编虫属 Genus *Spongoplegma* Haeckel, 1881

54. 南极柳编虫 *Spongoplegma antarcticum* Haeckel

55. 南极柳编虫（亲近种）*Spongoplegma* aff. *antarcticum* Haeckel

56. 方柳编虫（新种）*Spongoplegma quadratum* sp. nov.

胶球虫科 Family Collosphaeridae Müller, 1858

胶球虫属 Genus *Collosphaera* Müller, 1855

57. 百孔胶球虫 *Collosphaera confossa* Takahashi

58. 卵胶球虫 *Collosphaera elliptica* Chen and Tan

59. 胶球虫 *Collosphaera huxleyi* Müller

60. 大孔胶球虫 *Collosphaera macropora* Popofsky

61. 胶球虫（未定种 1）*Collosphaera* sp. 1

62. 胶球虫（未定种 2）*Collosphaera* sp. 2

尖球虫属 Genus *Acrosphaera* Haeckel, 1882

63. 锥尖球虫（新种）*Acrosphaera acuconica* sp. nov.

64. 蛛网尖球虫 *Acrosphaera arachnodictyna* Chen et al.

65. 松尖球虫 *Acrosphaera hirsuta* Perner

66. 神尖球虫？*Acrosphaera*? *mercurius* Lazarus

67. 刺尖球虫 *Acrosphaera spinosa* (Haeckel)

68. 刺尖球虫多刺亚种 *Acrosphaera spinosa echinoides* Haeckel

69. 尖球虫（未定种 1）*Acrosphaera* sp. 1

管球虫属 Genus *Siphonosphaera* Müller, 1858

70. 围织管球虫 *Siphonosphaera circumtexta* (Haeckel)

71. 杂管球虫（新种）*Siphonosphaera mixituba* sp. nov.

筒球虫属 Genus *Solenosphaera* Haeckel, 1887

72. 双管筒球虫（新种）*Solenosphaera bitubula* sp. nov.

73. 单管筒球虫 *Solenosphaera monotubulosa* (Hilmers) group

三管虫属 Genus *Trisolenia* Ehrenberg, 1860, emend. Bjørklund and Goll, 1979

74. 大光三管虫 *Trisolenia megalactis* Ehrenberg

75. 三管虫（未定种 1）*Trisolenia* sp. 1

六柱虫科 Family Hexastylidae Haeckel, 1881, emend. Petrushevskaya, 1975

矛球虫属 Genus *Lonchosphaera* Popofsky, 1908, emend. Dumitrică, 2014

76. 考勒矛球虫 *Lonchosphaera cauleti* Dumitrică

77. 粗糙矛球虫（新种）*Lonchosphaera scabrata* sp. nov.

78. 尖矛球虫 *Lonchosphaera spicata* Popofsky group

79. 矛球虫（未定种 1）*Lonchosphaera* sp. 1

三轴球虫科 Family Cubosphaeridae Haeckel, 1881, emend. Campbell, 1954

六矛虫属 Genus *Hexalonche* Haeckel, 1881

80. 君主六矛虫 *Hexalonche anaximensis* Haeckel

81. 内码六矛虫 *Hexalonche esmarki* Goll and Bjørklund

82. 冷六矛虫（新种）*Hexalonche gelidis* sp. nov.

83. 哲六矛虫 *Hexalonche philosophica* Haeckel

六枪虫属 Genus *Hexacontium* Haeckel, 1881

84. 阿奇六枪虫？*Hexacontium*? *akitaensis* (Nakaseko)

85. 三轴六枪虫 *Hexacontium axotrias* Haeckel

86. 内棘六枪虫 *Hexacontium enthacanthum* Jørgensen

87. 蜂巢六枪虫 *Hexacontium favosum* Haeckel

88. 主六枪虫 *Hexacontium hostile* Cleve

89. 蜜罩六枪虫 *Hexacontium melpomene* (Haeckel)

90. 厚棘六枪虫 *Hexacontium pachydermum* Jørgensen

91. 似六枪虫（新种）*Hexacontium parallelum* sp. nov.

92. 方角六枪虫（新种）*Hexacontium quadrangulum* sp. nov.

93. 方六枪虫 *Hexacontium quadratum* Tan

94. 杖六枪虫 *Hexacontium sceptrum* Haeckel

95. 棘六枪虫 *Hexacontium senticetum* Tan and Su

六葱虫属 Genus *Hexacromyum* Haeckel, 1881, emend. Chen et al., 2017

96. 美六葱虫 *Hexacromyum elegans* Haeckel

97. 稀六葱虫 *Hexacromyum rara* (Carnevale) group

星球虫科 Family Astrosphaeridae Haeckel, 1881

棘球虫属 Genus *Acanthosphaera* Ehrenberg, 1858

98. 恩玛提棘球虫 *Acanthosphaera enigmaticus* (Hollande and Enjumet)

99. 具缘棘球虫 *Acanthosphaera marginata* Popofsky

100. 极区棘球虫（新种）*Acanthosphaera polageota* sp. nov.

101. 棘球虫（未定种 1）*Acanthosphaera* sp. 1

日球虫属 Genus *Heliosphaera* Haeckel, 1862

102. 具齿日球虫 *Heliosphaera dentata* (Mast)

103. 大六角日球虫 *Heliosphaera macrohexagonaria* Tan

104. 小孔日球虫（新种）*Heliosphaera miniopora* sp. nov.

枝球虫属 Genus *Cladococcus* Müller, 1856

105. 枝球虫 *Cladococcus viminalis* Haeckel

海眼虫属 Genus *Haliomma* Ehrenberg, 1844

106. 棘动海眼虫 *Haliomma acanthophora* Popofsky

107. 星海眼虫（新种）*Haliomma asteranota* sp. nov.

108. 小海眼虫（新种）*Haliomma microlaris* sp. nov.

109. 卵海眼虫 *Haliomma ovatum* Ehrenberg

110. 柱海眼虫（新种）*Haliomma stylota* sp. nov.

111. 海眼虫（未定种 1）*Haliomma* sp. 1
112. 海眼虫（未定种 2）*Haliomma* sp. 2

小海眼虫属 Genus *Haliommetta* Haeckel, 1887, emend. Petrushevskaya, 1972

113. 粗针小海眼虫（新种）*Haliommetta hadraspina* sp. nov.
114. 中新世小海眼虫 *Haliommetta miocenica* (Campbell and Clark) group
115. 罗斯小海眼虫（新种）*Haliommetta rossina* sp. nov.

光眼虫属 Genus *Actinomma* Haeckel, 1862, emend. Nigrini, 1967

116. 北方光眼虫 *Actinomma boreale* Cleve
117. 耳蜗光眼虫 *Actinomma cocles* Renaudie and Lazarus
118. 怡光眼虫 *Actinomma delicatulum* (Dogiel)
119. 杜氏光眼虫（新种）*Actinomma dumitricanis* sp. nov.
120. 哥咯尼光眼虫 *Actinomma golownini* Petrushevskaya
121. 矛光眼虫 *Actinomma hastatum* (Haeckel)
122. 亨宁光眼虫 *Actinomma henningsmoeni* Goll and Bjørklund
123. 六针光眼虫 *Actinomma hexactis* Stöhr
124. 赫特光眼虫 *Actinomma holtedahli* Bjørklund
125. 粗糙光眼虫（新种）*Actinomma impolita* sp. nov.
126. 科固光眼虫 *Actinomma kerguelenensis* Caulet
127. 薄皮光眼虫（新种）*Actinomma laminata* sp. nov.
128. 瘦光眼虫 *Actinomma leptoderma* (Jørgensen)
129. 灰光眼虫 *Actinomma livae* Goll and Bjørklund
130. 神光眼虫（新种）*Actinomma magicula* sp. nov.
131. 海女光眼虫 *Actinomma medusa* (Ehrenberg)
132. 奇异光眼虫 *Actinomma mirabile* Goll and Bjørklund
133. 厚皮光眼虫 *Actinomma pachyderma* Haeckel
134. 宽光眼虫 *Actinomma plasticum* Goll and Bjørklund
135. 结实光眼虫（新种）*Actinomma solidula* sp. nov.
136. 优丝光眼虫 *Actinomma yosii* Nakaseko
137. 光眼虫（未定种 1）*Actinomma* sp. 1
138. 光眼虫（未定种 2）*Actinomma* sp. 2

海胆虫属 Genus *Echinomma* Haeckel, 1881

139. 波夫海胆虫 *Echinomma popofskii* Petrushevskaya

葱海胆虫属 Genus *Cromyechinus* Haeckel, 1881

140. 葱海胆虫（未定种 1）*Cromyechinus* sp. 1

绵林虫属 Genus *Spongodrymus* Haeckel, 1881

141. 鹿仁绵林虫 *Spongodrymus elaphococcus* Haeckel

根球虫属 Genus *Rhizosphaera* Haeckel, 1860

142. 近意根球虫 *Rhizosphaera paradoxa* Popofsky

143. 网根球虫 *Rhizosphaera reticulata* (Haeckel)

海绵球虫属 Genus *Spongosphaera* Ehrenberg, 1847

144. 无针海绵球虫 *Spongosphaera spongiosum* (Müller) group

145. 海绵球虫 *Spongosphaera streptacantha* Haeckel

灯球虫属 Genus *Lychnosphaera* Haeckel, 1881

146. 王灯球虫 *Lychnosphaera regina* Haeckel

147. 灯球虫（未定种 1）*Lychnosphaera* sp. 1

根编虫属 Genus *Rhizoplegma* Haeckel, 1881

148. 北方根编虫 *Rhizoplegma boreale* (Cleve)

梅子虫亚目 Suborder PRUNOIDEA Haeckel, 1883

空虫科 Family Ellipsidae Haeckel, 1882

空椭球虫属 Genus *Cenellipsis* Haeckel, 1887

149. 泊阁空椭球虫 *Cenellipsis bergontianus* Carnevale group

150. 莫尼卡空椭球虫 *Cenellipsis monikae* (Petrushevskaya)

针球虫科 Family Stylosphaeridae Haeckel, 1881

球剑虫属 Genus *Ellipsoxiphus* Dunikowski, 1882, emend. Haeckel, 1887

151. 纺锤球剑虫 *Ellipsoxiphus atractus* Haeckel

152. 薄球剑虫（新种）*Ellipsoxiphus leptodermicus* sp. nov.

153. 圆球剑虫（新种）*Ellipsoxiphus rotundalus* sp. nov.

154. 棘刺球剑虫（新种）*Ellipsoxiphus echinatus* sp. nov.

石芹虫属 Genus *Lithapium* Haeckel, 1887

155. 梨形石芹虫 *Lithapium pyriforme* Haeckel

156. 海咽石芹虫 *Lithapium halicapsa* Haeckel

轴梅虫属 Genus *Axoprunum* Haeckel, 1887

157. 十字轴梅虫 *Axoprunum stauraxoniurn* Haeckel

针球虫属 Genus *Stylosphaera* Ehrenberg, 1847

158. 冠针球虫光滑亚种 *Stylosphaera coronata laevis* Ehrenberg

159. 哥伦针球虫 *Stylosphaera goruna* Sanfilippo and Riedel

160. 米诺针球虫 *Stylosphaera minor* Clark and Campbell

161. 梨形针球虫 *Stylosphaera pyriformis* (Bailey)

162. 针球虫（未定种 1）*Stylosphaera* sp. 1

橄榄虫属 Genus *Druppatractus* Haeckel, 1887

163. 果橄榄虫（新种）*Druppatractus carpocus* sp. nov.

164. 矛橄榄虫 *Druppatractus hastatus* Blueford group

165. 不规则橄榄虫 *Druppatractus irregularis* Popofsky

166. 梨橄榄虫（新种）*Druppatractus pyriformus* sp. nov.

针蜓虫属 Genus *Stylatractus* Haeckel, 1887

167. 圣针蜓虫 *Stylatractus angelina* (Campbell and Clark)

168. 细针蜓虫（新种）*Stylatractus gracilus* sp. nov.
169. 不规则针蜓虫 *Stylatractus irregularis* (Takemura)
170. 外套针蜓虫 *Stylatractus palliatum* (Haeckel)
171. 富针蜓虫 *Stylatractus pluto* (Haeckel)

剑蜓虫属 Genus *Xiphatractus* Haeckel, 1887

172. 长针剑蜓虫（新种）*Xiphatractus longostylus* sp. nov.
173. 大轴谱剑蜓虫大轴谱亚种 *Xiphatractus megaxyphos megaxyphos* (Clark and Campbell)
174. 泡沫剑蜓虫 *Xiphatractus spumeus* Dumitrica
175. 剑蜓虫（未定种 1）*Xiphatractus* sp. 1
176. 剑蜓虫（未定种 2）*Xiphatractus* sp. 2

倍球虫属 Genus *Amphisphaera* Haeckel, 1881, emend. Petrushevskaya, 1975

177. 奥特倍球虫 *Amphisphaera aotea* Hollis
178. 裂蹼倍球虫 *Amphisphaera dixyphos* (Ehrenberg)
179. 尼氏倍球虫 *Amphisphaera nigriniae* Kamikuri
180. 辐射倍球虫 *Amphisphaera radiosa* (Ehrenberg) group
181. 光滑倍球虫（新种）*Amphisphaera sperlita* sp. nov.

针矛虫属 Genus *Stylacontarium* Popofsky, 1912

182. 阿克针矛虫 *Stylacontarium acquilonium* (Hays) group

海绵虫科 Family Sponguridae Haeckel, 1862, sensu Sanfilippo and Riedel, 1973

海绵虫属 Genus *Spongurus* Haeckel, 1862

183. 考勒海绵虫 *Spongurus cauleti* Goll and Bjørklund

木偶海绵虫属 Genus *Spongocore* Haeckel, 1887

184. 圆筒木偶海绵虫 *Spongocore cylindricus* Haeckel
185. 木偶海绵虫（未定种）*Spongocore* sp. 1

双月虫属 Genus *Amphymenium* Haeckel, 1881

186. 小玫瑰双月虫 *Amphymenium monstrosum* (Popofsky)
187. 蕨状双月虫（相似种）*Amphymenium* cf. *splendiarmatum* Clark and Campbell

盘虫亚目 Suborder DISCOIDEA Haeckel, 1862

镜盘虫科 Family Phacodiscidae Haeckel, 1881

镜盘虫属 Genus *Phacodiscus* Haeckel, 1881

188. 盾透镜盘虫 *Phacodiscus clypeus* Haeckel
189. 轮镜盘虫 *Phacodiscus rotula* Haeckel
190. 扁豆镜盘虫 *Phacodiscus lentiformis* Haeckel

透镜虫属 Genus *Astrophacus* Haeckel, 1881

191. 全织星透镜虫 *Astrophacus perplexus* (Clark and Campbell)

果盘虫科 Family Coccodiscidae Haeckel, 1862

圆石虫属 Genus *Lithocyclia* Ehrenberg, 1847

192. 小眼圆石虫 *Lithocyclia ocellus* Ehrenberg

孔盘虫科 Family Porodiscidae Haeckel, 1881

始盘虫亚科 Subfamily Archidiscida Haeckel, 1862

始盘虫属 Genus *Archidiscus* Haeckel, 1887

193. 十字始盘虫 *Archidiscus stauroniscus* Haeckel

洞盘虫亚科 Subfamily Trematodiscidae Haeckel, 1862

孔盘虫属 Genus *Porodiscus* Haeckel, 1881

194. 环孔盘虫 *Porodiscus circularis* Clark and Campbell

195. 小眼孔盘虫 *Porodiscus micromma* (Harting)

196. 周旋孔盘虫 *Porodiscus perispira* Haeckel

197. 四孔盘虫 *Porodiscus quadrigatus* Haeckel

膜包虫属 Genus *Perichlamydium* Ehrenberg, 1847

198. 不规则膜包虫（新种）*Perichlamydium irregularmus* sp. nov.

199. 具缘膜包虫 *Perichlamydium limbatum* Ehrenberg

200. 编膜包虫 *Perichlamydium praetextum* (Ehrenberg) group

201. 农神膜包虫 *Perichlamydium saturnus* Haeckel

针网虫亚科 Subfamily Stylodictyinae Haeckel, 1881

针网虫属 Genus *Stylodictya* Ehrenberg, 1847

202. 皮刺针网虫 *Stylodictya aculeata* Jørgensen

203. 转圈针网虫 *Stylodictya circularis* (Clark and Campbell)

204. 日旋针网虫 *Stylodictya heliospira* Haeckel

205. 小针网虫 *Stylodictya minima* Clark and Campbell

206. 强刺针网虫 *Stylodictya validispina* Jørgensen

207. 变形针网虫（新种）*Stylodictya variata* sp. nov.

208. 针网虫（未定种 1）*Stylodictya* sp. 1

209. 针网虫（未定种 2）*Stylodictya* sp. 2

十字网虫属 Genus *Staurodictya* Haeckel

210. 神女十字网虫 *Staurodictya medusa* Haeckel

211. 眼状十字网虫 *Staurodictya ocellata* (Ehrenberg)

双腕虫属 Genus *Amphibrachium* Haeckel, 1881

212. 双腕虫（未定种 1）*Amphibrachium* sp. 1

门盘虫科 Family Pylodiscidae Haeckel, 1887

六洞虫属 Genus *Hexapyle* Haeckel, 1881

213. 小刺六洞虫 *Hexapyle spinulosa* Chen and Tan

盘孔虫属 Genus *Discopyle* Haeckel, 1887

214. 盘孔虫（未定种 1）*Discopyle* sp. 1

海绵盘虫科 Family Spongodiscidae Haeckel, 1881

海绵盘虫属　Genus *Spongodiscus* Ehrenberg, 1854

215. 双凹海绵盘虫　*Spongodiscus biconcavus* Haeckel, emend. Chen et al.
216. 共同海绵盘虫　*Spongodiscus communis* Clark and Campbell group
217. 内凹海绵盘虫（新种）*Spongodiscus inconcavus* sp. nov.
218. 边带海绵盘虫（新种）*Spongodiscus perizonatus* sp. nov.
219. 平坦海绵盘虫（新种）*Spongodiscus planarius* sp. nov.
220. 放射海绵盘虫（新种）*Spongodiscus radialinus* sp. nov.
221. 海绵盘虫　*Spongodiscus resurgens* Ehrenberg group
222. 纹管海绵盘虫　*Spongodiscus trachodes* (Renz)
223. 海绵盘虫（未定种 1）*Spongodiscus* sp. 1
224. 海绵盘虫（未定种 2）*Spongodiscus* sp. 2

海绵轮虫属　Genus *Spongotrochus* Haeckel, 1860

225. 冰海绵轮虫　*Spongotrochus glacialis* Popofsky group
226. ?暗针海绵轮虫　?*Spongotrochus rhabdostyla* Bütschli
227. 异形海绵轮虫　*Spongotrochus vitabilis* Goll and Bjørklund
228. 海绵轮虫（未定种 1）*Spongotrochus* sp. 1

对臂虫属 Genus *Spongobrachium* Haeckel, 1881

229. 反对臂虫　*Spongobrachium antoniae* (O'Connor)

棒网虫属　Genus *Dictyocoryne* Ehrenberg, 1860

230. 截棒网虫　*Dictyocoryne truncatum* (Ehrenberg)
231. 棒网虫（未定种 1）*Dictyocoryne* sp. 1
232. 棒网虫（未定种 2）*Dictyocoryne* sp. 2

海绵星虫属　Genus *Spongaster* Ehrenberg, 1860

233. 海绵星虫　*Spongaster tetras* Ehrenberg
234. 海绵星虫（未定种 1）*Spongaster* sp. 1
235. 海绵星虫（未定种 2）*Spongaster* sp. 2

海绵门孔虫属　Genus *Spongopyle* Dreyer, 1889

236. 吻海绵门孔虫　*Spongopyle osculosa* Dreyer

炭篮虫亚目　Suborder LARCOIDEA Haeckel, 1887

炭篮虫科　Family Larcopylidae Dreyer, 1889

炭篮虫属　Genus *Larcopyle* Dreyer, 1889

237. 隐口炭篮虫　*Larcopyle adelstoma* (Kozlova and Gobovets)
238. 炭篮虫　*Larcopyle butschlii* Dreyer
239. 弗拉克炭篮虫　*Larcopyle frakesi* (Chen) group
240. 哈史炭篮虫　*Larcopyle hayesi* (Chen) group
241. 卵炭篮虫　*Larcopyle ovata* (Kozlova and Gorbovetz)
242. 多棘炭篮虫　*Larcopyle polyacantha* (Campbell and Clark)
243. 多棘炭篮虫巨人亚种　*Larcopyle polyacantha titan* Lazarus et al.

244. 巨人炭篮虫 *Larcopyle titan* (Campbell and Clark) group

245. 炭篮虫（未定种 1）*Larcopyle* sp. 1

箱虫科 Family Larnacidae Haeckel, 1883

壶箱虫属 Genus *Larnacalpis* Haeckel, 1887

246. 扁圆壶箱虫 *Larnacalpis lentellipsis* Haeckel

门孔虫科 Family Pyloniidae Haeckel, 1881

双腰带虫亚科 Subfamily Diplozonaria Haeckel, 1887, emend. Tan and Chen, 1990

四门孔虫属 Genus *Tetrapyle* Müller, 1858

247. 四叶四门孔虫 *Tetrapyle quadriloba* (Ehrenberg)

248. 塔四门孔虫 *Tetrapyle turrita* Haeckel

八门孔虫属 Genus *Octopyle* Haeckel, 1881

249. 六角八门孔虫 *Octopyle sexangulata* Haeckel

多门虫属 Genus *Pylonium* Haeckel, 1881

250. 古多门虫（新种）*Pylonium palaeonatum* sp. nov.

251. 新拟多门虫（新种）*Pylonium neoparatum* sp. nov.

门带虫属 Genus *Pylozonium* Haeckel, 1887

252. 石门带虫（新种）*Pylozonium saxitalum* sp. nov.

253. 门带虫（未定种 1）*Pylozonium* sp. 1

光眼虫科 Family Actinommidae Haeckel, 1862, sensu Riedel, 1967

梅孔虫属 Genus *Prunopyle* Dreyer, 1889

254. 南极梅孔虫 *Prunopyle antarctica* Dreyer

255. 原始梅孔虫（新种）*Prunopyle archaeota* sp. nov.

256. 小梅孔虫（新种）*Prunopyle minuta* sp. nov.

257. 口梅孔虫（新种）*Prunopyle opeocila* sp. nov.

258. 方形梅孔虫（新种）*Prunopyle quadrata* sp. nov.

259. 四毛梅孔虫 *Prunopyle tetrapila* Hays group

圆顶虫科 Family Tholonidae Haeckel, 1887

方顶虫属 Genus *Cubotholus* Haeckel, 1887

260. 规则方顶虫 *Cubotholus regularis* Haeckel

双口虫属 Genus *Dipylissa* Dumitrica, 1988

261. 本松双口虫 *Dipylissa bensoni* Dumitrica

三球虫属（新属）Genus *Triosphaera* gen. nov.

262. 三球虫（新种）*Triosphaera bulloidis* sp. nov.

石太阳虫科 Family Litheliidae Haeckel, 1862

包卷虫属 Genus *Spirema* Haeckel, 1881

263. 编织包卷虫 *Spirema flustrella* Haeckel

264. 冷特包卷虫 *Spirema lentellipsis* Haeckel

265. 苹果包卷虫 *Spirema melonia* Haeckel

266. 老包卷虫（新种）*Spirema vetula* sp. nov.
267. 包卷虫（未定种 1）*Spirema* sp. 1

石太阳虫属 Genus *Lithelius* Haeckel, 1862

268. 蜂房石太阳虫 *Lithelius alveolina* Haeckel
269. 空腔石太阳虫（新种）*Lithelius coelomalis* sp. nov.
270. 克林石太阳虫 *Lithelius klingi* Kamikuri
271. 小石太阳虫 *Lithelius minor* Jørgensen
272. 水手石太阳虫 *Lithelius nautiloides* Popofsky
273. 蜗牛石太阳虫 *Lithelius nerites* Tan and Su
274. 规则石太阳虫（新种）*Lithelius regularis* sp. nov.
275. 圆石太阳虫（新种）*Lithelius rotalarius* sp. nov.
276. 日石太阳虫 *Lithelius solaris* Haeckel
277. 螺石太阳虫 *Lithelius spiralis* Haeckel
278. 石太阳虫（未定种 1）*Lithelius* sp. 1

旋篮虫属 Genus *Larcospira* Haeckel, 1862

279. 多球旋篮虫 *Larcospira bulbosa* Goll and Bjørklund

石果虫属 Genus *Lithocarpium* Stöhr, 1880, emend. Petrushevskaya, 1975

280. 多棘石果虫 *Lithocarpium polyacantha* (Campbell and Clark) group

果虫属 Genus *Cromyodruppa* Haeckel, 1887

281. 同心葱果虫? *Cromyodruppa*? *concentrica* Lipman

旋壳虫科 Family Strebloniidae Haeckel, 1887

棘旋壳虫属 Genus *Streblacantha* Haeckel, 1887

282. 转棘旋壳虫 *Streblacantha circumyexta* (Jørgensen)

艇虫科 Family Phorticidae Haeckel, 1881

艇虫属 Genus *Phorticium* Haeckel, 1881

283. 艇虫 *Phorticium pylonium* Haeckel group
284. 艇虫（未定种 1）*Phorticium* sp. 1

果核虫属（新属）Genus *Carpodrupa* gen. nov.

285. 果核虫（新种）*Carpodrupa rosseala* sp. nov.
286. 果核虫（未定种 1）*Carpodrupa* sp. 1

罩笼虫目 Order NASSELLARIA Ehrenberg, 1875

编网虫亚目 Suborder PLECTOIDEA Haeckel, 1881

编网虫科 Family Plectaniidae Haeckel, 1881

棘编虫属 Genus *Plectacantha* Jørgensen, 1905

287. 房棘编虫 *Plectacantha oikiskos* Jørgensen

环骨虫亚目 Suborder STEPHOIDEA Haeckel, 1881

单环虫科 Family Stephanidae Haeckel, 1881

轭环虫属 Genus *Zygocircus* Bütschli, 1882

288. 轭环虫 *Zygocircus productus* (Hertwig)

篓虫亚目 Suborder SPYROIDEA Ehrenberg, 1847, emend. Petrushevskaya, 1971c

双眼虫科 Family Zygospyridae Haeckel, 1887

脊篮虫属 Genus *Liriospyris* Haeckel, 1881, emend. Goll, 1968

289. 光滑脊篮虫（新种）*Liriospyris glabra* sp. nov.

290. 脊篮虫（未定种 1）*Liriospyris* sp. 1

盔篮虫科 Family Tholospyridae Haeckel, 1887

盔篮虫属 Genus *Tholospyris* Haeckel, 1881

291. 南极盔篮虫 *Tholospyris antarctica* (Haecker) group

292. 小盔篮虫（新种）*Tholospyris peltella* sp. nov.

束篓虫属 Genus *Desmospyris* Haeckel, 1881

293. 头盔束篓虫（新种）*Desmospyris coryatis* sp. nov.

294. 弗塔束篓虫 *Desmospyris futaba* Kamikuri

295. 哈史束篓虫 *Desmospyris haysi* (Chen)

296. 多脚束篓虫（新种）*Desmospyris multichyparis* sp. nov.

297. 瑰篮束篓虫 *Desmospyris rhodospyroides* Petrushevskaya

298. 海绵束篓虫 *Desmospyris spongiosa* Hays

299. 三脚束篓虫（新种）*Desmospyris trichyparis* sp. nov.

葡萄虫亚目 Suborder BOTRYODEA Haeckel, 1881

管葡萄虫科 Family Cannobotryidae Haeckel, 1881, emend. Riedel, 1967

疑蜂虫属 Genus *Amphimelissa* Jørgensen, 1905

300. 方形疑蜂虫（新种） *Amphimelissa quadrata* sp. nov.

301. 棘刺疑蜂虫 *Amphimelissa setosa* (Cleve)

302. 疑蜂虫（未定种 1） *Amphimelissa* sp. 1

袋葡萄虫属 Genus *Botryopera* Haeckel, 1887

303. 同缘袋葡萄虫 *Botryopera conradae* (Chen)

304. 筒形袋葡萄虫（新种）*Botryopera cylindrita* sp. nov.

305. 三叶袋葡萄虫 *Botryopera triloba* (Ehrenberg) group

306. 袋葡萄虫（未定种 1）*Botryopera* sp. 1

葡萄门虫属 Genus *Botryopyle* Haeckel, 1881

307. 筛葡萄门虫 *Botryopyle cribrosa* (Ehrenberg)

囊篮虫属 Genus *Saccospyris* Haecker, 1908, emend. Petrushevskaya, 1965

308. 南极囊篮虫 *Saccospyris antarctica* Haecker

309. 同胸囊篮虫 *Saccospyris conithorax* Petrushevskaya

310. 网头囊篮虫 *Saccospyris dictyocephalus* (Haeckel) group

311. 囊篮虫（未定种 1）*Saccospyris* sp. 1

笼虫亚目 Suborder CYRTOIDEA Haeckel, 1862, emend. Petrushevskaya, 1971

三足壶虫科 Family Tripocalpidae Haeckel, 1887

原帽虫属 Genus *Archipilium* Haeckel, 1881

312. 直翼原帽虫 *Archipilium orthopterum* Haeckel
313. 直翼原帽虫（亲近种）*Archipilium* sp. aff. *A. orthopterum* Haeckel
314. 近似原帽虫（新种）*Archipilium vicinun* sp. nov.

三帽虫属 Genus *Tripilidium* Haeckel, 1881

315. 棒三帽虫? *Tripilidium*? *clavipes* Clark and Campbell

枝蓬虫属 Genus *Cladoscenium* Haeckel, 1881

316. 异地枝蓬虫 *Cladoscenium advena* (Clark and Campbell) group
317. 三胸枝蓬虫 *Cladoscenium tricolpium* (Haeckel)

小袋虫属 Genus *Peridium* Haeckel, 1881

318. 长棘小袋虫 *Peridium longispinum* Jørgensen
319. 小袋虫（未定种 1）*Peridium* sp. 1

袋虫属 Genus *Archipera* Haeckel, 1881

320. 双肋袋虫 *Archipera dipleura* Tan and Tchang

瓮笼虫科 Family Cyrtocalpidae Haeckel, 1887

小角虫属 Genus *Cornutella* Ehrenberg, 1838, emend. Nigrini, 1967

321. 加州小角虫 *Cornutella californica* Campbell and Clark
322. 格孔小角虫 *Cornutella clathrata* Ehrenberg
323. 圆锥小角虫 *Cornutella mitra* Ehrenberg
324. 深小角虫 *Cornutella profunda* Ehrenberg

始匣虫属 Genus *Archicapsa* Haeckel, 1881

325. 橄榄始匣虫（新种）*Archicapsa olivaeformis* sp. nov.

三肋笼虫科 Family Tripocyrtidae Haeckel, 1887, emend. Campbell, 1954

小孔帽虫亚科 Subfamily Sethopilinae Haeckel, 1881, emend. Campbell, 1954

网杯虫属 Genus *Dictyophimus* Ehrenberg, 1847

326. 棘刺网杯虫（新种）*Dictyophimus echinatus* sp. nov.
327. 燕网杯虫 *Dictyophimus hirundo* (Haeckel) group
328. 伊斯网杯虫 *Dictyophimus histricosus* Jørgensen
329. 光头网杯虫（新种）*Dictyophimus lubricephalus* sp. nov.
330. 宽头网杯虫 *Dictyophimus platycephalus* Haeckel

石蜂虫属 Genus *Lithomelissa* Ehrenberg, 1847

331. 强角石蜂虫（新种）*Lithomelissa arrhencorna* sp. nov.
332. 美丽石蜂虫（新种）*Lithomelissa callifera* sp. nov.
333. 头石蜂虫 *Lithomelissa capito* Ehrenberg
334. 挑战石蜂虫 *Lithomelissa challengerae* Chen
335. 强针石蜂虫（新种）*Lithomelissa cratospina* sp. nov.
336. 厄伦伯石蜂虫 *Lithomelissa ehrenbergii* Bütschli
337. 厄伦伯石蜂虫（亲近种）*Lithomelissa* sp. A aff. *L. ehrenbergi* Bütschli

338. 笑石蜂虫 *Lithomelissa gelasinus* O'Connor
339. 圆头石蜂虫（新种）*Lithomelissa globicapita* sp. nov.
340. 哈克石蜂虫（相似种）*Lithomelissa* cf. *haeckeli* Bütschli
341. 燕形石蜂虫（新种）*Lithomelissa hirundiforma* sp. nov.
342. 侧头石蜂虫 *Lithomelissa laticeps* Jørgensen
343. 小柱石蜂虫（新种）*Lithomelissa ministyla* sp. nov.
344. 厚壁石蜂虫（新种）*Lithomelissa pachyoma* sp. nov.
345. 轴神石蜂虫 *Lithomelissa poljanskii* (Petrushevskaya)
346. 前南极石蜂虫 *Lithomelissa preantarctica* (Pettrushevskaya)
347. 壮石蜂虫 *Lithomelissa robusta* Chen
348. 圆头石蜂虫 *Lithomelissa sphaerocephalis* Chen
349. 口刺石蜂虫（新种）*Lithomelissa stomaculeata* sp. nov.
350. 石蜂虫 *Lithomelissa thoracites* Haeckel
351. 三角石蜂虫 *Lithomelissa tricornis* Chen
352. 雅石蜂虫（新种）*Lithomelissa tryphera* sp. nov.
353. 石蜂虫（未定种 1）*Lithomelissa* sp. 1
354. 石蜂虫（未定种 2）*Lithomelissa* sp. 2

南极虫属 Genus *Antarctissa* Petrushevskaya, 1967

355. 代齿南极虫 *Antarctissa antedenticulata* Chen
356. 包头南极虫（新种）*Antarctissa cingocephalis* sp. nov.
357. 封口南极虫 *Antarctissa clausa* (Popofsky)
358. 同射南极虫 *Antarctissa conradae* Chen
359. 德弗兰南极虫 *Antarctissa deflandrei* (Petrushevskaya)
360. 小齿南极虫 *Antarctissa denticulata* (Ehrenberg) group
361. 小齿南极虫封口亚种 *Antarctissa denticulata clausa* Petrushevskaya
362. 长南极虫 *Antarctissa longa* (Popofsky)
363. 厚壁南极虫（新种）*Antarctissa pachyoma* sp. nov.
364. 翼南极虫（新种）*Antarctissa pterigyna* sp. nov.
365. 史科南极虫 *Antarctissa strelkovi* Petrushevskaya
366. 威特南极虫 *Antarctissa whitei* Bjørklund

海绵蜂虫属 Genus *Spongomelissa* Haeckel, 1887

367. 阔口海绵蜂虫（新种）*Spongomelissa chaenothoraca* sp. nov.
368. 小瓜海绵蜂虫 *Spongomelissa cucumella* Sanfilippo and Riedel

罩巾虫属 Genus *Velicucullus* Riedel and Campbell, 1952

369. 高罩巾虫 *Velicucullus altus* Abelmann
370. 奥古罩巾虫 *Velicucullus oddgurneri* Bjørklund
371. 海绵罩巾虫（新种）*Velicucullus spongiformus* sp. nov.

笠虫属 Genus *Helotholus* Jørgensen, 1905

372. 渐大笠虫 *Helotholus ampliata* (Ehrenberg)
373. 灯泡笠虫（新种）*Helotholus bulbosus* sp. nov.
374. 笠虫 *Helotholus histricosa* Jørgensen
375. 前方笠虫 *Helotholus praevema* Weaver
376. 万笠虫 *Helotholus vema* Hays
377. 小万笠虫（新种）*Helotholus vematella* sp. nov.

石囊虫属 Genus *Lithopera* Ehrenberg, 1847

378. 石囊虫（未定种 1）*Lithopera* sp. 1

双孔编虫属 Genus *Amphiplecta* Haeckel, 1881

379. 管双孔编虫（新种）*Amphiplecta siphona* sp. nov.
380. 双孔编虫（未定种 1）*Amphiplecta* sp. 1

灯犬虫属 Genus *Lychnocanoma* Haeckel, 1887, emend. Foreman, 1973

381. 美灯犬虫 *Lychnocanoma bellum* (Clark and Campbell)
382. 圆锥灯犬虫 *Lychnocanoma conica* (Clark and Campbell)
383. 秀灯犬虫（新种）*Lychnocanoma eleganta* sp. nov.
384. 大灯犬虫 *Lychnocanoma grande* (Campbell and Clark) group
385. 大灯犬虫皱亚种 *Lychnocanoma grande rugosum* (Riedel)
386. 大孔灯犬虫（新种）*Lychnocanoma macropora* sp. nov.

花篮虫科 Family Anthocyrtidae Haeckel, 1887

筛锥虫属 Genus *Sethopyramis* Haeckel, 1881

387. 方筛锥虫 *Sethopyramis quadrata* Haeckel
388. 筛锥虫（未定种 1）*Sethopyramis* sp. 1

裹锥虫属 Genus *Peripyramis* Haeckel, 1881

389. 围裹锥虫 *Peripyramis circumtexta* Haeckel

梯锥虫属 Genus *Bathropyramis* Haeckel, 1881

390. 分枝梯锥虫 *Bathropyramis ramosa* Haeckel
391. 扩口梯锥虫 *Bathropyramis tenthorium* (Haeckel)
392. 光面梯锥虫 *Bathropyramis trapezoides* Haeckel

盾虫属 Genus *Aspis* Nishimura, 1992

393. 小瓶盾虫（新种）*Aspis ampullarus* sp. nov.
394. 枝角盾虫（新种）*Aspis cladocerus* sp. nov.
395. 墙盾虫 *Aspis murus* Nishimura

格锥虫属 Genus *Cinclopyramis* Haeckel, 1881

396. 格锥虫（未定种 1）*Cinclopyramis* sp. 1

筛笼虫科 Family Sethocyrtida Haeckel, 1887

筛盔虫属 Genus *Sethocorys* Haeckel, 1881

397. 奥地苏筛盔虫 *Sethocorys odysseus* Haeckel

冠明虫属 Genus *Lophophaena* Ehrenberg, 1847, emend. Petrushevskaya, 1971

398. 头冠明虫（新种）*Lophophaena cephalota* sp. nov.
399. 圆头冠明虫（新种）*Lophophaena cyclocapita* sp. nov.
400. 圆筒冠明虫 *Lophophaena cylindrica* (Cleve)
401. 橄榄冠明虫（新种）*Lophophaena olivacea* sp. nov.
402. 球冠明虫？（新种）*Lophophaena? orbicularis* sp. nov.
403. 海绵冠明虫 *Lophophaena spongiosa* (Petrushevskaya)
404. 冠明虫（未定种 1）*Lophophaena* sp. 1

筛圆锥虫属 Genus *Sethoconus* Haeckel, 1881

405. 杜氏筛圆锥虫？*Sethoconus? dogieli* Petrushevskaya
406. 板筛圆锥虫 *Sethoconus tabulata* (Ehrenberg)
407. 筛圆锥虫（未定种 1）*Sethoconus* sp. 1

花笼虫属 Genus *Anthocyrtium* Haeckel, 1887

408. 拜洛花笼虫 *Anthocyrtium byronense* Clark
409. 五脚花笼虫（新种）*Anthocyrtium quinquepedium* sp. nov.

盔蜂虫属 Genus *Corythomelissa* Campbell, 1951

410. 钩盔蜂虫 *Corythomelissa adunca* (Sanfilippo and Riedel)
411. 头盔蜂虫（新种）*Corythomelissa corysta* sp. nov.
412. 粗糙盔蜂虫 *Corythomelissa horrida* Petrushevskaya
413. 盔蜂虫（未定种 1）*Corythomelissa* sp. 1

角笼虫属 Genus *Ceratocyrtis* Bütschli, 1882

414. 扩角笼虫 *Ceratocyrtis amplus* (Popofsky) group
415. 筒角笼虫（新种）*Ceratocyrtis cylindris* sp. nov.
416. 不规则角笼虫（新种）*Ceratocyrtis irregularis* sp. nov.
417. 阔角笼虫 *Ceratocyrtis manumi* Goll and Bjørklund
418. 雄角笼虫 *Ceratocyrtis mashae* Bjørklund
419. 围玛角笼虫（新种）*Ceratocyrtis perimashae* sp. nov.
420. 强壮角笼虫 *Ceratocyrtis robustus* Bjørklund
421. 角笼虫（未定种 1）*Ceratocyrtis* sp. 1

格头虫属 Genus *Dictyocephalus* Ehrenberg, 1860

422. 南方格头虫 *Dictyocephalus australis* Haeckel
423. 厚格头虫 *Dictyocephalus crassus* Carnevale
424. 乳格头虫 *Dictyocephalus papillosus* (Ehrenberg)
425. 塔格头虫（新种）*Dictyocephalus turritus* sp. nov.
426. 变形格头虫（新种）*Dictyocephalus variabilus* sp. nov.
427. 格头虫（未定种 1）*Dictyocephalus* sp. 1
428. 格头虫（未定种 2）*Dictyocephalus* sp. 2

果篮虫属 Genus *Carpocanium* Ehrenberg, 1847

429. 海绵果篮虫（新种）*Carpocanium spongiforma* sp. nov.

430. 果篮虫（未定种 1）*Carpocanium* sp. 1

果圆球虫属（新属） Genus *Carpoglobatus* gen. nov.

431. 管果圆球虫（新种）*Carpoglobatus tubulus* sp. nov.

足篮虫科 Family Podocyrtidae Haeckel, 1887

里曼虫属 Genus *Lipmanella* Loeblich and Tappan, 1961

432. 始弹里曼虫 *Lipmanella archipilium* (Petrushevskaya)

433. 网角里曼虫 *Lipmanella dictyoceras* (Haeckel)

434. 里曼虫（未定种 1）*Lipmanella* sp. 1

翼篮虫属 Genus *Pterocanium* Ehrenberg, 1847

435. 多门翼篮虫 *Pterocanium polypylum* Popofsky

436. 翼篮虫（未定种 1）*Pterocanium* sp. 1

假网杯虫属 Genus *Pseudodictyophimus* Petrushevskaya, 1971

437. 头盔假网杯虫 *Pseudodictyophimus galeatus* Caulet

438. 大针假网杯虫（新种）*Pseudodictyophimus gigantospinus* sp. nov.

439. 单薄假网杯虫 *Pseudodictyophimus gracilipes* (Bailey)

440. 单薄假网杯虫（亲近种）*Pseudodictyophimus* sp. aff. *P. gracilipes* (Bailey)

441. 纤细假网杯虫（新种）*Pseudodictyophimus tenellus* sp. nov.

辫篓虫科 Family Phormocyrtidae Haeckel, 1887

圆蜂虫属 Genus *Cycladophora* Ehrenberg, 1847, emend. Lombari and Lazarus, 1988

442. 双角圆蜂虫 *Cycladophora bicornis* (Popofsky) group

443. 锥圆蜂虫 *Cycladophora conica* Lombari and Lazarus

444. 戴维斯圆蜂虫 *Cycladophora davisiana* Ehrenberg group

445. 戴维斯圆蜂虫似角亚种 *Cycladophora davisiana cornutoides* (Petrushevskaya)

446. 假异圆蜂虫 *Cycladophora pseudoadvena* (Kozlova)

447. 棉胸圆蜂虫 *Cycladophora spongothorax* (Chen)

448. 长钟圆蜂虫（新种）*Cycladophora tinocampanula* sp. nov.

冠笼虫属 Genus *Lophocyrtis* Haeckel, 1887

449. 戈氏冠笼虫 *Lophocyrtis golli* Chen

窗袍虫属 Genus *Clathrocyclas* Haeckel, 1881, emend. Foreman, 1968

450. 君窗袍虫 *Clathrocyclas alcmenae* Haeckel

451. 车窗袍虫（相似种）*Clathrocyclas* cf. *C. danaes* Haeckel

宽口斯科虫属 Genus *Eurystomoskevos* Caulet, 1991

452. 匹楚宽口斯科虫 *Eurystomoskevos petrushevskaae* Caulet

453. 宽口斯科虫（未定种 1）*Eurystomoskevos* sp. 1

丽篮虫属 Genus *Lamprocyrtis* Kling, 1973

454. 射光丽篮虫 *Lamprocyrtis aegles* (Ehrenberg)

455. 丽篮虫（未定种 1）*Lamprocyrtis* sp. 1

神篓虫科 Family Theocyrtidae Haeckel, 1887, emend. Nigrini, 1967

冈瓦纳虫属 Genus *Gondwanaria* Petrushevskaya, 1975

456. 狄弗朗冈瓦纳虫 *Gondwanaria deflandrei* Petrushevskaya

457. 日本冈瓦纳虫 *Gondwanaria japonica* (Nakaseko) group

458. 米露冈瓦纳虫 *Gondwanaria milowi* (Riedel and Sanfilippo) group

459. 翼冈瓦纳虫（新种）*Gondwanaria pteroforma* sp. nov.

460. 隶董冈瓦纳虫 *Gondwanaria redondoensis* (Campbell and Clark)

461. 壮实冈瓦纳虫 *Gondwanaria robusta* (Abelmann)

462. 半滑冈瓦纳虫 *Gondwanaria semipolita* (Clark and Campbell)

463. 半滑冈瓦纳虫（相似种）*Gondwanaria* cf. *semipolita* (Clark and Campbell)

464. 瘦冈瓦纳虫（新种）*Gondwanaria tenuoria* sp. nov.

465. 冈瓦纳虫（未定种 1）*Gondwanaria* sp. 1

吊筐虫属 Genus *Artophormis* Haeckel, 1881

466. 简单吊筐虫（相似种）*Artophormis* sp. cf. *A. gracilis* Riedel

467. 吊筐虫（未定种 1）*Artophormis* sp. 1

468. 吊筐虫（未定种 2）*Artophormis* sp. 2

毛虫科 Family Podocampidae Haeckel, 1887

篮袋虫属 Genus *Cyrtopera* Haeckel, 1881

469. 小壶篮袋虫 *Cyrtopera laguncula* Haeckel

470. 巨篮袋虫（新种）*Cyrtopera magnifica* sp. nov.

石毛虫科 Family Lithocampidae Haeckel, 1887

石螺旋虫属 Genus *Lithostrobus* Bütschli, 1882

471. 长石螺旋虫（相似种）*Lithostrobus* cf. *longus* Grigorjeva

窄旋虫属 Genus *Artostrobus* Haeckel, 1887

472. 环窄旋虫 *Artostrobus annulatus* (Bailey)

473. 米斯窄旋虫 *Artostrobus missilis* (O'Connor)

474. 多节窄旋虫（新种）*Artostrobus multiartus* sp. nov.

475. 厚壁窄旋虫 *Artostrobus pachyderma* (Ehrenberg) group

476. 平板窄旋虫? *Artostrobus*? *pretabulatus* Petrushevskaya

477. 窄旋虫（未定种 1）*Artostrobus* sp. 1

细篮虫属 Genus *Eucyrtidium* Ehrenberg, 1847, emend. Nigrini, 1967

478. 环节细篮虫 *Eucyrtidium annulatum* (Popofsky)

479. 歪细篮虫（新种）*Eucyrtidium anomurum* sp. nov.

480. 丽转细篮虫 *Eucyrtidium calvertense* Martin

481. 芋口细篮虫 *Eucyrtidium cienkowskii* Haeckel group

482. 冰细篮虫（新种）*Eucyrtidium gelidium* sp. nov.

483. 颗粒细篮虫 *Eucyrtidium granulatum* (Petrushevskaya) group
484. 小孔细篮虫 *Eucyrtidium punctatum* (Ehrenberg) group
485. 洁雅细篮虫 *Eucyrtidium lepidosa* (Kozlova)
486. 细篮虫（未定种 1）*Eucyrtidium* sp. 1

石帽虫属 Genus *Lithomitra* Bütschli, 1881

487. 线石帽虫 *Lithomitra lineata* (Ehrenberg) group

管毛虫属 Genus *Siphocampe* Haeckel, 1881

488. 高奇剑管毛虫 *Siphocampe altamiraensis* (Campbell and Clark)
489. 隶杂管毛虫 *Siphocampe elizabethae* (Clark and Campbell)
490. 中突管毛虫（新种）*Siphocampe mesinflatis* sp. nov.
491. 方管毛虫 *Siphocampe quadrata* (Petrushevskaya and Kozlova)

石毛虫属 Genus *Lithocampe* Ehrenberg, 1838

492. 微缚石毛虫 *Lithocampe subligata* Stöhr group

旋篮虫属 Genus *Spirocyrtis* Haeckel, 1881, emend.

493. 丽旋篮虫（新种）*Spirocyrtis bellulis* sp. nov.
494. 圆梯旋篮虫？（亲近种）*Spirocyrtis*? aff. *gyroscalaris* Nigrini
495. 矩形旋篮虫（新种）*Spirocyrtis rectangulis* sp. nov.
496. 梯盘旋篮虫 *Spirocyrtis scalaris* Haeckel

陀螺虫科 Family Artostrobiidae Riedel, 1967, emend. Foreman, 1973

陀螺虫属 Genus *Artostrobium* Haeckel, 1887

497. 耳陀螺虫 *Artostrobium auritum* (Ehrenberg) group
498. 糙角陀螺虫 *Artostrobium rhinoceros* Sanfilippo and Riedel
499. 浪陀螺虫 *Artostrobium undulatum* (Popofsky)

旋葡萄虫属 Genus *Botryostrobus* Haeckel, 1887

500. 布拉旋葡萄虫 *Botryostrobus bramlettei* (Campbell and Clark)
501. 佐得旋葡萄虫 *Botryostrobus joides* Petrushevskaya

筐列虫属 Genus *Phormostichoartus* Campbell, 1951, emend. Nigrini, 1977

502. 管筐列虫 *Phormostichoartus fistula* Nigrini

第三节　分类系统及种类描述

泡沫虫目 Order SPUMELLARIA Ehrenberg, 1875

骨骼通常呈球形或球的变形，如椭球、圆盘、透镜、螺旋等。许多种类具有几个同心球壳，外壳称为皮壳，内壳常称为髓壳。

球虫亚目 Suborder SPHAEROIDEA Haeckel, 1887

具球形中央囊，壳为硅质的格孔状或海绵状球体，向三个轴向均等生长。

光滑球虫科 Family Liosphaeridae Haeckel, 1887, emend. Chen et al., 2017

球形壳，单体生活，一般表面光滑，也包含一些具棘刺或不规则骨针的类型。

空球虫属 Genus *Cenosphaera* Ehrenberg, 1854

具单一的格孔状球壳，有简单的壳孔和壳腔（无延长游离小管和内放射桁）。我们观察到此类群的一些标本壳内存在较为细弱的中心网状结构。

Haeckel（1887）认为该属是所有球形类中最为简单而古老的类群，既无外伸管，也无内放射桁，与 *Collosphaera* 的主要区别在于后者的壳体形态有时呈不规则状，并根据 *Cenosphaera* 的壁孔类型将该属分为 4 个亚属。显然，这是包含种类较为广泛的一个类群。

1. 原始南极空球虫（新种）*Cenosphaera archantarctica* sp. nov.

（图版 1，图 1–9）

壳体稍小，壳壁稍厚，表面光滑或略显粗糙，个别标本有细的棘突，具类圆形孔，大小相似或不等，亚规则或不规则排列，有六角形框架，一般孔径约为孔间桁宽 1–2 倍，横跨赤道有 6–9 个孔，有些标本的壁孔被充填或沉淀以致表面模糊，似为较老地层种类，分布于 Site 274 孔的底部，未见出现于较新的地层样品中。

标本测量：壳径 106–210μm，孔径 6–18μm。

模式标本：DSDP274-42-2 1（图版 1，图 4–5），来自南极罗斯海的 DSDP Leg 28 航次 Site 274 孔的 274-42-2 岩心样品中，保存在中国科学院南海海洋研究所沉积标本室。

地理分布 南极罗斯海，古新世。

该新种特征与 *Cenosphaera vesparia* Haeckel 较为相似，主要区别是后者的壁孔大小相同，排列规则，孔间桁较细，壳壁稍薄，且主要分布于较新的地层。

2. 坚空球虫 *Cenosphaera compacta* Haeckel

（图版 1，图 10–33）

Cenosphœra compacta Haeckel, 1887, p. 65, pl. 12, fig. 7.
Cenosphaera sp. A group 2, Lazarus and Pallant, 1989, p. 365, pl. 7, figs. 5–6.

该种与 *Cenosphaera nagatai* Nakaseko（Hays, 1965, p. 165, pl. 2, fig. 6）的主要区别是壳壁明显比后者厚，个体稍大，孔间桁稍宽。

地理分布 南极罗斯海，晚古新世—早上新世。

3. 冠空球虫 *Cenosphaera cristata* Haeckel

（图版 2，图 1–9）

Cenosphaera cristata Haeckel, 1887, p. 66；Riedel, 1958, p. 223, pl. 1, figs. 1–2；Petrushevskaya, 1967, pp. 10–11, figs. 7i–iv；Petrushevskaya, 1975, p. 567, pl. 1, figs. 3, 4, pl. 17, fig. 2；Nigrini and Moore, 1979, p. S41, pl. 4, figs. 2a–b；Keany, 1979, p. 51, pl. 1, figs. 1–2；陈木宏等，2017，83 页，图版 1，图 15–16，图版 2，图 1–5。
Cenosphaera aemiliana Carnevale, 1908, p. 7, pl. 1, fig. 1.

Cenosphaera subtilis Carnevale, 1908, p. 7, pl. 1, fig. 2.
Hexastyliidae gen. sp. indet., Petrushevskaya, 1975, p. 567, pl. 17, fig. 1.

地理分布　南极罗斯海，晚渐新世—现代。

4. 福地空球虫 *Cenosphaera elysia* Haeckel

（图版 2，图 10–11）

Cenosphæra elysia Haeckel, 1887, p. 64, pl. 12, fig. 8.

地理分布　南极罗斯海，晚始新世—晚中新世。

5. 巢空球虫 *Cenosphaera favosa* Haeckel

（图版 2，图 12–21；图版 3，图 1–4）

Cenosphæra favosa Haeckel, 1887, p. 62, pl. 12, fig. 10；Blueford, 1982, p. 193, pl. 1, figs. 1–2.
Cenosphaera sp., Chen, 1975, p. 453, pl. 6, fig. 9 (only).
Cenosphaera sp. D, Lazarus and Pallant, 1989, p. 365, pl. 7, fig. 2.
Cenosphaera sp., Petrushevskaya, 1967, p. 11, fig. 7V.

地理分布　南极罗斯海，早渐新世。

6. 果空球虫 *Cenosphaera mellifica* Haeckel

（图版 3，图 5–9）

Cenosphæra mellifica Haeckel, 1887, p. 62, pl. 12, fig. 9.
Cenosphaera sp. 1, Abelmann, 1992a, p. 378, pl. 1, fig. 4.

地理分布　南极罗斯海，晚始新世—早渐新世。

7. 南极小空球虫（新种）*Cenosphaera miniantarctica* sp. nov.

（图版 3，图 10–17）

Cenosphaera sp. D, Lazarus and Pallant, 1989, p. 365, pl. 7, fig. 1.

个体很小，壳壁相对较厚，表面光滑；壁孔圆形，规则排列，横跨赤道有 5–6 孔，具六角形框架，孔间桁较细，孔径是孔间桁宽的 3–4 倍，但孔径略小于壁厚（孔间桁高）。这是很特殊的小个体 *Cenosphaera* 类，在罗斯海主要分布于晚更新世地层。在较老的早中新世—渐新世地层中还有特征相近的标本，其特征接近 *Cenosphaera* sp. D（Lazarus and Pallant, 1989, p. 365, pl. 7, fig. 1），后者的个体略大（110–120μm），壳壁稍为粗糙，壁孔多些，横跨赤道有 8–9 孔以上，其他特征基本相同，因此将它们归为同一新种。

标本测量：50–95μm。

模式标本：DSDP274-2-3 1（图版 3，图 15），来自南极罗斯海的 DSDP Leg 28 航次 Site 274 孔的 274-2-3 岩心样品中，保存在中国科学院南海海洋研究所沉积标本室。

地理分布　南极罗斯海全新世，个别见于晚始新世地层中。

该新种的基本特征是个体极细小，同时以壁孔较少又规则分布而区别于其他已知种。

8. 娜荚空球虫 *Cenosphaera nagatai* Nakaseko

（图版 3，图 18–28）

Cenosphaera nagatai Nakaseko, 1959, pp. 6–7, pl. II, figs. la–b, 2；Hays, 1965, p. 165, pl. 2, fig. 6.
Actinomma sp. Gr. B, Lazarus and Pallant, 1989, p. 366, pl. 7, fig. 9.
Hexacromyum sexaculeatum (Stöhr), Petrushevskaya, 1975, p. 569, pl. 2, figs. 3–4.

地理分布 南极罗斯海，早渐新世—晚更新世。

9. 假空球虫（新种）*Cenosphaera pseudocoela* sp. nov.

（图版 3，图 29–32）

壳壁中等厚度，具类圆形或椭圆形孔，大小不等，无规律分布，孔径一般大于孔间桁宽，孔间桁稍细；壳表生出的棘刺末端分叉并相互连接，常可形成一个较薄的假外壳（外包壳），或看似双层壳，还有一些棘刺直接伸出外壳形成细锥形或柱状的骨针；表面粗糙。

标本测量：格孔壳直径 170–180μm，壳体外径 195–205μm，骨针长 15–30μm。

模式标本：DSDP274-23-5 1（图版 3，图 29–30），来自南极罗斯海的 DSDP Leg 28 航次 Site 274 孔的 274-23-5 岩心样品中，保存在中国科学院南海海洋研究所沉积标本室。

地理分布 南极罗斯海，晚始新世—晚渐新世。

该种特征与 *Cenosphaera solantarctica* sp. nov.接近，但后者的壁孔大小相近，排列规则，壳表无骨针，具六角形框架，表面平整。

10. 里德空球虫 *Cenosphaera riedeli* Blueford

（图版 4，图 1–6）

Cenosphaera riedeli Blueford, 1982, p. 193, pl. 1, figs. 7–9.
Thecosphaerella? *ptomatus* Sanfilippo and Riedel, Petrushevskaya, 1975, p. 571, pl. 1, figs. 1–2.

地理分布 南极罗斯海，晚古新世—早渐新世。

11. 太阳南极空球虫（新种）*Cenosphaera solantarctica* sp. nov.

（图版 4，图 7–21）

Cenosphaera sp., Chen, 1975, p. 453, pl. 7, figs. 1–2.
Cenosphaera sp., Keany, 1979, pl. 5, fig. 1.

壳体大小变化较大，一般圆球形，个别为椭球形；壳壁较厚，壁孔类圆形，大小相近，规则或亚规则排列，具六角形框架，孔径约为孔间桁的 2–3 倍，横跨赤道有 14–18 孔（个体小的壁孔也少）；壳体表面有许多棘刺，分别从各孔间桁节点上生出，呈角锥状，各刺长度基本相同，较长，在末端有分叉，这些棘刺在壳体外围形成一个较厚的放射刺状假外壳。

标本测量：壳径 125–340μm，孔径 9–15μm，刺长 15–25μm。

模式标本：DSDP274-29-3 2（图版 4，图 7–8），来自南极罗斯海的 DSDP Leg 28 航次 Site 274 孔的 274-29-3 岩心样品中，保存在中国科学院南海海洋研究所沉积标本室。

地理分布　南极罗斯海，早始新世—渐新世。

该新种特征与 *Cenosphaera favosa* Haeckel 较接近，主要区别在于后者的壳表棘刺呈圆锥状，末端削尖，不分叉，而且刺长相对较短。

12. 海绵空球虫（新种）*Cenosphaera spongiformis* sp. nov.

（图版 5，图 1–2）

单一格孔状壳体被厚密的棉絮状物所包裹。格孔壳中等厚度，壁孔呈类圆形或椭圆形，大小不等，孔径多为孔间桁宽的 2–3 倍，不规则排列，孔间桁多角形；格孔壳表的棘刺发育形成不规则藤架支撑着外围的棉絮状物，絮状物层的厚度约达格孔壳径的 1/6–1/4，有的标本可见个别棘刺伸出壳外。这是一种很特殊的 *Cenosphaera* 壳体结构。

标本测量：格孔壳直径 168–190μm，壳体外径 210–260μm，刺长 7–18μm。

模式标本：DSDP274-34-1 2（图版 5，图 1–2），来自南极罗斯海的 DSDP Leg 28 航次 Site 274 孔的 274-34-1 岩心样品中，保存在中国科学院南海海洋研究所沉积标本室。

地理分布　南极罗斯海，早始新世。

该新种具有圆球形的单一格孔状壳体，外围被厚密的棉絮状物所包裹，并形成一个厚密的絮状外壳，此特征明显地区别于其他种类。

13. 女神空球虫 *Cenosphaera veneris* Clark and Campbell

（图版 5，图 3–9）

Cenosphaera veneris Clark and Campbell, 1942, p. 20, pl. 4, figs. 6, 11, 13.
Cenosphaera (?) *oceanica* Clark and Campbell, Petrushevskaya, 1975, pl. 1, figs. 12–13.
cf. *Actinomma* (?) *californica* (Clark and Campbell), Dzinoridze et al., 1978, pl. 21, fig. 7.
?*Cenosphaera oceanica* Clark and Campbell, Lazarus and Pallant, 1989, p. 365, pl. 7, figs. 7–8.
Spumellina incertae sedis Forma B, Benson, 1966, pp. 284–285, pl. 19, figs. 9–11.

我们的标本特征与 Benson（1966）的标本非常相似。

地理分布　南极罗斯海，晚始新世—早渐新世。

14. 蜂空球虫 *Cenosphaera vesparia* Haeckel

（图版 5，图 10–25）

Cenosphæra vesparia Haeckel, 1887, p. 62, pl. 12, fig. 11.

我们的标本个体大小不等。

地理分布　南极罗斯海，早始新世—现代。

15. 剑空球虫（新种）*Cenosphaera xiphacantha* sp. nov.

（图版 5，图 26–27）

个体很大，单一圆球形壳，壁较厚，壁孔圆形，大小相近，排列规则，横跨赤道有 17–19 孔，具六角形框架，孔径是孔间桁宽的 2–3 倍，壳表的各节点上有小的锥形凸起，特别是从壳壁内侧直接向外发育出唯一一根角锥状的坚实骨针，骨针的前半段为三棱状，后半段角锥状，末端缩尖，形似一把短剑；壳表略粗糙，无其他针刺。

标本测量：格孔壳直径 310–325μm，骨针长 80–90μm。

模式标本：DSDP274-31-2 1（图版 5，图 26–27），来自南极罗斯海的 DSDP Leg 28 航次 Site 274 孔的 274-31-2 岩心样品，保存在中国科学院南海海洋研究所沉积标本室。

地理分布 南极罗斯海，晚始新世—早渐新世。

该新种的壳体特征与 *Cenosphaera vesparia* Haeckel 相似，主要区别是后者壳表无骨针，孔间桁在节点处不突起，表面光滑。

16. 空球虫（未定种 1）*Cenosphaera* sp. 1

（图版 6，图 1–2）

Cenosphaera sp. A, Abelmann, 1990, p. 691, pl. 3, fig. 12.
Acrosphaera (?) *mercurius* Lazarus, van de Paverd, 1995, p. 59, pl. 8, figs. 5–8.

地理分布 南极罗斯海，出现于中渐新世。

17. 空球虫（未定种 2）*Cenosphaera* sp. 2

（图版 6，图 3–4）

格孔壳壁较薄，外包壳壁较厚，壁孔类圆形，很小，大小相近，亚规则排列，孔间桁较宽，似有六角形框架，各节点有角锥形凸起，表面粗糙，壳外无骨针。

地理分布 南极罗斯海，出现于晚始新世。

18. 空球虫（未定种 3）*Cenosphaera* sp. 3

（图版 6，图 5–6）

个体较小，圆形孔具六角形框架，排列规则，壳表有一根大骨针和一些小骨针。

地理分布 南极罗斯海，出现于晚始新世。

19. 空球虫（未定种 4）*Cenosphaera* sp. 4

（图版 6，图 7–10）

壳表较薄，仅有一个皮壳，内为空腔；壁孔类圆形或多角形，分布无规律，孔径是孔间桁宽的 3–5 倍，孔间桁宽度变化不大，但排列不规则，甚至不在一个平面上，略有起伏，在各节点上还有一些突起或小枝。

地理分布 南极罗斯海，全新世。

果球虫属 Genus *Carposphaera* Haeckel, 1881

具一髓壳和一皮壳，二者由贯穿中央囊的放射桁相连。两壳之间的间距大于髓壳半径。

20. 角果球虫 *Carposphaera angulata* (Haeckel)

（图版 6，图 11–15）

Carposphaera angulata (Haeckel), van de Paverd, 1995, p. 80, pl. 14, figs. 1–5, pl. 15, figs. 1–3, pl. 16, figs. 1, 2, 4.

Acanthosphaera angulata Haeckel, 1887, p. 216, pl. 26, fig. 4.
Acanthosphaera reticulata Haeckel, 1887, p. 217, pl. 26, fig. 5.

地理分布　南极罗斯海，晚始新世—早上新世。

21. 异果球虫（新种）*Carposphaera anomala* sp. nov.

（图版 6，图 16–21）

皮壳椭球形或圆三角形，髓壳圆球形，大小之比为 2:1 或 3:1；皮壳壁孔较小而多，亚圆形或椭圆形，大小不等，不规则排列，具多角形孔间桁，横跨赤道有 8–15 孔；髓壳近圆球形，髓壳内有若干放射桁交汇在中心处；在皮壳的外面有一较薄易碎的外套包围，与皮壳之间由许多细放射桁连接，表面粗糙，无放射骨针。

标本测量：皮壳直径 100–120μm，髓壳直径 30–60μm。

模式标本：DSDP274-34-5 2（图版 6，图 20–21），来自南极罗斯海的 DSDP Leg 28 航次 Site 274 孔的 274-34-5 岩心样品中，保存在中国科学院南海海洋研究所沉积标本室。

地理分布　南极罗斯海，出现于中始新世。

该种特征与 *Carposphaera magnaporulosa* Clark and Campbell 较接近，主要区别是后者壳体呈圆球形，壁孔较大而少，而前者的壳体明显呈椭球形或圆三角形。

22. 圆果球虫 *Carposphaera globosa* Clark and Campbell

（图版 7，图 1–12）

Carposphaera globosa Clark and Campbell, 1945, p. 9, pl. 1, figs. 6–8；谭智源、宿星慧，1982，137 页，图版 1，图 3–4；Blueford, 1988, p. 247, pl. 3, fig. 6；谭智源、陈木宏，1999，125–126 页，图 5-18；陈木宏等，2017，85 页，图版 3，图 3–4。
Carposphaera buxiformis Clark and Campbell, 1942, p. 21, pl. 5, fig. 20.

地理分布　南极罗斯海，早始新世—晚中新世。

23. 大孔果球虫 *Carposphaera magnaporulosa* Clark and Campbell

（图版 7，图 13–27）

Carposphaera magnaporulosa Clark and Campbell, 1942, p. 21, pl. 5, figs. 15, 17, 21, 23.

个体很小，仅一髓壳和一皮壳，横跨赤道有约 5 个壁孔，表面光滑。罗斯海的标本基本符合这些特征，但壳表略显粗糙。

地理分布　南极罗斯海，晚古新世—早渐新世。

24. 结实果球虫（新种）*Carposphaera sterrmona* sp. nov.

（图版 7，图 28–29）

皮壳呈类圆球形或近立方形，壁较厚，皮壳壁孔较大，类圆形，大小不等，横跨赤道有约 6–7 孔，不规则排列，具六角形或不规则形框架；髓壳很小，圆球形，仅占皮壳的 1/5–1/4，髓壳壁孔很细，圆形，髓壳与皮壳之间由许多放射桁连接，有些放射桁在中间有分枝相连，这些分枝也连接到皮壳，与主放射桁共同形成放射骨针；壳表有约 50–60 根较为粗短的角锥形骨针，针基呈三棱状，针长约为皮壳径的 1/4；表面粗糙。

标本测量：皮壳直径 150–170μm，髓壳直径 25–30μm，骨针长 25–40μm。

模式标本：DSDP274-28-1 2（图版 7，图 28–29），来自南极罗斯海的 DSDP Leg 28 航次 Site 274 孔的 274-28-1 岩心样品中，保存在中国科学院南海海洋研究所沉积标本室。

地理分布 南极罗斯海，出现于早渐新世。

该新种特征与 *Carposphaera magnaporulosa* Clark and Campbell 接近，主要区别是后者个体较小，壳表一般光滑，无粗实的放射骨针。

25. 果球虫（未定种 1）*Carposphaera* sp. 1

（图版 7，图 30–31）

该未定种与 *Carposphaera polypora* (Haeckel)很相似，主要区别是前者壳表有若干三片棱柱状的骨针，很短（可能被折断）。

地理分布 南极罗斯海，出现于早渐新世。

26. 果球虫（未定种 2）*Carposphaera* sp. 2

（图版 7，图 32–33）

外壳近圆球形，似有一开口，口径小于壳径的一半，壁孔类六角形，较大，横跨赤道有约 6 孔，孔径为孔间桁的 4–5 倍，孔间桁较细，节点上有小棘突；内壳近葫芦形，其顶端封闭，伸至外壳开口处。

地理分布 南极罗斯海，出现于早渐新世。

偏球虫属（新属）Genus *Eccentrisphaera* gen. nov.

单皮壳有 2–3 个髓壳，或双皮壳有 1–2 个髓壳，各髓壳的壳体中心相互一致，但与皮壳的壳体中心不同，因此髓壳在皮壳内处于偏离中心的位置，靠近皮壳一侧内壁；壳表常有一些小棘凸或细骨针。

模式种：*Eccentrisphaera bimedullaris* sp. nov.

27. 双皮偏球虫（新种）*Eccentrisphaera biderma* sp. nov.

（图版 8，图 1–4）

具 2 个皮壳和 1–2 个髓壳；外皮壳壁稍薄，壁孔圆形，有六角形框架，排列规则，横跨赤道有 12–14 孔，表面光滑或有小棘凸；内皮壳较大，大小为外皮壳的 2/3，两壳的间距小于髓壳半径，壁孔圆形或类圆形，规则或不规则排列，横跨赤道有 9–10 孔；髓壳不在壳体中央，外髓壳紧贴内皮壳，内髓壳很小，位于外髓壳中心；4 个壳体的大小比例为 1∶3∶7∶10。

标本测量：外皮壳直径约 185–210μm，内皮壳孔径 130–145μm，外髓壳直径 62–75μm，内髓壳直径 22μm。

模式标本：DSDP274-12-5 1a（图版 8，图 1–2），来自南极罗斯海的 DSDP Leg 28 航次 Site 274 孔的 274-12-5 岩心样品中，保存在中国科学院南海海洋研究所沉积标本室。

地理分布 南极罗斯海，出现于晚中新世。

该新种区别于其他类似种类的主要特征是具有 2 个皮壳和 1–2 个髓壳，外髓壳的壳壁紧贴着皮壳的内壁，壳表光滑或骨针极少。

28. 双髓壳偏球虫（新种）*Eccentrisphaera bimedullaris* sp. nov.

（图版 8，图 5–20）

Actinomma cocles Renaudie and Lazarus, Renaudie, 2012, p. 36, pl. 4, fig. 6 (only).

壳体较小，皮壳壁薄或中等厚度，壁孔圆形，大小不一或近等，分布规则或亚规则，具六角形框架，孔间桁较细，孔径约为孔间桁宽的 4–5 倍，横跨赤道有 6–9 孔，一般壁薄的孔少，壁稍厚的相对孔较多；两个髓壳，不在壳体中心位置，外髓壳的一侧紧贴或靠近皮壳，内髓壳在外髓壳的中心。外髓壳壳径约为皮壳壳径的 1/3–1/2，内髓壳较小，约为外髓壳的 1/3；壳体表面常有少量的细小骨针。

标本测量：皮壳直径 106–143μm，皮壳孔径 13–25μm，骨针长 10–30μm。

模式标本：DSDP274-6-1 1（图版 8，图 8–9），来自南极罗斯海的 DSDP Leg 28 航次 Site 274 孔的 274-6-1 岩心样品中，保存在中国科学院南海海洋研究所沉积标本室。

地理分布 南极罗斯海，上新世—现代。

该新种有两个髓壳和一个皮壳，两个髓壳同心，但不是位于皮壳的中心，而是靠近皮壳的一侧，此特征明显区别于其他种类。

29. 小偏球虫 *Eccentrisphaera minima* (Clark and Campbell)

（图版 8，图 21–24）

Carposphaera minima Clark and Campbell, 1942, p. 21, pl. 4, figs. 8–9.

髓壳较小，约为皮壳的 1/3 或更小，在皮壳内处于非对称的位置（不在中心）。

地理分布 南极罗斯海，晚始新世—早渐新世。

30. 孔偏球虫（新种）*Eccentrisphaera porolaris* sp. nov.

（图版 8，图 25–26）

具两个皮壳和一个髓壳，髓壳紧贴内皮壳壳壁，三壳大小比例为 1∶2.5∶4；外皮壳圆球形或近椭球形，一端上有开口，口缘透明，很短，口径是壳径的 1/4–1/3，外皮壳的壁孔类圆形，大小近等或不等，亚规则排列，横跨赤道有 15–17 孔；内皮壳圆球形，壁孔近圆形，大小不等，具明显的六角形框架，亚规则排列，横跨赤道有 9–11 孔；髓壳不在皮壳中央，紧贴内皮壳一侧的内壁；壳表有很短的棘凸，无放射骨针。

标本测量：外皮壳直径约 200–230μm，内皮壳直径 115–120μm，髓壳直径 50–55μm。

模式标本：DSDP274-12-5 1b（图版 8，图 25–26），来自南极罗斯海的 DSDP Leg 28 航次 Site 274 孔的 274-12-5 岩心样品中，保存在中国科学院南海海洋研究所沉积标本室。

地理分布 南极罗斯海，出现于晚中新世。

该新种区别于其他类似种类的主要特征是具有两个皮壳和一个髓壳，外皮壳呈近椭球形，一端上有一个开口，其他相近种类的皮壳均为封闭的圆球形。

31. 三髓壳偏球虫（新种）*Eccentrisphaera trimedullaris* sp. nov.

（图版 8，图 27–29）

壳体较大，皮壳壁稍厚，孔较多，类圆形，大小近等，分布亚规则，有六角形框架，孔径约为孔间桁宽的 3–4 倍，横跨赤道有 15–17 孔；有三个髓壳，不在壳体中心位置，外髓壳的一侧紧贴皮壳，两个内髓壳在外髓壳的中心；四个壳体（三个髓壳和一个皮壳）的大小比例为 1∶3∶9∶18；壳表略显粗糙，有一些棘凸和个别小骨针。

标本测量：皮壳直径约 250μm，皮壳孔径 10–15μm，骨针长 8–15μm。

模式标本：DSDP274-7-1 1（图版 8，图 27–29），来自南极罗斯海的 DSDP Leg 28 航次 Site 274 孔的 274-7-1 岩心样品中，保存在中国科学院南海海洋研究所沉积标本室。

地理分布 南极罗斯海，晚上新世。

该新种区别于其他已有种类的主要特征是具有三个髓壳和一个皮壳，三个髓壳同心，但不与皮壳同一个中心，而是位于靠近皮壳的一侧，外髓壳的壳壁紧贴着皮壳的内壁。

光滑球虫属 Genus *Liosphaera* Haeckel, 1881

具两皮壳；两壳大小相近，壳间距小于内壳半径。

32. 果光滑球虫（新种）*Liosphaera carpolaria* sp. nov.

（图版 8，图 30–31）

双层壳，无髓壳；外壳稍薄，壁孔较大，多角形，大小不一，排列不规则，横跨赤道有 7–9 孔，孔间桁较细，孔径约为孔间桁宽的 3–7 倍，表面有一些小凸起；内壳壁孔近圆形，大小相近，亚规则排列，横跨赤道有 6–7 孔，具六角形框架，内壳与外壳由许多较细的放射桁相连；壳体表面略粗糙，无放射骨针。

标本测量：外壳直径 155–170μm，内壳孔径 100–106μm，两壳间距 25–35μm。

模式标本：DSDP274-10-5 2（图版 8，图 30–31），来自南极罗斯海的 DSDP Leg 28 航次 Site 274 孔的 274-10-5 岩心样品中，保存在中国科学院南海海洋研究所沉积标本室。

地理分布 南极罗斯海，早上新世。

该新种与 *Liosphaera hexagonia* Haeckel（1887, pl. 76, pl. 20, fig. 3）较相似，主要区别是后者的两个壳壁孔较多，六角形，横跨赤道均有约 20 孔，排列规则，两壳间距较小，外壳孔间桁呈细丝状，且壳表光滑。

33. 光滑球虫（未定种 1）*Liosphaera* sp. 1

（图版 8，图 32–33）

外壳近椭球形，内壳圆球形；双壳的壳壁较厚，壁孔类圆形，较小而多，亚规则排列，孔径是孔间桁宽的 2–3 倍，或孔径与孔间桁宽近等，两壳之间的放射桁很少。

地理分布 南极罗斯海，中中新世。

荚球虫属 Genus *Thecosphaera* Haeckel, 1881

具两髓壳与一皮壳。

34. 尖纹荚球虫 *Thecosphaera akitaensis* Nakaseko

（图版 9，图 1–13）

Thecosphaera akitaensis Nakaseko, 1971, p. 63, pl. 1, figs. 4a–b；Ling, 1975, p. 717, pl. 1, figs. 7–8；Sakai, 1980, p. 704, pl. 2, figs. 6a–b；陈木宏等，2017，88–89 页，图版 4，图 7–10。

地理分布　南极罗斯海，早始新世—中渐新世。

35. 内方荚球虫 *Thecosphaera entocuba* Chen et al.

（图版 9，图 14–21）

Thecosphaera entocuba Chen et al.，陈木宏等，2017，89 页，图版 4，图 11–14。
Spongoplegma sp., Chen, 1975, p. 454, pl. 22, figs. 1–2.

外髓壳呈正立方体形，与 *Joergensenium apollo* Kamikuri（2010, p. 99, pl. 3, figs. 1–3, 5, 6, pl. 4, figs. 17a–c）的主要区别在于后者的外髓壳呈七面体形，或不规则五角形，皮壳壁薄。

地理分布　南极罗斯海，晚渐新世。

36. 土豆荚球虫 *Thecosphaera glebulenta* (Sanfilippo and Riedel)

（图版 9，图 22–26）

Thecosphaerella glebulenta Sanfilippo and Riedel, 1973, p. 521, pl. 3, figs. 12–13, pl. 26, fig. 1；Kamikuri et al., 2013, figs. 6, 8a–b.

地理分布　南极罗斯海，早-中始新世。

37. 日本荚球虫 *Thecosphaera japonica* Nakaseko

（图版 10，图 1–4）

Thecosphaera japonica Nakaseko, 1971, p. 61, pl. 1, figs. 3a–b; Sakai, 1980, p. 704, pl. 2, figs. 5a–b; Motoyama, 1996, p. 252, pl. 2, figs. 3a–b；陈木宏等，2017，90 页，图版 4，图 19–20，图版 85，图 10–11。

地理分布　南极罗斯海，晚古新世—早渐新世。

38. 细孔荚球虫（新种）*Thecosphaera minutapora* sp. nov.

（图版 10，图 5–12）

具一个皮壳、两个髓壳；皮壳中等壁厚，壁孔类圆形，很小，较多，大小相近，横跨赤道有 12–14 孔，规则或亚规则排列，孔径是孔间桁宽的 1–2 倍，六角形框架较浅或无；两个髓壳均很小，有若干放射桁与皮壳相连，三壳之比为 1∶3∶12；壳体表面光滑，个别有很短的角锥状骨针。

标本测量：皮壳直径 170–210μm，外髓壳直径 45–60μm，内髓壳直径 15–20μm。

模式标本：DSDP274-30-6 1（图版 10，图 10–11），来自南极罗斯海的 DSDP Leg 28

航次 Site 274 孔的 274-30-6 岩心样品中，保存在中国科学院南海海洋研究所沉积标本室。

地理分布 南极罗斯海，出现于晚始新世。

该新种与 *Actinomma henningsmoeni* Goll and Bjørklund（1989, pl. 2, figs. 10–15）有些相似，但后者皮壳略有变形，不完全圆球形，壳表稍粗糙，髓壳稍大，约占壳体的 1/3。新种与 *Thecosphaera* 属其他各种的主要区别是新种的髓壳很小，仅占壳体的 1/4，皮壳壁孔较细小且数量多。

39. 显赫荚球虫 *Thecosphaera nobile* (Ehrenberg)

（图版 10，图 13–20）

Haliomma nobile Ehrenberg, Petrushevskaya, 1975, p. 568, pl. 17, fig. 16.

地理分布 南极罗斯海，主要出现于古新世。

40. 卵荚球虫（新种）*Thecosphaera ovata* sp. nov.

（图版 10，图 21–24）

皮壳卵形或椭球形，壁孔类圆形或多角形，大小相近，排列不规则，有或无六角形框架，横跨短轴赤道有 11–13 孔，孔径约为孔间桁宽的 2–4 倍，壳表光滑或粗糙，具六角形框架，有节点突起；外髓壳圆球形或椭球形，较大，壁厚；内髓壳均为圆球形，较小；两个髓壳与皮壳短轴的大小比例为 1∶2.5∶4，许多放射桁连接着三壳，表面无放射骨针。

标本测量：皮壳长轴直径 190–200μm，皮壳短轴直径 150–155μm，外髓壳直径 100–125μm，内髓壳直径 45–55μm。

模式标本：DSDP274-12-3 1（图版 10，图 23–24），来自南极罗斯海的 DSDP Leg 28 航次 Site 274 孔的 274-12-3 岩心样品中，保存在中国科学院南海海洋研究所沉积标本室。

地理分布 南极罗斯海，出现于晚中新世。

该新种以典型的椭球形壳体特征区别于 *Thecosphaera* 属的其他各个种类，壳体封闭。

41. 厚壁荚球虫（新种）*Thecosphaera pachycortica* sp. nov.

（图版 10，图 25–32）

具一个皮壳、两个髓壳；皮壳壁厚，壁孔类圆形，横跨赤道有 9–11 孔，有六角形框架，孔间桁较细，很高，桁宽是孔径的 1/4–1/3，从皮壳内壁到外壁的距离较大，似呈放射桁连接着皮壳内壁与外壁，形成近似两个壳层的皮壳；外髓壳呈葫芦形，其伸长的顶侧靠近皮壳，外髓壳的短直径是皮壳内径的 1/2，内髓壳圆球形，很小，大小约为外髓壳的 1/3；壳体表面光滑，有若干粗短的角锥状骨针。

标本测量：皮壳外径约 140–170μm，皮壳内径 100–113μm，外髓壳直径 44–50μm，内髓壳直径约 18μm，骨针长 6–18μm。

模式标本：DSDP274-23-5 1（图版 10，图 25–26），来自南极罗斯海的 DSDP Leg 28 航次 Site 274 孔的 274-23-5 岩心样品中，保存在中国科学院南海海洋研究所沉积标本室。

地理分布　南极罗斯海，晚古新世—早渐新世。

该新种特征与 *Thecosphaera sanfilippoae* Blueford 较相似，主要区别为后者皮壳壳壁较薄，外髓壳不呈葫芦形。

42. 帕维阿荚球虫 *Thecosphaera parviakitaensis* (Kamikuri)

（图版 11，图 1–4）

Hexacontium parviakitaensis Kamikuri, 2010, p. 98, pl. 1, figs. 9a–b, 13a–14b, pl. 4, fig. 15.
Thecosphaera akitaensis Nakaseko, Sakai, 1980, p. 704, pl. 2, figs. 6a–b.
Excentrodiscus japonicus (Nakaseko and Nishimura), Kamikuri, 2010, pp. 86, 94, pl. 2, figs. 2a–3b, 5a–6b, 8a–9b, 11a–b, 13a–14b, pl. 4, figs. 19a–c.

地理分布　南极罗斯海，始新世。

43. 网荚球虫（新种）*Thecosphaera reticularis* sp. nov.

（图版 11，图 5–19）

个体大小有变化，有的标本壳体较大；皮壳壳壁较厚，具类圆形孔，大小相近，亚规则排列，横跨赤道有 12–18 孔，有六角形框架，皮壳外表孔间桁突起末端有分枝，相互连接形成稀疏的细网状外表，有的标本孔间桁突起还进一步形成骨针，并在末端分叉相互交错连接形成海绵结构的外表；髓壳较小，两个髓壳与皮壳（内径）的大小比例约为 1∶3∶9，外髓壳壁孔很小，圆形，近规则排列，横跨赤道有 8–10 孔；壳体外表粗糙，有个别短小的放射骨针。

标本测量：皮壳直径 160–280μm，外髓壳直径 56–105μm，总壳（含外表网层）直径 180–375μm。

模式标本：DSDP274-36-1 2（图版 11，图 15–17），来自南极罗斯海的 DSDP Leg 28 航次 Site 274 孔的 274-36-1 岩心样品中，保存在中国科学院南海海洋研究所沉积标本室。

地理分布　南极罗斯海，早始新世—早渐新世。

该新种特征与 *Thecosphaera zittelii* Dreyer 较接近，主要区别在于后者皮壳外无末端分叉的突起或骨针，壳体外表无网状或海绵状组织。

44. 桑氏荚球虫 *Thecosphaera sanfilippoae* Blueford

（图版 12，图 1–10）

Thecosphaera sanfilippoae Blueford, 1982, pp. 198–199, pl. 5, figs. 5–6；陈木宏等，2017，90 页，图版 5，图 1–4。
Actinomma henningsmoeni Goll and Bjørklund, 1989, p. 728, pl. 2, figs. 10–15.
Thecosphaera inermis (Haeckel), Kamikuri et al., 2008, p. 171, pl. 2, figs. 5a–b.
Actinomma sp., Bjørklund, 1976, p. 1124, pl. 19, figs. 1–2.

地理分布　南极罗斯海，晚始新世—早渐新世。

45. 东芝荚球虫 *Thecosphaera tochigiensis* Nakaseko

（图版 12，图 11–20）

Thecosphaera tochigiensis Nakaseko, 1971, p. 62, pl. 1, figs. 5a–b.
Actinomma medusa (Ehrenberg), Abelmann, 1990, p. 690, pl. 1, fig. 6.

Amphisphaera aff. *magnasporulosa* (Clark and Campbell), Hollis, 2002, p. 286, pl. 1, fig. 1.
Hexactromyum delicatulum (Dogel) group, Petrushevskaya, 1975, p. 569, pl. 2, fig. 11.
Actinomma sp. B, Abelmann, 1990, p. 691, pl. 1, figs. 7A–B.
Actinomma sp., Bjørklund, 1976, p. 1124, pl. 19, figs. 1–2.

地理分布 南极罗斯海，中渐新世—晚中新世。

46. 兹特荚球虫 *Thecosphaera zittelii* Dreyer

（图版 12，图 21–28）

Thecosphaera zittelii Dreyer, 1890, p. 477, Taf. XVI, Fig. 5；陈木宏等，2017，90 页，图版 5，图 5–8。
Thecosphaera japonica Nakaseko, 1971, p. 61, pl. 1, figs. 3a–b；Motoyama, 1996, p. 252, pl. 2, figs. 3a–b.
Sphaeroidea gen. sp., Petrushevskaya, 1975, p. 630, pl. 17, fig. 17.

地理分布 南极罗斯海，晚始新世—晚渐新世。

47. 荚球虫（未定种 1）*Thecosphaera* sp. 1

（图版 13，图 1–2）

皮壳卵形或椭球形，壳长轴是短轴的 1.5 倍，两端略缩窄，壳壁孔及孔间桁很细，类圆形或不规则形孔，大小略有差异，排列不规则，横跨壳的短轴有 16–18 孔，孔径约为孔间桁宽的 1–3 倍，壳表平滑；两个髓壳均为圆球形，位于皮壳中心，髓壳上有许多放射桁与皮壳中部连接，两个髓壳的大小与皮壳短轴之比为 1∶3∶6；无放射骨针。

该未定种特征与 *Thecosphaera ovata* sp. nov.较接近，但后者的壁孔相对较大而少，髓壳较大，壳表稍粗糙或有棘凸。

地理分布 南极罗斯海，出现于中渐新世。

洋葱球虫属 Genus *Cromyosphaera* Haeckel, 1881

具四层壳，两个髓壳和两个皮壳，或为一个髓壳三个皮壳，各壳间由许多放射桁连接。

48. 四壳洋葱球虫 *Cromyosphaera quadruplex* Haeckel

（图版 13，图 3–5）

Cromyosphaera quadruplex Haeckel, 1887, p. 84, pl. 30, fig. 9.

四个壳体的大小比例为 1∶2∶4∶5。外皮壳光滑，具较大的六角形壁孔，孔径约为孔间桁宽的 10 倍；内皮壳的壁孔呈规则六角形或不规则多角形，孔径是孔间桁宽的 5 倍；两个髓壳的壁孔为规则圆形，孔径与桁宽近等。

罗斯海的标本特征与 Haeckel（1887）较相似，罗斯海标本的个体明显较大，壳壁较厚，尚有区别，但壳体形态与结构基本一致。暂定该种。

地理分布 南极罗斯海，出现于早渐新世。

49. 粗糙洋葱球虫（新种）*Cromyosphaera asperata* sp. nov.

（图版 13，图 6–7）

壳圆球形，具两个皮壳和两个髓壳，内皮壳的大小是外皮壳的一半，四个壳体的大小比例约为 1∶3∶8∶15，髓壳很小；外皮壳由稍粗的不规则孔间桁组成，呈海绵结构，壳壁较厚，壳表棘刺状；内皮壳与外髓壳的壳壁结构与外皮壳相似，但壳壁较薄，整个壳内的一些区域空间也有类似海绵组织的存在，内髓壳呈格孔状，壁孔很细；四个壳体之间由 17–22 根放射桁相连接，放射桁较粗，不伸出壳外，壳表粗糙，无放射骨针。

标本测量：外皮壳直径 400μm，内皮壳直径 220μm，外髓壳直径 90μm，内髓壳直径 30μm。

模式标本：DSDP274-25-1 2（图版 13，图 6–7），来自南极罗斯海的 DSDP Leg 28 航次 Site 274 孔的 274-25-1 岩心样品中，保存在中国科学院南海海洋研究所沉积标本室。

地理分布　南极罗斯海，出现于早渐新世。

该新种特征与 *Cromyosphaera quadruplex* Haeckel 较接近，主要区别是后者的四个壳均为格孔状，各壳的壁孔圆形或六角形，大小相等，排列规则，具六角形框架，无海绵结构，表面光滑。

编枝球虫属 Genus *Plegmosphaera* Haeckel, 1882

球壳呈海绵网状，中空，无髓壳。

50. 糙编枝球虫（新种）*Plegmosphaera asperula* sp. nov.

（图版 13，图 8–9；图版 14，图 1–14）

Plegmosphaera sp.，陈木宏等，2017，93 页，图版 6，图 4–6。

壳近圆球形或不规则球形、椭球形，个体较大，内部是一个大空腔；壳壁稍厚，由类似海绵组织的松散网格桁无规律交汇而成，一些平面上的节点处无突起；网孔为多类不规则形，形状各异，大小差异较大，散乱分布，各孔内缘圆滑；孔间桁一般较粗实，粗细略有差异，在一些部位呈三维方向发育；表面粗糙，或有凹凸，无放射骨针。

标本测量：壳直径约 250–460μm。

模式标本：DSDP274-25-3 2（图版 13，图 8–9），来自南极罗斯海的 DSDP Leg 28 航次 Site 274 孔的 274-25-3 岩心样品中，保存在中国科学院南海海洋研究所沉积标本室。

地理分布　南极罗斯海，早始新世—早渐新世。

新种与 *Plegmosphaera coelopila* Haeckel（1887, p. 88）的主要区别是后者的孔间桁较细和壳表平滑，而新种的壳体常呈不规则状，孔间桁较粗，壳表粗糙，有些凹凸不平。

51. 球编枝球虫（新种）*Plegmosphaera globula* sp. nov.

（图版 15，图 1–14）

单一球壳，圆球形或椭球形，壳壁较厚；壳壁呈复合海绵结构，壁孔类圆形或不规则形，大小不一，数量较多，孔间桁宽窄不一，无规律交错组成；壳体封闭，可见个别标本上有一个圆形小窗孔，位置不定；壳内常见有一个或更多的圆形小壳体，无规律地分布在放射虫壳体内的不同部位（不在壳的中心），疑为被吞噬或共生的小硅藻。壳体表面平整光滑或略粗糙，无放射骨针或棘凸。

标本测量：圆球形壳直径 190–270μm；椭球形壳长轴 210–350μm，短轴 150–290μm。

模式标本：DSDP274-26-1 2（图版 15，图 13–14），来自南极罗斯海的 DSDP Leg 28 航次 Site 274 孔的 274-26-1 岩心样品中，保存在中国科学院南海海洋研究所沉积标本室。

地理分布 南极罗斯海，早始新世—早渐新世。

该新种与 *Plegmosphaera petrushevia* sp. nov.的主要区别是后者的壳体均近似椭球形，无圆球形壳，壳壁较薄，壁孔与孔间桁较为细小，壳表被一些微型凸起的脊线分隔为一些大小不等的小区。

52. 陪氏编枝球虫（新种）*Plegmosphaera petrushevia* sp. nov.

（图版 15，图 15–20）

Plegmosphaera monikae Petrushevskaya, 1975, p. 655, pl. 30, figs. 1–7.
Cenosphaera sp. A, Abelmann, 1992a, pl. 1, fig. 6.

单一球壳，个体较大，类椭球形或卵形，形状常不太规则，壳壁较薄，由网状组织构成，近似海绵状，壁孔很小，近圆形，大小相似，不规则分布，但被壳表的一些微型凸起的脊线所分隔，呈现出许多相邻而大小不等的小区，每一小区中包含的壁孔数量不等，孔间桁较细，横跨赤道有 30–40 孔；壳内常见有一个或更多的圆形小壳体，无规律地分布在放射虫壳体内的不同部位（不在壳的中心），疑为被吞噬或共生的小硅藻。在壳体的一端可见一圆形的小窗孔，表面平整光滑，无骨针。

标本测量：壳直径 240–375μm。

模式标本：DSDP274-27-1 2（图版 15，图 15–16），来自南极罗斯海的 DSDP Leg 28 航次 Site 274 孔的 274-27-1 岩心样品中，保存在中国科学院南海海洋研究所沉积标本室。

地理分布 南极罗斯海，中渐新世。

新种与 *Plegmosphaera coelopila* Haeckel（1887, p. 88）的主要区别是后者的壳壁稍厚，壁孔较大而孔间桁很细（网孔径是孔间桁宽的 5–10 倍）。我们的标本与 *Plegmosphaera monikae* Petrushevskaya（1975）的标本特征相似，应属同一个种，但由于 Petrushevskaya（1975）在建立该新种时并没有任何文字的描述说明等，仅有照相图版及新种名称，完全不符合建种规范，认为是一个无效的种名。

53. 针编枝球虫（新种）*Plegmosphaera spiculata* sp. nov.

（图版 16，图 1–19）

单一球壳，圆球形、椭球形或长卵形，一般壁厚；壳壁海绵状、蜘蛛状或格孔状，壁孔类圆形或不规则形，较细小，分布无规律，孔间桁稍宽；在壳体一端常有一个浅开口，口缘具较粗的圆锥形齿，有的壳体表面上还可见一些粗短的圆锥形骨针或不规则形骨针；壳腔内常见个别圆形或条形的小壳体，无规律地分布在放射虫壳体内的不同部位（不在壳的中心），疑为被吞噬或共生的小硅藻；壳表粗糙。

标本测量：壳体长径 205–430μm，短径 130–340μm；口缘齿及骨针长 12–30μm，齿及针的基宽 10–30μm。

模式标本：DSDP274-33-5 1（图版 16，图 14–15），来自南极罗斯海的 DSDP Leg 28

航次 Site 274 孔的 274-33-5 岩心样品中，保存在中国科学院南海海洋研究所沉积标本室。

地理分布　南极罗斯海，晚始新世—中渐新世。

该种特征与 *Plegmosphaera globula* sp. nov.较接近，主要区别是后者的壳体无开口及口缘齿，壳表无任何骨针。

柳编虫属 Genus *Spongoplegma* Haeckel, 1881

外壳格孔近海绵架构，简单的格孔状髓壳被不规则的疏松海绵状结构组织所包裹，由放射桁将两壳相连。Haeckel（1887）认为 *Spongoplegma*= *Carposphaera*。

54. 南极柳编虫 *Spongoplegma antarcticum* Haeckel

（图版 17，图 1–16）

Spongoplegma antarcticum Haeckel, 1887, p. 90；Hays, 1965, pp. 165–167, pl. 1, fig. 1；Chen, 1975, p. 454, pl. 22, figs. 3–4.
Spongoplegma aff. *antarcticum* Haeckel, Takemura, 1992, p. 742, pl. 2, figs. 9–10.
Rhizosphaera antarcticum (Haeckel) group, Petrushevskaya, 1975, p. 571.
Diploplegma banzare Riedel, 1958, p. 223, pl. 1, figs. 3–4.
Thecosphaera antarctica Nakaseko, 1959, p. 8, pl. II, figs. 7, 10a–b.
Styptosphaera (?) *spumacea* Haeckel, Nigrini, 1970, p. 167, pl. 1, figs. 7–8；Nigrini and Moore, 1979, p. S71, pl. 8, figs. 6a–b；Kamikuri et al., 2008, p. 170, pl. 1, fig. 7.
Spongoplegma sp., Chen, 1975a, p. 454, pl. 22, figs. 1–2；Lazarus, 1992, p. 796, pl. 5, figs. 1–4.

地理分布　南极罗斯海，晚始新世—早上新世。

55. 南极柳编虫（亲近种）*Spongoplegma* aff. *antarcticum* Haeckel

（图版 17，图 17–18）

Spongoplegma antarcticum Haeckel, 1887, p. 90；Hays, 1965, pp. 165–167, pl. I, fig. 1.
Spongoplegma aff. *antarcticum* Haeckel, Takemura, 1992, p. 742, pl. 2, figs. 9–10.

地理分布　南极罗斯海，出现于中始新世。

56. 方柳编虫（新种）*Spongoplegma quadratum* sp. nov.

（图版 17，图 19–21）

个体较小，外形呈近正方形；外壳由疏松海绵结构或杂乱网络组成，包裹着髓壳，壁孔的形状、大小与分布无规律，较为疏松，孔间桁和放射桁宽近似；髓壳为简单圆球形，壁孔类圆形，孔间桁较细，髓壳直径约为外壳的 1/3–2/5；壳体表面粗糙，无放射骨针。

标本测量：外壳直径 118–150μm，内壳直径 40–45μm。

模式标本：DSDP274-1-5 1（图版 17，图 19–21），来自南极罗斯海的 DSDP Leg 28 航次 Site 274 孔的 274-1-5 岩心样品中，保存在中国科学院南海海洋研究所沉积标本室。

地理分布　南极罗斯海，全新世。

新种的壳体结构特征与 *Spongoplegma antarcticum* Haeckel 较相似，主要区别是后者的壳体较大，呈圆球形。

胶球虫科 Family Collosphaeridae Müller, 1858

营群体生活，各个体由气泡状的胶状体相连，伪足连成网状。每个中央囊具清楚、略呈不规则形的骨骼壳。

胶球虫属 Genus *Collosphaera* Müller, 1855

壳简单，内外平滑，无任何骨针或小管。

57. 百孔胶球虫 *Collosphaera confossa* Takahashi

（图版 18，图 1–6）

Collosphaera confossa Takahashi, 1991, p. 56, pl. 2, figs. 4–5；Kamikuri et al., 2008, p. 168, pl. 2, fig. 16.
Collosphaera spp., Kling, 1973, pl. 6, figs. 6–8.
Plegmosphaera monikae Petrushevskaya, 1975, pl. 30, figs. 1–7.
Cenosphaera? sp. aff. *Cenosphaera euganea* Squinabol, Petrushevskaya and Kozlova, 1972, p. 531, pl. 1, fig. 10, pl. 4, fig. 1.

地理分布 南极罗斯海，晚始新世—早渐新世。

58. 卵胶球虫 *Collosphaera elliptica* Chen and Tan

（图版 18，图 7–17）

Collosphaera elliptica Chen and Tan, 1989, p. 1, pls. 1–2；陈木宏、谭智源，1996，158 页，图版 2，图 11–12；陈木宏等，2017，93–94 页，图版 7，图 3–5。
Collosphaera reynoldsi Kamikuri, 2010, p. 97, pl. 3, figs. 18–25.
Collosphaera sp., Lazarus and Pallant, 1989, pp. 365–366, pl. 7, fig. 3.
Collosphaerid spp., Lazarus, 1992, pl. 5, figs. 9–10.

地理分布 属于世界性广布种，常见于各大洋的热带或温暖海域水体中，白令海和南极罗斯海也有发现，早始新世—现今。

59. 胶球虫 *Collosphaera huxleyi* Müller

（图版 18，图 18–27）

Collosphaera huxleyi Müller, 1855, p. 238；Müller, 1858, pp. 55–59, pl. 8, figs. 6–9；Haeckel, 1862, S. 534, pl. 34, figs. 1–11；Haeckel, 1887, p. 96；Cienkowski, 1871, S. 343, pl. 29, figs. 1–6；Popofsky, 1917, p. 241, text-figs. 2–3, pl. 13, figs. 1–9；斯特列尔科夫、列雪特尼阿克，1962，127–128 页，图 9；Strelkov and Reshetnyak, 1971, p. 332, pl. 4, figs. 21, 23, text-figs. 19–21；Boltovskoy and Riedel, 1980, p. 103, pl. 1, fig. 5；Takahashi, 1991, p. 56, pl. 2, figs. 8–11；van de Paverd, 1995, pp. 39, 43, pl. 1, fig. 4；陈木宏、谭智源，1996，157 页，图版 2，图 3–4，图版 36，图 6；谭智源、陈木宏，1999，130 页，图 5-23；Kamikuri et al., 2008, p. 168, pl. 2, fig. 23；陈木宏等，2017，94 页，图版 7，图 6–7。
Collosphaera spp., Kling, 1973, pl. 6, figs. 4–8.

地理分布 南极罗斯海，早渐新世—早始新世。

60. 大孔胶球虫 *Collosphaera macropora* Popofsky

（图版 18，图 28–29）

Collosphaera macropora Popofsky, 1917, p. 247, textfigs. 5–6, pl. 14, figs. 2a–c；Strelkov and Reshetnyak, 1971, p. 337, pl. 4, figs. 30–31；Johnson and Nigrini, 1980, p. 147, pl. 1, fig. 7, pl. 4, fig. 15；van de Paverd, 1995, p. 43, pl. 1, figs. 2, 5；Takahashi, 1991, p. 56, pl. 2, figs. 13–18；Kamikuri et al., 2008, p. 168, pl. 2, fig. 19.

Acrosphaera ? sp., O'Connor, 1997b, p. 111, pl. 3, fig. 1.

地理分布　南极罗斯海，出现于早上新世。

61. 胶球虫（未定种 1）*Collosphaera* sp. 1

（图版 18，图 30–31）

Collosphaerid spp., Lazarus, 1992, pl. 5, figs. 9–10.
Collosphaera spp., Kling, 1973, pl. 6, figs. 4–5.

地理分布　南极罗斯海，全新世。

62. 胶球虫（未定种 2）*Collosphaera* sp. 2

（图版 18，图 32–33）

壳表上少量孔的边缘有尖突。

地理分布　南极罗斯海，出现于中始新世。

尖球虫属 Genus *Acrosphaera* Haeckel, 1882

球壳简单，表面具放射状不规则分布的骨针。

63. 锥尖球虫（新种）*Acrosphaera acuconica* sp. nov.

（图版 19，图 1–4）

Hexastyliidae gen. sp. indet., Petrushevskaya, 1975, p. 567, pl. 17, fig. 1.

单一球壳，中等大小，壳壁稍厚，壳内为空腔；壳表在各孔间桁交会处有许多大小相近的尖角锥状凸起，呈放射骨针状；壳壁孔亚圆形或不规则形，大小不等，排列无规律，无六角形框架，横跨赤道有 9–11 孔，孔间桁较粗，孔径是孔间桁宽的 1–5 倍。

标本测量：壳体直径 200–215μm，壁孔直径 12–36μm，针长 18–25μm。

模式标本：DSDP274-25-3 2（图版 19，图 1–2），来自南极罗斯海的 DSDP Leg 28 航次 Site 274 孔的 274-25-3 岩心样品中，保存在中国科学院南海海洋研究所沉积标本室。

地理分布　南极罗斯海，晚始新世—早渐新世。

该未定种的壳体特征与 *Haliomma acanthophora* Popofsky（Takahashi, 1991, p. 68, pl. 9, fig. 3）较相似，但后者壳内中心有一个疏松网格状的小髓壳，有一些丝状放射细桁与外壳相连（我们的标本未见此特征）；与 *Elatomma penicillus* Haeckel（Takahashi, 1991, p. 69, pl. 9, fig. 10）也较相似，但后者有一髓壳。

64. 蛛网尖球虫 *Acrosphaera arachnodictyna* Chen et al.

（图版 19，图 5–14）

Acrosphaera arachnodictyna Chen et al.，陈木宏等，2017，95–96 页，图版 7，图 13–15，图版 8，图 1–4。

地理分布　南极罗斯海，中始新世—早上新世。

65. 松尖球虫 *Acrosphaera hirsuta* Perner

（图版 19，图 15–16）

Acrosphaera hirsuta Perner, 1892, p. 263, pl. 1, fig. 8；陈木宏等，2017，96 页，图版 9，图 3–5。

地理分布 南极罗斯海，出现于早渐新世。

66. 神尖球虫？*Acrosphaera*? *mercurius* Lazarus

（图版 19，图 17–22；图版 20，图 1–2）

Acrosphaera ? *mercurius* Lazarus, 1992, p. 794, pl. 1, figs. 11–16；van de Paverd, 1995, p. 59, pl. 8, figs. 1–3.

地理分布 南极罗斯海，中渐新世—晚中新世。

67. 刺尖球虫 *Acrosphaera spinosa* (Haeckel)

（图版 20，图 3–14）

Collosphaera spinosa Haeckel, 1862, S. 536, Taf. 34, Figs. 12–13.

Acrosphaera spinosa (Haeckel), Brandt, 1885, S. 263, Taf. 4, Figs. 33a–c；Haeckel, 1887, p. 100；Popofsky, 1917, S. 253, Abb. 14–16；Ling, 1972, p. 164, pl. 1, figs. 1–2；谭智源、宿星慧，1982，139 页，图版 1，图 1–2；van de Paverd, 1995, pp. 52–56, pl. 5, figs. 1–11；陈木宏、谭智源，1996，158 页，图版 2，图 14–16，图版 36，图 11–12；Boltovskoy, 1998, figs. 15–18；陈木宏等，2017，97 页，图版 9，图 7–8。

Polysolenia spinosa (Haeckel), Nigrini, 1967, p. 14, pl. 1, fig. 1；Nigrini and Moore, 1979, p. S19, pl. 2, fig. 5.

地理分布 南极罗斯海，中始新世—全新世。

68. 刺尖球虫多刺亚种 *Acrosphaera spinosa echinoides* Haeckel

（图版 20，图 15–23）

Acrosphaera echinoides Haeckel, 1887, p. 100, pl. 8, fig. 1；

Acrosphaera spinosa echinoides Haeckel, Bjørklund and Goll, 1979, pp. 1311–1312, pl. 1, figs. 12–13 (only), pl. 4, figs. 1–4, 7, 8；Lazarus, 1992, p. 794, pl. 5, figs. 5–8.

地理分布 南极罗斯海，中始新世—早渐新世。

69. 尖球虫（未定种 1）*Acrosphaera* sp. 1

（图版 20，图 24–25）

Acrosphaera sp., Abelmann, 1990, p. 690, pl. 1, fig. 1.

地理分布 南极罗斯海，出现于早渐新世。

管球虫属 Genus *Siphonosphaera* Müller, 1858

具一简单球壳，壳孔向外延伸成简单具硬壁的放射管，管口截平。

70. 围织管球虫 *Siphonosphaera circumtexta* (Haeckel)

（图版 21，图 1–4）

Clathrosphaera circumtexta Haeckel, 1887, p. 118, pl. 8, fig. 6.

Siphonosphaera circumtexta (Haeckel), van de Paverd, 1995, p. 60, pl. 8, figs. 1–3.

Acrosphaera circumtexta (Haeckel), Strelkov and Reshetnyak, 1971, p. 347, pl. 7, figs. 46, 47, 50, pl. 8, fig. 53.
Clathrosphaera arachnoides Haeckel, Takahashi, 1991, pl. 1, fig. 15.
Polysolenia sp., Riedel and Sanfilippo, 1971, pl. 1B, figs. 14, 17.
Polysolenia (?) sp., Riedel and Sanfilippo, 1977, pl. 18, fig. 5.

地理分布 南极罗斯海，早渐新世—晚中新世。

71. 杂管球虫（新种）*Siphonosphaera mixituba* sp. nov.

（图版 21，图 5–6）

单一球形壳，个体较大，壳壁厚。表面有许多壁孔，呈圆形、椭圆形、不规则形，大小不一，分布无规律，横跨赤道有 12–16 孔；在每一个壁孔边缘的壳壁向外伸出形成小管，较短，管的末端斜向截平，一侧较高尖凸，另一侧较低，靠近壁孔，这些特殊形状的不同小管汇集使壳体外围呈现许多杂乱的凸起状；管壁与壳壁相同，均为透明状，壳表无放射状骨针。

标本测量：球形壳直径 200–215μm，管宽（孔径）10–25μm，管长 15–25μm。

模式标本：DSDP274-30-4 2（图版 21，图 5–6），来自南极罗斯海的 DSDP Leg 28 航次 Site 274 孔的 274-30-4 岩心样品中，保存在中国科学院南海海洋研究所沉积标本室。

地理分布 南极罗斯海，出现于晚始新世。

该新种特征与 *Siphonosphaera cyathina* Haeckel（1887, p. 105, pl. 6, fig. 10）较接近，主要区别在于后者的壳壁较薄，小管较短，管端规则截平而没有斜向截面，壳体外围整洁；与部分 *Acrosphaera spinosa* (Haeckel)的主要区别是后者的壳表有细圆锥形的骨针。

筒球虫属 Genus *Solenosphaera* Haeckel, 1887

单一胶球壳，具由壳孔延伸形成的放射管，管壁上有窗孔，管口截平，光滑。与 *Siphonosphaera* 的主要区别是后者的管壁无小孔。

72. 双管筒球虫（新种）*Solenosphaera bitubula* sp. nov.

（图版 21，图 7–16）

个体较小，球壳有两个开口，位于球壳的两端，形成两个较短的口管，管宽约为壳径的一半，口端截平；壳壁较厚，透明，散布一些大小不等的类圆形壁孔，分布无规律；壳体表面光滑。

标本测量：球形壳直径 80–110μm，壳体长 120–150μm，管口宽 50–60μm，管长 18–30μm。

模式标本：DSDP274-1-5 2（图版 21，图 11–12），来自南极罗斯海的 DSDP Leg 28 航次 Site 274 孔的 274-1-5 岩心样品中，保存在中国科学院南海海洋研究所沉积标本室。

地理分布 南极罗斯海，晚上新世—晚更新世。

该新种以仅具两个口管为特征，区别于 *Siphonosphaera monotubulosa* Hilmers（1906, p. 82, pl. 1, fig. 5）及其他各个种类。

73. 单管筒球虫 *Solenosphaera monotubulosa* (Hilmers) group

（图版 21，图 17–33）

Siphonosphaera monotubulosa Hilmers, 1906, p. 82, pl. 1, fig. 5.
Otosphaera auriculata Petrushevskaya and Kozlova, 1972, pl. 9, fig. 2.

地理分布 南极罗斯海，早始新世—晚始新世。

三管虫属 Genus *Trisolenia* Ehrenberg, 1860, emend. Bjørklund and Goll, 1979

具格孔的胶球壳和放射管。

74. 大光三管虫 *Trisolenia megalactis* Ehrenberg

（图版 21，图 34–35）

Trisolenia megalactis Ehrenberg, 1872, pl. 8, fig. 19；Ehrenberg, 1873, p. 301；Bjørklund and Goll, 1979, pp. 1318, 1321, pl. 5, figs. 1–21；van de Paverd, 1995, pp. 70, 72, pl. 12, figs. 2–5, pl. 13, figs. 1–7.

地理分布 南极罗斯海，出现于早古新世。

75. 三管虫（未定种 1）*Trisolenia* sp. 1

（图版 21，图 36–37）

胶球壳，具 3–5 个放射管，管口截平，壳壁较厚，壁孔类圆形，大小不一，在孔的边缘有一近三角形的尖凸，使壳表呈现棘刺状。

地理分布 南极罗斯海，出现于早始新世。

六柱虫科 Family Hexastylidae Haeckel, 1881, emend. Petrushevskaya, 1975

初壳简单网格状（如呈现），位于中央囊内，皮壳为一球形格孔壳，表面有一些强壮的放射骨针。

矛球虫属 Genus *Lonchosphaera* Popofsky, 1908, emend. Dumitrică, 2014

壳内的放射桁常分叉，致延伸为外骨针数量增加，外部骨针数量不定（5–40）；皮壳的形状略有变化，孔不规则，赤道一面上有 10–20 孔，壳表有时可见小辅针。初壳为不规则多角形结构。分布于中新世？—现代。Petrushevskaya（1975）指出难以辨认该属与 *Hexastylus* 和 *Centrolonche* 的区别。

76. 考勒矛球虫 *Lonchosphaera cauleti* Dumitrică

（图版 21，图 38–39）

Lonchosphaera cauleti Dumitrică, 2014, p. 65, figs. 1l–n, 2d, e, 3d–h；陈木宏等，2017，100 页，图版 11，图 9–12。
Lonchosphaera sp. C, Petrushevskaya, 1975, p. 630, pl. 17, figs. 11–15.

地理分布 南极罗斯海，全新世。

77. 粗糙矛球虫（新种）*Lonchosphaera scabrata* sp. nov.

（图版 22，图 1–11）

Lonchosphaera sp. C, Petrushevskaya, 1975, p. 567, pl. 17, figs. 11–15.

皮壳壁厚，壁孔类圆形或椭圆形，大小不一，排列不规则，横跨赤道有 13–16 孔，孔径是孔间桁宽的 1–4 倍，孔间桁各节点在壳表上的突起发育形成许多大小与长度相近的尖角锥形或末端分叉的棱柱形的辅针，表面非常粗糙；初壳较模糊，放射桁很细，在壳腔内杂乱分布并连接到皮壳；壳体表面发育有 14–20 根较强壮的放射骨针，呈三棱柱状，骨针长一般约为皮壳半径的 1/2。

标本测量：皮壳直径 200–225μm，放射骨针长 40–105μm，辅针长 18–35μm。

模式标本：DSDP274-26-3 1（图版 22，图 5–6），来自南极罗斯海的 DSDP Leg 28 航次 Site 274 孔的 274-26-3 岩心样品中，保存在中国科学院南海海洋研究所沉积标本室。

地理分布　南极罗斯海，中始新世—早渐新世。

该新种特征与 *Lonchosphaera spicata* Popofsky 较相似，但后者壳壁稍薄，辅针不甚发育，壳体表面较为光滑。

78. 尖矛球虫 *Lonchosphaera spicata* Popofsky group

（图版 22，图 12–19）

Lonchosphaera spicata Popofsky, 1908, p. 218, pl. 24, fig. 2, pl. 25, figs. 2, 7；Petrushevskaya, 1975, p. 567, pl. 17, figs. 4–8.
Acanthosphaera sp., Hays, 1965, p. 169, pl. II, fig. 8；Chen, 1975, p. 450, pl. 22, fig. 5.
Heliosphaerea radiata Popofsky, Benson, 1966, pp. 160–161, pl. 5, figs. 1–2.
Acanthosphaera ? *haliformis* Petrushevskaya and Kozlova, 1972, pl. 9, fig. 16.
Lonchosphaera sp. B, Petrushevskaya, 1975, pl. 17, figs. 9–10.
Lonchosphaera sp. C, Petrushevskaya, 1975, p. 567, pl. 17, figs. 11–15.
Actinomma sp. C, Lazarus and Pallant, 1989, p. 366, pl. 7, fig. 20.
Actinomma ? *magnifenestra* Lazarus, 1992, p. 795, pl. 3, figs. 1–9.
Actinomma? *magnifenestra*? Lazarus, 1992, pl. 3, figs. 10–12.
Actinommid sp. P, Renaudie, 2012, p. 32, pl. 3, figs. 4a–b.

该种的网格状内壳常难以辨认，皮壳厚度和壳孔变化较大，表面光滑或在孔间桁节点上有小突起，不同个体标本的放射骨针长度有一定的差异。由于种征变化较大，我们将此类标本归为一个种群。

地理分布　南极罗斯海，早始新世—早渐新世。

79. 矛球虫（未定种 1）*Lonchosphaera* sp. 1

（图版 23，图 1–2）

个体稍小，类圆形孔，大小相近，横跨赤道有约 7–9 孔；壳表较厚，表面光滑，约 8–12 骨针，圆锥形，较为短小而细。

地理分布　南极罗斯海，出现于晚中新世。

三轴球虫科 Family Cubosphaeridae Haeckel, 1881, emend. Campbell, 1954

格孔壳简单或同心多层，六根放射状主针位于两平面上，相交呈直角。

六矛虫属 Genus *Hexalonche* Haeckel, 1881

具两个同心格孔球壳和六根大小相等的简单骨针。

80. 君主六矛虫 *Hexalonche anaximensis* Haeckel

（图版 23，图 3–22）

Hexalonche anaximensis Haeckel, 1887, p. 183, pl. 25, fig. 5.

我们的标本基本特征与 Haeckel（1887）较为接近，但 Haeckel（1887）认为该种的皮壳壁薄，直径是髓壳的两倍，而罗斯海的标本一般皮壳壁稍厚，壳径约为髓壳的三倍。因此，存在一定的差异。

地理分布 南极罗斯海，晚古新世—早渐新世。

81. 内码六矛虫 *Hexalonche esmarki* Goll and Bjørklund

（图版 23，图 23–27）

Hexalonche esmarki Goll and Bjørklund, 1989, p. 729, pl. 1, figs. 6–10.

该种的壁孔比 *Hexalonche anaximensis* Haeckel 多，表面常有一些粗辅针。

地理分布 南极罗斯海，中-晚始新世。

82. 冷六矛虫（新种）*Hexalonche gelidis* sp. nov.

（图版 23，图 28–29）

两层壳之间的间距较小，内壳较大，其直径是外壳直径的 3/4；壳壁较薄，外壳壁孔较小，类圆形，大小相近，排列不规则，横跨赤道有 14–16 孔，孔径约为孔间桁宽的 1–3 倍；内壳壁孔较大，近六角形，排列规则，横跨赤道有 4–5 孔，孔间桁较细，孔径是孔间桁的 5–6 倍；六个放射桁自内壳的壳壁生出，与皮壳连接，并伸出壳外形成六根放射骨针，呈细长圆锥形或针形，末端缩尖，针长接近壳体的半径；壳体表面光滑。

标本测量：皮壳直径 180μm，内壳直径 130μm，骨针长 80μm。

模式标本：DSDP274-2-3 2（图版 23，图 28–29），来自南极罗斯海的 DSDP Leg 28 航次 Site 274 孔的 274-2-3 岩心样品中，保存在中国科学院南海海洋研究所沉积标本室。

地理分布 南极罗斯海，出现于晚更新世。

该新种以具有六根对称的放射骨针和两个壳体的特征归入 *Hexalonche* 属，以其髓壳（内壳）的壳体较大和壁孔很大（六角形）的典型特征区别于其他各种。

83. 哲六矛虫 *Hexalonche philosophica* Haeckel

（图版 23，图 30–32；图版 24，图 1–2）

Hexalonche philosophica Haeckel, 1887, p. 186, pl. 22, fig. 4；陈木宏、谭智源，1996，165 页，图版 5，图 7–8，图版 38，图 5。

Hexacontium dionysus Kamikuri, 2010, p. 98, pl. 3, figs. 4, 7–11, 15.

地理分布　南极罗斯海，中始新世—晚中新世。

六枪虫属 Genus *Hexacontium* Haeckel, 1881

具三个同心格孔球壳和六根大小相等的简单骨针。

84. 阿奇六枪虫？*Hexacontium*? *akitaensis* (Nakaseko)

（图版 24，图 3–4）

Thecosphaera akitaensis Nakaseko, 1971, p. 63, pl. 1, figs. 4a–b.
Hexacontium akitaensis (Nakaseko), Kamikuri, 2010, pp. 98–99, pl. 1, figs. 7–8, 10–12, pl. 4, fig. 16.

我们的标本与 Kamikuri（2010）的标本特征有些相似，个体相对较小，壳表棘刺状，与 *Thecosphaera akitaensis* Nakaseko（1971）有明显区别。此类标本未见六根放射骨针，归入 *Hexacontium* 属不合理。罗斯海中未见 *Thecosphaera akitaensis* Nakaseko 标本。

地理分布　南极罗斯海，出现于中渐新世。

85. 三轴六枪虫 *Hexacontium axotrias* Haeckel

（图版 24，图 5–18）

Hexacontium axotrias Haeckel, 1887, p. 192, pl. 24, fig. 3.
Hexactonium pythagoraea (Haeckel), van de Paverd, 1995, pp. 129, 131, pl. 34, figs. 1–4, 6–7.
Hexalonche philosophica ? Haeckel, Lazarus, 1992, p. 795, pl. 4, figs. 1–3, 5.
Hexacontium pachydermum Jørgensen, Kamikuri et al., 2008, p. 169, pl. 2, figs. 3a–b.
Hexacontium hexacanthum Müller, van de Paverd, 1995, p. 129, pl. 32, figs. 1–2, pl. 38, fig. 2 (not pl. 33, figs. 1–7).

地理分布　南极罗斯海，中始新世—中渐新世。

86. 内棘六枪虫 *Hexacontium enthacanthum* Jørgensen

（图版 24，图 19–22）

Hexacontium enthacanthum Jørgensen, 1900, p. 52, pl. 2, fig. 14, pl. 4, fig. 20；Jørgensen, 1905, p. 115, pl. 8, figs. 30a–b；Schröder, 1909a, pl. XVII11, figs. 3a–d；Benson, 1966, pp. 149–150, pl. 3, figs. 13–14, pl. 4, figs. 1–3；Bjørklund, 1976, pl. 1, figs. 1–3；Cortese and Bjørklund, 1998b, pl. 1, figs. 12–20；陈木宏等，2017，103 页，图版 13，图 1–2。

地理分布　南极罗斯海，早始新世—中渐新世。

87. 蜂巢六枪虫 *Hexacontium favosum* Haeckel

（图版 25，图 1–10）

Hexacontium favosum Haeckel, 1887, p. 194, pl. 24, fig. 2.
Hexacontium hexacanthum (Müller), van de Paverd, 1995, p. 129, pl. 33, figs. 1–7 (not pl. 32, figs. 1–2, pl. 38, fig. 2).
Hexalonche philosophica ? Haeckel, Lazarus, 1992, pp. 795–796, pl. 4, figs. 1–3, 5.
Hexacontium minerva Kamikuri, 2010, pp. 97–98, pl. 3, figs. 12–14, 16–17.
Hexacontium sp., Bjørklund, 1976, p. 1124, pl. 19, fig. 5.

地理分布　南极罗斯海，中始新世—中渐新世。

88. 主六枪虫 *Hexacontium hostile* Cleve

（图版 25，图 11–14）

Hexacontium hostile Cleve, 1901, p. 9, pl. 6, fig. 4.
Hexacontium cf. *enthacanthum* Jørgensen, Abelmann, 1990, p. 691, pl. 2, figs. 2A–B.

地理分布 南极罗斯海，中渐新世。

89. 蜜罩六枪虫 *Hexacontium melpomene* (Haeckel)

（图版 25，图 15–22）

Stylosphaera melpomene Haeckel, 1887, p. 135, pl. 16, fig. 1.
Hexalonche octocolpa Haeckel, 1887, p. 183, pl. 22, figs. 6, 6a.
Hexacontium melpomene (Haeckel), van de Paverd, 1995, pp. 127–128, pl. 31, figs. 2, 7–8.

该种壳表较粗糙，或有一些细小的辅针。

地理分布 南极罗斯海，晚古新世—全新世零星出现。

90. 厚棘六枪虫 *Hexacontium pachydermum* Jørgensen

（图版 25，图 23–31）

Hexacontium pachydermum Jørgensen, 1900, p. 53；Jørgensen, 1905, pl. 8, figs. 31a–b；Bjørklund, 1976, p. 1124, pl. 1, figs. 4–9；Cortese and Bjørklund, 1998b, pp. 166–167, pl. 1, figs. 1–7, 9–11.
Hexacontium pythagoraea (Haeckel), van de Paverd, 1995, pp. 129, 131, pl. 34, figs. 1–4, 6–7.
Hexacontium spp., Abelmann, 1990, p. 691, pl. 2, figs. 1A–B.

地理分布 南极罗斯海，中始新世—早上新世。

91. 似六枪虫（新种）*Hexacontium parallelum* sp. nov.

（图版 26，图 1–6）

Hexacontid juveniles van de Paverd, 1995, pl. 32, figs. 3, 6.

皮壳不完整，一般仅见放射骨针上的侧分枝或侧刺及其衍生发育的部分皮壳，皮壳可能易碎而不能完整保存；常见两个完整的髓壳，壳体较小，外髓壳中等壁厚，具类圆形孔，大小相近，分布无规律，孔径与孔间桁宽近等，横跨赤道有 6–8 个，由若干细放射桁与内髓壳相连，内髓壳很小，格孔状或网格状；三壳大小比例为 1∶3∶6–9；从外髓壳长出的六根三棱柱状放射桁延伸为三棱角锥状的放射骨针，骨针末端缩尖，壳表面光滑，无辅针。

标本测量：皮壳直径 115–180μm，外髓壳直径 60–75μm，内髓壳直径 25–30μm，骨针长 30–60μm。

模式标本：DSDP274-21-1 2（图版 26，图 5–6），来自南极罗斯海的 DSDP Leg 28 航次 Site 274 孔的 274-21-1 岩心样品中，保存在中国科学院南海海洋研究所沉积标本室。

地理分布 南极罗斯海，中渐新世—早上新世。

壳体特征与 *Actinomma boreale* Cleve（1899, pl. 1, fig. 5a）相似，但后者放射骨针在七根以上，且常有完整的外壳；与标本 *Hexacontid juveniles*（van de Paverd, 1995, pl. 32, figs. 3, 6）基本一致，作者仅将其作为幼体而尚未完全确认。综合不同标本类型的特征组成，认为我们与 van de Paverd（1995）的标本均应归入 *Hexacontium* 属，可视为一个新种。

92. 方角六枪虫（新种）*Hexacontium quadrangulum* sp. nov.

（图版 26，图 7–11）

具一个皮壳、两个髓壳，三壳大小之比为 1∶3∶8；皮壳中等壁厚，壁孔类圆形，大小相近，亚规则排列，横跨赤道有 10–13 孔，孔间桁宽变化不大，无六角形框架，孔径是孔间桁宽的 2–4 倍；外髓壳呈立方形，壁孔近圆形，较细，大小近等，亚规则排列，每条边上有 7–8 孔；内髓壳圆球形，很小，由若干放射桁与外髓壳相连；六条主放射桁自立方形的外髓壳顶角生出连接到皮壳，三棱柱形，延伸至皮壳外形成六根相互垂直对称的放射状骨针，骨针三棱角锥状，较短，针长小于皮壳半径；壳表光滑，无辅针或棘凸。

标本测量：皮壳直径 180–235μm，外髓壳直径 75–90μm，内髓壳直径 30–40μm，骨针长 45–60μm。

模式标本：DSDP274-32-1 1（图版 26，图 9–11），来自南极罗斯海的 DSDP Leg 28 航次 Site 274 孔的 274-32-1 岩心样品中，保存在中国科学院南海海洋研究所沉积标本室。

地理分布 南极罗斯海，晚始新世—中渐新世。

该种的壳体特征与 *Joergensenium apollo* Kamikuri（2010, pl. 3, figs. 1–3, 5, 6）较相似，具有方形的外髓壳，但后者的放射桁不伸出皮壳形成六根放射骨针，且个体略小。新种与 *Hexacontium senticetum* Tan and Su 的主要区别是后者的髓壳呈圆球形；与 *Hexacontium quadratum* Tan 的区别是后者的皮壳呈近立方形，而髓壳为圆球形。

93. 方六枪虫 *Hexacontium quadratum* Tan

（图版 26，图 12–15）

Hexacontium quadratum Tan，谭智源，1993，192 页，图版 9，图 2–3；谭智源、陈木宏，1999，157 页，图版 6，图 10–11；陈木宏等，2017，104 页，图版 13，图 8–14。

Hexacontium melpomene (Haeckel), van de Paverd, 1995, pp. 127–128, pl. 31, figs. 2, 7–8.

地理分布 南极罗斯海，中渐新世—晚更新世。

94. 杖六枪虫 *Hexacontium sceptrum* Haeckel

（图版 26，图 16–19）

Hexacontium sceptrum Haeckel, 1887, p. 194, pl. 24, fig. 1.

地理分布 南极罗斯海，中-晚始新世。

95. 棘六枪虫 *Hexacontium senticetum* Tan and Su

（图版 26，图 20–23）

Hexacontium senticetum Tan and Su，谭智源、宿星慧，1982，145 页，图版 5，图 5–10；陈木宏、谭智源，1996，167 页，图版 6，图 5–7。

Hexacontium pachydermum Jørgensen, Kamikuri et al., 2008, p. 169, pl. 2, figs. 3a–b.

地理分布 南极罗斯海，中始新世。

六葱虫属 Genus *Hexacromyum* Haeckel, 1881, emend. Chen et al., 2017

壳体四层，两个皮壳和两个髓壳，具六根或更多对称分布的大小相同的放射骨针。

96. 美六葱虫 *Hexacromyum elegans* Haeckel

（图版 27，图 1–2）

Hexacromyum elegans Haeckel, 1887, p. 201, pl. 24, fig. 9；Takahashi and Honjo, 1981, p. 148, pl. 3, fig. 15；Takahashi, 1991, p. 73, pl. 13, figs. 4–5, 7；陈木宏等，2017，104 页，图版 13，图 15–16。

地理分布 南极罗斯海，出现于晚始新世。

97. 稀六葱虫 *Hexacromyum rara* (Carnevale) group

（图版 27，图 3–4）

Stauroloncherara Carnevale, 1908, p. 15, pl. 2, fig. 8.
Hexacromyum rara (Carnevale) group, Petrushevskaya, 1975, p. 569, pl. 2, figs. 1–2.

地理分布 南极罗斯海，出现于早中新世。

星球虫科 Family Astrosphaeridae Haeckel, 1881

格孔球壳简单或多重，具 8 根或更多（常为 20–60 根）的放射针。

棘球虫属 Genus *Acanthosphaera* Ehrenberg, 1858

具一简单的格孔球壳，放射骨针简单、同形。

98. 恩玛提棘球虫 *Acanthosphaera enigmaticus* (Hollande and Enjumet)

（图版 27，图 5–6）

Porococcus enigmaticus Hollande and Enjumet, 1960, p. 113, pl. 58, figs. 1–3.
Diplosphaera lychnosphaera (juveniles), van de Paverd, 1995, pp. 94–95, pl. 21, figs. 3–4 (not pl. 22, fig. 4).

简单格孔壳，类圆形孔，大小相近，亚规则排列，横跨赤道有约 8–10 孔；放射骨针三棱角锥状，末端缩尖，无侧刺或分叉，针长近等或大于壳径。

地理分布 南极罗斯海，出现于早上新世。

99. 具缘棘球虫 *Acanthosphaera marginata* Popofsky

（图版 27，图 7–18）

Acanthosphaera marginata Popofsky, 1912, p. 96, pl. 1, fig. 2.
Arachnosphaera dichotoma Jørgensen, Bjørklund, 1976, p. 1124, pl. 3, figs. 5–9.
Acanthosphaera actinota (Haeckel), Boltovskoy and Riedel, 1980, p. 97, pl. 1, fig. 7；Abelmann, 1992a, p. 374, pl. 1, fig. 10.

地理分布 南极罗斯海，晚更新世—全新世。

100. 极区棘球虫（新种）*Acanthosphaera polageota* sp. nov.

（图版 27，图 19–22）

Acanthosphaera sp., Hays, 1965, p. 169, pl. 2, fig. 8.

单一球形壳，壁厚有变化，表面粗糙；壁孔圆或亚圆形，大小不同，有多角形或六角形框架，孔径是孔间桁宽的 2–3 倍，横跨赤道有 10–15 孔，常见 10 孔左右；具 6–9 根长度相近的放射主骨针，呈三棱角锥状，主骨针长约为壳直径的 1/4，也有一些标本的主骨针长略大于壳半径，通常是无规律地分布在壳体表面，或成对分布；有些标本还可见在孔间桁节点上生出的针状辅针或角锥形凸起。

标本测量：皮壳直径 115–125μm，骨针长 30–40μm。

模式标本：DSDP274-16-1 1（图版 27，图 19–20），来自南极罗斯海的 DSDP Leg 28 航次 Site 274 孔的 274-16-1 岩心样品中，保存在中国科学院南海海洋研究所沉积标本室。

地理分布　南极罗斯海，中古新世—晚渐新世。Hays（1965）认为该种出现于南极老地层岩心，偶尔发现于极峰以北岩心顶部，但未见于现代的极峰以南。

该新种以壁孔和骨针小而多，分布无规律区别于其他种类。

101. 棘球虫（未定种 1）*Acanthosphaera* sp. 1

（图版 27，图 23–24）

个体较大，壳孔较多，横跨赤道有约 16–20 个，圆形，大小相近，具六角形框架，排列规则；三棱角锥状放射骨针较多，一般较短，无辅针。

地理分布　南极罗斯海，出现于早上新世。

日球虫属 Genus *Heliosphaera* Haeckel, 1862

具一简单格孔球壳，壳表有两类不同形的主骨针和小辅针。

102. 具齿日球虫 *Heliosphaera dentata* (Mast)

（图版 27，图 25–34）

Heteracantha dentata Mast, 1910, p. 37, pl. 5, fig. 47；Bjørklund, 1976, p. 1124, pl. 14, figs. 10–12.

Anomalacantha denata (Mast), Benson, 1966, p. 170, pl. 5, figs. 10–11；Nigrini and Moore, 1979, p. S37, pl. 4, fig. 4；Lazarus and Pallant, 1989, p. 366, pl. 7, figs. 23–24；Kamikuri et al., 2008, p. 168, pl. 1, figs. 2–3.

单一格孔壳较小，壳壁较厚，横跨赤道一般仅有 4–5 孔。该种与 *Lychnosphaera regina* Haeckel（1887, p. 277, pl. 11, figs. 1–4）的沉积物标本（具纤细的海绵状皮壳，皮壳破损或消失）较为相似，主要区别是后者保存的格孔髓壳相对稍大，横跨赤道有 6–8 孔。

地理分布　南极罗斯海，晚始新世—中渐新世。

103. 大六角日球虫 *Heliosphaera macrohexagonaria* Tan

（图版 28，图 1–2）

Heliosphaera macrohexagonaria Tan，谭智源，1998，151 页，图 139；谭智源、陈木宏，1999，165 页，图 5-73；陈木宏等，2017，106 页，图版 14，图 6–7。

Lychnosphaera regina Haeckel，陈木宏等，2017，122 页，图版 26，图 8–9，12–13（其他不是）。

该种壳表较薄，放射骨针较为细小。

地理分布 南极罗斯海，出现于晚始新世。

104. 小孔日球虫（新种）*Heliosphaera miniopora* sp. nov.

（图版 28，图 3–6）

格孔壳壁稍厚，表面粗糙，壁孔较小，大小相近，类圆形，亚规则排列，横跨赤道有 7–9 个，孔间桁较宽，与孔径近等，有六角形框架；放射骨针 12–18 根，三片棱柱状，较为粗长，骨针侧缘光滑无侧刺，长度大于壳的直径，骨针初端与末端的宽度基本不变，不缩尖；壳表的一些节点上有小凸或形成小辅针。

标本测量：壳直径 105–120μm，骨针长 95–130μm。

模式标本：DSDP274-16-1 1（图版 28，图 3–4），来自南极罗斯海的 DSDP Leg 28 航次 Site 274 孔的 274-16-1 岩心样品中，保存在中国科学院南海海洋研究所沉积标本室。

地理分布 南极罗斯海，出现于晚渐新世。

该新种以壁孔小而多、壁稍厚及骨针无侧刺的特征区别于 *Heliosphaera dentata* (Mast)和 *Heliosphaera macrohexagonaria* Tan。

枝球虫属 Genus *Cladococcus* Müller, 1856

单一格孔壳，放射骨针有侧枝。

105. 枝球虫 *Cladococcus viminalis* Haeckel

（图版 28，图 7–16）

Cladococcus viminalis Haeckel, 1862, p. 369, taf. xiv, figs. 2, 3；Haeckel, 1887, p. 226；Bjørklund, 1976, p. 1124, pl. 1, figs. 10–12.
Hexastylus solonis Haeckel, van de Paverd, 1995, p. 92, pl. 20, figs. 1–4.

球形格孔壳，圆形孔，具六角形框架；放射骨针三棱柱状，较长，侧缘有齿或侧分枝，沉积物中的标本骨针均已折断，仅保留近距针基的部分。

地理分布 南极罗斯海，晚始新世—早上新世。

海眼虫属 Genus *Haliomma* Ehrenberg, 1844

具一髓壳和一皮壳，二者由放射梁相连。壳表覆盖有同类简单的放射骨针。Petrushevskaya, 1975 将该属修订为具三层壳且无壳表骨针，在此不予采纳。

106. 棘动海眼虫 *Haliomma acanthophora* Popofsky

（图版 28，图 17–18）

Haliomma acanthophora Popofsky, 1912, pp. 101–102, text-fig. 13；陈木宏等，2017，107–108 页，图版 14，图 9–12，图版 15，图 1–2。
Carposphaera acanthophora Benson, 1966, pp. 127–131, pl. 2, figs. 8–10.
Actinosphaera acanthophora Dumitrica, 1973, p. 832, pl. 20, figs. 1–2.
Haliomma erinaceum Renz, 1976, p. 101, pl. 2, figs. 4a–b.

Actinosphaera cristata(?) Benson, 1983, pp. 499–500.

地理分布　南极罗斯海，出现于中渐新世。

107. 星海眼虫（新种）*Haliomma asteranota* sp. nov.

（图版 28，图 19–22）

Spumellarian gen. et sp. indet ＃1 and ＃2, Lazarus and Pallant, 1989, p. 367, pl. 7, figs. 25–28.

个体较小，皮壳多角形或亚球形，格孔状，壁孔类圆形，不规则排列，孔径是孔间桁宽的 1–3 倍，横跨赤道有 6–8 孔；髓壳不规则形，孔间桁杂乱网格状，稀疏，大小接近皮壳的一半；壳体表面光滑，具 8–14 根放射骨针，主要从多角形皮壳的各顶角长出，较为粗壮，三棱角锥状，末端缩尖，针长接近皮壳直径，无辅针。

标本测量：皮壳直径 60–75μm，髓壳直径 30–45μm，骨针长 50–65μm。

模式标本：DSDP274-24-3 1（图版 28，图 21–22），来自南极罗斯海的 DSDP Leg 28 航次 Site 274 孔的 274-24-3 岩心样品中，保存在中国科学院南海海洋研究所沉积标本室。

地理分布　南极罗斯海，早渐新世；格陵兰海区、北大西洋，渐新世。

该种特征与 *Haliomma microlaris* sp. nov.较近，主要区别是后者的放射骨针呈小圆柱状，较短，末端截平，髓壳较小。罗斯海的标本与 Lazarus 和 Pallant（1989, p. 367, pl. 7, figs. 25–28）鉴定的 Spumellarian gen. et sp. indet 标本基本一致。

108. 小海眼虫（新种）*Haliomma microlaris* sp. nov.

（图版 28，图 23–32；图版 29，图 1–2）

壳体很小，圆球形，两壳之比近 3∶1；皮壳壁孔类圆形或六角形，大小相近，规则或不规则排列，孔径是孔间桁宽的 3–4 倍，横跨赤道有 5–7 孔；髓壳呈多角形网格状，由若干细桁连接组成；壳体表面在一些孔间桁节点上生出小圆柱形的放射骨针，骨针末端截平，针长约为壳径的 1/5–1/3。

标本测量：皮壳直径 56–81μm，髓壳直径 19–30μm，骨针长 13–40μm。

模式标本：DSDP274-40-1 1（图版 28，图 27–28），来自南极罗斯海的 DSDP Leg 28 航次 Site 274 孔的 274-40-1 岩心样品中，保存在中国科学院南海海洋研究所沉积标本室。

地理分布　南极罗斯海，早—晚古新世。

该新种以其个体细小、髓壳网格状和具若干小圆柱形放射骨针的特征明显区别于其他已知种类，为南极环境的特有类型。

109. 卵海眼虫 *Haliomma ovatum* Ehrenberg

（图版 29，图 3–4）

Haliomma ovatum Ehrenberg, 1844, p. 83, tab. 19, figs. 48–49；Ehrenberg, 1874, p. 236；陈木宏等，2017，109 页，图版 15，图 16–17。

地理分布　南极罗斯海，出现于晚中新世。

110. 柱海眼虫（新种）*Haliomma stylota* sp. nov.

（图版 29，图 5–20）

两层壳，一个皮壳和一个髓壳；皮壳格孔状，中等壁厚，具圆形或类圆形孔，较小，亚规则排列，横跨赤道有 9–11 孔，有浅的六角形框架，孔径是孔间桁宽的 2–3 倍；髓壳圆球形或梨形，大小约为皮壳的 1/3，壁孔较少；约 6–12 根放射桁连接髓壳与皮壳，延伸至壳外发育成较粗壮的三片棱柱状放射骨针，末端不缩尖，较长，有侧刺或侧齿；壳表一般光滑，无辅针。

标本测量：皮壳直径 120–140μm，髓壳直径 50–60μm，骨针长 65–135μm。

模式标本：DSDP274-28-3 1（图版 29，图 11–12），来自南极罗斯海的 DSDP Leg 28 航次 Site 274 孔的 274-28-3 岩心样品中，保存在中国科学院南海海洋研究所沉积标本室。

地理分布 南极罗斯海，中始新世—中渐新世。

新种与 *Stylosphaera megaxyphos tetraxyphos* Clark and Campbell（1942, pl. 6, figs. 1, 8）和 *Actinomma* sp. (ancestor to *A. golownini* ?)（Lazarus, 1992, pl. 4, figs. 1, 4, 9, 13）的主要区别是后者的放射骨针无侧齿或刺，且为相对较短的三棱角锥状。

111. 海眼虫（未定种 1）*Haliomma* sp. 1

（图版 29，图 21–22）

具一个髓壳和一个皮壳；皮壳壁薄，壁孔较大，孔间桁较细，具六角形框架，髓壳特征与皮壳相似；放射骨针呈三片棱柱状，有侧刺。该未定种与 *Haliomma stylota* sp. nov. 的主要区别是壳壁较薄，壁孔较大，孔间桁较细。

地理分布 南极罗斯海，出现于中始新世。

112. 海眼虫（未定种 2）*Haliomma* sp. 2

（图版 29，图 23–26）

个体很小，具一个皮壳和一个髓壳，均格孔状；皮壳亚球形或椭球形，髓壳圆球形，二壳大小之比为 1/4–1/3；皮壳上生出六根放射骨针，近对称分布，呈细柱状，较长。未定种特征与 *Haliomma microlaris* sp. nov.较接近，但后者的个体稍大，髓壳一般呈网格状，放射骨针较多、较短，且不规则分布。

地理分布 南极罗斯海，晚更新世—全新世。

小海眼虫属 Genus *Haliommetta* Haeckel, 1887, emend. Petrushevskaya, 1972

具一皮壳和二髓壳。皮壳孔大小近等。主骨针无内杆与内部的髓壳相连，偶有小骨针。

113. 粗针小海眼虫（新种）*Haliommetta hadraspina* sp. nov.

（图版 29，图 27–32）

具一个皮壳、两个髓壳，三壳大小比例为 1∶3∶6；皮壳壁厚，壁孔类圆形，大小相近或不等，不规则或亚规则排列，横跨赤道有 6–9 孔，有六角形框架；外髓壳近圆球形，较模糊，内髓壳圆球形，很小，有规则排列的细孔；壳表一般粗糙，在孔间桁节点上有棘凸，发育有6–9根较粗的放射骨针，呈三棱柱状，很短，针基宽，无辅针。

标本测量：皮壳直径150–165μm，外髓壳直径65–100μm，内髓壳直径25–38μm，骨针长15–25μm。

模式标本：DSDP274-30-4 2（图版29，图31–32），来自南极罗斯海的DSDP Leg 28航次Site 274孔的274-30-4岩心样品中，保存在中国科学院南海海洋研究所沉积标本室。

地理分布　南极罗斯海，始新世。

该种与*Haliommetta rossina* sp. nov.的主要区别是后者个体略小，皮壳壁稍薄，孔间桁较细，放射骨针较多和长，三棱角锥状，有辅针。

114. 中新世小海眼虫 *Haliommetta miocenica* (Campbell and Clark) group

（图版30，图1–11）

Heliosphaera miocenica Campbell and Clark, 1944a, p. 16, pl. 2, figs. 10–14.

Halimmetta miocenica (Campbell and Clark) group, Petrushevskaya and Kozlova, 1972, pp. 517–519, pl. 9, figs. 8–9；Chen, 1975, p. 453, pl. 20, fig. 14；陈木宏、谭智源，1996，172–173页，图版8，图2–7；陈木宏等，2017，110页，图版16，图2–5。

Echinomma quadrisphaera Dogiel, Petrushevskaya, 1969, p. 138, fig. 1(4).

Hexacontium inerme Haeckel, van de Paverd, 1995, p. 125, pl. 31, figs. 3, 6.

Actinomma popofskii (Petrushevskaya), Kamikuri et al., 2008, p. 167, pl. 1, fig. 31.

该种皮壳特征有一定的变化范围，如在罗斯海较新地层中的标本皮壳壁稍薄，而在较老地层中的标本皮壳壁较厚，且壳表放射骨针或辅针不太发育，后人对不同区域标本的鉴定特征与Campbell和Clark（1944a）存在着一定差异，因此Petrushevskaya和Kozlova（1972）将此类标本归为一个种群是合适的。

较老地层中的标本皮壳壁厚且壁孔细而多，与*Actinomma henningsmoeni* Goll and Bjørklund（1989, p. 728, pl. 2, figs. 10–15）和*Actinomma medusa* (Ehrenberg)（Abelmann, 1990, pl. 1, fig. 6）有明显的区别。

地理分布　南极罗斯海，中始新世—全新世。

115. 罗斯小海眼虫（新种）*Haliommetta rossina* sp. nov.

（图版30，图12–15）

具一个皮壳、两个髓壳，三壳大小比例为1∶3∶8；皮壳壁厚中等，壁孔类圆形或椭圆形，大小相近，亚规则排列，具六角形框架，孔径是孔间桁宽的3–4倍，横跨赤道有7–8孔；两个髓壳圆球形，均为格孔壳，外髓壳壁孔类圆形，大小相近，有六角形框架，内外髓壳之间有一些细放射桁相连，但不直接连到皮壳；自外髓壳连接到皮壳的放射桁较粗，延伸至壳外形成10–16根放射主骨针，放射主骨针呈三角棱锥状，末端缩尖，侧缘无刺；壳体表面粗糙，有一些小辅针，

标本测量：皮壳直径125–138μm，外髓壳直径60–70μm，内髓壳直径25–30μm，骨针长60–105μm。

模式标本：DSDP274-30-6 2（图版30，图14–15），来自南极罗斯海的DSDP Leg 28航次Site 274孔的274-30-6岩心样品中，保存在中国科学院南海海洋研究所沉积标本室。

地理分布 南极罗斯海，始新世。

新种与 *Haliomma stylota* sp. nov.的特征较相近，区别在于前者具有两个髓壳，壁孔较少，一般壳表稍粗糙，有较短小的辅针或突起，*Haliomma stylota* sp. nov.的主骨针呈三片棱柱状而不是角锥状，有侧刺。

光眼虫属 Genus *Actinomma* Haeckel, 1862, emend. Nigrini, 1967

具三个同心格孔球壳，放射骨针很多，形状多数简单、相似，或有小辅针。

116. 北方光眼虫 *Actinomma boreale* Cleve

（图版 30，图 16–24）

Actinomma boreale Cleve, 1899, p. 26, pl. 1, fig. 5；Cortese and Bjørklund, 1997, pl. 1, figs. 1–10；Bjørklund et al., 1998, pl. 1, figs. 6–7；Dolven, 1998, pl. 1, figs. 1–6；Cortese and Bjørklund, 1998a, pl. 1, figs. 1–18, pl. 3, figs. 1–3, 6；Itaki, 2003, pl. 1, figs. 13–17；Itaki and Ikehara, 2004, pl. 1, figs. 11–13；陈木宏等，2017，111 页，图版 16，图 9–14，图版 17，图 1–2。

Chromyechinus borealis Jørgensen, 1905, pp. 117–118, pl. 8, fig. 35, pl. 9, figs. 36–37；Bjørklund, 1976, p. 1124, pl. 2, figs. 7–15；Molina-Cruz, 1991, fig. 2(3).

地理分布 南极罗斯海，晚始新世—中中新世。

117. 耳蜗光眼虫 *Actinomma cocles* Renaudie and Lazarus

（图版 30，图 25–30）

Actinomma cocles Renaudie and Lazarus, Renaudie, 2012, p. 36, pl. 4, figs. 3–4, 6, 8A–B.

地理分布 南极罗斯海，全新世。

118. 怡光眼虫 *Actinomma delicatulum* (Dogiel)

（图版 31，图 1–2）

Heliosoma delicatulum Dogiel, Dogiel and Reshetnjak, 1952, pp. 7–8, fig. 2.

Echinomma delicatulum (Dogiel), Petrushevskaya, 1968, pp. 18–20, fig. 11；Chen, 1975, p. 453, pl. 24, fig. 1.

Echinomma sp., Riedel, 1958, p. 225, pl. 1, fig. 6.

地理分布 南极罗斯海，出现于中上新世。

119. 杜氏光眼虫（新种）*Actinomma dumitricanis* sp. nov.

（图版 31，图 3–15）

具三个格孔壳，三壳之间近等间距，大小比例 1∶3∶5；外皮壳壁较厚，壁孔类圆形，较小，亚规则或不规则分布，孔间桁较宽，孔径近等或略小于孔间桁宽，横跨赤道有 14–20 孔；内皮壳和髓壳的壳壁比外皮壳略薄，壁孔结构与分布三壳相似，各壳之间由许多放射桁相连；壳体表面光滑或有锥凸，可见个别较短的圆锥形放射骨针，也常可见外皮壳上有一圆形开孔，孔径约为壳体直径的 1/4。

标本测量：外皮壳直径 170–250μm，内皮壳直径 110–140μm，髓壳直径 40–55μm，骨针长 15–40μm，开孔孔径 40–60μm。

模式标本：DSDP274-24-5 1（图版 31，图 8–10），来自南极罗斯海的 DSDP Leg 28

航次 Site 274 孔的 274-24-5 岩心样品中，保存在中国科学院南海海洋研究所沉积标本室。

地理分布　南极罗斯海，晚古新世—晚上新世。

Bjørklund（1976, p. 1121, pl. 20, figs. 8, 9）描述 *Actinomma holtedahli* Bjørklund 的髓壳为海绵状，较小，两个皮壳的间距很近，外皮壳比内皮壳壁薄。我们的罗斯海标本结构虽然与该种较相似，但髓壳一般呈格孔球状，尤其是外皮壳一般壁很厚，而内皮壳相对壁薄，两个皮壳间距较大，因此，两者存在明显的差异。Abelmann（1990, p. 690, pl. 1, figs. 4A–B）的 *Actinomma* cf. *holtedahli* Bjørklund，以及 Caulet（1991, p. 531, pl. 1, figs. 1–2）的 *Actinomma kerguelenensis* Caulet 标本均外壁略厚，还有一些较细或粗的三棱角锥状放射骨针，骨针较长，而且两个皮壳的间距很近，与 *Actinomma holtedahli* Bjørklund 有明显的特征区别。

120. 哥咯尼光眼虫 *Actinomma golownini* Petrushevskaya

（图版 31，图 16–17）

Actinomma golownini Petrushevskaya, 1975, p. 569, pl. 2, fig. 16；Abelmann, 1990, p. 691, pl. 1, fig. 8；Abelmann, 1992a, p. 776, pl. 1, figs. 7–8.

Petrushevskaya（1975）在描述中提及最外面有第四层壳，但已几乎消失，在 Abelmann（1990, 1992a）的标本中也未见第四壳，但我们的罗斯海标本却较好地保留有第四壳特征（很薄）。

地理分布　南极罗斯海，出现于早渐新世。

121. 矛光眼虫 *Actinomma hastatum* (Haeckel)

（图版 31，图 18–23）

Heliosoma hastatum Haeckel, 1887, p. 241, pl. 28, fig. 4.
Actinomma hastatum (Haeckel), van de Paverd, 1995, p. 134, pl. 35, fig. 5.
Echinomma toxopneustes Haeckel, 1887, p. 259, pl. 29, fig. 1.
Cromyomma circumtextum Haeckel, 1887, p. 262, pl. 30, fig. 4.
Cromyomma icosacanthus Haeckel, 1887, p. 263, pl. 30, fig. 1.
Cromyechinus dodecacanthus Haeckel, 1887, p. 264, pl. 30, fig. 3.
Actinomma sp., Ling and Lazarus, 1990, p. 355, pl. 1, fig. 3, pl. 4, fig. 7.

地理分布　南极罗斯海，中始新世—中渐新世。

122. 亨宁光眼虫 *Actinomma henningsmoeni* Goll and Bjørklund

（图版 31，图 24–33）

Actinomma henningsmoeni Goll and Bjørklund, 1989, p. 728, pl. 2, figs. 10–15.
Actinomma sp. B, Abelmann, 1990, p. 691, pl. 1, figs. 7A–B；Abelmann, 1992a, p. 776, pl. 1, figs. 3–4.

地理分布　南极罗斯海，晚古新世—中上新世。

123. 六针光眼虫 *Actinomma hexactis* Stöhr

（图版 32，图 1–7）

Actinomma hexactis Stöhr, 1880, p. 91, pl. 2, fig. 7；Blueford, 1982, pl. 4, figs. 7–8；陈木宏等，2017，112 页，图版 17，图 7–8。

Hexacontium hexactis Haeckel, 1887, p. 192.
Hexacontium (*Hexacontosa*) *nipponicum* Nakaseko, 1955, pl. 3, figs. 5a–c.

罗斯海的标本个体较大。

地理分布 南极罗斯海，中始新世—中渐新世。

124. 赫特光眼虫 *Actinomma holtedahli* Bjørklund

（图版 32，图 8–13）

Actinomma holtedahli Bjørklund, 1976, p. 1121, pl. 20, figs. 8–9.
Actinomma sp., Bjørklund, 1976, p. 1124, pl. 20, fig. 7.
Haliommoidea gen. sp. indet., Petrushevskaya, 1975, p. 568, pl. 1, fig. 10.
Actinomma beroes (Ehrenberg) group, Petrushevskaya, 1975, p. 569, pl. 2, fig. 15.

外壳壁薄，无放射骨针。我们的一些标本个体偏小。

地理分布 南极罗斯海，中-晚始新世。

125. 粗糙光眼虫（新种）*Actinomma impolita* sp. nov.

（图版 32，图 14–25）

个体较大，三壳大小比例为 1∶3∶7.5；皮壳壁较厚，壁孔类圆形、椭圆形或不规则形，大小不同，排列不规则，具多角形框架，孔间桁宽窄不一，孔径是孔间桁宽的 1–5 倍，横跨赤道有 12–16 孔；外髓壳壁厚中等，壁孔类圆形，孔径是孔间桁宽的 2–3 倍，横跨赤道有 7–9 孔；内髓壳很小，格孔状；约 12–20 根放射桁连接着髓壳与皮壳，延伸至壳外形成放射主骨针，主骨针三棱角锥状，较粗短，长度一般仅为皮壳直径的 1/4；壳体表面粗糙，在孔间桁节点上有凸起，或可发育为角锥形或棘刺状的辅针。

标本测量：壳直径 220–310μm，外髓壳直径 106–125μm，内髓壳直径 40–50μm，骨针长 40–110μm。

模式标本：DSDP274-33-5 2（图版 32，图 18–19），来自南极罗斯海的 DSDP Leg 28 航次 Site 274 孔的 274-33-5 岩心样品中，保存在中国科学院南海海洋研究所沉积标本室。

地理分布 南极罗斯海，中始新世—中渐新世。

该种特征与 *Actinomma laminata* sp. nov.较接近，主要区别在于后者的壳壁较薄，壳表粗糙，有棘凸或辅针，主骨针较粗短。

126. 科固光眼虫 *Actinomma kerguelenensis* Caulet

（图版 33，图 1–8）

Actinomma kerguelenensis Caulet, 1991, p. 531, pl. 1, figs. 1–2.
Actinomma cf. *holtedahli* Bjørklund, Abelmann, 1990, p. 690, pl. 1, figs. 4a–b.

有四层壳，外壳呈密集网状结构，多角形孔，壳表具一些放射骨针。该种与 *Actinomma livae* Goll and Bjørklund 很相似，主要区别是后者壳表较粗糙，壁稍厚，常只见三个壳。

地理分布 南极罗斯海，早始新世—早渐新世。

127. 薄皮光眼虫（新种）*Actinomma laminata* sp. nov.

（图版 33，图 9–20）

个体较大，三壳大小比例为 1∶2.2∶4.5；皮壳壁较薄，壁孔类圆形，大小相近，亚规则排列，有六角形或多角形框架，孔间桁较细，孔径是孔间桁宽的 3–6 倍，横跨赤道有 12–14 孔；外髓壳壁厚中等，壁孔类圆形，孔径是孔间桁宽的 2–3 倍，横跨赤道有 7–9 孔；内髓壳很小，格孔状；约 10–16 根放射桁连接着髓壳与皮壳，延伸至壳外形成三棱角锥状的放射骨针，不同的标本放射骨针长度不同，一般小于皮壳半径，但最长者可接近皮壳直径；壳体表面光滑，无辅针。

标本测量：皮壳直径 210–250μm，外髓壳直径 110–130μm，内髓壳直径 50–60μm，骨针长 50–220μm。

模式标本：DSDP274-23-5 2（图版 33，图 9–10），来自南极罗斯海的 DSDP Leg 28 航次 Site 274 孔的 274-23-5 岩心样品中，保存在中国科学院南海海洋研究所沉积标本室。

地理分布　南极罗斯海，中始新世—中渐新世。

该种特征与 *Actinomma leptoderma* (Jørgensen)较接近，主要区别在于后者的个体较小，壁孔较少，放射骨针也较为短小。

128. 瘦光眼虫 *Actinomma leptoderma* (Jørgensen)

（图版 34，图 1–18）

Echinomma leptodermum Jørgensen, 1900, p. 57；Jørgensen, 1905, p. 116, pl. 8, figs. 33a–c；Bjørklund, 1976, pl. 1, figs. 10–12, pl. 2, figs. 1–6；Hays, 1965, p. 169, pl. 1, fig. 2.

Actinomma leptoderma (Jørgensen)，陈木宏、谭智源，1996，173 页，图版 8，图 13–14，图版 40，图 7–8；Cortese and Bjørklund, 1998b, p. 153, pl. 2, figs. 1–14, pl. 3, figs. 4, 5, 9, 10, 16；Kamikuri et al., 2008, p. 167, pl. 1, fig. 30；陈木宏等，2017，113 页，图版 17，图 9–13，图版 18，图 1–2。

Actinomma leptoderma ? (Jørgensen), Nigrini and Moore, 1979, p. S35, pl. 3, fig. 7.

Actinomma leptoderma leptoderma (Jørgensen), Kruglikova et al., 2007, pl. 1, figs. 1–10.

Actinomma sp., Benson, 1966, pp. 164–166, pl. 5, figs. 5–6.

Actinomma sp. B, Kruglikova et al., 2007, pl. 1, figs. 21–30.

地理分布　南极罗斯海，早渐新世—全新世。

129. 灰光眼虫 *Actinomma livae* Goll and Bjørklund

（图版 34，图 19–28）

Actinomma livae Goll and Bjørklund, 1989, pp. 728–729, pl. 1, figs. 1–5；陈木宏等，2017，113–114 页，图版 20，图 1–8。

Actinomma sp., Bjørklund, 1976, pl. 20, fig. 7.

该种与 *Actinomma kerguelenensis* Caulet 较相似，可有两个髓壳和两个皮壳，而且均具放射骨针，主要区别是前者的两个皮壳结构相似，均为格孔状，而后者的内皮壳为格孔状，外皮壳呈密集网格状，使壳表有时不平坦或有些起伏。根据罗斯海标本特征，认为该种以外皮壳稍薄、无网状结构且表面较平滑而别于后者。

地理分布　南极罗斯海，早始新世—早渐新世。

130. 神光眼虫（新种）*Actinomma magicula* sp. nov.

（图版 35，图 1–13）

具两个球形格孔壳，圆球形或椭球形，中等壁厚，两壳的间距较小，大小比例为 3∶4；外壳壁孔类圆形或不规则形，孔径约为孔间桁宽的 2–4 倍，不规则排列，孔间桁较细，横跨赤道有 7–9 孔；内壳壁孔类圆形，大小相近，亚规则排列，孔径是孔间桁宽的 2–3 倍，横跨赤道有 6–7 孔；在内壳腔里常有一些稀疏的不规则状网桁交错充填着部分或整个内腔；壳表光滑或略粗糙，有个别很短的锥形骨针。

标本测量：外壳直径 105–180μm，内壳直径 75–120μm，骨针长 10–20μm。

模式标本：DSDP274-24-5 1（图版 35，图 1–3），来自南极罗斯海的 DSDP Leg 28 航次 Site 274 孔的 274-24-5 岩心样品中，保存在中国科学院南海海洋研究所沉积标本室。

地理分布 南极罗斯海，中古新世—中更新世。

该种特征与 *Actinomma dumitricanis* sp. nov.有些接近，主要区别在于后者具三个格孔壳、个体较大，外壳壳壁较厚，壁孔很小，数量较多，而且有一个较大的圆形开孔。本新种的不同标本个体大小差异明显，越老地层的壳体越小。

131. 海女光眼虫 *Actinomma medusa* (Ehrenberg)

（图版 35，图 14–15）

Haliomma medusa Ehrenberg, 1844a, p. 83；Ehrenberg, 1854, taf. 22, fig. 33 (only).
Actinomma medusa (Ehrenberg) Petrushevskaya, 1975, p. 568, pl. 2, figs. 6–8；Abelmann, 1990, pp. 690–691, pl. 1, fig. 6.
Actinomma okurai Nakaseko and Nishimura, 1971, p. 68, pl. 1, figs. 3–5 (only).

地理分布 南极罗斯海，出现于晚始新世。

132. 奇异光眼虫 *Actinomma mirabile* Goll and Bjørklund

（图版 35，图 16–27）

Actinomma mirabile Goll and Bjørklund, 1989, p. 729, pl. 2, figs. 1–3, pl. 4, figs. 33–34.

地理分布 南极罗斯海，晚古新世—早上新世。

133. 厚皮光眼虫 *Actinomma pachyderma* Haeckel

（图版 35，图 28–29）

Actinomma pachyderma Haeckel, 1887, p. 254, pl. 29, figs. 4, 5；陈木宏等，2017，115 页，图版 21，图 7–14。
Actinomma sp., Benson, 1966, pp. 164–166, pl. 5, fig. 6 (only).

地理分布 南极罗斯海，晚上新世。

134. 宽光眼虫 *Actinomma plasticum* Goll and Bjørklund

（图版 35，图 30–33；图版 36，图 1–28）

Actinomma plasticum Goll and Bjørklund, 1989, p. 729, pl. 2, figs. 4–9.

地理分布 南极罗斯海，早始新世—中渐新世。

135. 结实光眼虫（新种）*Actinomma solidula* sp. nov.

（图版 36，图 29–32）

Amphisphaera kina Hollis, 1993, p. 318, pl. 1, figs. 3–4；Hollis, 2002, p. 277, pl. 1, figs. 6a–b.

具两个格孔壳，个体较小；外壳壁厚，具圆形孔和六角形框架，孔径是孔间桁宽的 2–3 倍，壁孔大小相近，排列规则，横跨赤道有 7–9 孔；内壳较小，直径约为外壳的 1/4，格孔状，壁孔圆形或六角形，孔间桁很细，横跨赤道有 5–7 孔；壳体外表光滑，各节点上有锥形凸起，有 8–12 根放射骨针，很短，呈角锥形，无辅针。

标本测量：外壳直径 120–125μm，内壳直径 35–45μm，骨针长 12–25μm，针基宽 10–13μm。

模式标本：DSDP274-22-3 2（图版 36，图 29–30），来自南极罗斯海的 DSDP Leg 28 航次 Site 274 孔的 274-22-3 岩心样品中，保存在中国科学院南海海洋研究所沉积标本室。

地理分布　南极罗斯海，中始新世—中渐新世。

我们的标本看似与 *Amphisphaera kina* Hollis（1993, p. 318, pl. 1, figs. 3–4；Hollis, 2002, p. 277, pl. 1, figs. 6a–b）有些相似，但后者的骨针集中在两极周围，外壳壳壁稍薄，内壳则较大。Hollis（1993）与 Hollis（2002）的标本特征存在一定差异，相对于前者，后者皮壳较薄且骨针较长（长短不一），还具有两个髓壳，是否为同一种类仍待商榷。Hollis 等（2020, pl. 1, figs. 1a–b）的标本特征与 Hollis（1993）和 Hollis（2002）的完全不同，仅见一很短的极针，壳壁厚，内壳不详。*Amphisphaera* 属的特征是有三或四个同心格孔状球形壳，仅两根大小相近、形状相似、长度近等或不同的极针。因此，*Amphisphaera kina* Hollis 种的特征及其归属仍不清楚。此罗斯海新种与 *Actinomma plasticum* Goll and Bjørklund 的主要区别是后者有十根放射片状骨针及辅针，四层壳，常见三层，外壳壁薄，多角形孔大小相近，排列规则，孔间桁很细，因此两者之间的差异较大。

136. 优丝光眼虫 *Actinomma yosii* Nakaseko

（图版 37，图 1–6）

Actinomma yosii Nakaseko, 1959, p. 10, pl. 2, figs. 8a–b, 9a–b；Weaver, 1976, p. 573, pl. 1, fig. 1.
Hexacromyum sexaculeatum (Stöhr), Petrushevskaya, 1975, p. 569, pl. 2, fig. 5.
Actinomma eldredgei Renaudie and Lazarus, Renaudie, 2012, pp. 32, 34, pl. 4, figs. 1–2, 5.

该种个体很小，皮壳孔圆形，具六角形框架，排列规则；放射骨针较为细小，很短，三棱片状；壳表常有一些小凸，略显粗糙。与 *Actinomma plasticum* Goll and Bjørklund 较相似，主要区别在于此种的标本个体较小，放射骨针较细、较短。

地理分布　南极罗斯海，中渐新世。

137. 光眼虫（未定种 1）*Actinomma* sp. 1

（图版 37，图 7–10）

个体很小，三层壳，壁孔稀疏，为 *Actinomma* 类的古老或初始类型。

地理分布　南极罗斯海，中古新世。

138. 光眼虫（未定种 2）*Actinomma* sp. 2

（图版 37，图 11–12）

个体较大，皮壳壁薄，表面光滑，孔间桁较宽，放射骨针为三棱角锥状。

地理分布 南极罗斯海，早渐新世。

海胆虫属 Genus *Echinomma* Haeckel, 1881

三个同心格孔壳，具两种类型的放射骨针——较大的主针和小辅针。

139. 波夫海胆虫 *Echinomma popofskii* Petrushevskaya

（图版 37，图 13–14）

Echinomma popofskii Petrushevskaya, 1967, pp. 20–22, figs. 12I–III.

地理分布 南极罗斯海，中-晚渐新世。

葱海胆虫属 Genus *Cromyechinus* Haeckel, 1881

具四个同心格孔球壳和简单的两类骨针——大主针和小辅针。

140. 葱海胆虫（未定种 1）*Cromyechinus* sp. 1

（图版 37，图 15–16）

第一外壳不完整，第二外壳壁较厚，孔小而多；第一内壳呈正方形，第二内壳为很小的圆形。放射骨针较多，大小不一。

地理分布 南极罗斯海，中渐新世。

绵林虫属 Genus *Spongodrymus* Haeckel, 1881

球形壳为坚实海绵状，无格孔状髓壳，壳表放射骨针无分叉。

141. 鹿仁绵林虫 *Spongodrymus elaphococcus* Haeckel

（图版 37，图 17–20）

Spongodrymus elaphococcus Haeckel, 1887, p. 272, pl. 18, fig. 9；van de Paverd, 1995, p. 110, pl. 28, figs. 1a–b, pl. 29, fig. 3.
Spongoplegma rugosa Hollande and Enjumet, 1960, pl. 46, fig. 3, pl. 48, fig. 6.

地理分布 南极罗斯海，早上新世。

根球虫属 Genus *Rhizosphaera* Haeckel, 1860

具两个同心的髓壳和一个格孔状外皮壳，外有海绵状骨骼网，放射状骨针很多。

142. 近意根球虫 *Rhizosphaera paradoxa* Popofsky

（图版 37，图 21–22）

Rhizosphaera paradoxa Popofsky, 1912, p. 111, pl. 8, fig. 3.

壳体中心区有一明显的杂乱结构，由桁条无规律交接组成，由其节点处伸出放射桁至近壳表处有侧枝相互连接形成外壳，桁枝及放射桁较粗。罗斯海标本的个体较小。

地理分布 南极罗斯海，全新世。

143. 网根球虫 *Rhizosphaera reticulata* (Haeckel)

（图版 38，图 1–8）

Cenosphaera reticulata Haeckel, 1862, p. 349, pl. 11, fig. 2；Clark and Campbell, 1945, pl. 1, fig. 5；Hollande and Enjumet, 1960, pp. 87–88, pl. 31, figs. 1–5, pl. 32, fig. 1, pl. 33, figs. 1–2.

Rhizosphaera reticulata (Haeckel), van de Paverd, 1995, p. 102, pl. 24, figs. 1, 7.

地理分布 南极罗斯海，晚始新世—中中新世。

海绵球虫属 Genus *Spongosphaera* Ehrenberg, 1847

球形壳，两个格孔髓壳被外围的疏松海绵所包围，若干放射桁自髓壳长出并形成放射骨针。

144. 无针海绵球虫 *Spongosphaera spongiosum* (Müller) group

（图版 38，图 9–12；图版 39，图 1–23）

Dictyosoma spongiosum Müller, 1858, p. 31, pl. 2, figs. 9–11.

Dictyoplegma spongiosum (Müller), Haeckel, 1862, p. 458.

Spongodictyon spongiosum (Müller), Haeckel, 1887, p. 90.

Tetrasphaera spongiosa Popofsky, 1912, p. 112, text-fig. 23.

Spongosphaera spongiosum (Müller), van de Paverd, 1995, p. 114, pl. 28, fig. 3.

两个髓壳很小，位于球形壳的中心，外围的海绵组织一般较为疏松，也有一些个体在靠近中央区较为致密，甚至出现内部的密层状结构，放射骨针不发育或不明显。种类特征有一定的变化范围，因此归定为一个种群。

地理分布 南极罗斯海，晚古新世—晚渐新世。

145. 海绵球虫 *Spongosphaera streptacantha* Haeckel

（图版 40，图 1–8）

Spongosphaera streptacantha Haeckel, 1860, p. 840；Haeckel, 1862, S. 455, Taf. 26, Figs. 1–3； Haeckel, 1887, p. 282；Popofsky, 1912, S. 109, Abb. 22, Taf. 8, Fig. 4；谭智源、张作人，1976，234 页，图 11；陈木宏、谭智源，1996，176 页，图版 10，图 5–7，图版 41，图 1–2。

Spongosphaera polyacantha forma *streptacantha* van de Paverd, 1995, p. 114, pl. 28, figs. 2, 5–7.

地理分布 南极罗斯海，晚始新世—中渐新世。

灯球虫属 Genus *Lychnosphaera* Haeckel, 1881

髓壳简单、球形、格孔状，生出各放射幅针并由较粗的放射主针与海绵状的皮壳相连接。

146. 王灯球虫 *Lychnosphaera regina* Haeckel

（图版 40，图 9–22）

Lychnosphaera regina Haeckel, 1887, p. 277, pl. 11, figs. 1–4；陈木宏等，2017，122 页，图版 26，图 1–7，10–11（不含其他）。
Cladococcus lychnosphaerae Hollande and Enjumet, 1960, p. 115, pl. 55, figs. 1–2.
Thalassoplegma tenuis (Mast) Hollande and Enjumet, 1960, p. 118, pl. 50, fig. 6.
Arachnosphaera dichotoma Jørgensen, 1900, p. 60, pl. 3, fig. 18.

地理分布 南极罗斯海，早始新世—早渐新世。

147. 灯球虫（未定种 1）*Lychnosphaera* sp. 1

（图版 40，图 23–24）

单一格孔壳，壁孔较大，亚规则排列，孔间桁较细，放射骨针三棱角锥状，与壳半径近等长，较粗壮。

地理分布 南极罗斯海，晚更新世。

根编虫属 Genus *Rhizoplegma* Haeckel, 1881

髓壳简单、球形、格孔状，无辅针，以坚实的放射状主针与海绵状皮壳相连。

148. 北方根编虫 *Rhizoplegma boreale* (Cleve)

（图版 41，图 1–3）

Hexadoras borealis Cleve, 1899, pl. 2, fig. 4.
Rhizoplegma boreale (Cleve), Jørgensen, 1900, pp. 61, 62；Jørgensen, 1905, p. 118, pl. 9, fig. 38, pl. 10, fig. 38；Schröder, 1909b, fig. 23；Petrushevskaya, 1967, pp. 12–14, fig. 8；Ling et al., 1971, p. 710, fig. 3, pl. 1, figs. 2–3；Bjørklund, 1976, pl. 3, figs. 10–16, pl. 4, figs. 1–3；Abelmann, 1992b, pl. 1, fig. 13, tab. 1；Itaki and Takahashi, 1995, fig. 10b, tab. 1；Bjørklund et al., 1998, pl. 1, fig. 8；Dolven, 1998, pl. 9, figs. 2–3；陈木宏等，2017，122–123 页，图版 27，图 1–14。
Rhizoplegma boreale var. *antarctica* Popofsky, 1908, pp. 216–217, pl. 24, fig. 1.

地理分布 南极罗斯海，中渐新世—中上新世。

梅子虫亚目 Suborder PRUNOIDEA Haeckel, 1883

中央囊椭圆或柱状，单轴延伸；具硅质的椭圆或柱状壳体，有窗孔，单轴生长常在主轴两端形成轴极。

空虫科 Family Ellipsidae Haeckel, 1882

具简单的椭球格孔壳，无赤道缢，无内包的髓壳，中央囊椭圆或圆柱状，无环状赤道缢。

空椭球虫属 Genus *Cenellipsis* Haeckel, 1887

具简单椭球壳，无放射针和极管。

149. 泊阁空椭球虫 *Cenellipsis bergontianus* Carnevale group

（图版 41，图 4–5）

Cenellipsis bergontianus Carnevale, 1908, p. 19, pl. 3, figs. 5–7.
Cenellipsis bergontianus Carnevale group, Dzinoridze et al., 1978, pl. 22, fig. 10.

地理分布 南极罗斯海，中渐新世。

150. 莫尼卡空椭球虫 *Cenellipsis monikae* (Petrushevskaya)

（图版 41，图 6–12；图版 42，图 1–10）

Plegmosphaera sp. aff. *P. lepticali* Renz, Takahashi, 1991, p. 61, pl. 5, fig. 11.
Prunopyle frakesi Chen, 1975, p. 454, pl. 10, figs. 1–3.
Cenosphaera ? *oceanica* Clark and Campbell, Petrushevskaya, 1975, pl. 1, figs. 12–13.
Plegmosphaera monikae Petrushevskaya, 1975, pl. 30, figs. 1–7.
Lithocarpium monikae Petrushevskaya, 1975, p. 572, pl. 4, figs. 6–10.

个体较大，宽椭球形，壁稍薄，壁孔小或细而密集，壳表光滑，有一个圆形的小窗孔。此类壳内常见有一个或更多的圆形小壳体，无规律地分布在放射虫壳体内的不同部位，疑为被吞噬或共生的小硅藻。此类标本的内部结构呈现不同类型，*Lithocarpium monikae* Petrushevskaya（1975）的图版 4，图 8–9 为内空，图版 4 中的其他图却在壳内中心有髓壳，而且在建立该新种时还包括了图版 30，图 1–7（不同归属）。*Prunopyle frakesi* Chen（1975）的一些特征与 Petrushevskaya（1975）较接近，但又增加了口孔及缘齿的观测，可能两者为同物，又同时发表，应为同一个种。

地理分布 南极罗斯海，早始新世—中渐新世。

针球虫科 Family Stylosphaeridae Haeckel, 1881

球形壳一个或多个，壳表的两极上有对称的两根骨针。

球剑虫属 Genus *Ellipsoxiphus* Dunikowski, 1882, emend. Haeckel, 1887

具单一的椭球形壳，主轴在两极延伸为两根对应的强壮骨针，其大小相同、形状相似。

151. 纺锤球剑虫 *Ellipsoxiphus atractus* Haeckel

（图版 42，图 11–14）

Ellipsoxiphus atractus Haeckel, 1887, p. 298, pl. 14, fig. 1；Kamikuri et al., 2008, p. 169, pl. 2, fig. 14
Stylosphaera radiosa Ehrenberg, Abelmann, 1990, p. 692, pl. 2, figs. 4b–c (only).

壳壁较厚，壁孔中有不规则的类圆形网孔；两根极针较粗短，三棱角锥状。

地理分布 南极罗斯海，全新世。

152. 薄球剑虫（新种）*Ellipsoxiphus leptodermicus* sp. nov.

（图版 43，图 1–6）

单一椭球形壳，壁较薄，具大小近等的圆形壁孔，无网孔，排列规则，横跨赤道有

6–7 孔，孔径是孔间桁宽的 4–6 倍，孔间桁较细，具浅的六角形框架；两根极针大小相似，较细，长圆锥或圆柱形（末端缩尖），不同标本的极针长度变化较大，一般小于壳体短轴，也有较长者大于壳体的长轴；壳体表面光滑，无辅针或明显的棘凸。

标本测量：壳长轴 100–180μm，壳短轴 85–160μm，骨针长 50–120μm。

模式标本：DSDP274-1-3 1（图版 43，图 3–4），来自南极罗斯海的 DSDP Leg 28 航次 Site 274 孔的 274-1-3 岩心样品中，保存在中国科学院南海海洋研究所沉积标本室。

地理分布 南极罗斯海，早上新世—全新世。

该新种特征与 *Xiphostylus alcedo* Haeckel（1887, p. 127, pl. 13, fig. 4）较接近，主要区别是后者壳壁较厚，具深的六角形框架，两根极针大小或长短不同，呈三棱角锥状，较粗壮，长针长度是短针的 4–5 倍，极针针基较宽。

153. 圆球剑虫（新种）*Ellipsoxiphus rotundalus* sp. nov.

（图版 43，图 7–9）

单一球壳，圆球形，壳壁很厚，壁孔圆形，具六角形框架，规则排列，横跨赤道有 11–13 个；孔间桁较细，约为孔径的 1/4–1/3，内壁与外表形态一致，内外间距较大，形成假双层壳；两根极针长短不一，相差较大，呈三片棱柱状，无侧刺；壳体表面光滑，无辅针与棘突。

标本测量：壳外径 210μm，壳内径 160μm，壳厚 25μm，骨针长 45–160μm。

模式标本：DSDP274-36-1 2（图版 43，图 7–9），来自南极罗斯海的 DSDP Leg 28 航次 Site 274 孔的 274-36-1 岩心样品中，保存在中国科学院南海海洋研究所沉积标本室。

地理分布 南极罗斯海，早始新世。

该新种特征与 *Xiphostylus alcedo* Haeckel（1887, p. 127, pl. 13, fig. 4）的主要区别是后者的壳体呈椭球形，壁孔较少，极针为三棱角锥状。

154. 棘刺球剑虫（新种）*Ellipsoxiphus echinatus* sp. nov.

（图版 43，图 10–11）

单一椭球壳，壳壁较薄，类圆形孔，横跨赤道有 7–8 孔，亚规则排列，孔间桁较细，孔径是孔间桁宽的 4–6 倍；壳体表面粗糙，在各节点上有棘刺，或形成辅针；两根极针近针形，长度不同，极针的初段三棱状，稍粗，末段细圆锥状，迅速削尖；壳表一些节点上发育有一些角锥状的辅针。

标本测量：壳长轴 110–120μm，壳短轴 105–110μm，极针长 50–60μm。

模式标本：DSDP274-1-5 2（图版 43，图 10–11），来自南极罗斯海的 DSDP Leg 28 航次 Site 274 孔的 274-1-5 岩心样品中，保存在中国科学院南海海洋研究所沉积标本室。

地理分布 南极罗斯海，全新世。

该种特征与 *Ellipsoxiphus leptodermicus* sp. nov.的主要区别是后者的壁孔排列规则，壳表光滑无棘刺或辅针，极针完全为细长的圆锥或圆柱状，不变形。

石芹虫属 Genus *Lithapium* Haeckel, 1887

具单一的椭球形或梨形壳，仅有一极针出现在壳长轴的一端。

155. 梨形石芹虫 *Lithapium pyriforme* Haeckel

（图版 43，图 12–13）

Lithapium pyriforme Haeckel, 1887, p. 303, pl. 14, fig. 9.

圆形孔，大小相同，具六角形框架，节点上有小凸起，极针短小。

地理分布　南极罗斯海，中上新世。

156. 海咽石芹虫 *Lithapium halicapsa* Haeckel

（图版 43，图 14–15）

Lithapium halicapsa Haeckel, 1887, pp. 303–304, pl. 14, fig. 8.

壳壁孔类圆形或心形，大小不同，孔缘向内有一小尖突，无六角形框架，无节点凸起，壳表光滑，极针短小。

地理分布　南极罗斯海，中上新世。

轴梅虫属 Genus *Axoprunum* Haeckel, 1887

单一椭球形壳，腔内有四根相互垂直的放射棒近等长地自内壁向中心伸长，末端游离，无交集。两根极针的大小与形状相似。

157. 十字轴梅虫 *Axoprunum stauraxoniurn* Haeckel

（图版 43，图 16–19）

Axoprunum stauraxoniurn Haeckel, 1887, p. 298, pl. 48, fig. 4；Hays, 1965, p. 176, pl. 1, fig. 3；Kamikuri et al., 2008, p. 168, pl. 2, figs. 13a–b.

该种标本的内棒末端或许在腔的中央分叉游离部分连接，形成一个模糊的假内壳。

地理分布　南极罗斯海，中渐新世。

针球虫属 Genus *Stylosphaera* Ehrenberg, 1847

具两个同心格孔球壳，主轴两极上两根骨针大小相等，形状相似。

158. 冠针球虫光滑亚种 *Stylosphaera coronata laevis* Ehrenberg

（图版 44，图 1–6）

Stylosphaera laevis Ehrenberg, 1873, p. 259; Ehrenberg, 1876, pl. 25, fig. 6; Sanfilippo and Riedel, 1973, pp. 520–521, pl. 1, fig. 19, pl. 25, figs. 5–6.

Stylosphaera coronata laevis Ehrenberg, Chen, 1975, p. 455, pl. 5, fig. 3.

Stylosphaera sp. C, Petrushevskaya, 1975, pl. 2, figs. 12–13.

Druppatractus sp., Chen, 1975, p. 505, pl. 20, figs. 11–12；Weaver, 1976, p. 573, pl. 5, figs. 1–2, pl. 6, fig. 1.

Druppatractus hastatus Blueford group, Abelmann, 1990, p. 691, pl. 2, fig. 3.

地理分布 南极罗斯海，中-晚中新世。

159. 哥伦针球虫 *Stylosphaera goruna* Sanfilippo and Riedel

（图版 44，图 7–10）

Stylosphaera goruna Sanfilippo and Riedel, 1973, p. 521, pl. 1, figs. 20–22, pl. 25, figs. 9–10.
Amphisphaera goruna (Sanfilippo and Riedel) Hollis, 2002, pl.1, fig. 5.

两个壳，除两根极针外，还有若干较粗大的辅针，壳壁厚或薄。

地理分布 南极罗斯海，早渐新世—全新世

160. 米诺针球虫 *Stylosphaera minor* Clark and Campbell

（图版 44，图 11–26；图版 45，图 1–24）

Stylosphaera minor Clark and Campbell, 1942, p. 27, pl. 5, figs. 1, 2, 2a, 12；Hollis, 2002, p. 289, pl. 1, fig. 13.
Stylatractus ? sp. A, O'Connor, 1997b, p. 116, pl. 3, fig. 11.
Axoprunum bispiculum (Popofsky), Takemura, 1992, pp. 741–742, pl. 1, figs. 1–2.
Axoprunum pierinae (Clark and Campbell), Hollis et al., 1997, p. 44, pl. 1, figs. 9, 11–13 (only).

地理分布 南极罗斯海，晚古新世—晚渐新世。

161. 梨形针球虫 *Stylosphaera pyriformis* (Bailey)

（图版 45，图 25–26）

Haliomma pyriformis Bailey, 1856, p. 1, pl. 1, fig. 29.
Stylosphaera pyriformis (Bailey), van de Paverd, 1995, p. 122, pl. 31, fig. 5.
Stylosphaera sp. B, Boltovskoy and Jankilevich, 1985, pl. 2, fig. 12.

地理分布 南极罗斯海，晚上新世。

162. 针球虫（未定种 1）*Stylosphaera* sp. 1

（图版 45，图 27–28）

具一个皮壳和两个髓壳，三壳大小比例为 1∶3∶9；皮壳呈宽椭球形，壁厚，壁孔与孔间桁近等宽，在孔间桁的节点处有明显的圆锥形突起；髓壳较小，壳体的表面粗糙，有棘凸；极针比壳径短，为圆柱形，末端缩尖。

地理分布 南极罗斯海，晚始新世。

橄榄虫属 Genus *Druppatractus* Haeckel, 1887

具简单的椭球状皮壳和髓壳。主轴两极有两个相对的极针，大小与形状均不相同。

163. 果橄榄虫（新种）*Druppatractus carpocus* sp. nov.

（图版 46，图 1–4）

两壳均近圆球形，大小之比为 3∶5，壁孔较大，壳间距较小；外壳壁厚中等，壁孔类圆形，大小相似，排列规则，有六角形框架，孔间桁较细，孔径是孔间桁宽的 4–5 倍，横跨赤道有 6–7 孔；内壳较大，不同标本壁孔六角形或类圆形，排列规则或不规则，大

小相近或不等，横跨赤道有 5–7 孔；许多放射桁连接着两壳，其中一根伸出壳外发育为极针，细圆锥形或柱形，末端略缩尖，另一极针不发育，不同标本的针长有一定差异；壳表光滑无刺。

标本测量：外壳直径 110–120μm，内壳直径 70–75μm，极针长 25–80μm。

模式标本：DSDP274-10-3 1（图版 46，图 3–4），来自南极罗斯海的 DSDP Leg 28 航次 Site 274 孔的 274-10-3 岩心样品中，保存在中国科学院南海海洋研究所沉积标本室。

地理分布　南极罗斯海，早上新世。

该种与 *Druppatractus hastatus* Blueford 的主要区别在于后者的内壳较小，仅为外壳的约 1/3，壁孔小而多，横跨赤道有 11–13 孔，极针较长，针长大于孔径。

164. 矛橄榄虫 *Druppatractus hastatus* Blueford group

（图版 46，图 5–9）

Druppatractus hastatus Blueford, 1982, p. 206, pl. 6, figs. 3–4；Lazarus and Pallant, 1989, p. 363, pl. 6, figs. 13–15.
Druppatractus hastatus Blueford group, Abelmann, 1990, p. 691, pl. 2, fig. 3.
Stylosphaera spp., Nigrini and Lombari, 1984, p. S31, pl. 4, figs. 4a–b.

该种壳体外形与 *Xiphosphaera gaea* Haeckel（1887, p. 123, pl. 14, fig. 5）较相似，但后者仅具一个壳（无内壳）。

地理分布　南极罗斯海，中始新世—早渐新世。

165. 不规则橄榄虫 *Druppatractus irregularis* Popofsky

（图版 46，图 10–11）

Druppatractus irregularis Popofsky, 1913, pp. 114–115, textfigs. 24–26；Benson, 1966, pp. 180–182, pl. 7, figs. 7–11.
Dorydruppa bensoni Takahashi, 1991, p. 78, pl. 15, figs. 11–14.

地理分布　南极罗斯海，出现于中上新世。

166. 梨橄榄虫（新种）*Druppatractus pyriformus* sp. nov.

（图版 46，图 12–14）

个体较小，外壳椭球形，壳壁稍厚；内壳近梨形，两端缩窄；两壳大小之比为 1∶2；壁孔圆形，大小近等，具六角形框架，排列规则，孔径是孔间桁宽的 3–4 倍，横跨赤道有约 6–7 孔；壳表粗糙，有棘凸；极针为针状，无侧棱，两针的发育程度不等。

标本测量：壳长轴 85–95μm，壳短轴 80–90μm，极针长 20–80μm。

模式标本：DSDP274-11-3 1（图版 46，图 12–14），来自南极罗斯海的 DSDP Leg 28 航次 Site 274 孔的 274-11-3 岩心样品中，保存在中国科学院南海海洋研究所沉积标本室。

地理分布　南极罗斯海，晚中新世。

该种特征与 *Druppatractus irregularis* Popofsky 较接近，主要区别是后者的皮壳壁很薄，壁孔亚规则排列，无六角形框架，髓壳的壁孔较多，圆形，排列规则，壳表极为光滑。

针蜓虫属 Genus *Stylatractus* Haeckel, 1887

具简单椭球形皮壳和两个髓壳，主轴上有两根对称的极针，大小相等，形状相似。

167. 圣针蜓虫 *Stylatractus angelina* (Campbell and Clark)

（图版 46，图 15–16）

Stylosphaera angelina Campbell and Clark, 1944a, p. 12, pl. 1, figs. 14–20.
Axoprunum angelinum (Campbell and Clark), Kling, 1973, p. 634, pl. 1, figs. 13–16, pl. 6, figs. 14–18.
Amphistylus angelinus (Campbell and Clark) Chen, 1975, p. 453, pl. 21, figs. 3–4; Takemura, 1992, p. 741, pl. 1, figs. 8–9.
Stylatractus sp., Hays, 1965, p. 167, pl. 1, fig. 6.
Stylatractus universus Hays, 1970, p. 215, pl. 1, figs. 1–2; Kling, 1971, p. 1086, pl. l, fig. 7; Lazarus, 1990, p. 717, pl. 6, figs. 9–11; Nigrini and Sanfilippo, 2001, pp. 410–411.

地理分布 南极罗斯海，晚中新世。

168. 细针蜓虫（新种）*Stylatractus gracilus* sp. nov.

（图版 46，图 17–23；图版 47，图 1–19）

Amphisphaera sp., Chen, 1975, p. 453, pl. 6, figs. 1–2.
Axoprunum bispiculum (Popofsky), Takemura, 1992, p. 741, pl. 1, figs. 1–2; Chen, 1975, p. 454, pl. 21, figs. 1–2; Hollis, 2002, pl. 1, figs. 7–9.
Axoprunum liostylum (Ehrenberg) group, Petrushevskaya, 1975, p. 571, pl. 2, fig. 22.
Stylacontarium sp. aff. *S. bispiculum*, Kling, 1973, p. 634, pl. 6, figs. 19–23, pl. 14, figs. 5–8.
Stylacontarium bispiculum Popofsky, Kling, 1973, pl. 15, figs. 11–14; Benson, 1983, p. 508, pl. 1, figs. 7–10.
Axoprunum pierinae (Clark and Campbell), Hollis et al., 1997, p. 44, pl. 1, figs. 9–10, 14 (only).

壳近圆球形或椭球形，有一个皮壳和两个髓壳，三壳大小比例为 1∶2∶6；皮壳一般壳壁较厚，也有一些标本壳壁稍薄，壁孔类圆形或椭圆形，大小相近或不同，亚规则或不规则排列，具六角形框架，横跨赤道一般有 10–12 孔，孔径是孔间桁宽的 2–3 倍，也有个别壁孔大小悬殊的标本仅有 8 孔；两个格孔状髓壳很小，壁孔类圆形，较小；两根极针形状相似，细圆锥形或针形；多数壳体表面光滑，有的标本孔间桁节点上有很小的棘突，壳表略显粗糙。该种的皮壳厚度与壁孔均存在一些变化，但基本出现于相同的地层层位中。

标本测量：壳长轴 110–150μm，壳短轴 95–145μm，极针长 60–110μm。

模式标本：DSDP274-20-5 1（图版 47，图 1–2），来自南极罗斯海的 DSDP Leg 28 航次 Site 274 孔的 274-20-5 岩心样品中，保存在中国科学院南海海洋研究所沉积标本室。

地理分布 南极罗斯海，早始新世—晚渐新世。

该种与 *Stylacontarium bispiculum* Popofsky（1912, p. 91, pl. 2, fig. 2）有明显的区别，后者壳壁较薄，壁孔较大，约为孔间桁宽的3–4 倍，节点上无突起，壳表光滑，Kling（1973）、Chen（1975）和 Takemura（1992）等将此类标本定为 *Stylacontarium bispiculum* 实属有误，应是不同的另一种。该种与 *Stylosphaera liostylus* Ehrenberg（1873, p. 259；Ehrenberg, 1876, pl. 25, fig. 7）较为相似，但后者壁孔排列规则，表面有许多明显的圆锥形棘刺，末端尖，极针圆锥状，较粗短。

169. 不规则针蜓虫 *Stylatractus irregularis* (Takemura)

（图版 48，图 1–10）

Axoprunum (?) *irregularis* Takemura, 1992, p. 742, pl. 3, figs. 8–11.

地理分布　南极罗斯海，晚始新世—中渐新世。

170. 外套针蜓虫 *Stylatractus palliatum* (Haeckel)

（图版 48，图 11–21）

Ellipsoxiphus elegans var. *palliatus* Haeckel, 1887, p. 296, pl. 14, fig. 7.
Ellipsoxiphium palliatum Haecker, 1908a, p. 441, pl. 84, fig. 587；Takahashi, 1991, p. 75, pl. 14, figs. 11–17.
Druppatractus acquilonius Hays, Takahashi and Honjo, 1981, p. 147, pl. 3, fig. 5.

该种具有一皮壳和二髓壳，主要特征是壳表由各节点生出的辅针末端相互连接形成一个网状的假外壳，两极针相似。Haeckel（1887）建立此种就存在疑问，认为仅具一个简单的壳体（无内壳），因此归入 *Ellipsoxiphium* 属，而 Haecker（1908a）的标本与 Haeckel（1887）有一定差异，主要是前者的极针呈针棒形，后者的极针却有侧棱。Takahashi（1991）的标本与我们的标本特征基本一致，有明显的两个髓壳，但壳表特征与 Haeckel（1887）相同，皮壳孔圆形，具六角形框架，而我们标本的极针却与 Haecker（1908a）一样为针棒状，无侧棱。综上所述，此类标本定为 *Stylatractus palliatum* (Haeckel)较合适。

171. 富针蜓虫 *Stylatractus pluto* (Haeckel)

（图版 49，图 1–6）

Amphisphaera pluto Haeckel, 1887, p. 144, pl. 17, figs. 7–8.
Stylatractus neptunus Haeckel, 1887, p. 328, pl. 17, fig. 6；Riedel, 1958, p. 226, pl. 1, fig. 9.
Xiphatractus pluto (Haeckel), Benson, 1966, p. 184, pl. 7, figs. 14–17；Takahashi, 1991, p. 77, pl. 15, figs. 1–3.
Xiphatractus cronos (Haeckel), Benson, 1966, p. 182, pl. 7, figs. 12–13.
? *Stylatractus* sp., Nigrini and Moore, 1979, p. 555, pl. 7, figs. 1a–b.
Stylosphaera hispida Ehrenberg, van de Paverd, 1995, p. 122, pl. 31, fig. 1.

极针圆锥形，三壳之比为 1∶3∶6，皮壳壁孔横跨赤道有 8–9 个。
地理分布　南极罗斯海，早—晚上新世。

剑蜓虫属 Genus *Xiphatractus* Haeckel, 1887

具一简单的椭球形皮壳和两个髓壳，主轴上的两根极针大小相异、形状不同。

172. 长针剑蜓虫（新种）*Xiphatractus longostylus* sp. nov.

（图版 49，图 7–10）

壳呈椭球形或近圆球形，具一个皮壳和两个髓壳，三壳大小比例为 1∶2.5∶5；皮壳中等壁厚，壁孔类圆形，大小相近，排列规则或亚规则，横跨赤道有 6–7 孔，具六角形框架，孔径约为孔间桁宽的三倍，在各节点上有角锥形凸起，末端尖；外髓壳椭球形，内髓壳圆球形，两个髓壳格孔状，壳壁略薄，壁孔类圆形，较少；两根极针较长，接近皮壳直径，大小与形状相似，在针基处为三棱状，其余部分呈细柱状，末端稍缩尖；壳体表面粗糙。

标本测量：皮壳长轴 120–140μm，皮壳短轴 110–125μm，外髓壳长轴 70–75μm，外髓壳短轴 55–60μm，内髓壳直径 25μm，极针长 60–110μm。

模式标本：DSDP274-10-5 1（图版 49，图 7–8），来自南极罗斯海的 DSDP Leg 28 航次 Site 274 孔的 274-10-5 岩心样品中，保存在中国科学院南海海洋研究所沉积标本室。

地理分布 南极罗斯海，晚中新世。

该种特征与 *Xiphatractus spumeus* Dumitrica 较相似，主要区别是后者的壳壁稍厚，极针很短，呈三棱角锥状，被 Dumitrica（1973）报道出现在北大西洋海区，这两个种可能代表明显不同的区域生物地理环境特征。罗斯海标本与 Takahashi（1991）的 *Xiphatractus* sp. B 标本非常相似，但后者的极针较短，与 *Xiphatractus spumeus* Dumitrica 基本相同。

173. 大轴谱剑蜓虫大轴谱亚种 *Xiphatractus megaxyphos megaxyphos* (Clark and Campbell)

（图版 49，图 11–16）

Stylosphaera megaxyphos megaxyphos Clark and Campbell, 1942, p. 25, pl. 5, figs. 5, 6, 14, pl. 6, fig. 6.

该类标本有两个髓壳，而 Clark 和 Campbell（1942）的标本仅见一个髓壳。极针一般两根，也可更多或不发育，极针圆柱形或有侧棱的角锥状。壳壁较厚，孔类圆形，大小不一，不规则或亚规则排列。

地理分布 南极罗斯海，早-中渐新世。

174. 泡沫剑蜓虫 *Xiphatractus spumeus* Dumitrica

（图版 49，图 17–18）

Xiphatractus spumeus Dumitrica, 1973, p. 833, pl. 20, fig. 9.
Xiphatractus sp. B, Takahashi, 1991, p. 77, pl. 15, figs. 6–7.

地理分布 南极罗斯海，早上新世。

175. 剑蜓虫（未定种 1）*Xiphatractus* sp. 1

（图版 49，图 19–20）

外皮壳呈变形的椭球形，壁孔较小而多，分布不规律；内髓壳有点模糊；极针三棱角锥状，较短，长针是短针的三倍长。

地理分布 南极罗斯海，早上新世。

176. 剑蜓虫（未定种 2）*Xiphatractus* sp. 2

（图版 49，图 21–22）

三壳之比为 1∶3∶9，球形壳，皮壳壁较厚，孔类圆形，大小不一，不规则分布，孔间桁稍宽；极针多于两根，不对称分布。此类标本特征较特殊，不属 *Actinomma* 类，暂归此属。

地理分布 南极罗斯海，早上新世。

倍球虫属 Genus *Amphisphaera* Haeckel, 1881, emend. Petrushevskaya, 1975

具三或四个同心格孔状球形壳，两根骨针大小相近、形状相似、长度近等或不同。

177. 奥特倍球虫 *Amphisphaera aotea* Hollis

（图版 50，图 1–2）

Amphisphaera aotea Hollis, 1993, p. 316, pl. 1, figs. 1–2.

我们的标本与 Hollis（1993）标本的特征完全一致。

地理分布 南极罗斯海，早上新世。

178. 裂蹼倍球虫 *Amphisphaera dixyphos* (Ehrenberg)

（图版 50，图 3–4）

Haliomma dixyphos Ehrenberg, 1844, p. 83；Ehrenberg, 1854, pl. 22, fig. 31.
Amphisphaera dixyphos (Ehrenberg), Petrushevskaya, 1975, p. 570, pl. 2, fig. 17.
Amphisphaera santhaennae (Campbell and Clark), Petrushevskaya, 1975, p. 570, pl. 2, fig. 21.
Stylosphaera dixyphos Ehrenberg, Abelmann, 1990, p. 692, pl. 2, fig. 7.

地理分布 南极罗斯海，出现于早渐新世。

179. 尼氏倍球虫 *Amphisphaera nigriniae* Kamikuri

（图版 50，图 5–9）

Amphisphaera nigriniae Kamikuri, 2010, p. 97, pl. 2, figs. 1a–b, 4a–b, 7a–b, 10a–b, 12a–b, pl. 4, figs. 18a–b.

外壳椭球形，壳壁稍厚，外表略显粗糙，壁孔较多，横跨赤道有约 12–15 孔，三壳之比为 1∶3∶6，极针较细，长圆锥形。该种与 *Stylacontarium bispiculum* Popofsky（1912, p. 91, pl. 2, fig. 2）的主要区别在于后者外壳壁薄光滑，壁孔较大，孔间桁稍宽，横跨赤道仅有 8–9 孔，内壳较小，三壳之比约为 1∶3∶10。

地理分布 南极罗斯海，早上新世—全新世。

180. 辐射倍球虫 *Amphisphaera radiosa* (Ehrenberg) group

（图版 50，图 10–23）

Stylosphaera radiosa Ehrenberg, 1854, p. 256；Ehrenberg, 1876, taf. 24, fig. 5；Abelmann, 1990, p. 692, pl. 2, figs. 4a–c, 7.
Stylatractus radiosus (Ehrenberg), Petrushevskaya and Kozlova, 1972, p. 520.
Amphisphaera radiosa (Ehrenberg) group, Petrushevskaya, 1975, p. 570, pl. 2, figs. 18–20.
(?) *Druppatractus agostnellii* Carnevale, 1908, p. 20, pl. 3, fig. 10.
Lithatractus santaennae pusillus Campbell and Clark, 1944a, p. 19, pl. 2, figs. 23–25.
Stylatractus santaennae (Campbell and Clark), Petrushevskaya and Kozlova, 1972, p. 520, pl. 11, fig. 10；Lazarus, 1992, p. 796, pl. 4, figs. 6–8.

地理分布 南极罗斯海，晚始新世—晚中新世。

181. 光滑倍球虫（新种）*Amphisphaera sperlita* sp. nov.

（图版 50，图 24–25）

具三个格孔壳；外壳椭球形，壁很薄，光滑，圆形孔较大，横跨赤道有 5–6 孔，具六角形框架，孔间桁较细，孔径是孔间桁宽的 4–5 倍；中间壳椭球形，形状与外壳相似，壳壁比外壳略厚，壁孔较大；内壳近圆球形，不太规则，壁孔较大；极针较长，细棒状，末端缩尖；壳壁光滑，无棘刺。

标本测量：皮壳长轴 140μm，皮壳短轴 120μm，中壳长轴 95μm，中壳短轴 80μm，内壳直径 50μm，极针长 70–110μm。

模式标本：DSDP274-10-3 2（图版 50，图 24–25），来自南极罗斯海的 DSDP Leg 28 航次 Site 274 孔的 274-10-3 岩心样品中，保存在中国科学院南海海洋研究所沉积标本室。

地理分布 南极罗斯海，出现于早上新世。

该种特征与 *Lithatractus santaennae* Campbell and Clark（1944b, p. 19, pl. 2, figs. 20–22；Petrushevskaya, 1975, p. 570, pl. 2, fig. 21）较接近，但后者壳壁稍厚，极针较短，为三棱角锥状。

针矛虫属 Genus *Stylacontarium* Popofsky, 1912

具三个同心格孔状球壳，两根极针大小或形状不同。

182. 阿克针矛虫 *Stylacontarium acquilonium* (Hays) group

（图版 51，图 1–16）

Druppatractus acquilonius Hays, 1970, p. 214, pl. 1, figs. 4–5；Kling, 1971, p. 1086, pl. 1, figs. 5–6；Morley, 1985, pl. 4, figs. 1a–b.

Stylacontarium acquilonium (Hays), Kling, 1973, p. 634, pl. 1, figs. 17–20, pl. 14, figs. 1–4；Ling, 1973, p. 777, pl. 1, figs. 6–7.

Axoprunum bispiculum (Popofsky), Hollis et al., 1997, pp. 43–44, pl. 1, fig. 14.

Hays（1970）描述该种的髓壳为椭球形或不规则状，由疏松网状物构成，认为该种主要分布于北半球高纬度海区，近赤道消失。实际上，该种特征变化较大，皮壳很厚或较薄均有，髓壳为椭球形、圆球形、不规则形、菱形或四方形，不同作者的判断依据存在差异。一般来说，如髓壳呈菱形或四方形则归属该种无疑。至今为止，由于应用中该种特征变化较大，定为种群较合适。

地理分布 南极罗斯海，晚始新世—中渐新世。

海绵虫科 Family Sponguridae Haeckel, 1862, sensu Sanfilippo and Riedel, 1973

壳为海绵状椭球体或圆柱状，由全部或部分的海绵结构组成，无赤道收缩缢，有或无一个被包围的髓壳。

海绵虫属 Genus *Spongurus* Haeckel, 1862

壳椭球或圆柱形，有时呈三节状，实心海绵结构，无内部空腔和格孔状髓壳。无极刺和格孔外膜。

183. 考勒海绵虫 *Spongurus cauleti* Goll and Bjørklund

（图版 51，图 17–22）

Spongurus cauleti Goll and Bjørklund, 1989, p. 730, pl. 3, figs. 4–9.

该种两端可能有棘刺。

地理分布　南极罗斯海，早—晚始新世。

木偶海绵虫属 Genus *Spongocore* Haeckel, 1887

壳呈椭圆形或圆柱形（可分为三节），由实心的海绵组织构成，无内部空腔和格孔髓壳。无极针。壳体外围有一环绕的格孔覆盖物（外包膜），由放射骨针与海绵壳相连。

184. 圆筒木偶海绵虫 *Spongocore cylindricus* Haeckel

（图版 52，图 1–16）

Spongurus cylindricus Haeckel, 1862, p. 465, pl. 27, fig. 1；van de Paverd, 1995, pp. 149, 151, pl. 40, figs. 1–9.
Spongocore puella Haeckel, 1887, p. 347, pl. 48, fig. 6；Benson, 1966, p. 187, pl. 8, figs. 1–3；Nigrini, 1970, p. 168, pl. 2, fig. 3；Kling, 1973, p. 635, pl. 7, figs. 18–22；陈木宏、谭智源，1996，178–179 页，图版 11，图 10–11，图版 41，图 6–7。
Spongocore polyacantha Popofsky, 1912, S. 116, Abb. 27；谭智源、宿星慧，1982，150 页，图版 VII，图 10。
Spongocore puer Campbell and Clark, 1944a, p. 22, pl. 3, figs. 7–9.
Ommatogramma dumitrikai Petrushevskaya, 1975, p. 577, pl. 37, figs. 4–5.
Spongocore spp., Abelmann, 1990, p. 692, pl. 3, fig. 1.

该种的哑铃状海绵壳的外部体型和内部结构有一定变化，两端一般呈增大的椭球形，也有的末尾缩小或增大并截平（酒杯状），中间棒体部分一般为纯海绵组织构成，也有的呈层状结构，似乎可将它们分为 2–3 个种。由于它们几乎同时出现于各相同的地层层位中，因此将这些存在差异特征的标本归为同一个种类，以方便鉴定与应用。

地理分布　南极罗斯海，早始新世—中渐新世。

185. 木偶海绵虫（未定种 1）*Spongocore* sp. 1

（图版 53，图 1–9）

壳体近似 *Spongocore cylindricus* Haeckel，但海绵组织已被改造变异成近格孔状，孔细小，类圆形，无规律分布，壳体也常不完整，仅保存一半，也未见包膜和骨针等。出现于 Site 274 的最底部，明显属于古老地层类型。

地理分布　南极罗斯海，早-中古新世。

双月虫属 Genus *Amphymenium* Haeckel, 1881

两个简单不分开的室臂对称地分布在一条轴线上，由侧翼连接。

186. 小玫瑰双月虫 *Amphymenium monstrosum* (Popofsky)

（图版 53，图 10–15）

Amphibrachium monstrosum Popofsky, 1912, p. 135, pl. 4, fig. 2.

海绵臂的中心有环形结构，两端分别由 4–6 个臂室组成。罗斯海的标本个体较小。*Amphibrachium* 的属征是没有臂室，因此该种归入 *Amphymenium* 属较为合适。

地理分布 南极罗斯海，早古新世—中渐新世。

187. 蕨状双月虫（相似种）*Amphymenium* cf. *splendiarmatum* Clark and Campbell

（图版 53，图 16–27）

Amphymenium cf. *splendiarmatum* Clark and Campbell, Hollis, 1997, pp. 49–50, pl. 3, fig. 9.

Amphymenium splendiarmatum Clark and Campbell, 1942, p. 46, pl. 1, figs. 12, 14；Sanfilippo and Riedel, 1973, p. 524, pl. 11, figs. 6–8, pl. 28, figs. 6–8.

Amphymenium (?) *splendiarmatum* Clark and Campbell, Petrushevskaya, 1975, p. 577, pl. 7, fig. 1, pl. 37, figs. 1–3.

地理分布 南极罗斯海，中始新世—中渐新世。

盘虫亚目 Suborder DISCOIDEA Haeckel, 1862

中央囊和外壳均为铁饼状或透镜状，壳体硅质，有窗孔。单一轴向缩小生长。

镜盘虫科 Family Phacodiscidae Haeckel, 1881

皮壳为透镜状格孔壳，髓壳球形，一个或两个，由放射桁与皮壳连接，壳缘无赤道腰带。

镜盘虫属 Genus *Phacodiscus* Haeckel, 1881

壳体为圆形透镜状，双髓壳，边缘无赤道腰带和放射骨针。

188. 盾透镜盘虫 *Phacodiscus clypeus* Haeckel

（图版 54，图 1–6）

Phacodiscus clypeus Haeckel, 1887, p. 425, pl. 35, figs. 6, 9.

Phacodiscus testatus Kozlov, Dzinoridze et al., 1978, pl. 23, fig. 12.

皮壳边缘缩尖，无其他结构物。

地理分布 南极罗斯海，晚始新世—中渐新世。

189. 轮镜盘虫 *Phacodiscus rotula* Haeckel

（图版 54，图 7–8）

Phacodiscus rotula Haeckel, 1887, p. 424, pl. 35, fig. 7.

皮壳边缘很厚，呈截面状，近似一腰带。

地理分布 南极罗斯海，出现于晚始新世。

190. 扁豆镜盘虫 *Phacodiscus lentiformis* Haeckel

（图版 54，图 9–13）

Phacodiscus lentiformis Haeckel, 1887, p. 425, pl. 35, fig. 8.

Heliostylus sp., Hollis, 2002, pl. 1, fig. 15.

皮壳边缘缩尖，边缘环的表面上有规则排列的小脊，无放射骨针。

透镜虫属 Genus *Astrophacus* Haeckel, 1881

具双髓壳，壳盘边缘有 10–12 个或更多的放射骨针，数量不定，分布无规律。该属与 *Heliodiscus* 较相似，主要区别是后者仅具一个髓壳。

191. 全织星透镜虫 *Astrophacus perplexus* (Clark and Campbell)

（图版 54，图 14–22；图版 55，图 1–22）

Heliodiscus perplexus Clark and Campbell, 1942, pp. 40–41, pl. 3, fig. 12.
Periphaena decora Ehrenberg, Hollis, 1997, p. 46, pl. 1, figs. 21–24.
Heliodiscus sp. A, Abelmann, 1990, p. 692, pl. 3, fig. 2.
Heliodiscus linckiaformis Clark and Campbell, 1942, p. 49, pl. 3, fig. 13.
Periphaena dupla (?) group, Petrushevskaya, 1975, pl. 1, figs. 14–15.
Heliostylus sp., Hollis, 2002, pl. 1, fig. 15.

具两个髓壳，皮壳边缘有一环状物，放射骨针较短小，数量不等。

该种与 *Periphaena decora* Ehrenberg（1873, p. 246；Sanfilippo and Riedel, 1973, p. 523, pl. 8, figs. 8–10, pl. 27, figs. 2–5；Takemura, 1992, p. 743, pl. 6, fig. 8）的主要区别是后者髓壳简单（仅一个髓壳），边缘有一实心的赤道腰带（薄片状），无放射骨针。

地理分布 南极罗斯海，晚古新世—中渐新世。

果盘虫科 Family Coccodiscidae Haeckel, 1862

壳呈中心双凸的透镜圆盘状，髓壳为简单或双重的同心圆室，其外围的皮壳由许多放射桁连接的一个以上同心赤道环室组成。

圆石虫属 Genus *Lithocyclia* Ehrenberg, 1847

壳盘简单圆形，盘缘无放射骨针或附属物，髓壳简单。

192. 小眼圆石虫 *Lithocyclia ocellus* Ehrenberg

（图版 56，图 1）

Lithocyclia ocellus Ehrenberg, 1854, pl. 36, fig. 30；Ehrenberg, 1873, p. 240；Riedel and Sanfilippo, 1971, p. 1588, pl. 3A, fig. 6；Kamikuri et al., 2013, fig. 6(11).

地理分布 南极罗斯海，出现于早渐新世。

孔盘虫科 Family Porodiscidae Haeckel, 1881

具扁平的盘状壳。中央室简单、球形，围绕着同心的室环。盘两面有筛板。

始盘虫亚科 Subfamily Archidiscida Haeckel, 1862

格孔壳的中央室为简单球形或透镜形，单一同心外室被放射桁分隔成若干小室。

始盘虫属 Genus *Archidiscus* Haeckel, 1887

具一个中央房室和单一的外围环室，环室被放射桁分隔成 2–6 个或更多的分室，壳缘无放射骨针。

193. 十字始盘虫 *Archidiscus stauroniscus* Haeckel

（图版 56，图 2–3）

Archidiscus stauroniscus Haeckel, 1887, p. 487, pl. 48, fig. 9.

个体很小，近四方盘形。

地理分布 南极罗斯海，全新世。

洞盘虫亚科 Subfamily Trematodiscidae Haeckel, 1862

壳盘无放射状附属物（盘缘具心针或室壁或特别的口吻），盘由 2–4 个或较多的同心环组成。

孔盘虫属 Genus *Porodiscus* Haeckel, 1881

具简单的圆盘，盘由数环构成，盘缘无放射状附属物或特别的口吻。

194. 环孔盘虫 *Porodiscus circularis* Clark and Campbell

（图版 56，图 4–5）

Porodiscus circularis Clark and Campbell, 1942, p. 42, pl. 2, figs. 2, 6, 10.

地理分布 南极罗斯海，出现于中渐新世。

195. 小眼孔盘虫 *Porodiscus micromma* (Harting)

（图版 56，图 6–7）

Flustrella micromma Harting, 1863, p. 16, pl. 3, fig. 47.
Porodiscus micromma (Harting), Takahashi and Honjo, 1981, p. 149, pl. 5, figs. 7–8；Takahashi, 1991, p. 82, pl. 20, figs. 13–14.

地理分布 南极罗斯海，全新世。

196. 周旋孔盘虫 *Porodiscus perispira* Haeckel

（图版 56，图 8–9）

Porodiscus perispira Haeckel, 1887, p. 495, pl. 41, fig. 2.

地理分布 南极罗斯海，早-中渐新世。

197. 四孔盘虫 *Porodiscus quadrigatus* Haeckel

（图版 56，图 10–11）

Porodiscus quadrigatus Haeckel, 1887, p. 494, pl. 41, fig. 3.

地理分布　南极罗斯海，中渐新世。

膜包虫属 Genus *Perichlamydium* Ehrenberg, 1847

具一简单的圆盘，无放射针及室壁，盘缘环绕着一个薄的有孔的赤道腰带。

198. 不规则膜包虫（新种）*Perichlamydium irregularmus* sp. nov.

（图版 56，图 12–19）

海绵盘呈圆形、类四方形或不规则形，有一定变化；室环较少，仅 2–4 个，主要集中在中央区，外围区域室环不连续或呈杂乱结构，不太规则；盘的边缘有一条加厚的边带，使边界清晰平滑，边带常不完全连续与封闭，盘体也有缺失或不完整；盘缘一般无放射骨针，仅个别标本在不完整的盘缘处有细小的刺。

标本测量：海绵盘直径 140–300μm，边缘带宽 6–12μm。

模式标本：DSDP274-23-1 2（图版 56，图 14–15），来自南极罗斯海的 DSDP Leg 28 航次 Site 274 孔的 274-23-1 岩心样品中，保存在中国科学院南海海洋研究所沉积标本室。

地理分布　南极罗斯海，早-中渐新世。

该种与 *Perichlamydium limbatum* Ehrenberg 的主要区别在于后者的室环较多，排列较有规律，而且没有盘边缘的加厚带。

199. 具缘膜包虫 *Perichlamydium limbatum* Ehrenberg

（图版 57，图 1–16）

Perichlamydium limbatum Ehrenberg, Ehrenberg, 1854, pl. 22, fig. 20；Haeckel, 1887, p. 514；Petrushevskaya, 1975, p. 575, pl. 6, fig. 11, pl. 39, figs. 1–4.

Perichlamydium sp. aff. *P. limbatum* Ehrenberg, Petrushevskaya, 1975, pl. 6, fig. 12.

Ommatodiscus sp., Benson, 1966, p. 210, pl. 10, figs. 2, 7；Molina-Cruz, 1982, pl. 2, fig. 1.

地理分布　南极罗斯海，中始新世—中渐新世。

200. 编膜包虫 *Perichlamydium praetextum* (Ehrenberg) group

（图版 58，图 1–11）

Flustrella praetextum Ehrenberg, 1844, p. 81.

Perichlamydium praetextum (Ehrenberg), Ehrenberg, 1854, pl. 22, fig. 21；Haeckel, 1887, p. 499；陈木宏等，2017，140 页，图版 37，图 12–14，图版 85，图 14–15.

Perichlamydium praetextum (Ehrenberg) group, Petrushevskaya, 1975, p. 575, pl. 6, fig. 10.

该种与 *Perichlamydium limbatum* Ehrenberg 较为相似，Petrushevskaya（1975）认为两者的主要区别是后者室环较少、较宽而个体较大，实际上许多标本中还是难以区别。该种与 *Perichlamydium saturnus* Haeckel 的区别也不太清晰，根据 Haeckel（1887）对这两个种的描述及图像，它们之间的区别应该是将室环不连续或被分隔的类型定为 *P. saturnus* 种，其他的均归为 *Perichlamydium praetextum* (Ehrenberg) group 较合适，此种群的室环特征变化较大。

地理分布　南极罗斯海，早-中渐新世。

201. 农神膜包虫 *Perichlamydium saturnus* Haeckel

（图版 58，图 12–16）

Perichlamydium saturnus Haeckel, 1887, p. 499, pl. 41, fig. 5.
Stylochlamydium venustum (Bailey), Takahashi, 1991, p. 82, pl. 20, fig. 11.

地理分布 南极罗斯海，早-中渐新世。

针网虫亚科 Subfamily Stylodictyinae Haeckel, 1881

盘缘具实心放射针，无室臂或边缘开孔。

针网虫属 Genus *Stylodictya* Ehrenberg, 1847

盘具实心放射针（5 根或更多，常为 8–12 根），规则或不规则地排列在圆形或多角形盘缘上。盘缘简单，无赤道腰带。

202. 皮刺针网虫 *Stylodictya aculeata* Jørgensen

（图版 58，图 17–23）

Stylodictya aculeata Jørgensen, 1905, S. 119, Taf. 10, Fig. 41；Petrushevskaya, 1967, p. 35, pl. 17, figs. 1–3；陈木宏、谭智源，1996，183 页，图版 14，图 3。
Stylodictya gracilis Ehrenberg, Petrushevskaya, 1975, p. 576, pl. 7, fig. 12.

地理分布 南极罗斯海，中渐新世—全新世。

203. 转圈针网虫 *Stylodictya circularis* (Clark and Campbell)

（图版 59，图 1–5）

Porodiscus circularis Clark and Campbell, 1942, p. 42, pl. 2, figs. 2, 6, 10.
Xiphospira sp. cf. *X. circularis* (Clark and Campbell), Kling, 1973, pl. 2, figs. 1–3, pl. 7, figs. 11–17.

地理分布 南极罗斯海，早古新世—晚始新世。

204. 日旋针网虫 *Stylodictya heliospira* Haeckel

（图版 59，图 6–13）

Stylodictya heliospira Haeckel, 1887, p. 512, pl. 41, fig. 8.
Spongodiscus sp., Lazarus and Pallant, 1989, p. 368, pl. 8, fig. 25.
Plectodiscus circularis (Clark and Campbell), Hollis, 2002, p. 288, pl. 2, figs. 21–22.

室环围绕中心向外螺旋状发育，盘缘有小的放射骨针。

地理分布 南极罗斯海，早始新世—早渐新世。

205. 小针网虫 *Stylodictya minima* Clark and Campbell

（图版 59，图 14–18）

Stylodictya minima Clark and Campbell, 1942, pp. 45–46, pl. 2, fig. 9.

地理分布 南极罗斯海，晚始新世—早渐新世。

206. 强刺针网虫 *Stylodictya validispina* Jørgensen

（图版 59，图 19–26；图版 60，图 1–8）

Stylodictya validispina Jørgensen, 1905, p. 119, pl. 10, figs. 40a–c；Petrushevskaya, 1967, p. 33, fig. 17IVV；Nigrini and Moore, 1979, p. S103, pl. 13, figs. 5a–b；Weaver, 1983, p. 678, pl. 1, fig. 9；Abelmann, 1990, p. 693, pl. 3, fig. 10；陈木宏、谭智源，1996，183 页，图版 14，图 1–2；陈木宏等，2017，142 页，图版 38，图 8–13。
Stylodictya tenuispina Jørgensen, 1905, pp. 118–119, pl. 10, figs. 39a–c.
Xiphospira sp. cf. *X. circularis* (Clark and Campbell), Kling, 1973, p. 635, pl. 2, figs. 1–3, pl. 7, figs. 11–14 (only).
Stylodictya stellata Bailey group, Petrushevskaya, 1975, p. 576, pl. 6, fig. 9.
Stylospira dujardini (Haeckel), Dzinoridze et al., 1978, pl. 36, figs. 15–16.
Stylodictya sp. cf. *validispina* Jørgensen，谭智源，1993，212 页，图版 6，图 4–6。

地理分布　南极罗斯海，晚渐新世—早上新世。

207. 变形针网虫（新种）*Stylodictya variata* sp. nov.

（图版 60，图 9–12）

个体较大，圆盘形，海绵组织较疏松，室环间距稍大，有 4–5 个室环；中心室环圆形，往外室环间距逐渐增大，第 1–2 室环呈椭圆形，第 3–4 室环近圆形或不规则形，外围室环常不完整或不连续，壁孔大小参差不齐，海绵组织分布不均匀；海绵盘上的放射桁发育不明显；盘缘不整齐，凹凸不平，有一些小棘刺，或盘缘破损状。

标本测量：海绵盘直径 150–240μm。

模式标本：DSDP274-25-5 1（图版 60，图 11–12），来自南极罗斯海的 DSDP Leg 28 航次 Site 274 孔的 274-25-5 岩心样品中，保存在中国科学院南海海洋研究所沉积标本室。

地理分布　南极罗斯海，早-中渐新世。

该种海绵盘特征与 *Stylodictya heliospira* Haeckel 有些相似，主要区别是后者室环较规则，呈螺旋状排列，间距较小，有放射桁，海绵组织较为均匀致密，边缘无棘刺。

208. 针网虫（未定种 1）*Stylodictya* sp. 1

（图版 60，图 13–14）

两个标本的海绵盘均为圆盘形；一个标本的室环螺旋形，海绵组织疏松；另一个标本的室环椭圆形，间距较小，海绵组织致密。

地理分布　南极罗斯海，早-中渐新世。

209. 针网虫（未定种 2）*Stylodictya* sp. 2

（图版 60，图 15）

海绵盘呈四方形，海绵组织较为疏松，室环螺旋状，往外渐变宽。

地理分布　南极罗斯海，出现于晚渐新世。

十字网虫属 Genus *Staurodictya* Haeckel

海绵盘圆形或方形，有四根放射桁与骨针相互垂直交叉。

210. 神女十字网虫 *Staurodictya medusa* Haeckel

（图版 60，图 16；图版 61，图 1–14）

Staurodictya medusa Haeckel, 1887, p. 506, pl. 42, fig. 3.
Stylodictya tenuispina Jørgensen, 1905, pp. 118–119, pl. 10, fig. 39.
Staurodictya targaeformis Clark and Campbell, 1942, p. 43, pl. 3, fig. 6.
Stylodictya targaeformis (Clark and Campbell), Petrushevskaya, 1975, pl. 6, figs. 7–8.

壳盘方形，同心环基本连续状，四根放射桁及骨针较为细小。

地理分布 南极罗斯海，早始新世—中渐新世。

211. 眼状十字网虫 *Staurodictya ocellata* (Ehrenberg)

（图版 61，图 15–21）

Stylodictya ocellata Ehrenberg, 1876, p. 84, pl. 23, fig. 7.
Staurodictya ocellata (Ehrenberg) Haeckel, 1887, p. 508.
Tholodiscus ocellatus (Ehrenberg) Kozlova, Petrushevskaya and Kozlova, 1972, p. 525, pl. 18, figs. 1–2.
Xiphospira ocellata (Ehrenberg), Petrushevskaya, 1975, p. 576, pl. 7, fig. 11.

壳盘方形或不规则形，同心环方形或椭圆形，常被断开而不连续，放射骨针较为粗壮。

地理分布 南极罗斯海，晚始新世—晚渐新世。

双腕虫属 Genus *Amphibrachium* Haeckel, 1881

具两个简单不分叉的室臂，相对位于一轴上，无翼膜。

212. 双腕虫（未定种 1）*Amphibrachium* sp. 1

（图版 61，图 22）

Amphibrachium sp.，陈木宏等，2017，143 页，图版 39，图 10–13，图版 85，图 16。

罗斯海标本个体较小，骨针不发育。

地理分布 南极罗斯海，出现于早上新世。

门盘虫科 Family Pylodiscidae Haeckel, 1887

壳呈扁盘形，其简单球形中央室由 1–2 个同心三射型腰带所环绕，每个腰带具三门孔，有三个简单的臂室使之分开。盘面具三孔或格状门孔。

六洞虫属 Genus *Hexapyle* Haeckel, 1881

三个臂室包围着一个三洞型髓壳，臂间凹槽形成门洞，由一赤道腰带相连。

213. 小刺六洞虫 *Hexapyle spinulosa* Chen and Tan

（图版 61，图 23–28）

Hexapyle spinulosa Chen and Tan，陈木宏、谭智源，1989，3–4 页，图版 1，图 8；陈木宏、谭智源，1996，188 页，图版 17，图 6–7；陈木宏等，2017，143 页，图版 40，图 1–2。

Hexapyle sp.，谭智源、宿星慧，1982，156 页，图版 X，图 1。

地理分布　南极罗斯海，早上新世—全新世。

盘孔虫属 Genus *Discopyle* Haeckel, 1887

具三门形髓壳和门盘形皮壳，被一赤道室环所包围。在壳盘的边缘有一孔口，被一刺冠环绕。

214. 盘孔虫（未定种 1）*Discopyle* sp. 1

（图版 61，图 29–30）

Discopyle sp.，陈木宏等，2017，144–145 页，图版 40，图 6–7。

地理分布　南极罗斯海，全新世。

海绵盘虫科 Family Spongodiscidae Haeckel, 1881

中央室简单，海绵状网架覆盖其上。无筛板。

海绵盘虫属 Genus *Spongodiscus* Ehrenberg, 1854

海绵盘呈圆盘形或圆饼形，简单，无赤道腰带或室臂，一般表面无刺。

215. 双凹海绵盘虫 *Spongodiscus biconcavus* Haeckel, emend. Chen et al.

（图版 62，图 1–13）

Spongodiscus biconcavus Haeckel, 1887, p. 577；Popofsky, 1912, p. 143, pl. 6, fig. 2；Benson, 1966, pp. 214–215, pl. 11, fig. 1, text-fig. 14；Benson, 1983, p. 508；谭智源、张作人, 1976, pp. 255–256, text-fig. 25；Takahashi, 1991, p. 84, pl. 19, figs. 4–6；van de Paverd, 1995, pp. 155–156, pl. 41, figs. 1, 5, 7–8；Takahashi et al., 2003, p. 189；Kunitomo et al., 2006, p. 145, figs. 1c, d；Okazaki et al., 2008, p. 81；Odette et al., 2008, p. 86, pl. 3, fig. 1；Itaki et al., 2009, p. 46, pl. 8, figs. 11–12；Chen et al., 2014, p. 102, figs. 4a–k；陈木宏等，2017，145–146 页，图版 40，图 8–11。

Spongodiscus sp., Ling, 1973, p. 778, pl. 1, figs. 9–10；Ling, 1975, p. 725, pl. 4, fig. 5；Ling, 1980, p. 368, pl. 1, fig. 7；Sakai, 1980, p. 709, pl. 6, fig. 5；Matul et al., 2002, p. 30, figs. 4: 3–4；Ikenoue et al., 2016, p. 40, pl. 2, figs. 1–20.

Spongodiscus sp. 3, Renz, 1974, p. 796, pl. 15, fig. 11.

Spongodiscus americanus Kuzlova，陈木宏、谭智源，1996，190 页，图版 17，图 15–16，图版 44，图 3。

Schizodiscus japonicus Matsuzaki and Suzuki, Matsuzaki et al., 2014, pp. 209, 211, pl. 2, figs. 27–30；Matsuzaki et al., 2015, p. 27, figs. 4: 5–6.

地理分布　南极罗斯海，中渐新世—早上新世。

216. 共同海绵盘虫 *Spongodiscus communis* Clark and Campbell group

（图版 63，图 1–13）

Spongodiscus communis Clark and Campbell, 1942, p. 47, pl. 2, figs. 1, 11, 13, 14, pl. 3, figs. 1, 4.

地理分布　南极罗斯海，晚始新世—中中新世。

217. 内凹海绵盘虫（新种）*Spongodiscus inconcavus* sp. nov.

（图版 64，图 1–6）

圆形海绵盘，较大，中心区很薄，易碎，常破损形成一个破洞；自中央区往外海绵盘渐增厚，盘的双面在中央区呈现凹陷形，盘缘处略减薄；盘体形态与 *Spongodiscus biconcavus* Haeckel 的中央区增厚正好相反；盘的边缘有或无一个 V 形缺口，无放射骨针和放射桁。

标本测量：海绵盘直径 205–340μm。

模式标本：DSDP274-10-3 1（图版 64，图 2），来自南极罗斯海的 DSDP Leg 28 航次 Site 274 孔的 274-10-3 岩心样品中，保存在中国科学院南海海洋研究所沉积标本室。

地理分布　南极罗斯海，中渐新世—早上新世。

该种与 *Spongodiscus biconcavus* Haeckel 的主要区别在于后者的海绵盘在中央区增厚，两者的壳体结构正好相反。

218. 边带海绵盘虫（新种）*Spongodiscus perizonatus* sp. nov.

（图版 64，图 7–12）

海绵盘较大，呈不规则形或类圆形，边界线弯曲状，盘体边缘被一增厚或加粗的围壁所封闭，形成一条明显的条带，轮廓清晰；海绵组织在中部较厚和密集，往外稍变薄和稀疏；无放射骨针和放射桁，表面光滑。

标本测量：海绵盘直径 220–275μm。

模式标本：DSDP274-23-1 1（图版 64，图 9–10），来自南极罗斯海的 DSDP Leg 28 航次 Site 274 孔的 274-23-1 岩心样品中，保存在中国科学院南海海洋研究所沉积标本室。

地理分布　南极罗斯海，晚始新世—中渐新世。

该种区别于此类其他各种的主要特征在于海绵盘的边缘具有一条加粗或增厚的围带，边界清晰，盘体常呈不规则形。

219. 平坦海绵盘虫（新种）*Spongodiscus planarius* sp. nov.

（图版 65，图 1–2）

圆形海绵盘，较大，稍薄，盘面平坦，无增厚区，整个壳体的海绵组织基本相同，较为均一；盘孔清晰，近圆形，大小相似，均匀分布，孔径是孔间桁宽的 1–2 倍，均较细；无环室结构，无放射桁，盘面无凹凸特征；盘缘有个别短而细小的放射骨针。

标本测量：海绵盘直径 290–310μm，骨针长 9–13μm。

模式标本：DSDP274-23-1 1（图版 65，图 1–2），来自南极罗斯海的 DSDP Leg 28 航次 Site 274 孔的 274-23-1 岩心样品中，保存在中国科学院南海海洋研究所沉积标本室。

地理分布　南极罗斯海，出现于中渐新世。

该种与 *Spongodiscus biconcavus* Haeckel 的主要区别是后者的海绵盘在中央区有增厚，海绵组织有变化而不是均一的。

220. 放射海绵盘虫（新种）*Spongodiscus radialinus* sp. nov.

（图版 65，图 3–6）

壳体圆盘形，双面平坦无凹凸，海绵组织在中央较致密，往边缘渐疏松，无中心环室结构；从中心向外发育有 35–40 条放射桁，近等间距放射状分布，各桁之间仅有一排或叠加的网孔，孔径随桁距增加而变大；海绵盘边缘无附属物包围，边界轮廓参差不齐；无放射骨针或棘刺。

标本测量：海绵盘直径 165–215μm。

模式标本：DSDP274-1-5 1（图版 65，图 5），来自南极罗斯海的 DSDP Leg 28 航次 Site 274 孔的 274-1-5 岩心样品中，保存在中国科学院南海海洋研究所沉积标本室。

地理分布 南极罗斯海，全新世。

该种与 *Spongodiscus trachodes* (Renz)的主要区别是后者的海绵盘具中心环室，无放射桁，整个壳体的海绵组织较为致密、细小，均匀分布。

221. 海绵盘虫 *Spongodiscus resurgens* Ehrenberg group

（图版 65，图 7–23）

Spongodiscus resurgens Ehrenberg, 1854, S. 54; Stöhr, 1880, S. 117, Taf. 6, Fig. 11; Haeckel, 1887, p. 577; Boltovskoy and Riedel, 1987, p. 98, pl. 2, fig. 24；陈木宏、谭智源，1996，189 页，图版 17，图 14；Kamikuri et al., 2008, p. 170, pl. 1, fig. 15.

Schizodiscus favus (Ehrenberg) *maxima* (Popofsky), Petrushevskaya, 1975, p. 574, pl. 5, figs. 6–7, pl. 34, figs. 1–2.

地理分布 南极罗斯海，早古新世—全新世。

222. 纹管海绵盘虫 *Spongodiscus trachodes* (Renz)

（图版 66，图 1–10）

Spongocyclia trachodes Renz, 1974, pl. 4, figs. 1–4, pl. 10, fig. 13；Ling and Lazarus, 1990, p. 356, pl. 1, fig. 7.

Stylospongia elliptica (Carnevale) subsp. *spiralis* Bjørklund and Kellogg, Dzhinoridze et al., 1978, pl. 25, figs. 7, 10.

海绵盘的中央区有 3–4 个发育不很完整的同心螺旋环，外围环纹一般较凌乱，壳盘平整，海绵组织均匀，盘缘清晰，无放射骨针。

地理分布 南极罗斯海，早始新世—晚渐新世。

223. 海绵盘虫（未定种 1）*Spongodiscus* sp. 1

（图版 66，图 11–13）

Spongodiscus anomalum (Popofsky), van de Paverd 1995, p. 156, pl. 41, figs. 6, 9–10.

Spongodiscus sp., Bjørklund, 1976, p. 1124, pl. 20, fig. 1.

我们的标本特征与 van de Paverd（1995, p. 156, pl. 41, figs. 6, 9–10）相似，均是海绵盘中央有一个较小的增厚区，约占整个海绵盘的 1/4，其外围为疏松的海绵组织，而且向边缘空隙渐变大。Popofsky（1912, p. 136, pl. 3, fig. 4）建立了新种 *Amphieraspedum anomalum* Popofsky，该种的海绵盘中有一增厚的长条状室臂，中心有 3–4 个同心环室，室臂的另一端也发育有约 3 个同心环，在该室臂的外围海绵盘较薄，同样具有条带状的围室。因此，van de Paverd（1995）鉴定的 *Spongodiscus anomalum* (Popofsky)完全不具备该种的基本特征，而是一个未定种 *Spongodiscus* sp.。

地理分布 南极罗斯海，晚始新世—全新世。

224. 海绵盘虫（未定种 2）*Spongodiscus* sp. 2

（图版 66，图 14）

海绵盘近椭圆形，壳体海绵组织一般较均匀，在长轴一端有两根粗短的三棱柱状骨针，相隔独生，盘缘上无其他放射骨针或辅针。

地理分布 南极罗斯海，出现于中渐新世。

海绵轮虫属 Genus *Spongotrochus* Haeckel, 1860

海绵盘圆形，具很多实心放射骨针（5–10 根或更多），散布在整个盘面和盘缘，或规则地分布在盘两边。

225. 冰海绵轮虫 *Spongotrochus glacialis* Popofsky group

（图版 67，图 1–8；图版 68，图 1–11）

Spongotrochus glacialis Popofsky, 1908, S. 228, Taf. 26, Fig. 8, Taf. 27, Fig. 1, Taf. 28, Fig. 2；Riedel, 1958, p. 227, pl. 2, figs. 1–2, text-fig. 1；Casey, 1971a, p. 331, pl. 23, figs. 4–5；Chen, 1975, p. 455, pl. 24, figs. 5–6；谭智源、张作人，1976，256 页，图版 I，图 3–4；陈木宏、谭智源，1996，191 页，图版 18，图 4–5，图版 44，图 5–6；Motoyama and Nishimura, 2005, p. 111, figs. 9.8–9.9；Kamikuri et al., 2008, p. 170, pl. 1, fig. 14；陈木宏等，2017，147 页，图版 41，图 6，图版 42，图 1–2。

Spongotrochus cf. *glacialis* Popofsky, Benson, 1966, p. 218, pl. 11, fig. 4.

Spongotrochus (?) *glacialis* Popofsky, Kling, 1973, pl. 2, figs. 4–6.

Spongotrochus glacilialis Popofsky group, Petrushevskaya, 1975, p. 575, pl. 5, fig. 8, pl. 35, figs. 1–6.

地理分布 南极罗斯海，中渐新世—晚上新世。

226. ?暗针海绵轮虫 ?*Spongotrochus rhabdostyla* Bütschli

（图版 69，图 1–8）

?*Spongotrochus rhabdostyla* Bütschli, 1882, pl. 26, fig. 2.

Larcopyle pylomaticus (Riedel), Kamikuri et al., 2008, pl. 1, fig. 27.

海绵盘为长方形，有许多细的同心环。未能找到?*Spongotrochus rhabdostyla* Bütschli 的文字描述，罗斯海标本的海绵盘形状和结构与 Bütschli（1882）和 Kamikuri 等（2008）（作者鉴定明显有误）的种类图版很相似，但 Bütschli（1882）的图示中有一较粗的不规则形骨针，这一特征在罗斯海标本中一般不见，或为 Bütschli（1882）的图示有误。

地理分布 南极罗斯海，全新世。

227. 异形海绵轮虫 *Spongotrochus vitabilis* Goll and Bjørklund

（图版 69，图 9–11）

Spongotrochus vitabilis Goll and Bjørklund, 1989, p. 730, pl. 3, figs. 1–3；陈木宏等，2017，147 页，图版 42，图 3。

Spongotrochus sp.，谭智源、张作人，1976，257 页，图 26；谭智源、陈木宏，1999，234–235 页，图 5-145。

海绵结构在中央区较致密，向盘缘逐渐变疏松，边缘有一些细小的放射骨针。

地理分布　南极罗斯海，晚始新世—中渐新世。

228. 海绵轮虫（未定种 1）*Spongotrochus* sp. 1

（图版 69，图 12–13）

海绵盘近蝴蝶形，中部收窄，两侧对称，类圆形，中部海绵组织略致密，向两外侧渐疏松，放射骨针自近中部生长，并延伸至海绵盘两侧末端之外。

地理分布　南极罗斯海，全新世

对臂虫属 Genus *Spongobrachium* Haeckel, 1881

壳体有一轴拉长，海绵盘中有两个相对的海绵臂，臂间或边缘有翼膜。

229. 反对臂虫 *Spongobrachium antoniae* (O'Connor)

（图版 69，图 14–16）

Spongotrochus antoniae O'Connor, 1997b, pp. 103–104, pl. 1, figs. 1–4, pl. 4, figs. 1–6.

海绵盘近似长椭圆形，长轴上有一加厚的对称臂，边缘基本被翼膜包覆，侧缘略加宽。

地理分布　南极罗斯海，早始新世。

棒网虫属 Genus *Dictyocoryne* Ehrenberg, 1860

盘为圆形或三角形，边缘具三条海绵状臂，臂间有一翼膜。

230. 截棒网虫 *Dictyocoryne truncatum* (Ehrenberg)

（图版 69，图 17–18）

Rhopalodicytum truncatum Ehrenberg, 1862, p. 240.

Dictyocoryne truncatum (Ehrenberg), van de Paverd, 1995, p. 160, pl. 3, figs. 2–5, 7, pl. 44, figs. 6, 7, 9, 10–13；陈木宏、谭智源，1996，191 页，图版 19，图 2–4，图版 44，图 9。

该种南极的标本与其他海区的标本有一定的差异，主要区别是南极标本的三个海绵臂较薄，除了中央区凸起之外，臂的表面较平坦，内部似有同心环，臂端边缘清晰。

地理分布　南极罗斯海，出现于晚始新世。

231. 棒网虫（未定种 1）*Dictyocoryne* sp. 1

（图版 69，图 19）

海绵盘近灯泡形，两端大小差异较大，两侧对称，盘内有三叉状的海绵棒，伸向小端的海绵棒较粗长，另两个棒相对较细小，其相互之间夹角也较小，为与长棒之间分别夹角的一半，三棒之间发育有较完整的翼膜。

地理分布　南极罗斯海，出现于晚始新世。

232. 棒网虫（未定种 2）*Dictyocoryne* sp. 2

（图版 69，图 20）

Hymeniastrum sp., O'Connor, 1997b, p. 112, pl. 3, fig. 2.

海绵盘三角形，三臂较短，中央盘相对较大，翼膜发育不完整。

地理分布 南极罗斯海，出现于晚始新世。

海绵星虫属 Genus *Spongaster* Ehrenberg, 1860

海绵盘呈四方形或圆形，有四个海绵臂，臂间由不同结构的海绵翼膜相连。

233. 海绵星虫 *Spongaster tetras* Ehrenberg

（图版 69，图 21–22）

Spongaster tetras Ehrenberg, 1872, S. 299, Taf. 6(3), Fig. 8；Haeckel, 1887, p. 597；Casey, 1971b, p. 331, pl. 23, figs. 18–19；谭智源、张作人，1976，258 页，图版 2，图 5–6；陈木宏、谭智源，1996，192 页，图版 20，图 1–3，图版 44，图 11–12。

Histiastrum quadratum Popofsky, 1912, S. 142, Taf. 6, Fig. 4, Abb. 59.

地理分布 南极罗斯海，晚始新世—中渐新世。

234. 海绵星虫（未定种 1）*Spongaster* sp. 1

（图版 69，图 23–24）

海绵状壳盘呈四方形，在中央有一明显的凸起增厚区，大小约占壳体的 1/3，该中央增厚区的海绵组织较为致密，其外围的海绵组织相对疏松，孔隙较大，孔径约为孔间桁的 2–3 倍。该未定种的壳体形状与 *Spongaster tetras* Ehrenberg 较相似，主要区别是后者的壳盘海绵组织较为致密均匀，孔隙较小，中央增厚区不太明显。

地理分布 南极罗斯海，晚始新世。

235. 海绵星虫（未定种 2）*Spongaster* sp. 2

（图版 69，图 25）

海绵盘近四方形，但四个海绵臂的发育不明显，盘内似有环形结构，在四方形的一边中部边缘上有一个三角形缺口，缺口深度约为边长的 1/3。盘缘无放射骨针。

地理分布 南极罗斯海，晚始新世。

海绵门孔虫属 Genus *Spongopyle* Dreyer, 1889

壳海绵状，常呈圆形，边缘上有一个或更多门孔。

236. 吻海绵门孔虫 *Spongopyle osculosa* Dreyer

（图版 70，图 1–2）

Spongopyle osculosa Dreyer, 1889, S. 42, Taf. 11, Figs. 99, 100；Riedel, 1958, p. 226, pl. 1, fig. 12；Abelmann, 1990, p. 693, pl. 3, fig. 11；陈木宏、谭智源，1996，192 页，图版 20，图 5；陈木宏等，2017，148 页，图版 43，图 2–8。

Spongopyle resurgens osculosa (Dreyer), Petrushevskaya and Kozlova, 1972, p. 528, pl. 21, fig. 4；van de Paverd, 1995, p. 154, pl. 41, fig. 3.

地理分布　南极罗斯海，中渐新世。

炭篮虫亚目 Suborder LARCOIDEA Haeckel, 1887

中央囊和外壳均为扁豆形，具硅质的扁豆状壳体，从三个相互垂直的轴向上不均等地生长。

炭篮虫科 Family Larcopylidae Dreyer, 1889

壳主轴极上有口。

炭篮虫属 Genus *Larcopyle* Dreyer, 1889

壳内一般呈旋转或包覆结构，外壁具一极口。

237. 隐口炭篮虫 *Larcopyle adelstoma* (Kozlova and Gobovets)

（图版 70，图 3–23）

Prunopyle adelstoma Kozlova and Gorbovets, 1966, p. 67, pl. 10, figs. 3–4；Hollis, 2002, p. 289, pl. 2, figs. 5–8.
Prunopyle ovata Kozlova and Gorbovetz, 1966, p. 67, pl. 10, figs. 5–8；Hollis et al., 1997, p. 48, pl. 2, figs. 5–8.
Prunopyle spp., Hollis, 1997, p. 45, pl. 7, figs. 12–13.

地理分布　南极罗斯海，早古新世—早上新世。

238. 炭篮虫 *Larcopyle butschlii* Dreyer

（图版 71，图 1–36）

Larcopyle butschlii Dreyer, 1889, S. 1124, Taf. 10, Fig. 70；Benson, 1966, p. 280, pl. 19, figs. 3–5；Nigrini and Moore, 1979, p. S131, pl. 17, figs. 1a–b；谭智源、宿星慧，1982，159 页，图版 XII，图 1–4；Abelmann, 1990, p. 24, pl. 4, fig. 4；陈木宏、谭智源，1996，193 页，图版 20，图 6–10，图版 45，图 1–3；谭智源、陈木宏，1999，239 页，图 5-148；Lazarus et al., 2005, pp. 102, 106–107, pl. 1, figs. 10–14；Kamikuri et al., 2008, p. 169, pl. 1, fig. 32；陈木宏等，2017，149 页，图版 44，图 1–5。
Larcopyle group Mullineaux and Smith, 1986, p. 66, pl. 2, fig. 3.
Tholospironium cervicorne Haeckel, van de Paverd, 1995, pp. 185–186, pl. 54, figs. 1–4, 6–12, pl. 58, fig. 7, pl. 59, fig. 6.

地理分布　南极罗斯海，早古新世—全新世。

239. 弗拉克炭篮虫 *Larcopyle frakesi* (Chen) group

（图版 72，图 1–19；图版 73，图 1–21）

Lithocarpium monikae Petrushevskaya, 1975, p. 572, pl. 4, figs. 6–10.
Prunopyle sp. D, Abelmann, 1990, p. 693, pl. 4, figs. 1A–B.

地理分布　南极罗斯海，早始新世—全新世。

240. 哈史炭篮虫 *Larcopyle hayesi* (Chen) group

（图版 74，图 1–9；图版 75，图 1–21）

Prunopyle hayesi Chen, 1975, p. 454, pl. 9, figs. 3–5；Lazarus, 1990, p. 717, pl. 5, figs. 5–8；Takemura, 1992, p. 742, pl. 1, figs. 13–14；Hollis, 1997, p. 48, pl. 2, figs. 12–14.

Prunopyle hayesi Chen group, Abelmann, 1990, p. 693, pl. 3, fig. 14.
Larcopyle hayesi (Chen), Lazarus et al., 2005, p. 119, pl. 11, figs. 1–21.
Ommatodiscus haeckeli Stöhr, Petrushevskaya, 1975, p. 572, pl. 3, figs. 12–16, pl. 32. figs. 1–7.

地理分布 南极罗斯海，早古新世—中中新世。

241. 卵炭篮虫 *Larcopyle ovata* (Kozlova and Gorbovetz)

（图版 76，图 1–27）

Prunopyle ovata Kozlova and Gorbovetz, 1966, p. 67, pl. 10, figs. 5–8；Hollis, 1997, p. 48, pl. 2, figs. 6–10.

地理分布 南极罗斯海，晚古新世—晚渐新世。

242. 多棘炭篮虫 *Larcopyle polyacantha* (Campbell and Clark)

（图版 77，图 1–10）

Larnacanthap polyacantha Campbell and Clark, 1944a, p. 30, pl. 5, figs. 4–7；Petrushevskaya, 1975, p. 572, pl. 29, fig. 6 (only).
Prunopyle titan Campbell and Clark, Hays, 1965, p. 173, pl. 2, fig. 4.

地理分布 南极罗斯海，早始新世—晚中新世。

243. 多棘炭篮虫巨人亚种 *Larcopyle polyacantha titan* Lazarus et al.

（图版 77，图 11–32）

Larcopyle polyacantha titan Lazarus et al., 2005, p. 108, pl. 3, figs. 1–12.
Lithocarpium polyacantha (Campbell and Clark) group, Petrushevskaya, 1975, p. 572, pl. 3, figs. 6–8；Abelmann, 1990, p. 24, pl. 4, fig. 2.
Lithocarpium ? sp., Petrushevskaya, 1975, pl. 3, fig. 9.
Prunopyle titan Chen, 1975, p. 454, pl. 23, figs. 1–2.
Prunopyle titan Chen, Lazarus, 1990, p. 717, pl. 5, figs. 1, 3, 4.
Prunopyle polycantha (Campbell and Clark), Hollis, 1997, p. 48, pl. 2, figs. 25–27.

Lazarus 等（2005）集合 *Larnacanthap polyacantha* Campbell and Clark 和 *Prunopyle titan* Chen 两个种及其过渡类型的特征建立了此亚种，该亚种与后两者的主要区别是其外壳一般具有极盖或多层结构，外表不完全平整。

地理分布 南极罗斯海，中始新世—中渐新世。

244. 巨人炭篮虫 *Larcopyle titan* (Campbell and Clark) group

（图版 78，图 1–30）

Prunopyle titan Campbell and Clark, 1944a, p. 20, pl. 3, figs. 1–3；Abelmann, 1990, p. 693, pl. 3, fig. 16.
Larcopyle titan (Campbell and Clark), Lazarus et al., 2005, p. 111, pl. 4, figs. 11–17.
Lithocarpium titan (Campbell and Clark), Petrushevskaya, 1975, p. 572, pl. 4, fig. 5.
Prunopyle sp. B group, Abelmann, 1990, p. 693, pl. 4, figs. 3A–B.
Prunopyle occidentalis Clark and Campbell, Hollis, 2002, p. 289, pl. 2, fig. 9.
Prunopyle ovata Kozlova, Hollis, 1997, p. 48, pl. 2, figs. 6–7 (only).
Middourium regulare (Borissenko), Hollis, 2006, pl. 3, fig. 9.

地理分布 南极罗斯海，早始新世—早上新世。

245. 炭篮虫（未定种 1）*Larcopyle* sp. 1

（图版 78，图 31–32）

Prunopyle sp., O'Connor, 1997b, p. 116, pl. 3, fig. 9.

个体较小，橄榄形，具两个髓壳和一个皮壳，表面略粗糙，有一缩窄的开口。未定种与 *Prunopyle* sp.（O'Connor, 1997b）较相似，但后者的壳体为卵形。

地理分布 南极罗斯海，晚更新世。

箱虫科 Family Larnacidae Haeckel, 1883

具完整格孔的椭球形外壳，无开放的门孔和收缩环，外皮壳与内髓壳均为三带型结构，相互交叉。

壶箱虫属 Genus *Larnacalpis* Haeckel, 1887

皮壳简单，无放射骨针，髓壳双层，箱形。

246. 扁圆壶箱虫 *Larnacalpis lentellipsis* Haeckel

（图版 79，图 1–4）

Larnacalpis lentellipsis Haeckel, 1887, p. 620, pl. 50, fig. 2.

个体较小，皮壳简单，表面光滑无刺，壁孔类圆形或近六角形。

地理分布 南极罗斯海，早中新世—早上新世。

门孔虫科 Family Pyloniidae Haeckel, 1881

外皮壳格孔状，具 2–4 个或更多对称的门孔。同心腰带系统 1–3 个（每个系统具 3 条腰带）。

双腰带虫亚科 Subfamily Diplozonaria Haeckel, 1887, emend. Tan and Chen, 1990

格孔状腰带构成两个同心系，位于两个同心椭圆透镜面，三带型髓壳被皮壳腰带包覆构成三个相似面；背面观髓壳中等大小，其门孔呈柿形，皮壳侧腰带呈两翼形或椭圆形；顶面观髓壳大，具柿形门孔，其两端突出于皮壳侧腰带中央。

四门孔虫属 Genus *Tetrapyle* Müller, 1858

髓壳呈椭圆透镜状，三带型，外绕两个十字交叉的格孔状腰带，赤道腰带（初腰带）较小，侧腰带（次腰带）较大。四门孔简单，位于两腰带之间，无矢隔。

247. 四叶四门孔虫 *Tetrapyle quadriloba* (Ehrenberg)

（图版 79，图 5–27）

Schizomma quadrilobum Ehrenberg, 1860, S. 815；Ehrenberg, 1872, Taf. 10, Figs. 12–14.

Tetrapyle quadriloba (Ehrenberg), Haeckel, 1862, S. 436；Haeckel, 1887, p. 645；Popofsky, 1912, S. 150, Abb. 70–72；谭智源、宿星慧，1982，160 页，图版 XII，图 9；陈木宏、谭智源，1996，193 页，图版 20，图 11，14–15，图版 45，图 4–6。

Tetrapyle octacantha Müller, Benson, 1966, pp. 245–250, pl. 15, figs. 3, 7–8 (only).

需要讨论的是目前有文献将 *Tetrapyle*、*Octopyle* 和 *Pylonium* 等不同属征的种类归为 *Tetrapyle octacantha* Müller group，首先是 Benson（1966）持此观点，其后 Itaki 等（2009）也持相同观点（尽管未在同物异名表列出 Benson 的相关文献），两者的鉴定标准基本一致，这一结果还出现在“Radiolaria.org”网页上的“Species list”中。实际上，这三个属各自有着基本不同的定义特征，其中包含了许多不同的种类，这些种类有着明显不同的特征，它们较为严格地遵循放射虫分类的系统描述与组成，早已被广泛认可与应用，因此将之笼统归为一个种群是缺乏依据的，不符合分类规范，尽管它们的壳体三带型结构在不同视角会造成判断上的困惑，但前人已经有针对性地对各个属与种进行了描述说明和图件展示。建议为了避免分类与认知上的混乱，还是尽量尊重较为严格的属种分类与鉴定原则和遵守生物命名法规。

地理分布 南极罗斯海，早古新世—全新世。

248. 塔四门孔虫 *Tetrapyle turrita* Haeckel

（图版 79，图 28–29）

Tetrapyle turrita Haeckel, 1887, p. 649, pl. 9, fig. 10.

表面光滑，有十根圆锥形骨针，其中两根在壳体主轴两极的顶端上，其他八根对称分布在门孔两侧，壳长是壳宽的两倍。

地理分布 南极罗斯海，全新世。

八门孔虫属 Genus *Octopyle* Haeckel, 1881

壳三带型，椭圆透镜状，两个门孔分别被一矢隔一分为二。

249. 六角八门孔虫 *Octopyle sexangulata* Haeckel

（图版 79，图 30–31）

Octopyle sexangulata Haeckel, 1887, p. 653, pl. 9, fig. 12.

壳体六角形或椭圆形，表面有棘刺，具六根较粗的圆锥形骨针，分别在门孔（两极）顶端的外侧和腰带的两侧。

地理分布 南极罗斯海，出现于中渐新世。

多门虫属 Genus *Pylonium* Haeckel, 1881

具三带型椭圆透镜状髓壳，被三个垂直交叉的格孔腰带状皮壳包覆。较小的皮壳为横腰带（初壳），较大的为侧腰带（第 2 壳）和矢腰带（第 3 壳）。

250. 古多门虫（新种）*Pylonium palaeonatum* sp. nov.

（图版 79，图 32–33）

壳呈椭圆透镜状，壳壁较厚，壁孔小，类圆形，孔间桁稍粗实；三带型的外髓壳较大，呈椭圆箱形，壳长与皮壳横腰带长（皮壳壳宽）近等，两者壳壁相互紧贴，有许多放射桁与内髓壳连接，内髓壳较小，约为外髓壳的1/3；皮壳的侧腰带在两端封闭，形成四个半月形的门孔，门孔上有若干较粗的桁枝连接着外髓壳与皮壳；壳体表面无放射骨针或棘刺，有小锥凸。

标本测量：皮壳长140μm、宽110μm；外髓壳长100μm、宽75μm；内髓壳长45μm、宽30μm。

模式标本：DSDP274-40-1 1（图版79，图32–33），来自南极罗斯海的DSDP Leg 28航次Site 274孔的274-40-1岩心样品中，保存在中国科学院南海海洋研究所沉积标本室。

地理分布　南极罗斯海，晚古新世。

该种特征与*Pylonium quadricorne* Haeckel（1887, p. 655, pl. 9, fig. 14）较相似，但后者的壳表有许多棘刺，壁薄孔大，门孔肾形。显然，前者属较老地层种，后者为近现代种。

地理分布　南极罗斯海，出现于晚古新世。

251. 新拟多门虫（新种）*Pylonium neoparatum* sp. nov.

（图版79，图34–39）

皮壳的侧腰带在两端开放，形成阔开口，口径约为壳宽的3/4，壁孔类圆形，较小，排列规则；三带型髓壳较小，椭圆形或近圆形，外髓壳大小仅为皮壳宽度（横腰带长）的一半，内髓壳大小是外髓壳的1/3；壳表光滑无刺。

标本测量：皮壳长110–145μm、宽95–115μm；外髓壳长55–65μm、宽50–55μm；内髓壳长20–25μm、宽15–20μm。

模式标本：DSDP274-30-4 1（图版79，图34–35），来自南极罗斯海的DSDP Leg 28航次Site 274孔的274-30-4岩心样品中，保存在中国科学院南海海洋研究所沉积标本室。

地理分布　南极罗斯海，晚始新世。

该种与*Pylonium palaeonatum* sp. nov.的主要区别在于前者的髓壳较小，与皮壳之间存在较大的空间，皮壳两端开放，而后者的皮壳为封闭状，部分壳体与髓壳相贴。该种在Stie 274地层中的出现层位在*Pylonium palaeonatum* sp. nov.的层位之上，年代相对较新，可能是由后者演化而来。

门带虫属　Genus *Pylozonium* Haeckel, 1887

三带型髓壳，被双格孔状皮壳所包围；内皮壳和外皮壳均为门孔型，由三个相互完全交叉的腰带组成，它们是横腰带、侧腰带和矢腰带。

252. 石门带虫（新种）*Pylozonium saxitalum* sp. nov.

（图版79，图40–53）

三壳形状相同，为椭圆透镜状，有一些放射桁相连；三壳之比为1∶3∶8，壳壁稍厚；外壳壁孔类圆形，大小相近，排列规则，多个门孔较清晰；内壳很小，轮廓清楚，但结构不详；壳体表面粗糙或有小棘突，无放射骨针。仅见于罗斯海早新生代地层。

标本测量：皮壳长120–140μm、宽105–125μm；外髓壳长50–75μm、宽40–55μm。

模式标本：DSDP274-42-2 1（图版 79，图 44–45），来自南极罗斯海的 DSDP Leg 28 航次 Site 274 孔的 274-42-2 岩心样品中，保存在中国科学院南海海洋研究所沉积标本室。

地理分布 南极罗斯海，早古新世。

该种的壳体结构比 *Pylozonium* sp. 1 更加完善和清晰，应是后者的早期发育类型，与 *Pylozonium novemcinctum* Haeckel（1887, p. 659）的基本特征较接近，但后者个体明显较大且四个门孔发育更加清晰。

253. 门带虫（未定种 1）*Pylozonium* sp. 1

（图版 80，图 1–14）

个体较小，三壳形状相似，近圆形或椭圆形透镜状，三壳之比为 1∶2∶3，外壳表面粗糙，无放射骨针。该未定种可能是 *Pylozonium* 类群的初始类型，仅见于新生代早期的同一层中，在地层底部的标本结构较模糊，似已被充填或改造。

地理分布 南极罗斯海，早-晚古新世。

光眼虫科 Family Actinommidae Haeckel, 1862, sensu Riedel, 1967

泡沫虫中特殊的一类，具球形或椭球形的格孔壳，常无内骨针。

梅孔虫属 Genus *Prunopyle* Dreyer, 1889

由两个以上的椭球形格孔壳组成，在外壳的一端有一个大极孔。Dreyer（1889）将其分为 *Sphaeropyle* 和 *Prunopyle* 两个属，但两者之间的一些种类特征很相近，因此难以区别这两个属的基本特征，相关种类的归属问题有待于进一步分析。

254. 南极梅孔虫 *Prunopyle antarctica* Dreyer

（图版 80，图 15–16）

Prunopyle antarctica Dreyer, 1889, pp. 100–101, pl. 10, fig. 75；Riedel, 1958, p. 225, pl. 1, figs. 7–8；Chen, 1975, p. 454, pl. 23, figs. 5–6；Nakaseko and Nishimura, 1982, p. 102, pl. 58, figs. 1–3, 5；Nishimura, 2003, pp. 197–200, pl. 1, figs. 1–12；陈木宏等，2017，152–153 页，图版 44，图 17–18，图版 45，图 1–5。

Cromyechinus antarctica (Dreyer), Petrushevskaya, 1967, pp. 22–26, figs. 13I–VI, 14I–VII；Itaki et al., 2009, p. 44, pl. 2, figs. 10a–b.

地理分布 南极罗斯海，晚更新世。

255. 原始梅孔虫（新种）*Prunopyle archaeota* sp. nov.

（图版 80，图 17–33）

个体较小，壳体不太规则，近椭球形或圆球形，壳壁与桁枝较为粗实；一个皮壳表面光滑或有一些棘凸，有一开口，口端收缩或开阔；具两个髓壳，圆球形、椭球形，或内部呈杂乱状，结构模糊，髓壳形态不清，似为发育不完善。该种似为 *Prunopyle* 的初始类型，分布于新生代地层的底部。

标本测量：外皮壳长 65–100μm、宽 55–85μm；外髓壳长 30–50μm、宽 30–45μm，外皮壳开口宽 25–65μm。

模式标本：DSDP274-43-3 1（图版 80，图 21–22），来自南极罗斯海的 DSDP Leg 28 航次 Site 274 孔的 274-43-3 岩心样品中，保存在中国科学院南海海洋研究所沉积标本室。

地理分布　南极罗斯海，早古新世。

该种以个体较小，内部结构形态较为模糊不清而区别于该类群的其他各种。由于具有明显的地层限制与意义，应属罗斯海古近系的一个新种。

256. 小梅孔虫（新种）*Prunopyle minuta* sp. nov.

（图版 80，图 34–57）

Prunopyle sp., O'Connor, 1997b, p. 116, pl. 3, fig. 9.

壳体较小，椭球形或卵形，有 3–4 层，1–2 个皮壳和 2 个髓壳；外皮壳的一端有开口，口缘有少量棘刺，壁孔较小，类圆形；内皮壳一般紧贴外皮壳，或基本不分开，呈封闭状；外髓壳一般椭球形或长卵形，个别为不规则形，大小是皮壳的 1/2–3/5；内髓壳常呈圆球形，也有长椭球形，很小；壳体表面光滑无刺。

标本测量：外皮壳长 80–120μm、宽 60–75μm，外髓壳长 60–70μm、宽 35–50μm，内髓壳直径 20–30μm，外皮壳开口宽 15–30μm。

模式标本：DSDP274-32-1 1（图版 80，图 40–41），来自南极罗斯海的 DSDP Leg 28 航次 Site 274 孔的 274-32-1 岩心样品中，保存在中国科学院南海海洋研究所沉积标本室。

地理分布　南极罗斯海，晚古新世—晚上新世。

该种基本特征是个体较小，与 *Prunopyle tetrapila* Hays 的主要区别是后者个体相对较大，仅有一个皮壳，两个髓壳均呈圆球形，位于皮壳中心。

257. 口梅孔虫（新种）*Prunopyle opeocila* sp. nov.

（图版 81，图 1–4）

三层壳；外壳椭球形，或不太规则，壁孔类圆形，有六角形框架，孔径与孔间桁宽近等，表面光滑，无放射骨针，极孔（开口）较大，口径约占壳径的 1/2，口缘上无齿或骨针；两个内壳均为圆球形，略偏离外壳内的中心位置，或接近开口处，结构不太规则；各壳之间的放射桁发育不连续，三壳之比约为 1∶2.5∶4。

标本测量：外壳长 140–200μm、宽 125–180μm，外髓壳径 90–110μm，内髓壳径 40–50μm，开口宽 65–70μm。

模式标本：DSDP274-10-1 2（图版 81，图 3–4），来自南极罗斯海的 DSDP Leg 28 航次 Site 274 孔的 274-10-1 岩心样品中，保存在中国科学院南海海洋研究所沉积标本室。

地理分布　南极罗斯海，早上新世。

该种与 *Prunopyle tetrapila* Hays 的主要区别在于后者的内壳处于外壳内的中心位置，结构规则，极孔较小，一般小于孔径的 1/3，口缘上有齿或骨针。

258. 方形梅孔虫（新种）*Prunopyle quadrata* sp. nov.

（图版 81，图 5–32）

个体较小，壳有 4–5 层；外皮壳一般呈圆筒形或箱形，侧视为长方形，两侧呈直线形，个别标本为椭球形，壳壁较厚，壁孔类圆形，较小，致密，孔间桁粗实，在壳体一端有较宽的开口，口缘上常有一些小齿，无口缘环；内皮壳呈椭球形或箱形，紧贴外皮壳；髓壳有 2–3 个，外髓壳一般椭球形，内髓壳圆球形或椭球形，髓壳与皮壳的大小之比一般为 1∶3∶5∶6，不同标本有一定的变化；壳体表面光滑，无刺，阔开口处常有小齿。

标本测量：外皮壳长 115–150μm、宽 80–105μm，外髓壳长 75–85μm、宽 70–80μm，外皮壳开口宽 40–75μm。

模式标本：DSDP274-24-1 2（图版 81，图 9–10），来自南极罗斯海的 DSDP Leg 28 航次 Site 274 孔的 274-24-1 岩心样品中，保存在中国科学院南海海洋研究所沉积标本室。

地理分布 南极罗斯海，早古新世—中渐新世。

这是发现于南极罗斯海较老地层中的特有种类，其特征与 *Prunopyle minuta* sp. nov. 有些相似，但后者的个体更小，呈椭球形，两端圆弧状，侧视椭圆形，壳层略少；而此种的壳体一般两端几乎截平，与侧面直角交会，侧面直线形，整个壳体侧视呈长方形。

259. 四毛梅孔虫 *Prunopyle tetrapila* Hays group

（图版 82，图 1–19）

Prunopyle tetrapila Hays, 1965, p. 172, pl. 2, fig. 5; Chen, 1975, p. 454, pl. 23, figs. 3–4; Keany, 1979, pl. 5, fig. 3; Takemura, 1992, p. 742, pl. 2, figs. 1–2.

Prunopyle tetrapila Hays group, Abelmann, 1990, p. 693, pl. 3, fig. 13.

Prunopyle trypopyren Caulet, 1991, p. 533, pl. 1, figs. 6–7 (not fig. 5).

Sphaeropyle robusta Kling, 1973, p. 634, pl. 1, figs. 11–12, pl. 6, figs. 9–13, pl. 13, figs. 1–5.

Sphaeropyle langii Dreyer, Kling, 1973, pl. 1, figs. 5–10, pl. 13, figs. 6–8.

地理分布 南极罗斯海，早渐新世—早更新世。

圆顶虫科 Family Tholonidae Haeckel, 1887

具壳孔规则排列的完整皮壳，皮壳由 2–6 个或更多的半球形或帽状的拱顶构成。这些拱顶两两相对地位于壳的三个轴极上。拱顶之间有缢勒，中央室简单或箱形。

方顶虫属 Genus *Cubotholus* Haeckel, 1887

皮壳简单，六个半球形的拱顶成对位于相互垂直的三个轴极上，并包覆着一个立方箱形中央室。

260. 规则方顶虫 *Cubotholus regularis* Haeckel

（图版 83，图 1–2）

Cubotholus regularis Haeckel, 1887, p. 680, pl. 10, fig. 14；Renz, 1976, p. 113, fig. 18；谭智源，1993，214 页，图版Ⅱ，图 10–11；陈木宏、谭智源，1996，196 页，图版 22，图 5–6；陈木宏等，2017，156 页，图版 47，图 3–4。

地理分布 南极罗斯海，出现于晚古新世。

双口虫属 Genus *Dipylissa* Dumitrica, 1988

壳体由沿极轴的各个单一帽状房室交互围绕排列形成，各室间的旋转角度近 90°，旋壳的末端有开口。我们认为 Dumitrica（1988）将其归入门孔虫科（Pyloniidae Haeckel, 1881）存疑，两者之间应有明显的差别。

261. 本松双口虫 *Dipylissa bensoni* Dumitrica

（图版 83，图 3–6）

Spirema sp., Benson, 1966, pp. 268–269, pl. 18, figs. 9–10；Benson, 1983, p. 508, pl. 6, figs. 3–4.

Dipylissa bensoni Dumitrica, 1988, pp. 190, 192, pl. 3, figs. 1–7, pl. 4, figs. 11–15, pl. 6, figs. 1–15；Boltovskoy, 1998, fig. 15.83；陈木宏等，2017，157 页，图版 47，图 7–14。

地理分布 南极罗斯海，全新世。

三球虫属（新属）Genus *Triosphaera* gen. nov.

皮壳由三个夹角为 120° 的类球形房室组成，各房室之间自壳中心往外发育有明显的勒缢；髓壳很小，位于皮壳内近中央处，类球形或不规则形。

模式种：*Triosphaera bulloidis* sp. nov.

262. 三球虫（新种）*Triosphaera bulloidis* sp. nov.

（图版 83，图 7–10）

个体较小，呈圆三角形；皮壳的三个大小相近的圆球形房室相互紧靠排列在一个平面上，相互之间呈 120° 的夹角，房室之间有勒缢，壳体呈封闭状；各房室的壳壁内有一些生自壳中心的桁樑，桁樑随室壁发育而弯曲支撑着壳壁；壁孔类圆形或椭圆形，不规则分布，孔径与孔间桁宽相近，或为孔间桁宽的 1–2 倍，具六角形或多角形框架；壳壁较厚，表面光滑，在壳表的勒缢附近有个别圆锥形的骨针；髓壳简单或不规则状。

标本测量：壳直径 75–90μm，房室宽 50–65μm。

模式标本：DSDP274-31-2 1（图版 83，图 9–10），来自南极罗斯海的 DSDP Leg 28 航次 Site 274 孔的 274-31-2 岩心样品中，保存在中国科学院南海海洋研究所沉积标本室。

地理分布 南极罗斯海，晚始新世。

该种的壳体形态结构与浮游有孔虫 *Globigerina* 或 *Globigerinoides* 较为相似，但后者的壳质为碳酸钙，末房室有一开口，壳壁上没有壁孔和孔间桁框架。

石太阳虫科 Family Litheliidae Haeckel, 1862

壳呈对称螺旋形，并可由一螺旋面分成两对称叶（全部螺旋卷曲位于此面），初室简单或呈箱形。

包卷虫属 Genus *Spirema* Haeckel, 1881

髓壳简单球形或亚球形，皮壳椭球形或亚球形，具螺旋式结构，表面光滑或棘刺状，无放射骨针。

263. 编织包卷虫 *Spirema flustrella* Haeckel

（图版 83，图 11–14）

Spirema flustrella Haeckel, 1887, p. 692.

壳卵形或类球形，螺旋圈三层，壳表棘刺状，甚至有些圆锥形放射骨针，壁孔类圆形，不规则分布。

地理分布 南极罗斯海，晚始新世。

264. 冷特包卷虫 *Spirema lentellipsis* Haeckel

（图版 83，图 15–26）

Spirema lentellipsis Haeckel, 1887, p. 692.

个体稍大，椭球形或近圆球形，螺旋壳层多而密集，一般四层或更多，壳表光滑或略粗糙，无放射骨针。

地理分布 南极罗斯海，早始新世—中渐新世。

265. 苹果包卷虫 *Spirema melonia* Haeckel

（图版 84，图 1–44）

Spirema melonia Haeckel, 1887, p. 692, pl. 49, fig. 1；陈木宏等，2017，158 页，图版 47，图 15–23。
Lithelius spiralis Haeckel, Bjørklund, 1976, p. 1124, pl. 5, fig. 1.
Lithelius nautiloides Popofsky, Kamikuri et al., 2008, p. 169, pl. 1, fig. 37.

个体略小，近圆球形，螺旋壳圈较少，不超过三层，壳表光滑，无放射骨针。

地理分布 南极罗斯海，早古新世—全新世。

266. 老包卷虫（新种）*Spirema vetula* sp. nov.

（图版 85，图 1–10）

壳呈圆球形，内部有不对称的螺旋结构，自近球形的中心初室长出，至外表形成一个完全包覆的封闭外壳，外壳壁较厚，有许多较小的类圆形孔，不规则分布，孔间桁或具六角形框架，或为无序相连；初室很小；壳表面光滑，无针刺。

标本测量：外壳直径 95–160μm，髓壳直径 10–20μm。

模式标本：DSDP274-42-2 1（图版 85，图 5–6），来自南极罗斯海的 DSDP Leg 28 航次 Site 274 孔的 274-42-2 岩心样品中，保存在中国科学院南海海洋研究所沉积标本室。

地理分布 南极罗斯海，早古新世—晚始新世。

该新种与 *Spirema melonia* Haeckel 相似，主要区别是后者的壳体呈椭球形或不规则形，壳表有出口，无封闭的外壳。

267. 包卷虫（未定种 1）*Spirema* sp. 1

（图版 85，图 11–34）

壳体近球形，个体小，外壳常不完整，内部旋转结构多样，壳表略粗糙，或有小刺。可能是 *Spirema* 类的一种原始的古老类型。

地理分布　南极罗斯海，早—晚古新世。

石太阳虫属 Genus *Lithelius* Haeckel, 1862

具简单球形或亚球形髓壳和椭圆透镜状或亚球形螺旋状结构的外皮壳。壳表有很多简单或分枝的放射状骨针。

268. 蜂房石太阳虫 *Lithelius alveolina* Haeckel

（图版 85，图 35–42）

Lithelius alveolina Haeckel, 1862, S. 520, Taf. 27, Figs. 8–9；Haeckel, 1887, p. 694；Renz, 1976, pp. 190–191, pl. 1, fig. 16；陈木宏、谭智源，1996，197 页，图版 22，图 12；陈木宏等，2017，158–159 页，图版 48，图 1–8。

Lithelius sp. A, Lazarus and Pallant, 1989, p. 368, pl. 8, fig. 12 (only).

Lithelius aff. *foremanae* Sanfilippo and Riedel, Hollis, 1997, pl. 2, figs. 3–4.

Spirema sp., Kling, 1973, p. 635, pl. 7, figs. 23–25.

地理分布　南极罗斯海，中渐新世—晚上新世。

269. 空腔石太阳虫（新种）*Lithelius coelomalis* sp. nov.

（图版 85，图 43–46）

壳体近圆球形；初室很大，直径约为壳径的 2/3–3/5，圆球形，内为空腔，初室的壳壁格孔状，壁孔类圆形，大小相近，亚规则或不规则排列，孔径是孔间桁宽的 2–3 倍，有六角形或多角形框架，横跨赤道有 10–14 孔；外围的旋转壳仅有 1–2 个，旋圈之间的距离很小，近等间距，旋壳的壳壁网格状，壁孔大小相近，杂乱重叠分布；壳体表面粗糙，有许多细小的棘突，无放射骨针。

标本测量：壳长 140–165μm、宽 120–145μm，初室直径 90–100μm，旋圈间距 13–15μm。

模式标本：DSDP274-26-1 2（图版 85，图 45–46），来自南极罗斯海的 DSDP Leg 28 航次 Site 274 孔的 274-26-1 岩心样品中，保存在中国科学院南海海洋研究所沉积标本室。

地理分布　南极罗斯海，中渐新世。

该种以具有较大圆球形的初室和较少的旋壳区别于 *Lithelius* 的其他各个种类。

270. 克林石太阳虫 *Lithelius klingi* Kamikuri

（图版 86，图 1–8）

Lithelius klingi Kamikuri, 2010, p. 95, pl. 4, figs. 9–14.

Spirema sp., Kling, 1973, p. 635, pl. 7, figs. 23–25.

地理分布　南极罗斯海，全新世。

271. 小石太阳虫 *Lithelius minor* Jørgensen

（图版 86，图 9–22）

Lithelius minor Jørgensen, 1900, S. 65, Taf. 5, Fig. 24；Benson, 1966, p. 262, pl. 17, figs. 5–7；Nigrini and Moore, 1979, p. S135, pl. 17, figs. 3, 4a–b；陈木宏、谭智源，1996，197–198 页，图版 14，图 2，图版 22，图 15–16；陈木宏等，2017，159 页，图版 48，图 9–19。

地理分布 南极罗斯海，早—晚古新世、全新世。

272. 水手石太阳虫 *Lithelius nautiloides* Popofsky

（图版 86，图 23–28）

Lithelius nautiloides Popofsky, 1908, S. 230, Taf. 27, Fig. 4; Riedel, 1958, p. 228, pl. 2, fig. 3, textfig. 2; Petrushevskaya, 1967, p. 53, figs. 27, 28I, 29I；Chen, 1975, p. 455, pl. 24, fig. 7；Nigrini and Moore, 1979, p. S137, pl. 17, fig. 15；陈木宏、谭智源，1996，198 页，图版 22，图 17–18；陈木宏等，2017，159 页，图版 49，图 1–11。

Lithelius (?) *nautiloides* Popofsky, Petrushevskaya, 1975, pl. 3, figs. 1, 3, 5, pl. 33, fig. 34.

Lithelius nautiloides group Popofsky, Abelmann, 1990, p. 694, pl. 4, fig. 5.

地理分布 南极罗斯海，晚上新世—全新世。

273. 蜗牛石太阳虫 *Lithelius nerites* Tan and Su

（图版 86，图 29–32）

Lithelius nerites Tan and Su，谭智源、宿星慧，1982，162 页，图版 12，图 5；陈木宏、谭智源，1996，197 页，图版 22，图 11；陈木宏等，2017，159 页，图版 49，图 12–13。

地理分布 南极罗斯海，中中新世—早上新世。

274. 规则石太阳虫（新种）*Lithelius regularis* sp. nov.

（图版 86，图 33–38）

个体中等大小，椭球形或近箱形，壳体短轴约为长轴的 1/2–2/3。初室很小，圆球形；旋圈 3–4 个，看似单旋，各圈层间距相近，连接的放射桁很多，分布较均匀；壳壁网格状，孔小而密集，孔间桁不规则分布；壳体表面粗糙，有一些很短的小棘刺，无放射骨针。

标本测量：壳长轴 135–170μm，短轴 100–130μm，旋圈间距 12–18μm。

模式标本：DSDP274-25-3 1（图版 86，图 33–34），来自南极罗斯海的 DSDP Leg 28 航次 Site 274 孔的 274-25-3 岩心样品中，保存在中国科学院南海海洋研究所沉积标本室。

地理分布 南极罗斯海，早始新世—中渐新世。

该种的壳体结构与 *Lithelius nautiloides* Popofsky 有些相似，主要区别是后者一般呈近圆球形，圈层间距渐增大，常有放射骨针。

275. 圆石太阳虫（新种）*Lithelius rotalarius* sp. nov.

（图版 87，图 1–9）

Lithelius sp. E, Petrushevskaya, 1975, p. 573, pl. 3, fig. 2.

壳呈圆球形，外围封闭，个体较大；具双螺旋结构，3–4 个旋圈，旋距自内向外渐增大；髓壳（初室）圆球形，位于壳中心，大小是壳体的1/5；壳壁网格状，壁孔类圆形，大小相近，杂乱分布，较为密集，孔径是孔间桁宽的1–3倍；壳表面粗糙，孔间桁节点上有小棘凸，还可见少量细圆柱形或角锥形的小骨针。

标本测量：壳直径205–240μm，骨针长20–45μm。

模式标本：DSDP274-36-1 2（图版87，图7–9），来自南极罗斯海的DSDP Leg 28航次Site 274孔的274-36-1岩心样品中，保存在中国科学院南海海洋研究所沉积标本室。

地理分布　南极罗斯海，早始新世—中渐新世。

该种特征与 *Lithelius nautiloides* Popofsky 较接近，主要区别是后者的壳体外围不封闭，末圈有出口，壳壁较光滑或粗糙，常有放射状骨针。

罗斯海的标本特征与 *Lithelius* sp. E（Petrushevskaya, 1975）基本相同，被Petrushevskaya（1975）报道出现于亚南极的始新世地层中。

276. 日石太阳虫 *Lithelius solaris* Haeckel

（图版87，图10–13）

Lithelius solaris Haeckel, 1887, p. 695, pl. 49, fig. 2；谭智源、张作人，1976，263页，图34；陈木宏、谭智源，1996，196页，图版22，图7。

地理分布　南极罗斯海，全新世。

277. 螺石太阳虫 *Lithelius spiralis* Haeckel

（图版87，图14–32）

Lithelius spiralis Haeckel, 1862, S. 519, Taf. 27, Fig. 6–7；谭智源、宿星慧，1982，162页，图版13，图9–11；陈木宏、谭智源，1996，196页，图版22，图8，图版45，图13；陈木宏等，2017，160页，图版50，图1–4。

地理分布　南极罗斯海，全新世。

278. 石太阳虫（未定种1）*Lithelius* sp. 1

（图版88，图1–8）

壳体类球形，个体较小，中心区结构模糊，外部的1–2层较清晰，旋距很小，壳表面光滑或有小刺。

地理分布　南极罗斯海，早古新世。

旋篮虫属 Genus *Larcospira* Haeckel, 1862

外皮壳亚球形或椭圆透镜形，由单或复螺旋形赤道腰带构成。螺旋壳环绕一主轴旋转。

279. 多球旋篮虫 *Larcospira bulbosa* Goll and Bjørklund

（图版88，图9–10）

Larcospira bulbosa Goll and Bjørklund, 1989, p. 729, pl. 2, figs. 16–21.

地理分布　南极罗斯海，出现于晚古新世。

石果虫属 Genus *Lithocarpium* Stöhr, 1880, emend. Petrushevskaya, 1975

壳体椭球形，具极管，约 10 个同心壳（或为密集螺旋），壳层之间距离小于 10μm。近极处的壳层发育不完整，间距达 15–30μm，壳表有外套。

280. 多棘石果虫 *Lithocarpium polyacantha* (Campbell and Clark) group

（图版 88，图 11–20）

Larnacantha polyacantha Campbell and Clark, 1944a, p. 30, pl. 5, figs. 4–7.
Lithocarpium polyacantha (Campbell and Clark) group, Petrushevskaya, 1975, p. 572, pl. 3, figs. 6–8, pl. 29, fig. 6；Abelmann, 1990, p. 694, pl. 4, fig. 2；O'Connor, 1993, p. 37, pl. 2, figs. 12–13；陈木宏等，2017，161 页，图版 51，图 1–9。
Porodiscus bassanii Principi, 1909, p. 12, taf. 1, fig. 31.
Prunopyle titan Campbell and Clark, Hays, 1965, p. 173, pl. 2, fig. 4；Bandy et al., 1971, pl. 1, figs. 7–9 (only).

地理分布 南极罗斯海，早古新世—全新世。

果虫属 Genus *Cromyodruppa* Haeckel, 1887

具四个以上的同心壳（两个髓壳和两个以上的皮壳），髓壳圆球形或卵形，皮壳一般呈卵形或椭球形，无骨针和极管。Haeckel（1887）建立该属时将其归入 Druppulida Haeckel, 1882 科中，但该科的壳体为格孔状，无海绵结构（Haeckel, 1887, p. 306）。因此，本书的 *Cromyodruppa*? *concentrica* Lipman 与其有区别，可能为另一新属，前人建立和使用该种时均已质疑其归属。

281. 同心葱果虫? *Cromyodruppa*? *concentrica* Lipman

（图版 88，图 21–22）

Cromyodruppa (?) *concentrica* Lipman, Kozlova and Gorbovetz, 1966, p. 62, pl. 1, figs. 1–4；Foreman, 1978, p. 742, pl. 2, fig. 18；Ling and Lazarus, 1990, p. 355, pl. 1, figs. 11–14；Ling, 1991, p. 319, pl. 1, fig. 4.

壳体稍大，近橄榄形或长筒形，具 5–7 个同心壳，各壳层的间距相似，壳壁由海绵组织构成，两端似有极盖。

地理分布 南极罗斯海，早始新世。

旋壳虫科 Family Strebloniidae Haeckel, 1887

具不对称的螺旋形多室壳，由数量不定的圆形房室组成，呈上升螺旋状，壳的两半不对等。初房简单或箱形。

棘旋壳虫属 Genus *Streblacantha* Haeckel, 1887

具球形、亚球形或椭圆透镜形的简单初室，自初室发育有螺旋上升的小室。壳表有放射骨针。

282. 转棘旋壳虫 *Streblacantha circumyexta* (Jørgensen)

（图版 88，图 23–31）

Sorolarcus circumyextus Jørgensen, 1900, p. 65.
Streblacantha circumyexta (Jørgensen), Jørgensen, 1905, S. 121, Taf. 11, Fig. 46a–c；Schröder, 1909b, S. 60, Textfig. 37a–b；Bjørklund, 1976, p. 1124, pl. 5, figs. 9–12；谭智源、宿星慧，1982，163–164 页，图版 13，图 12–14；陈木宏、谭智源，1996，199 页，图版 23，图 6–9，图版 46，图 4–6；陈木宏等，2017，162 页，图版 51，图 10–17。
Phorticium circumtextum ? (Jørgensen), van de Paverd 1995, pl. 59, fig. 2.
Phorticium octopyle (Haeckel), van de Paverd 1995, p. 194, pl. 57, figs. 1–4, 6–8.
Tholospira cervicornis Haeckel, Itaki et al., 2009, p. 48, pl. 11, figs. 15–17.

地理分布　南极罗斯海，早古新世—晚上新世。

艇虫科 Family Phorticidae Haeckel, 1881

具很不规则的单室壳，由原始的透镜状格孔壳不规则变化而成。不规则的皮壳包围着一个规则或亚规则的透镜状或三带型的髓壳。

艇虫属 Genus *Phorticium* Haeckel, 1881

外皮壳格孔状，内包一椭圆透镜状的 *Larnacilla* 型髓壳。

283. 艇虫 *Phorticium pylonium* Haeckel group

（图版 89，图 1–39）

Phorticium pylonium Haeckel, 1887, p. 709, pl. 49, fig. 10；Cleve, 1899, p. 31, pl. 3, fig. 2；Jørgensen, 1905, S. 120, Taf. 10, Fig. 42a–b, Taf. 11, Fig. 42e, f, 43–45；Benson, 1966, pp. 252–256, pl. 16, figs. 5–9, pl. 17, figs. 1–3；谭智源、张作人，1976，266 页，图 38a–b；陈木宏、谭智源，1996，199 页，图版 23，图 10–11。
Phorticium pylonium Haeckel group, Benson, 1983, p. 506, pl. 7, figs. 15–16.
Phorticium polycladum Tan and Tchang，谭智源、张作人，1976，267 页，图 39a–b；陈木宏、谭智源，1996，199 页，图版 23，图 12–13；Kamikuri et al., 2008, pl. 2, fig. 15.
Pylospira octopyle (Haeckel)? Molina-Cruz, 1977, p. 335, pl. 3, fig. 9；Nigrini and Moore, 1979, p. S139–S140, pl. 17, figs. 6a–c.

地理分布　南极罗斯海，早古新世—全新世。

284. 艇虫（未定种 1）*Phorticium* sp. 1

（图版 89，图 40–47）

Larcospira minor (Jørgensen), Bjørklund, 1976, p. 1124, pl. 5, figs. 2, 4–6 (only).

个体很小，髓壳椭圆透镜状或箱形。
地理分布　南极罗斯海，中中新世—全新世。

果核虫属（新属）Genus *Carpodrupa* gen. nov.

壳箱形或椭球形，壳体有 2–3 层，呈非螺旋状的同心结构，各壳层之间的距离近等，形状相似，无开口。

模式种：*Carpodrupa rosseala* sp. nov.

285. 果核虫（新种）*Carpodrupa rosseala* sp. nov.

（图版 90，图 1–8）

Amphitholus (?) sp. A, Lazarus and Pallant, 1989, p. 367, pl. 8, figs. 4, 10.

椭球形或箱形壳，较小，壳体封闭，有三层非螺旋的同心结构，互不连串，壳层的间距近等或略加大，各壳层的形状相似；髓壳很小，圆球形，一个或数个堆积；壳壁网格状，较薄，壁孔多类型，较小，孔间桁叠网状，杂乱；壳体表面粗糙，有小棘刺，无放射骨针，无开口。

标本测量：壳长轴 125–145μm，短轴 100–115μm，壳层间距 10–19μm。

模式标本：DSDP274-25-1 1（图版 90，图 1–2），来自南极罗斯海的 DSDP Leg 28 航次 Site 274 孔的 274-25-1 岩心样品中，保存在中国科学院南海海洋研究所沉积标本室。

地理分布 南极罗斯海，晚始新世—中渐新世。

该种特征与 *Lithelius regularis* sp. nov.较相似，主要区别在于后者的壳体呈螺旋结构，末圈房室有一个开口。

286. 果核虫（未定种 1）*Carpodrupa* sp. 1

（图版 90，图 9–12）

个体稍大，椭球形或近圆球形，有 2–3 层同心壳体，最外层壳常破损不完整；髓壳很小，圆球形，有 3–4 个聚集在一起；表面略显粗糙，有少量小棘凸。

地理分布 南极罗斯海，晚渐新世。

罩笼虫目 Order NASSELLARIA Ehrenberg, 1875

壳的两极不同，常两侧对称，自一条中央棒生出一根骨针及数条放射桁，具一个环（常为 D 形），或呈帽形结构，常分若干壳节单列排序。

编网虫亚目 Suborder PLECTOIDEA Haeckel, 1881

壳为发育不全的原始三脚形，由长自中央点或棒的放射骨针组成，骨针简单或分叉，在分叉的末端可汇合成一个疏松枝编状的不完整格孔壳。骨骼中无环形物。

编网虫科 Family Plectaniidae Haeckel, 1881

具一编织状的骨架，由放射骨针的分叉交汇连接组成，所有骨针由一中央点或中央棒生出。

棘编虫属 Genus *Plectacantha* Jørgensen, 1905

具四根初始骨针——矢向的、背部的、基部的和 D 环，并发育出具五角形孔的三大网格——腹网、左侧网和右侧网的壳体。

287. 房棘编虫 *Plectacantha oikiskos* Jørgensen

（图版 91，图 1–2）

Plectacantha oikiskos Jørgensen, 1905, pp. 131–132, pl. 13, fig. 57；Benson, 1966, pp. 353–355, pl. 23, figs. 18–19 (not 20).
Phormacantha hystrix (Jørgensen), Petrushevskaya, 1971d, p. 129–130, figs. 68I–V.
Plectacantha ? sp., Benson, 1966, pp. 356–357, pl. 23, figs. 21–24.

地理分布　南极罗斯海，全新世。

环骨虫亚目 Suborder STEPHOIDEA Haeckel, 1881

无完整的格孔壳。骨骼由一个或以上的简单环组成，其间可由疏松并被大开口分隔的网状物相连。有一个矢环，控制着壳体的两侧。

单环虫科 Family Stephanidae Haeckel, 1881

具一简单矢环，无任何格孔状骨骼网。

轭环虫属 Genus *Zygocircus* Bütschli, 1882

环具简单的肋或两侧对称的翼，平滑或有棘刺，无分枝骨针和基足。

288. 轭环虫 *Zygocircus productus* (Hertwig)

（图版 91，图 3–10）

Lithocircus productus Hertwig, 1879, p. 197, Taf. 7, Fig. 4.
Zygocircus productus (Hertwig), Bütschli, 1882, p. 496, Taf. 28, Fig. 9；Haeckel, 1887, p. 948；Petrushevskaya, 1971d, p. 281, pl. 145, figs. 10–11；Abelmann, 1992a, p. 384, pl. 3, fig. 6；陈木宏等，2017，166–167 页，图版 53，图 8–11。
Zygocircus productus capulosus Popofsky, Goll, 1979, p. 381, pl. 2, figs. 4–5 (not 6–9).

罗斯海的标本个体很小。

地理分布　南极罗斯海，晚上新世—全新世。

篓虫亚目 Suborder SPYROIDEA Ehrenberg, 1847, emend. Petrushevskaya, 1971c

壳具完整格架，头部被一矢缢分为双叶室，有一矢环。

双眼虫科 Family Zygospyridae Haeckel, 1887

无头盔与胸部，壳仅由双室头及其骨突构成。

脊篮虫属 Genus *Liriospyris* Haeckel, 1881, emend. Goll, 1968

壳无胸，仅由一个双室的头部及其骨突组成。有六个基脚和三个棒状顶角，或已退化。

289. 光滑脊篮虫（新种）*Liriospyris glabra* sp. nov.

（图版 91，图 11–26）

Liriospyris sp.，陈木宏等，2017，169 页，图版 54，图 9–10。

Lophospyris pentagona (Ehrenberg) *hyperborea* (Jørgensen) emend. Gol, Takahashi, 1991, p. 103, pl. 29, fig. 8 (only).
Lophospyris stabilis stabilis (Goll), van de Paverd, 1995, p. 210, pl. 63, fig. 20 (only).

个体较小，双室壳近长椭球形、肾形，矢缢明显，壁厚中等，两端圆弧；壁孔类圆形或六角形，较小，大小相近或不等，排列不规则，在两个室壳上呈不对称分布，孔间桁稍宽，似有六角形框架；顶角和基脚均较短小或基本消失；表面光滑，仅见个别骨针。该种不同标本的壁孔形状与孔间桁宽度、壁厚以及个体大小均有一定的变化。

标本测量：壳（头）宽 70–135μm，壳高 55–115μm。

模式标本：DSDP274-12-3 1（图版 91，图 13–14），来自南极罗斯海的 DSDP Leg 28 航次 Site 274 孔的 274-12-3 岩心样品中，保存在中国科学院南海海洋研究所沉积标本室。

地理分布 南极罗斯海，晚古新世—全新世。

该种与 *Ceratospyris pentagona* Ehrenberg 和 *Lophospyris stabilis stabilis* (Goll)的主要区别是后二者双室上的壁孔很大，数量较少，呈五角形，壳壁具刺，基脚较长；与 *Lophospyris pentagona* (Ehrenberg) *hyperborea* (Jørgensen)的主要区别是后者在矢环两边的壁孔较大，边缘壁孔较小，常有棘刺，基脚较长。

290. 脊篮虫（未定种 1）*Liriospyris* sp. 1

（图版 91，图 27–30）

个体较小，近椭圆形或扁肾形，矢缢较浅，壳表较薄；壁孔六角形或类圆形，大小不一，排列不规则，孔间桁较细；顶角和基脚均较短小或基本消失，壳表较光滑。

地理分布 南极罗斯海，晚古新世—全新世。

盔篮虫科 Family Tholospyridae Haeckel, 1887

壳盔帽形，无胸，由双室状的头部组成。

盔篮虫属 Genus *Tholospyris* Haeckel, 1881

具三个基脚和一个顶角，格孔状壳。

291. 南极盔篮虫 *Tholospyris antarctica* (Haecker) group

（图版 91，图 31–38）

Phormospyris antarctica Haecker, 1907, p. 124, fig. 9.
Triceraspyris antarctica (Haecker), Haecker, 1908b, pp. 445–446, pl. 84, fig. 586；Riedel, 1958, p. 230, pl. 2, figs. 6–7；Chen, 1975, p. 456, pl. 15, fig. 6；Vigour and Lazarus, 2002, pl. 2, fig. 10.
Triceraspyris antarctica (Haecker) group, Petrushevskaya, 1975, p. 593, pl. 8, fig. 1, pl. 27, figs. 1–2.
Triceraspyris antarctica?, Vigour and Lazarus, 2002, pl. 2, figs. 11–14.
Triceraspyris coronate Vigour and Lazarus, 2002, pl. 2, figs. 15–16.
Tripospyris bicornis Popofsky, 1908, pp. 269–270, pl. XXX, fig. 6.
Tripospyris biloculata Popofsky, 1908, p. 269, pl. XXX, fig. 7.
Trissospyris sp. B, Bjørklund, 1976, pl. 18, figs. 12–17.
Clathrospyris vogti Goll and Bjørklund, 1989, p. 731, pl. 4, figs. 25, 26, 32.
Trssocyclid sp., Abelmann, 1992a, pl. 3, fig. 10.

该种壳体 D 环在顶部形成缩缢，双室形状肺形或猪笼形，基脚有时变宽或分叉，壳表光滑或有刺，种的基本特征相似。因此，归为种群。

地理分布 南极罗斯海，早渐新世—早上新世。

292. 小盔篮虫（新种）*Tholospyris peltella* sp. nov.

（图版 91，图 39–54）

Trissocyclid sp., Abelmann, 1992b, pl. 3, fig. 10.
Trissospyris sp. A, Bjørklund, 1976, pl. 18, figs. 7–11.

个体很小，双室形状为肺形或猪笼形，矢缢较深；双室的壁孔类圆形或不规则形，大小不一，数目不定，分布不规律，有的标本在 D 环两侧有 1–2 个较大的孔，不同标本的孔间桁宽窄不同；顶角很短，底下的基脚末端常内聚并相互连接，在双室底部中间凹陷处形成一个封闭的近半圆形或三角形的小壳室。壳体表面光滑，无骨针或棘刺。

标本测量：壳（头）宽 65–85μm，壳高（不含基脚室）45–55μm。

模式标本：DSDP274-33-3 2（图版 91，图 39–40），来自南极罗斯海的 DSDP Leg 28 航次 Site 274 孔的 274-33-3 岩心样品中，保存在中国科学院南海海洋研究所沉积标本室。

地理分布 南极罗斯海，中古新世—晚始新世。

该种特征与 *Tholospyris antarctica* (Haecker)较为接近，主要区别在于后者的个体较大，各基脚向下或外倾，不汇聚形成小室，壳壁稍粗糙或有小棘。

束篓虫属 Genus *Desmospyris* Haeckel, 1881

无头角，若干基脚或相互连接为格孔状壳。

293. 头盔束篓虫（新种）*Desmospyris coryatis* sp. nov.

（图版 91，图 55–56）

个体很小，仅一个头室，呈头盔形；头室壳壁由 2–3 层的藤状（孔间桁）结构组成，壁孔相互叠加，类圆形，大小不一，孔间桁较细，不规则交叉连接，各连接点在壳表形成许多棘突；头室之下开口近截平，有一口沿，仅在中间处发育一个不完整的基脚，较短，且有侧棘，在中下部还有 2–3 条侧枝连接到头室下方；表面显粗糙，无骨针和顶针。

标本测量：壳（头）宽 55–65μm，头室高 65–70μm，基脚长 18–25μm。

模式标本：DSDP274-1-3 1（图版 91，图 55–56），来自南极罗斯海的 DSDP Leg 28 航次 Site 274 孔的 274-1-3 岩心样品中，保存在中国科学院南海海洋研究所沉积标本室。

地理分布 南极罗斯海，全新世。

该种的壳体形态与大小与 *Desmospyris multichyparis* sp. nov.较接近，主要区别是后者壳壁为简单格孔形，具有五个没有侧刺的基脚，各自独立均匀地分布在头室下缘，互不连接。

294. 弗塔束篓虫 *Desmospyris futaba* Kamikuri

（图版 92，图 1–6）

Dendrospyris futaba Kamikuri, 2010, p. 100, pl. 5, figs. 17, 18, 23–25.

头室间的矢隔较深，双室顶部拱起；壳壁薄，光滑，壁孔小，孔间桁稍宽；胸部开口处常有一些齿状物。

地理分布　南极罗斯海，早-中渐新世。

295. 哈史束篓虫 *Desmospyris haysi* (Chen)

（图版 92，图 7–24）

Dendrospyris haysi Chen, 1974, p. 482, pl. 2, figs. 3–5；Chen, 1975, p. 455, pl. 15, figs. 3–5；Weaver, 1976, p. 579, pl. 2, figs. 7–9, pl. 7, fig. 4；Lazarus, 1990, p. 716, pl. 5, fig. 9.
Desmospyris anthocyrtoides (Bütschli), Benson, 1966, p. 332, pl. 23, figs. 6–8.
Phormospyris stabilis stabilis (Goll), Kling, 1979, p. 309, pl. 1, fig. 18；Takahashi, 1991, p. 104, pl. 30, figs. 2–5.
Lophospyris stabilis stabilis (Goll), van de Paverd, 1995, p. 210, pl. 63, figs. 13–16 (only).
Gorgospyris ? spp., O'Connor, 1997b, p. 112, pl. 3, figs. 13–15.

地理分布　南极罗斯海，晚古新世—中中新世。

296. 多脚束篓虫（新种）*Desmospyris multichyparis* sp. nov.

（图版 92，图 25–26）

个体很小，仅一个头室，伸长钟罩形，自顶部往下逐渐扩大，开口端不缩窄；头室壁孔类圆形或多角形，大小不同，分布无规律，孔间桁较细，发育有些杂乱，在壳体表面有明显节点突起，呈角锥状；口端仅截平，无口缘，向下生出五个较细的近圆柱形基脚，基脚末端略内倾，各脚间无附属连接物；壳表粗糙，顶部的角锥形凸起较明显。

标本测量：壳（头）宽 50–55μm，头室高 55–65μm，基脚长 20–25μm。

模式标本：DSDP274-1-3 1（图版 92，图 25–26），来自南极罗斯海的 DSDP Leg 28 航次 Site 274 孔的 274-1-3 岩心样品中，保存在中国科学院南海海洋研究所沉积标本室。

地理分布　南极罗斯海，全新世。

该种与 *Desmospyris trichyparis* sp. nov.的主要区别是后者的头部为双室，个体稍大，基脚仅三个且有格孔壁相互连接，壳表光滑。

297. 瑰篮束篓虫 *Desmospyris rhodospyroides* Petrushevskaya

（图版 92，图 27–46）

Rhodospyris sp. A, Petrushevskaya and Kozlova, 1972, p. 531, pl. 38, fig. 11.
Desmospyris rhodospyroides Petrushevskaya, 1975, p. 593, pl. 10, figs. 27–29, 31, 32；Lazarus, 1990, p. 716, pl. 5, figs. 11–12；Lazarus, 1992, p. 797, pl. 7, figs. 3–4.

该种与 *Desmospyris haysi* (Chen)很相似，主要区别是前者头胸之间的缩缢较浅或不明显，两者壳宽近等，头室较规整；而后者头胸之间的缩缢较深，头部两室常呈泡状，形态有一定变化。

地理分布　南极罗斯海，晚始新世—全新世。

298. 海绵束篓虫 *Desmospyris spongiosa* Hays

（图版 93，图 1–20）

Desmospyris spongiosa Hays, 1965, p. 173, pl. 2, fig. 1；Petrushevskaya, 1973b, pl. 3, fig. 22；Petrushevskaya, 1975, p. 593, pl. 10, figs. 33–34, pl. 27, fig. 8；Keany, 1979, p. 54, pl. 2, fig. 12, pl. 3, fig. 1；Lazarus, 1990, p. 716, pl. 4, figs. 9–11.

地理分布　南极罗斯海，中中新世—早上新世。

299. 三脚束篓虫（新种）*Desmospyris trichyparis* sp. nov.

（图版 93，图 21–23）

Trissocyclid sp. A, Abelmann, 1990, p. 695, pl. 5, figs. 5A–C.
Trissocyclidae gen. et sp. indet., Hollis, 2002, p. 293, pl. 3, fig. 3 (only).

个体中等大小，壳壁稍厚；头部呈双室，被 D 形矢环分隔，壁孔圆形或亚圆形，大小相差不大，一般在靠近隔环处略大，分布较均匀；三个角锥状基脚较坚实，向下伸长，基脚之间有部分的格孔结构相互连接形成不完整的格孔壁，壁孔与头室壁孔相似，类圆形或椭圆形，大小不等，数量较少，分布无规律，孔间桁较宽；壳表光滑无刺。

标本测量：壳（头）宽 105–115μm，头室高 70–80μm，基脚长 50–70μm。

模式标本：DSDP274-11-3 1（图版 93，图 21），来自南极罗斯海的 DSDP Leg 28 航次 Site 274 孔的 274-11-3 岩心样品中，保存在中国科学院南海海洋研究所沉积标本室。

地理分布　南极罗斯海，晚中新世。

该种壳体特征与 *Desmospyris spongiosa* Hays 较接近，主要区别是后者的基脚已完全消失，形成一个完整的筒状腹部，开口近截平，壳壁有海绵或网格组织。

葡萄虫亚目 Suborder BOTRYODEA Haeckel, 1881

具完整的格孔壳，头部多室，呈一叶状，由缩缢分为三个或更多的头叶。

管葡萄虫科 Family Cannobotryidae Haeckel, 1881, emend. Riedel, 1967

壳为叶状的头部，头与胸界线不明显；或具一大一小头室，胸呈亚圆柱形。

疑蜂虫属 Genus *Amphimelissa* Jørgensen, 1905

无胸和腹，该属外表与 *Lithomelissa* 相似，但其结构基本不同。四根初始骨针自 D 刺生出，一矢针、二侧针和一背针，形成一个不完全的内格架头部，其外包围了一个格架壳，向下延续为胸。

300. 方形疑蜂虫（新种） *Amphimelissa quadrata* sp. nov.

（图版 93，图 24–25）

个体稍大，呈方块形；头部较大，被相互垂直交叉的两个矢向缩缢分隔为四个大小相近的类球形室房，其下部阔开口；胸部发育不完整，或仅存很短小的残壁，阔开口；壳壁孔类圆形，大小不同，分布不规则，具六角形框架，孔径是孔间桁的 3–4 倍；表面粗糙，有一些小棘刺。此标本的头部结构与形态较为特殊，暂定为该属。

标本测量：壳宽 120–140μm，壳高 125–150μm。

模式标本：DSDP274-10-3 2（图版 93，图 24–25），来自南极罗斯海的 DSDP Leg 28 航次 Site 274 孔的 274-10-3 岩心样品中，保存在中国科学院南海海洋研究所沉积标本室。

地理分布　南极罗斯海，早上新世。

该新种与同类的其他种主要区别是具有独特的两个相互垂直的缩缢，将头部分为四个大小相似的头室，壳体呈近正方形。

301. 棘刺疑蜂虫 *Amphimelissa setosa* (Cleve)

（图版 94，图 1–19）

Botryopyle setosa Cleve, 1899, p. 27, Taf. I, Figs. 10a–b；谭智源、张作人，1976，272 页，图 45；谭智源、陈木宏，1999，286 页，图 5-207；陈木宏等，2017，173 页，图版 55，图 18–23。

Lithomelissa setosa (Cleve), Jørgensen, 1900, p. 81, pl. 4, figs. 21–22.

Lithomelissa setosa Jørgensen, 1905, pp. 135–136, pl. 16, figs. 81–83, pl. 18, figs. 108a–b；Bjørklund, 1973, pl. 2, fig. j；Bjørklund, 1974, fig. 8；Bjørklund, 1976, pl. 8, figs. 1–13, pl. 11, figs. 19–23；Takahashi, 1991, p. 97, pl. 25, figs. 16–22；陈木宏等，2017，189 页，图版 66，图 6–8。

Amphimelissa setosa (Cleve), Jørgensen, 1905, p. 137, pl. 18, figs. 109a–b；Bjørklund and Swanberg, 1987, pp. 245–254, figs. 3A–K, 4A–Y；Schröder-Ritzrau, 1995, pl. 7, figs. 1–3；Bjørklund et al., 1998, pl. 2, figs. 12–14, 30–33；Dolven, 1998, pl. 12, fig. 1；Matul and Abelmann, 2005, figs. 1: 1–12；Itaki et al., 2009, p. 55, pl. 23, figs. 40–41.

Botryocampe cf. *inflata* Bailey, Petrushevskaya, 1971d, fig. 79v.

Botryopyle (?) *dionisii* Petrushevskaya, 1975, p. 589, pl. 13, fig. 18, pl. 26, fig. 10.

地理分布 南极罗斯海，早始新世—晚渐新世。

该种名实际上包含两个不同的种类，首先定名的是 *Botryopyle setosa* Cleve（1899）的标本，头部明显分为 2–4 叶，而后 Jørgensen（1900）在鉴定另一类头部基本不分叶的标本时仍使用种名 *Lithomelissa setosa* (Cleve)（承认种名作者，并有文献引用），但在 1905 年的文献中却改为 *Lithomelissa setosa* Jørgensen，此后造成了引用易被混淆的状况。Cleve（1899）的种类特征常见于南北半球的高纬度海区，属典型的冷水种，罗斯海的标本也具此类特征。实际上，*Amphimelissa setosa* (Cleve)与 *Lithomelissa setosa* Jørgensen 的主要区别是前者的头部呈叶状，2–3 个纵向缩缢将其分为 3–4 个头室，而后者头部简单，仅有一个头室。因此，根据生物命名法规原则，Jørgensen（1900, 1905）的 *Lithomelissa setosa* Jørgensen 种类名称属于无效名，其标本种类应属另一种或需更改种名，以免由于命名的混乱造成使用上的混淆。

302. 疑蜂虫（未定种 1） *Amphimelissa* sp. 1

（图版 94，图 20–21）

该未定种与 *Amphimelissa* setosa (Cleve)较相似，主要区别是未定种的头部分叶不明显，无头胸之间的勒缢（无胸？）。

地理分布 南极罗斯海，出现于早上新世。

袋葡萄虫属 Genus *Botryopera* Haeckel, 1887

壳简单，仅由一个叶状头组成，无管状或放射状的附属物。该属形态特征与 *Antarctissa* 属较相似，但常呈近圆筒状，胸部伸长，拱形勒缢位于头腔下部的 1/3–1/2 之间。

303. 同缘袋葡萄虫 *Botryopera conradae* (Chen)

（图版 95，图 1–4）

Antarctissa conradae Chen, 1974, p. 484, pl. 3, figs. 1–3.
Botryopera conradae (Chen), Petrushevskaya, 1971d, p. 193, fig. 2(6).

地理分布 南极罗斯海，全新世。

304. 筒形袋葡萄虫（新种）*Botryopera cylindrita* sp. nov.

（图版 95，图 5–10）

Antarctissa denticulata (?) (Ehrenberg) var. *faceata* Petrushevskaya, 1967, pp. 86–88, figs. 50II–III.
Lithomelissa sp. P, Petrushevskaya, 1973, p. 1044, fig. 3(12).

壳分二节，呈圆筒形，中等壁厚；头部有一较大的主头室，类圆球形或椭球形，在其下侧围绕有若干小头叶，直接连接着胸壳（第二壳节）；胸壳完全为长圆筒形，侧缘直或略弯曲，口端微缩窄，截平，无口缘环或齿，有的标本口部残缺状；壁孔类圆形，大小不等，排列不规则，但较均匀分布；壳表光滑，无骨针。

标本测量：主头室长 30–60μm、宽 35–55μm，胸壳长 70–85μm、宽 45–75μm。

模式标本：DSDP274-12-5 1（图版 95，图 7–8），来自南极罗斯海的 DSDP Leg 28 航次 Site 274 孔的 274-12-5 岩心样品中，保存在中国科学院南海海洋研究所沉积标本室。

地理分布 南极罗斯海，中中新世—晚上新世。

该种特征与 *Antarctissa denticulata* (Ehrenberg)较接近，主要区别是后者壳体常呈圆三角形，胸壳较短，口端圆弧形，壳表有刺或骨针等附属物。

305. 三叶袋葡萄虫 *Botryopera triloba* (Ehrenberg) group

（图版 95，图 11–30）

Lithobotrys triloba Ehrenberg, 1854, pl. 19, fig. 35；pl. 22, fig. 30a.
Botryopera triloba (Ehrenberg) group, Petrushevskaya, 1975, p. 591, pl. 11, figs. 27–29, 36–39, pl. 20, figs. 3–4.
Lithomelissa boreale (Ehrenberg) Petrushevskaya, 1967, p. 83, pl. 48, figs. 1–4.
Lithomelissa sp. B, Petrushevskaya, 1967, p. 82, fig. 47.
Trisulcus borealis (Ehrenberg) Petrushevskaya, 1971b, p. 145, pl. 74.

地理分布 南极罗斯海，中更新世—全新世。

306. 袋葡萄虫（未定种 1）*Botryopera* sp. 1

（图版 95，图 31–34）

个体较小，头脑丘状，结构略显复杂，半球形，头胸之间勒缢明显；胸呈长圆筒状，壁孔类圆形，亚规则排列；壳表光滑。

地理分布 南极罗斯海，晚始新世—中渐新世。

葡萄门虫属 Genus *Botryopyle* Haeckel, 1881

头部分叶，无头管，胸部简单，口端开放，无腹壳。

307. 筛葡萄门虫 *Botryopyle cribrosa* (Ehrenberg)

（图版 95，图 35–50）

? *Lithobotrys cribrosa* Ehrenberg, 1873, p. 237；Ehrenberg, 1876, pl. 3, fig. 20.
Botryocella cribrosa (Ehrenberg) group, Hollis, 1997, pp. 53–54, pl. 4, figs. 1–4.
Botryopyle multicellaris (Haeckel), van de Paverd, 1995, p. 256, pl. 77, figs. 13–16, 18.

地理分布 南极罗斯海，中始新世—中渐新世。

囊篮虫属 Genus *Saccospyris* Haecker, 1908, emend. Petrushevskaya, 1965

无顶针，壳口缘有细齿冠。

308. 南极囊篮虫 *Saccospyris antarctica* Haecker

（图版 152，图 22–27）

Saccospyris antarctica Haecker, 1907, p. 124, figs. 10a–b; Haecker, 1908a, pp. 447–448, pl. 84, figs. 584, 589–590; Petrushevskaya, 1965, pp. 96–98, fig. 10；Petrushevskaya, 1968, pp. 149–150, fig. 85II；Petrushevskaya, 1975, p. 589, pl. 13, fig. 21；Weaver, 1976a, p. 582, pl. 1, fig. 6；Weaver, 1983, p. 678, pl. I, fig. 4；Abelmann, 1992a, p. 777；陈木宏等，2017，233–234 页，图版 84，图 9。
Botryopyle antarctica (Haecker), Riedel, 1958, pp. 224–226, text-fig. 13, pl. 4, fig. 12 (in part).
Botryopyle (?) *antarctica* (Haecker), Chen, 1975, pl. 18, fig. 10.
Botryocampe inflata Bailey group, Petrushevskaya, 1975, pl. 13, figs. 24–25.

地理分布 南极罗斯海，晚上新世—全新世。

309. 同胸囊篮虫 *Saccospyris conithorax* Petrushevskaya

（图版 152，图 28–29）

Saccospyris conithorax Petrushevskaya, 1965, p. 98, fig. 11.
Botryocampe conithorax (Petrushevskaya), Petrushevskaya, 1971b, p. 156, pl. 79, fig. 4.
Botryocampe conithorax (Petrushevskaya) group, Petrushevskaya, 1975, p. 588, pl. 13, fig. 27.

地理分布 南极罗斯海，出现于中渐新世。

310. 网头囊篮虫 *Saccospyris dictyocephalus* (Haeckel) group

（图版 152，图 30–31）

Botryopyle dictyocephalus Haeckel, 1887, p. 1113, pl. 96, fig. 6.
Botryopyle dictyocephalus Haeckel group, Riedel and Sanfilippo, 1971, p. 1602, pl. 1J, figs. 21–26, pl. 2J, figs. 16–18, pl. 3F, figs. 9–12；Takemura, 1992, pp. 743–744, pl. 3, fig. 7.

地理分布 南极罗斯海，出现于中渐新世。

311. 囊篮虫（未定种 1）*Saccospyris* sp. 1

（图版 152，图 32–34）

Lithocorythium sp., Petrushevskaya, 1975, pl. 13, fig. 23.

地理分布 南极罗斯海，全新世。

笼虫亚目 Suborder CYRTOIDEA Haeckel, 1862, emend. Petrushevskaya, 1971

壳呈圆锥形或帽形，多节壳的各节单一直线形排列。

三足壶虫科 Family Tripocalpidae Haeckel, 1887

壳简单，不分节，具一个头部和三个辐射脚。

原帽虫属 Genus *Archipilium* Haeckel, 1881

具三条侧肋或翼，口端截平，无末端脚，无顶角。

312. 直翼原帽虫 *Archipilium orthopterum* Haeckel

（图版 96，图 1–2）

Archipilium orthopterum Haeckel, 1887, p. 1139, pl. 98, fig. 7；陈木宏等，2017, 173–174 页，图版 55，图 24。

地理分布　南极罗斯海，出现于晚始新世。

313. 直翼原帽虫（亲近种）*Archipilium* sp. aff. *A. orthopterum* Haeckel

（图版 96，图 3–5）

Archipilium sp. aff. *A. orthopterum* Haeckel, p. 114, pl. 36, figs. 5, 7.

地理分布　南极罗斯海，早上新世。

314. 近似原帽虫（新种）*Archipilium vicinun* sp. nov.

（图版 96，图 6–7）

Archipilium sp., Caulet, 1991, pl. 2, fig. 7.
Archipilium sp., O'Connor, 1997b, p. 111, pl. 3, fig. 17.

壳体较小，圆锥形，顶部有一半球形的小拱，往下迅速扩大，阔开口，口缘不平整，呈残缺状；壁孔类圆形或椭圆形，分布很不规则，孔径大小悬殊，是孔间桁宽的 1–4 倍，孔间桁宽窄不一；三条侧肋从半球形拱顶之下开始，延伸至壳中部向外发育成为斜下自由生长的侧翼，侧翼呈长圆柱状，末端逐渐缩尖，有的略内倾，侧翼长大于壳长；壳体表面光滑无刺，无顶角。

标本测量：壳长 70–75μm、宽 55–70μm，侧翼长 80–95μm。

模式标本：DSDP274-10-5 1（图版 96，图 6–7），来自南极罗斯海的 DSDP Leg 28 航次 Site 274 孔的 274-10-5 岩心样品中，保存在中国科学院南海海洋研究所沉积标本室。

地理分布　南极罗斯海，出现于早上新世；南印度洋，新西兰（Kaipara Harbour），出现于渐新世—中中新世。

该新种区别于同类群其他各种的基本特征是口部不平整，无口缘或缘环，常有一小拱顶，壁孔类型差异较大。罗斯海标本特征与 Caulet（1991）和 O'Connor（1997b）的标本均非常相似。

三帽虫属 Genus *Tripilidium* Haeckel, 1881

无侧肋，有三个简单或分叉的末端脚，一个顶角。

315. 棒三帽虫? *Tripilidium*? *clavipes* Clark and Campbell

（图版 96，图 8–11）

Tripilidium (?) *clavipes* Clark and Campbell, Abelmann, 1990, p. 695, pl. 5, fig. 1.
Tripilidium (?) *clavipes advena* Clark and Campbell, 1945, p. 34, pl. 7, figs. 31–33.

地理分布 南极罗斯海，晚始新世—早上新世。

枝蓬虫属 Genus *Cladoscenium* Haeckel, 1881

一个游离分叉的内柱延伸形成一个顶角，三个基脚游离，无侧翼。

316. 异地枝蓬虫 *Cladoscenium advena* (Clark and Campbell) group

（图版 96，图 12–13）

Tripilidium clavipes advena Clark and Campbell, 1945, p. 34, pl. 7, fig. 31 (only).
Tripodiscium clavipes (Clark and Campbell) Petrushevskaya, 1971b, fig. 34VI.
Caldoscenium (?) *advena* (Clark and Campbell) group, Petrushevskaya, 1975, p. 589, pl. 11, fig. 16, pl. 24, fig. 7.
Cladoscenium cf. *tricolpum* (Haeckel), Benson, 1966, p. 387, pl. 25, figs. 10–11.
Pseudodictyophimus amundseni Goll and Bjørklund, 1989, p. 732, pl. 5, figs. 1–4.
Lithomelissa sp., Hollis, 2002, p. 297, pl. 4, fig. 2.

地理分布 南极罗斯海，出现于中渐新世。

317. 三胸枝蓬虫 *Cladoscenium tricolpium* (Haeckel)

（图版 96，图 14–17）

Euscenium tricolpium Haeckel, 1887, p. 1147, pl. 53, fig. 12.
Cladoscenium tricolpium (Haeckel), Jørgensen, 1900, pp. 78–79; Jørgensen, 1905, p. 134, pl. 15, figs. 71–73; Schröder, 1914, pp. 93–95, figs. 40–43; Bjørklund, 1976, p. 1124, pl. 7, figs. 5–8.

地理分布 南极罗斯海，中渐新世。

小袋虫属 Genus *Peridium* Haeckel, 1881

壳具简单的腔，无自由的柱。有三个自由的基足，一个顶角。

318. 长棘小袋虫 *Peridium longispinum* Jørgensen

（图版 96，图 18–38）

Peridium longispinum Jørgensen, 1900, pp. 75–76；Jørgensen, 1905, p. 135, pl. 15, figs. 75–79, pl. 16, fig. 80；Bjørklund, 1976, pl. 7, figs. 9–15；Schröder-Ritzrau, 1995, pl. 6, figs. 3–4；Bjørklund et al., 1998, pl. 2, figs. 26–27；Dolven, 1998, pl. 12, fig. 5；Bjørklund and Kruglikova, 2003, p. 245 (not figured)；陈木宏等，2017，175–176 页，图版 57，图 2–9。
Arachnocorallium sp., Petrushevskaya, 1975, p. 592, pl. 9, figs. 17–18 (only).
Peromelissa thoracites Haeckel, van de Paverd, 1995, pp. 222–223, pl. 65, figs. 12–14, 16 (not others).
Plagoniid gen. et sp. indet., Takemura, 1992, p. 744, pl. 1, fig. 10.

该种的形态结构与 *Psilomelissa calvata* Haeckel（1887, p. 1209, pl. 56, fig. 3）较相似，主要区别是前者壳表有许多棘刺，而后者的壳表光滑无刺。

地理分布　南极罗斯海，晚始新世—全新世。

319. 小袋虫（未定种 1）*Peridium* sp. 1

（图版 96，图 39–40）

头呈盔形，有一个细柱状顶角，具三个基脚。

地理分布　南极罗斯海，出现于晚始新世。

袋虫属 Genus *Archipera* Haeckel, 1881

壳单节，无内柱（或内杆），具两个或多个顶角。

320. 双肋袋虫 *Archipera dipleura* Tan and Tchang

（图版 96，图 41–46）

Archipera dipleura Tan and Tchang，谭智源、张作人，1976，274 页，图 48a–f；谭智源、陈木宏，1999，291 页，图 5-215；陈木宏等，2017，177–178 页，图版 58，图 4–13。

地理分布　南极罗斯海，早渐新世—早上新世。

瓮笼虫科 Family Cyrtocalpidae Haeckel, 1887

具一简单而无节的壳，有一简单的头，无放射脊骨突。

小角虫属 Genus *Cornutella* Ehrenberg, 1838, emend. Nigrini, 1967

壳圆锥形，向着口的一边逐渐扩大，有头角。

321. 加州小角虫 *Cornutella californica* Campbell and Clark

（图版 97，图 1–6）

Cornutella californica Campbell and Clark, 1944b, pp. 22–23, pl. 7, figs. 33, 34, 42, 43；Foreman, 1968, pp. 21–22, pl. 3, figs. la–c；Ling and Lazarus, 1990, p. 355, pl. 3, fig. 17；Ling, 1991, p. 319, pl. 1, fig. 14.

Cornutella sp. aff. *C. californica* Clark and Campbell, Bjørklund, 1976, p. 1124, pl. 23, figs. 23–24.

Cornutella stiligera Ehrenberg group, Petrushevskaya and Kozlova, 1972, p. 551, pl. 30, figs. 14–15.

头部有一较长的顶角。

地理分布　南极罗斯海，晚始新世—中渐新世。

322. 格孔小角虫 *Cornutella clathrata* Ehrenberg

（图版 97，图 7–13）

Cornutella clathrata Ehrenberg, 1838, p. 129；Ehrenberg, 1844, p. 77；Ehrenberg, 1847, p. 42；Ehrenberg, 1854, pl. 22, figs. 39a–c；Ehrenberg, 1854, p. 1183；Abelmann, 1990, p. 696, pl. 8, fig. 8.

Cornutella clathrata Ehrenberg, 1858, p. 31.

Cornutella curvata Haeckel, 1887, p. 1183.

地理分布 南极罗斯海，早始新世—中中新世。

323. 圆锥小角虫 *Cornutella mitra* Ehrenberg

（图版 97，图 14–17）

Cornutella mitra Ehrenberg, 1874, p. 221；Ehrenberg, 1876, p. 68, pl. 2, fig. 8.

Zealithapium mitra (Ehrenberg), O'Connor, 1999, pp. 5–6, pl. 9, fig. 47；Pascher et al., 2015, Taxonomic Notes, p. 4, pl. 1, fig. 8；Hollis et al., 2020, pl. 1, fig. 10.

地理分布 南极罗斯海，早上新世。

324. 深小角虫 *Cornutella profunda* Ehrenberg

（图版 97，图 18–27；图版 98，图 1–11）

Cornutella clathrata profunda Ehrenberg, 1854, S. 241, Taf. 35b, Fig. 21.

Cornutella profunda Ehrenberg, 1858, S. 31；Riedel, 1958, p. 232, pl. 3, figs. 1–2；Benson, 1966, pp. 439–432, pl. 29, figs. 7–8；Nigrini, 1967, pp. 60–63, pl. 6, figs. 5a–c；Kling, 1973, pp. 635–636, pl. 3, figs. 1–4, pl. 9, figs. 8–17；Petrushevskaya, 1975, p. 587, pl. 13, figs. 32–33；Weaver, 1983, p. 675, pl. 1, fig. 6；陈木宏、谭智源，1996，207–208 页，图版 27，图 7，图版 49，图 1；陈木宏等，2017，180–181 页，图版 59，图 21–25，图版 60，图 1–6。

Cornutella hexagona Haeckel, 1887, p. 1180, pl. 54, fig. 9.

Cornutella sethoconus Haeckel, 1887, p. 1180, pl. 54, fig. 11.

Sethoconus orthoceras (Haeckel), Haeckel, 1887, p. 1294, pl. 54, fig. 11.

Sethoconus bimarginatus (Haeckel), Haeckel, 1887, p. 1295, pl. 54, fig. 12.

Cornutella californica (Campbell and Clark), Hollis, 2002, p. 287, pl. 6, fig. 5 (only).

地理分布 南极罗斯海，早古新世—全新世。

始匣虫属 Genus *Archicapsa* Haeckel, 1881

壳不分节，头简单，无顶角，口部封闭（有格板）。

325. 橄榄始匣虫（新种）*Archicapsa olivaeformis* sp. nov.

（图版 98，图 12–13）

壳近橄榄形，中间最宽，两端缩窄，个体较小；壳壁稍厚，壁孔类圆形或椭圆形，大小不等，亚规则排列，孔径是孔间桁宽的 1–2 倍；头呈半球形，无顶角，头与胸之间无缩缢；胸的末端封闭无开口，无具孔的筛板；壳体表面光滑，无刺。

标本测量：壳长 110–115μm、宽 75–80μm。

模式标本：DSDP274-34-1 2（图版 98，图 12–13），来自南极罗斯海的 DSDP Leg 28 航次 Site 274 孔的 274-34-1 岩心样品中，保存在中国科学院南海海洋研究所沉积标本室。

地理分布 南极罗斯海，晚始新世。

该新种特征与 *Archicapsa quadriforis* Haeckel（1887, p. 1192）较为接近，主要区别在于后者壳体呈卵形，较胖，壳长仅略大于壳宽，口端基盘上为具四孔的筛板，壳表粗糙。

三肋笼虫科 Family Tripocyrtidae Haeckel, 1887, emend. Campbell, 1954

具两节壳，由一个横缢分为头和胸，有三条放射肋。（部分有不完整的第三节壳）

小孔帽虫亚科 Subfamily Sethopilinae Haeckel, 1881, emend. Campbell, 1954

壳的基部为阔开口。

网杯虫属 Genus *Dictyophimus* Ehrenberg, 1847

具三条完整的胸肋，在口缘上延长变成三个实心而分散的脚，头具一顶角。

326. 棘刺网杯虫（新种）*Dictyophimus echinatus* sp. nov.

（图版 98，图 14–29）

Ceratocyrtis sp., Petrushevskaya, 1975, p. 590, pl. 19, fig. 1.

Cyrtoidae gen. sp., Petrushevskaya, 1967, p. 91, fig. 52III.

壳呈钟罩形，自上而下渐扩大；头部小球形，部分陷入胸内，壁孔很小，有一个细角锥形的顶角；胸壳中等壁厚，壁孔类圆形或椭圆形，大小不一，排列不规则，一般在靠近出口处壁孔较大，孔径是孔间桁宽的 1–4 倍，孔间桁较窄；胸壁上的三条主胸肋较发育，至口缘处生长为细锥形基脚；胸口完全开放，在口缘处还常伸出 3–5 个细长骨针，形态和大小与三个基脚相类似；壳表常粗糙，有一些不同形状的骨针或刺。

标本测量：壳长 75–150μm、宽 95–170μm，头径 40–50μm，顶角长 30–55μm，基脚长 30–95μm。

模式标本：DSDP274-20-5 1（图版 98，图 20–21），来自南极罗斯海的 DSDP Leg 28 航次 Site 274 孔的 274-20-5 岩心样品中，保存在中国科学院南海海洋研究所沉积标本室。

地理分布　南极罗斯海，晚始新世—全新世。

该种与 *Dictyophimus mawsoni* Riedel（1958, p. 234, pl. 3, figs. 6–7）特征较为接近，主要区别是后者的三个实心末端脚较为强壮，具有一个粗大的顶角。

327. 燕网杯虫 *Dictyophimus hirundo* (Haeckel) group

（图版 99，图 1–2）

Pterocorys hirundo Haeckel, 1887, p. 1318, pl. 71, fig. 4；Riedel, 1958, p. 238, pl. 3, fig. 11, pl. 4, fig. 1, textfig. 9；Petrushevskaya, 1967, p. 115, pl. 67, figs. 1–5；Petrushevskaya, 1971b, pl. 111, figs. 4–5；陈木宏、谭智源，1996，217 页，图版 31，图 2–4，图版 52，图 2。

Dictyophimus hirundo (Haeckel) group, Petrushevskaya, 1975, p. 583；Nigrini and Moore, 1979, p. N35, pl. 22, figs. 2–4；Abelmann, 1992b, p. 380, pl. 4, fig. 8；陈木宏等，2017，185 页，图版 62，图 10–16。

Dictyophimus sp. aff. *D. hirundo* (Haeckel), Petrushevskaya and Kozlova, 1972, p. 553, pl. 27, figs. 16–17.

地理分布　南极罗斯海，中中新世—全新世。

328. 伊斯网杯虫 *Dictyophimus histricosus* Jørgensen

（图版 99，图 3–12）

Dictyophimus histricosus Jørgensen, 1905, p. 138, pl. 16, fig. 89；陈木宏等，2017，186 页，图版 63，图 1。

Helotholus histricosa Jørgensen, Benson, 1966, p. 462, pl. 31, fig. 6 (not figs. 4, 5, 7, 8).

Helotholus histricosa Jørgensen group, Benson, 1983, p. 504, pl. 8, fig. 1 (not figs. 2–3).

Dictyophimus cf. *platycephalus* Haeckel, Abelmann, 1990, p. 696, pl. 7, fig. 11.

Jørgensen（1905）同时建立了两个新种 *Dictyophimus histricosus* Jørgensen 和 *Helotholus histricosa* Jørgensen，它们的种名实际上是相同的，因此容易在名称上引起混淆。它们分别归为两个不同的属，前者具有三个末端脚，而后者无末端脚。

地理分布 南极罗斯海，晚始新世—中渐新世。

329. 光头网杯虫（新种）*Dictyophimus lubricephalus* sp. nov.

（图版 99，图 13–14）

Dictyophimus sp. 4, Abelmann, 1992a, p. 380, pl. 4, fig. 6.

个体较小，壳近圆筒形，顶部圆弧形，侧面垂直，主要由头部构成，胸不明显，无缩缢；壳壁较薄，孔较大，类圆形，有六角形框架，孔间桁较细，孔径是孔间桁宽的 2–5 倍；顶针不发育或很小，仅呈小锥状凸起；三个实心侧翼为头部侧骨针斜下生长直接发育形成，呈三棱角锥状，直挺，末端缩尖；壳表光滑，无棘刺。

标本测量：壳长 50μm、宽 75μm，侧翼长 30–45μm。

模式标本：DSDP274-20-3 1（图版 99，图 13–14），来自南极罗斯海的 DSDP Leg 28 航次 Site 274 孔的 274-20-3 岩心样品中，保存在中国科学院南海海洋研究所沉积标本室。

地理分布 南极罗斯海，出现于晚渐新世。

该新种特征与 *Dictyophimus histricosus* Jørgensen 接近，主要区别是后者的壳体呈圆锥形或钟罩形，有一头部和顶角，侧翼内弯，胸口有一细隔环，可见腹壳的残部。

330. 宽头网杯虫 *Dictyophimus platycephalus* Haeckel

（图版 99，图 15–16）

Dictyophimus platycephalus Haeckel, 1887, p. 1198, pl. 60, figs. 4–5.
Dictyophimus (?) *platycephalus* (?) Haeckel, Petrushevskaya, 1967, p. 71, pl. 42, figs. 1–3.

地理分布 南极罗斯海，出现于早上新世。

石蜂虫属 Genus *Lithomelissa* Ehrenberg, 1847

具三个自胸壁生出的游离侧翼或实心骨针。无末端足。头具一个或多个角。

331. 强角石蜂虫（新种）*Lithomelissa arrhencorna* sp. nov.

（图版 99，图 17–20）

个体较小，近拱顶圆筒形，头与胸的高度和长度相近，壳壁较厚；头为半球形，顶角三棱角锥状，较粗壮；胸呈短圆筒形，三个侧翼较小，细角锥状，开放的口缘有一些孔间桁形成的锯齿；壁孔类圆形或椭圆形，大小不等，不规则分布，孔径与孔间桁宽近等，孔间桁较宽，各节点上有凸起，壳表不太平整，无棘刺。

标本测量：壳长 105–140μm，壳宽（头与胸）90–120μm，头长 45–70μm，胸长 60–80μm，顶角长 50–80μm，侧翼长 15–25μm。

模式标本：DSDP274-23-3 1（图版 99，图 17–18），来自南极罗斯海的 DSDP Leg 28 航次 Site 274 孔的 274-23-3 岩心样品中，保存在中国科学院南海海洋研究所沉积标本室。

地理分布 南极罗斯海，早始新世—中渐新世。

该种特征与 *Lithomelissa ministyla* sp. nov.较接近，主要区别是后者壳壁较薄，顶针细小，有口缘环，无锯齿，壳表较平滑。

332. 美丽石蜂虫（新种）*Lithomelissa callifera* sp. nov.

（图版 99，图 21–25）

Lophophaenoma sp., Petrushevskaya, 1971d, p. 117, figs. 63III–IV.

头近圆球形，内部骨针结构发育，一根垂直向上形成细柱状或角锥状的顶针，一根向下延伸至壳口，三根侧向形成胸肋；胸壳近圆筒形，胸肋向外成细柱状骨针，各肋之间的壳壁略膨凸，口部开阔，有的标本口端收缩，并由桁枝连接成穿孔状，似有三个基脚；整个壳体壁薄，壁孔较大，类圆形或椭圆形，不规则排列，壳表很光滑，无棘刺。

标本测量：壳长 105–160μm、宽 95–140μm，头径 50–80μm，胸长 55–95μm，顶针长 40–85μm，侧翼长 25–40μm。

模式标本：DSDP274-10-3 1（图版 99，图 23–24），来自南极罗斯海的 DSDP Leg 28 航次 Site 274 孔的 274-10-3 岩心样品中，保存在中国科学院南海海洋研究所沉积标本室。

地理分布 南极罗斯海，早上新世。

该新种区别于其他各种的主要特征为具有完善发达的头部骨针结构及其衍生物，胸肋之间的壳壁有膨凸，表面光滑无刺。

333. 头石蜂虫 *Lithomelissa capito* Ehrenberg

（图版 99，图 26–27）

Lithomelissa capito Ehrenberg, 1874, p. 240；Ehrenberg, 1876, p. 78, pl. 3, fig. 14.
Dictyocephalus ? *obtusus* Ehrenberg, Bütschli, 1882, pl. 31, figs. 10a–b.

地理分布 南极罗斯海，出现于中渐新世。

334. 挑战石蜂虫 *Lithomelissa challengerae* Chen

（图版 100，图 1–6）

Lithomelissa challengerae Chen, 1975, p. 457, pl. 8, fig. 3；Takemura, 1992, p. 744, pl. 4, figs. 11–12.

地理分布 南极罗斯海，晚始新世—中渐新世。

335. 强针石蜂虫（新种）*Lithomelissa cratospina* sp. nov.

（图版 100，图 7–17）

Lithomelissa sphaerocephalis Chen, Hollis et al., 1997, p. 52, pl. 3, figs. 12–14.
Botryometra poljanskii Petrushevskaya, 1975, p. 588, pl. 26, fig. 13.

壳分两节，头近圆球形，胸钟罩形，胸与头间缩缢较深或有一颈，胸宽略大于头宽（直径），胸较短，口部开放，残缺状，无口缘；顶针呈三棱角锥状，较粗实强壮；壁孔类圆形，大小接近，分布不规则，孔径多数与孔间桁宽近等；孔间桁的节点上有小棘凸，无刺，壳表略显粗糙。

标本测量：壳长 105–170μm、宽 110–130μm，头径 80–95μm，胸长 60–100μm，顶针长 65–75μm。

模式标本：DSDP274-10-3 2（图版 100，图 11–13），来自南极罗斯海的 DSDP Leg 28 航次 Site 274 孔的 274-10-3 岩心样品中，保存在中国科学院南海海洋研究所沉积标本室。

地理分布 南极罗斯海，晚始新世。

该新种特征与 *Lithomelissa sphaerocephalis* Chen 较接近，主要区别在于后者的头部壁孔不明显，顶针较为短小，而且壳表有海绵状覆盖物。

336. 厄伦伯石蜂虫 *Lithomelissa ehrenbergii* Bütschli

（图版 101，图 1–23）

Lithomelissa ehrenbergii Bütschli, 1882, p. 519, pl. 33, figs. 21a–b；Haeckel, 1887, p. 1204；Hollis, 1997, pp. 51–52, pl. 3, figs. 17–20.

Lithomelissa(?) *ehrenbergii* Dumitrica, 1973, p. 837, pl. 25, figs. 6–7.

Lithomelissa aff. *mitra* Bütschli, Chen, 1975, p. 458, pl. 8, figs. 4–5.

Gondwanaria japonica (Nakaseko) group, Petrushevskaya, 1975, pl. 21, figs. 4–5 (only).

Lithomelissa sp., Hollis, 2002, pl. 4, fig. 2.

该种的鉴定特征存在较大疑惑，一般胸近圆筒状，顶角和侧翼呈三棱角锥状，壁薄，表面光滑；头部球形或半球形，颈部明显或无，顶角和侧翼长或短，形态上有一定的变化，似乎与其他相近种的特征区别也不大。此类标本需要进一步的种类归纳与厘清。

地理分布 南极罗斯海，晚始新世—中渐新世。

337. 厄伦伯石蜂虫（亲近种）*Lithomelissa* sp. A aff. *L. ehrenbergi* Bütschli

（图版 101，图 24–25）

Lithomelissa sp. A aff. *L. ehrenbergi* Bütschli, Chen, 1975, p. 458, pl. 11, figs. 1–2.

地理分布 南极罗斯海，出现于早上新世。

338. 笑石蜂虫 *Lithomelissa gelasinus* O'Connor

（图版 102，图 1–4）

Lithomelissa gelasinus O'Connor, 1997a, p. 69, text-fig. 4, pl. 2, figs. 3–6, pl. 6, figs. 6–9；O'Connor, 2000, p. 206, pl. 1, figs. 7–9.

地理分布 南极罗斯海，中渐新世。

339. 圆头石蜂虫（新种）*Lithomelissa globicapita* sp. nov.

（图版 102，图 5–12）

Lithomelissa sp., Petrushevskaya, 1971d, p. 90, figs. 46I–V.

壳体较小，中等壁厚，表面光滑；头部很大，近圆球形，壁孔均较小，类圆形，无规律散布，孔间桁较宽，头部的顶针不发育，或有个别较小的侧向骨针；胸部很短，呈残缺状，与头壳近等宽，两者间缩缢较深，壁孔特征与头部壁孔相似，口部开放，不平整。

标本测量：壳长 110–130μm，头长 90–100μm、宽 80–100μm，胸长 20–50μm、宽 80–90μm。

模式标本：DSDP274-33-3 2（图版 102，图 11–12），来自南极罗斯海的 DSDP Leg 28 航次 Site 274 孔的 274-33-3 岩心样品中，保存在中国科学院南海海洋研究所沉积标本室。

地理分布 南极罗斯海，晚始新世。

该新种特征与 *Lithomelissa thoracites* Haeckel 较接近，主要区别在于后者的胸宽明显大于头部，壁孔较大或孔间桁较细，壳表略粗糙。罗斯海的标本特征与 Petrushevskaya（1971d, p. 90, figs. 46I–V）完全一致。

340. 哈克石蜂虫（相似种）*Lithomelissa* cf. *haeckeli* Bütschli

（图版 102，图 13–18）

Lithomelissa cf. *haeckeli* Bütschli, Petrushevskaya, 1971d, p. 89, fig. 45II.
Lithomelissa sp., Petrushevskaya, 1971d, p. 89, figs. 45III–VI.
Lithomelissa sp. aff. *L. haeckeli* Bütschli, Petrushevskaya, 1975, p. 592, pl. 9, figs. 24–26, pl. 24, fig. 8.

地理分布 南极罗斯海，晚始新世—中渐新世。

341. 燕形石蜂虫（新种）*Lithomelissa hirundiforma* sp. nov.

（图版 102，图 19–22）

个体较小，壳呈圆锥形；头为半球形，较小，头与胸之间的缩缢不明显，有一个较短的角锥形顶角；胸为圆锥形或下部近圆筒形，肩部下斜，三个胸肋在胸壳的中下部继续斜下延伸为三个较为短小的三棱角锥状侧翼，胸口开放，近截平，有一细环状口缘；整个壳体壁稍厚，壁孔类圆形或椭圆形，较小，靠近开口处略增大，分布不规则，孔间桁稍粗，有节点凸起，壳壁稍显粗糙，无棘刺。

标本测量：壳长 105–125μm、宽 85–100μm，头宽 40–50μm，胸长 80–95μm，顶针长 15–25μm，侧翼长 40–50μm。

模式标本：DSDP274-12-3 2（图版 102，图 21–22），来自南极罗斯海的 DSDP Leg 28 航次 Site 274 孔的 274-12-3 岩心样品中，保存在中国科学院南海海洋研究所沉积标本室。

地理分布 南极罗斯海，晚中新世。

该种与 *Lithomelissa ehrenbergii* Bütschli 的主要区别是后者个体及头部较大，头近圆球形，顶针和侧翼均较为粗壮，胸呈圆筒状，无口缘。

342. 侧头石蜂虫 *Lithomelissa laticeps* Jørgensen

（图版 102，图 23–26）

Lithomelissa laticeps Jørgensen, 1905, p. 136, pl. 16, figs. 84a–b.
Trisulcus nana (Popofsky), Petrushevskaya, 1971d, p. 144, pl. 73, figs. I–III.

地理分布 南极罗斯海，晚中新世—全新世。

343. 小柱石蜂虫（新种）*Lithomelissa ministyla* sp. nov.

（图版 102，图 27–35）

壳分两节，头壳呈宽半球形，胸为短圆筒形，两者宽度接近；头与胸之间的外表有缩缢，内壁有隔环，在交界处的内部中心发育数条横向和一条垂直向上的头部骨针，垂向骨针穿过头壁形成顶针，骨针与顶针均为细柱状；整个壳体壁孔相似，类圆形或椭圆形，大小不一，分布不规则，孔间桁较宽；壳表平滑，部分壳表孔缘凸起并相连成矮墙状；胸壳开口处常有一个口缘环，或已破碎不平整。

标本测量：壳长 95–145μm，头长 30–65μm、宽 65–95μm，胸长 50–80μm、宽 70–110μm。

模式标本：DSDP274-31-6 1（图版 102，图 32–33），来自南极罗斯海的 DSDP Leg 28 航次 Site 274 孔的 274-31-6 岩心样品中，保存在中国科学院南海海洋研究所沉积标本室。

地理分布 南极罗斯海，晚始新世—中渐新世。

该新种特征与 *Lithomelissa poljanskii* (Petrushevskaya)较相似，主要区别是后者的头呈圆球形，较高，顶针为三棱角锥状或细角锥状，壁孔较均匀，无口缘环。

344. 厚壁石蜂虫（新种）*Lithomelissa pachyoma* sp. nov.

（图版 103，图 1–4）

Lithomelissa sp., Petrushevskaya, 1971d, p. 89, figs. 45III–VI.

壳体较小，主要为头部，胸部不完整，壳壁较厚；头呈圆球形，较大，壁孔小而模糊，孔间桁宽厚，头腔内的垂向骨针发育形成顶针，顶针较短小，呈细柱或角锥形；胸近圆筒状，胸壳长度小于或大于头壳半径，宽度略小于或接近头的直径，壁孔类圆形，大小相似，口部近截平；整个壳体的各孔间桁节点上有较粗的锥形凸起，形成很厚的壳壁。

标本测量：壳长 100–115μm，头长 65–75μm、宽 65–80μm，胸长 25–45μm、宽 55–65μm。

模式标本：DSDP274-25-5 1（图版 103，图 1–2），来自南极罗斯海的 DSDP Leg 28 航次 Site 274 孔的 274-25-5 岩心样品中，保存在中国科学院南海海洋研究所沉积标本室。

地理分布 南极罗斯海，中渐新世—晚更新世。

该新种的头部骨针和顶针特征等与 *Lithomelissa ministyla* sp. nov.较接近，主要区别在于后者的头壳呈半球形，壳壁较薄，各孔间桁节点上无明显的锥形凸起。罗斯海的标本特征与 *Lithomelissa* sp.（Petrushevskaya, 1971d, p. 89, figs. 45III–VI）非常相似。

345. 轴神石蜂虫 *Lithomelissa poljanskii* (Petrushevskaya)

（图版 103，图 5–10）

Botryometra poljanskii Petrushevskaya, 1975, p. 590, pl. 13, figs. 9–10, pl. 21, fig. 7, pl. 26, fig. 13.

地理分布 南极罗斯海，晚始新世—早渐新世。

346. 前南极石蜂虫 *Lithomelissa preantarctica* (Pettrushevskaya)

（图版 103，图 11–20）

Saccospyris preantarctica Petrushevskaya, 1975, p. 589, pl. 13, figs. 19–20.
Ceratocyrtis sp. R, Petrushevskaya, 1975, p. 590, pl. 11, fig. 12, pl. 18, figs. 1–3, pl. 19, fig. 1.

地理分布　南极罗斯海，晚始新世—早渐新世。

347. 壮石蜂虫 *Lithomelissa robusta* Chen

（图版 103，图 21–33）

Lithomelissa robusta Chen, 1975, p. 457, pl. 9, figs. 1–2；Abelmann, 1990, p. 695, pl. 5, figs. 2A–B.
Lychnocanium (?) sp., Bjørklund, 1976, p. 1124, pl. 24, figs. 10–11.
Lophophaena mugaica (Grigorjeva), Hollis, 2002, p. 297, pl. 4, figs. 3–6.

地理分布　南极罗斯海，晚始新世—中渐新世。

348. 圆头石蜂虫 *Lithomelissa sphaerocephalis* Chen

（图版 104，图 1–7）

Lithomelissa sphaerocephalis Chen, 1975, p. 457, pl. 8, figs. 1–2；Takemura, 1992, p. 744, pl. 4, figs. 8–9；Hollis, 1997, p. 52, pl. 3, figs. 12–14.
Lithomelissa gelasinus O'Connor, 1997b, p. 71, pl. 2, figs. 3–6, pl. 6, figs. 6–9.

头部无孔，顶针长为头长的一半，壳表有海绵状覆盖物。

地理分布　南极罗斯海，中渐新世。

349. 口刺石蜂虫（新种）*Lithomelissa stomaculeata* sp. nov.

（图版 104，图 8–12）

个体稍大，中等壁厚；头近圆球形，壁孔较大，有一垂直向上的三棱角锥状顶针，还有个别侧向的骨针或刺，头与胸之间有较深缩缢；胸呈钟罩形，往下略扩大，壳体的最宽处在胸口，壳壁孔大小不一，类圆形或椭圆形，分布无规律，在胸的中部发育有三个侧翼，侧翼基部具窗孔，末端为细柱形或角锥形，口部开阔，具许多由孔间桁向下延伸形成的角锥形末端骨针，呈刺状，较长；壳表光滑或有一些小棘刺。

标本测量：壳长 150–180μm、宽 125–145μm，头径 50–60μm，胸长 90–130μm，顶针长 20–40μm，侧翼长 30–70μm，末端骨针长 20–50μm。

模式标本：DSDP274-10-3 2（图版 104，图 8–9），来自南极罗斯海的 DSDP Leg 28 航次 Site 274 孔的 274-10-3 岩心样品中，保存在中国科学院南海海洋研究所沉积标本室。

地理分布　南极罗斯海，早上新世。

该新种形态特征与 *Lithomelissa stigi* Bjørklund（1976）的差异较大，主要区别在于后者的头部和胸壳均较小，侧翼不太发育，口端骨针较少，很短。

350. 石蜂虫 *Lithomelissa thoracites* Haeckel

（图版 104，图 13–28）

Lithomelissa thoracites Haeckel, 1862, p. 301, pl. 6, figs. 2–8；Hertwig, 1879, p. 204, pl. 8, fig. 1；Popofsky, 1913, pp. 337–338, text-figs. 44–47；Benson, 1966, pp. 366–369, pl. 24, figs. 10–13；陈木宏、谭智源，1996，209 页，图版 28，图 1–4；陈木宏等，2017，190 页，图版 66，图 9–14。

Lithomelissa monoceras Popofsky, 1913, p. 335, text-fig. 43, pl. 32, fig. 7.

Antarctissa (?) sp., Bjørklund, 1976, pl. 19, figs. 10–12.

罗斯海的标本壳壁相对稍厚，形态有一定的变化。

地理分布 南极罗斯海，早始新世—中渐新世。

351. 三角石蜂虫 *Lithomelissa tricornis* Chen

（图版 105，图 1–24）

Lithomelissa tricornis Chen, 1975, p. 458, pl. 8, figs. 6–7；Abelmann, 1990, p. 695, pl. 5, fig. 3；Takemura, 1992, p. 744, pl. 2, figs. 5–6.

Lithomelissa maureenae O'Connor, 1997b, p. 72, pl. 2, figs. 7–10, pl. 6, figs. 10–11, pl. 7, figs. 1–6.

地理分布 南极罗斯海，晚始新世—早上新世。

352. 雅石蜂虫（新种）*Lithomelissa tryphera* sp. nov.

（图版 105，图 25–33）

壳钟罩形，壁薄，壁孔不规则分布，壳表或有次生网状组织；头较大，半球形，常呈透明状，壁孔模糊，头与胸之间无缩缢或很浅，但有内隔环；胸壳自上而下逐渐扩大，胸壁孔类圆形，大小不一，不规则排列，孔间桁较宽，表面平滑；头顶有一很小的突起角，或为三棱角锥状顶角，也可见三个实心侧翼；口开阔，无口缘，常不平整。该种类个体稍小，结构特征有些变化。

标本测量：壳长 115–160μm、宽 125–145μm，头长 50–60μm、宽 80–110μm，顶针长 10–35μm，侧翼长 10–30μm。

模式标本：DSDP274-28-3 1（图版 105，图 27–28），来自南极罗斯海的 DSDP Leg 28 航次 Site 274 孔的 274-28-3 岩心样品中，保存在中国科学院南海海洋研究所沉积标本室。

地理分布 南极罗斯海，晚始新世—中渐新世。

该新种的形态特征与 *Lithomelissa tricornis* Chen 较接近，主要区别在于后者的头部明显具有三个顶角，头与胸之间的缩缢较深。

353. 石蜂虫（未定种 1）*Lithomelissa* sp. 1

（图版 105，图 34–35）

壳钟罩形，头部有一顶针。壳壁较厚，壁孔小，类圆形，无规律分布，孔间桁较宽；壳表光滑无刺，口缘无骨针。

地理分布 南极罗斯海，出现于早上新世。

354. 石蜂虫（未定种 2）*Lithomelissa* sp. 2

（图版 105，图 36–37）

头较大，顶角三棱角锥状，较短，胸呈喇叭状，阔开口，不完整。

地理分布　南极罗斯海，出现于中渐新世。

南极虫属 Genus *Antarctissa* Petrushevskaya, 1967

壳分两节，头比胸略小，胸的上部常包覆了头的下部，两者紧密相扣连接。

355. 代齿南极虫 *Antarctissa antedenticulata* Chen

（图版 106，图 1–4）

Antarctissa antedenticulata Chen, 1975, p. 456, pl. 18, figs. 1–2.

地理分布　南极罗斯海，中渐新世—中中新世。

356. 包头南极虫（新种）*Antarctissa cingocephalis* sp. nov.

（图版 106，图 5–6）

壳呈子弹形，外表由胸壳及上部较厚的壳壁连续往上包覆组成，头壳完全被封闭在里面；头的上部圆球状，下部缩窄为颈，头长与颈长接近，头宽约为颈宽的两倍、胸宽的一半，头与颈一起的长度与胸长近等；胸圆筒状，肩部圆弧形，开口近截平，无口缘；头与胸的壁孔相似，均为大小不等的类圆形或椭圆形，不规则排列，孔间桁较宽，板片状；整个壳体的外表被不规则的网格结构外套包裹为一体，无缩缢，外包层的条形孔间桁较粗，较为疏松，壁孔大小不等，可见内部结构；壳体表面光滑，无棘刺。

标本测量：壳长 150μm、宽 100μm，头径 55μm，颈宽 30μm，胸长 80μm，胸宽 90μm。

模式标本：DSDP274-16-1 1（图版 106，图 5–6），来自南极罗斯海的 DSDP Leg 28 航次 Site 274 孔的 274-16-1 岩心样品中，保存在中国科学院南海海洋研究所沉积标本室。

地理分布　南极罗斯海，出现于早中新世。

该新种区别于其他种类的主要特征是具有一个完全封闭的外包层，将整个头部及胸上部包覆在内。

357. 封口南极虫 *Antarctissa clausa* (Popofsky)

（图版 106，图 7–18）

Helotholus histricosa Jørgensen var. *clausa* Popofsky, 1908, p. 281, pl. 33, fig. 1, pl. 34, fig. 1.
Antarctissa denticulata (Ehrenberg) *clausa* Popofsky, Petrushevskaya, 1967, p. 87, pl. 49, fig. 5；Petrushevskaya, 1971b, pl. 64, fig. 2.
Antarctissa clausa (Popofsky), Petrushevskaya, 1975, p. 591, pl. 24, fig. 9.

地理分布　南极罗斯海，中渐新世—全新世。

358. 同射南极虫 *Antarctissa conradae* Chen

（图版 106，图 19–28）

Antarctissa conradae Chen, 1974, p. 484, pl. 3, figs. 1–3；Chen, 1975, p. 457, pl. 17, figs. 1–5.

Botryopera conradae (Chen), Petrushevskaya, 1986, p. 193, fig. 2(6).
Lithomelissa sp. P, Petrushevskaya, 1973, p. 1044, fig. 3(12).

该种的头部有一顶针。

地理分布 南极罗斯海，晚中新世。

359. 德弗兰南极虫 *Antarctissa deflandrei* (Petrushevskaya)

（图版 106，图 29–42）

Antarctissa deflandrei (Petrushevskaya), Lazarus, 1990, p. 713, pl. 3, figs. 18–19；Abelmann, 1990, p. 694, pl. 4, figs. 10A–B.
Botryopera deflandrei Petrushevskaya, 1975, p. 592, pl. 11, figs. 30–31.
Antarctissa denticulata faceata Petrushevskaya, 1967, pp. 86–88, fig. 50IV.
Lithomelissa setosa Jørgensen, Takahashi, 1991, p. 97, pl. 25, figs. 16–19, 21–22.

该种的头部无顶针或欠发育。

地理分布 南极罗斯海，晚中新世—全新世。

360. 小齿南极虫 *Antarctissa denticulata* (Ehrenberg) group

（图版 107，图 1–50）

Lithobotrys denticulata Ehrenberg, 1844, p. 203.
Lithopera denticulata (Ehrenberg), Ehrenberg, 1873, pl. 12, fig. 7；Haeckel, 1887, p. 1083；Haecker, 1907, pp. 123–124, fig. 8.
Helotholus histricosa var. *clausa* Popofsky, 1908, pp. 281–282, pl. XXXIII, fig. 1.
Antarctissa denticulata (Ehrenberg), Petrushevskaya, 1968, pp. 84–86, figs. 49I–IV；Petrushevskaya, 1971d, p. 122, fig. 64I；Chen, 1975, p. 457, pl. 18, figs. 3–8；Lazarus, 1990, p. 713, pl. 3, figs. 1–4；Abelmann, 1992b, p. 378, pl. 3, figs. 17–18.
Antarctissa ewingi Chen, 1974, p. 486, pl. 3, figs. 4–6；Chen, 1975, p. 457, pl. 16, figs. 5–9.
Antarctissa cylindrica Petrushevskaya, 1975, p. 591, pl. 11, figs. 19–20；Lazarus, 1990, p. 713, pl. 3, figs. 8–12.
Antarctissa robusta Petrushevskaya, 1975, p. 591, pl. 11, figs. 21–22；Lazarus, 1990, p. 714, pl. 3, figs. 6–7.
Antarctissa longa (Popofsky), Keany, 1979, p. 55, pl. 3, figs. 5–6, pl. 5, figs. 11–12.

该种群包括已知的若干种，其共同特征是二节壳，头常部分陷入胸中，被后者所包裹，壳体呈圆三角形或圆柱形，壳表具刺、类海绵状或平滑，口部圆弧形或截平，一般开放，或有口盖封闭。种群的个体形态变化较大。由于各自建种的定义含糊不清或与标本不符，造成各种间的界线不确定，相互交叉，因此难以界定种间的区别，加之同为南极生态类型，归为一个种群较合适，以便鉴定、统计与应用。

地理分布 南极罗斯海，早中新世—全新世。

361. 小齿南极虫封口亚种 *Antarctissa denticulata clausa* Petrushevskaya

（图版 108，图 1–17）

Antarctissa denticulata clausa Petrushevskaya, 1971d, p. 122, figs. 64II.

地理分布 南极罗斯海，晚上新世—全新世。

362. 长南极虫 *Antarctissa longa* (Popofsky)

（图版 108，图 18–40）

Helotholus longus Popofsky, 1908, pp. 282–283, pl. 34, fig. 2.
Antarctissa longa (Popofsky) Petrushevskaya, 1967, p. 91, pl. 51, fig. 1；Petrushevskaya, 1968, p. 90, fig. 51I；Petrushevskaya, 1975,

p. 591, pl. 11, figs. 8–10, pl. 18, fig. 6；Petrushevskaya, 1986, p. 191, pl. 1, fig. 11.
Antarctissa strelkovi Petrushevskaya, Lazarus, 1990, p. 713, pl. 3, figs. 13–15.

地理分布　南极罗斯海，早渐新世—全新世。

363. 厚壁南极虫（新种）*Antarctissa pachyoma* sp. nov.

（图版 109，图 1–11）

二节壳，呈子弹头形，壳壁较厚；头部较大，近圆球形、半球形或椭球形，与胸宽近等，下部常被胸壁包覆，两者间在壳内有缩缢，但外表却平滑过渡；胸壳较短，胸长小于头长和胸宽，口部开放，基本截平，但无口缘环和齿；壁孔类圆形，大小不一，分布无规律，头部与胸部的壁孔特征基本一致，无明显变化；整个壳体表面平滑无刺，无顶角，无肋翼。

标本测量：壳长 125–175μm、宽 100–130μm，头径 85–110μm，胸长 30–60μm。

模式标本：DSDP274-10-3 1（图版 109，图 6–7），来自南极罗斯海的 DSDP Leg 28 航次 Site 274 孔的 274-10-3 岩心样品中，保存在中国科学院南海海洋研究所沉积标本室。

地理分布　南极罗斯海，早渐新世—早上新世。

该种特征与 *Antarctissa antedenticulata* Chen 较接近，主要区别在于后者个体稍小，不呈子弹形，壳壁较薄，外表上头与胸之间存在较深的缩缢，开口处有一个较细的口缘环。

364. 翼南极虫（新种）*Antarctissa pterigyna* sp. nov.

（图版 109，图 12–15）

Antarctissa (?) sp., Bjørklund, 1976, pl. 19, figs. 10–12.

个体稍大，壳壁较厚，无顶针；头呈圆球形，较大，头径与上部胸宽近等，孔间桁较粗，有些叠交，壁孔模糊不清或类圆形，较小，各孔间桁节点上有小锥形凸起；头与胸间有领隔或较深的缩缢，胸部较短，近圆筒形，上部较窄，往下略变宽，在胸上部发育有三个具窗孔的三角形侧翼，胸壁格孔状，壁孔类圆形或不规则形，大小不同，分布无规律，孔间桁较宽，部分不规则，表面光滑，壳体最宽处在口端，口部开放，残缺不平整；壳体表面无棘刺。

标本测量：壳长 170–190μm、宽 115–125μm，头直径 85–100μm，胸长 80–115μm。

模式标本：DSDP274-32-3 2（图版 109，图 14–15），来自南极罗斯海的 DSDP Leg 28 航次 Site 274 孔的 274-32-3 岩心样品中，保存在中国科学院南海海洋研究所沉积标本室。

地理分布　南极罗斯海，晚始新世。

该新种特征与 *Antarctissa antedenticulata* Chen 较接近，主要区别在于后者胸部无侧翼，口部截平，部分有筛板，壳表平滑。

365. 史科南极虫 *Antarctissa strelkovi* Petrushevskaya

（图版 109，图 16–29）

Antarctissa strelkovi Petrushevskaya, 1967, pp. 88–90, figs. 51III–VI.

Peromelissa denticulata (Ehrenberg), Haecker, 1908a, pp. 448–452, pl. 84, figs. 582, 583, 591；Riedel, 1958, p. 236, pl. 3, fig. 9.

口部封闭，壳表有刺。

地理分布 南极罗斯海，早—晚上新世。

366. 威特南极虫 *Antarctissa whitei* Bjørklund

（图版 109，图 30–33）

Antarctissa whitei Bjørklund, 1976, p. 1125, pl. 13, figs. 9–14.

地理分布 南极罗斯海，全新世。

海绵蜂虫属 Genus *Spongomelissa* Haeckel, 1887

三个实心侧翼从胸节的侧面长出，无末端脚，有一个或更多头角。

367. 阔口海绵蜂虫（新种）*Spongomelissa chaenothoraca* sp. nov.

（图版 109，图 34–35）

Spongomelissa sp., Chen, 1975, p. 458, pl. 10, fig. 4；Bjørklund, 1976, p. 1124, pl. 22, figs. 6–9.
Lithomelissa sp., Hollis, 2002, p. 297, pl. 4, figs. 1–2.

壳近钟罩形，有一个顶角和三个侧翼；头半球形，较为宽大，顶角自头腔内的中心垂直骨针延伸形成，呈三棱角锥状；胸近圆筒形，口部开阔，常破损为残留状，自胸壳上部生出的侧翼为实心状，其基部有窗孔；壳壁较薄或稍厚，壁孔类圆形，大小接近或不等，排列不规则，孔间桁稍细；壳表光滑或有小棘刺，不同海区的标本特征略有差异。

标本测量：壳长大于 125μm，壳宽 115μm，头长 70–75μm、宽 95–110μm，胸长不完整，顶角长 45–55μm。

模式标本：DSDP274-20-3 1（图版 109，图 34–35），来自南极罗斯海的 DSDP Leg 28 航次 Site 274 孔的 274-20-3 岩心样品中，保存在中国科学院南海海洋研究所沉积标本室。

地理分布 南极罗斯海，中渐新世；西南太平洋，晚古新世—始新世；挪威海，渐新世。

该新种的基本特征与 *Spongomelissa cucumella* Sanfilippo and Riedel 较接近，主要区别在于后者的个体较小，口部收缩，侧翼较长。

368. 小瓜海绵蜂虫 *Spongomelissa cucumella* Sanfilippo and Riedel

（图版 110，图 1–28）

Spongomelissa cucumella Sanfilippo and Riedel, 1973, p. 530, pl. 19, figs. 6–7, pl. 34, figs. 7–10；陈木宏、谭智源，1996，210 页，图版 28，图 8；陈木宏等，2017，191 页，图版 66，图 21–24。
Dictyophimus cf. *platycephalus* Haeckel, Abelmann, 1990, p. 696, pl. 7, fig. 11.
Pseudodictyophimus bicornis (Ehrenberg), Okazaki et al., 2005b, p. 2250, pl. 9, fig. 6.

地理分布 南极罗斯海，晚始新世—晚更新世。

罩巾虫属 Genus *Velicucullus* Riedel and Campbell, 1952

胸呈阔钟形或盘形，边缘板状或表面海绵状；头部分若干脑叶。

369. 高罩巾虫 *Velicucullus altus* Abelmann

（图版 110，图 29–30）

Velicucullus altus Abelmann, 1990, p. 698, pl. 8, figs. 5A–C；Abelmann, 1992a, p. 777, pl. 3, fig. 8；Takemura and Ling, 1997, p. 114, pl. 1, fig. 6.

个体较大，盘缘扁平，有海绵组织。

地理分布　南极罗斯海，出现于早上新世。

370. 奥古罩巾虫 *Velicucullus oddgurneri* Bjørklund

（图版 111，图 1–13）

Velicucullus oddgurneri Bjørklund, 1976, p. 1126, pl. 19, figs. 6–9.
Lampromitra coronata Keany, 1979, p. 56, pl. 4, fig. 10, pl. 5, fig. 14.
Velicucullus cf. *oddgurneri* (Bjørklund), Abelmann, 1990, p. 698, pl. 8, fig. 6；Abelmann, 1992a, p. 777, pl. 3, figs. 9–10.
Squinabolella putahensis Pessagno, Ling, 1991, p. 320, pl. 1, fig. 13.

地理分布　南极罗斯海，晚始新世—早上新世。

371. 海绵罩巾虫（新种）*Velicucullus spongiformus* sp. nov.

（图版 112，图 1–5）

壳呈钟形或斗笠形，整个壳体均由海绵组织构成；头呈半球形或圆筒形，顶部开窗，或有一个半球形状的格孔盖，无顶针与棘刺；胸呈阔圆锥形，自上而下迅速扩大，阔开口，在口缘有一些实心的放射骨针；壳体表面无棘刺。

标本测量：头长 50–80μm、宽 45–125μm，胸长 50–100、宽 90–260μm。

模式标本：DSDP274-29-1 2（图版 112，图 4–5），来自南极罗斯海的 DSDP Leg 28 航次 Site 274 孔的 274-29-1 岩心样品中，保存在中国科学院南海海洋研究所沉积标本室。

地理分布　南极罗斯海，晚古新世—晚始新世。

该新种的壳体形态与 *Velicucullus altus* Abelmann（1992a, p. 777, pl. 3, fig. 8）较接近，但后者的壳体中-上部为格孔壳，仅外围部分为海绵结构，且口缘无骨针。

笠虫属 Genus *Helotholus* Jørgensen, 1905

具一头棘和四胸棘，胸腔内有一中央杆。

372. 渐大笠虫 *Helotholus ampliata* (Ehrenberg)

（图版 115，图 3–6）

Cornutella ampliata Ehrenberg, 1873, p. 221；Ehrenberg, 1876, pl. 2, fig. 5.
Sethoconus galea Cleve, 1899, p. 33, pl. 4, fig. 3.

地理分布 南极罗斯海，晚始新世。

373. 灯泡笠虫（新种）*Helotholus bulbosus* sp. nov.

（图版 115，图 7–8）

壳呈灯泡形，上部膨大，下部缩小；头呈圆球形，较大，无顶针，壳内的头骨针发育，在头与胸的交界处形成多孔形骨架，个别斜向伸出壳外形成细圆柱形的头壳侧针，中心骨针（中央杆）垂直向下延长直接伸出胸口发育成三棱角锥状的唯一“基脚”；胸呈圆筒形，较短，胸宽度仅为头径的一半，胸长是胸宽的 1/3，口部开放，无挡板，口近截平，口缘不平整；整个壳体壁稍薄，壁孔为小类圆形，不规则分布，孔间桁较宽，孔径与桁宽近等；壳表略显起伏不平，无棘刺。

标本测量：壳长 155–165μm，头直径 125–135μm，胸长 25–30μm、宽 75–85μm，基脚长 40–45μm。

模式标本：DSDP274-23-5 1（图版 115，图 7–8），来自南极罗斯海的 DSDP Leg 28 航次 Site 274 孔的 274-23-5 岩心样品中，保存在中国科学院南海海洋研究所沉积标本室。

地理分布 南极罗斯海，出现于中渐新世。

该种与 *Lophophaena*? *orbicularis* sp. nov. 的主要区别在于后者的胸壳发育不完善，头部骨针不发育，壳表平滑。

374. 笠虫 *Helotholus histricosa* Jørgensen

（图版 115，图 9–12）

Helotholus histricosa Jørgensen, 1905, p. 137, pl. 16, figs. 86–88.
Ceratocyrtis sp. aff. *C. histricosus* (Jørgensen), Bjørklund, 1976, p. 1124, pl. 15, figs. 6–8, pl. 18, figs. 18–21.

地理分布 南极罗斯海，晚始新世。

375. 前方笠虫 *Helotholus praevema* Weaver

（图版 115，图 13–27）

Helotholus praevema Weaver, 1983, pp. 677–678, pl. 3, figs. 1, 5–15.
Ceratocyrtis amplus (Popofsky), Petrushevskaya, 1975, p. 590, pl. 11, figs. 3–6, 13, pl. 19, fig. 2.

地理分布 南极罗斯海，中渐新世—晚更新世。

376. 万笠虫 *Helotholus vema* Hays

（图版 116，图 1–18）

Helotholus vema Hays, 1965, p. 176, fig. A, pl. 2, fig. 3；Chen, 1975, p. 497, pl. 16, figs. 1–4；Weaver, 1976, p. 580, pl. 1, figs. 10–11.
Trisulcus nana (Popofsky), Petrushevskaya, 1971d, p. 144, pl. 73, figs. I–III.
Trisulcus borealis (Ehrenberg), Petrushevskaya, 1971d, p. 146, figs. 74II–IV (only).
Pseudocubus vema (Hays), Keany and Kennett, 1972, p. 539, fig. 4(10–11).
Helotholus ? *hays* Lazarus, 1992, p. 797, pl. 8, figs. 1–17.

地理分布 南极罗斯海，早上新世—全新世。

377. 小万笠虫（新种）*Helotholus vematella* sp. nov.

（图版 116，图 19–26）

个体较小；头较小，呈半球形，部分陷入胸壳内，缩缢较深，有 2–3 个较短的角锥形顶针与侧针；胸近圆筒形，口部略收缩，胸壁上有个别较短的角锥形骨针；头腔内的一根中心骨针延伸至胸口处，形成一个水平状的多孔骨架封盖胸口，因此开口处不完全开放，口端近截平；整个壳壁较薄，壁孔较小，类圆形或椭圆形，大小不等，不规则排列，孔径与孔间桁宽近等或较小，孔间桁较宽。

标本测量：壳长 95–120μm，头长 25–40μm、宽 40–60μm，胸长 55–75μm、宽 95–115，口宽 50–65μm。

模式标本：DSDP274-28-5 1（图版 116，图 19–20），来自南极罗斯海的 DSDP Leg 28 航次 Site 274 孔的 274-28-5 岩心样品中，保存在中国科学院南海海洋研究所沉积标本室。

地理分布　南极罗斯海，晚始新世—早渐新世。

该种标本特征与 *Helotholus vema* Hays 有些接近，主要区别在于后者个体相对较大，壳壁较厚，壁孔较大，口部无收缩，可能分布于较新的地层中。

石囊虫属 Genus *Lithopera* Ehrenberg, 1847

头的下部陷入胸的顶部，有一顶角，三个放射肋仅发育在胸腔内，壳口封闭（胸的底部为格孔封闭物）。

378. 石囊虫（未定种 1）*Lithopera* sp. 1

（图版 116，图 27–28）

壳体较小，头圆球形，部分陷入胸腔，胸近球形或椭球形，口部封闭；壳壁稍厚，壁孔较小，类圆形，大小相近，亚规则排列，孔径略小于孔间桁宽，壳表光滑无其他附属物。该未定种壳体结构简单，个体小，可能是 *Lithopera* 属的早期类型。

地理分布　南极罗斯海，出现于晚始新世。

双孔编虫属 Genus *Amphiplecta* Haeckel, 1881

具三条内胸肋，被包围在胸部的网格内。头顶为一大开口，由一冠状刺所环绕。

379. 管双孔编虫（新种）*Amphiplecta siphona* sp. nov.

（图版 116，图 29–38）

两节壳均呈圆筒形；头呈短筒形，顶部开放，近截平，常残缺不平整，开口处的缘壁上无向上的骨针；胸近圆筒形，肩部圆弧状，两侧平直不扩大，胸口开放，口缘残缺状，不平整；头与胸的壳壁稍厚，格孔状，壁孔类圆形或不规则形，大小不一，排列不规则，孔间桁也宽窄不一，因不同标本而异；壳体表面一般平滑，但有的标本壁孔边缘生出一些角锥状小刺，使表面显粗糙。

标本测量：壳长 125–190μm，头长 55–80μm、宽 55–110μm，胸长 65–125μm、宽 125–150μm。

模式标本：DSDP274-32-1 1（图版 116，图 29–30），来自南极罗斯海的 DSDP Leg 28 航次 Site 274 孔的 274-32-1 岩心样品中，保存在中国科学院南海海洋研究所沉积标本室。

地理分布 南极罗斯海，晚始新世。

该新种特征与 *Amphiplecta acrostoma* Haeckel（1887, p. 1223, pl. 97, fig. 10）有些接近，但后者的头壁开口处有一些向上的骨针，胸壳呈圆锥形，胸口扩展；与 *Amphiplecta* sp.（Petrushevskaya, 1971d, p. 104, fig. 54）的区别是后者胸壳呈扩大的圆锥盘形。

380. 双孔编虫（未定种 1）*Amphiplecta* sp. 1

（图版 116，图 39–40）

个体较大，壳壁呈海绵状。

地理分布 南极罗斯海，出现于晚始新世。

灯犬虫属 Genus *Lychnocanoma* Haeckel, 1887, emend. Foreman, 1973

具三个实心的末端脚，无胸肋，头有一顶角，三个基脚向下延伸，轻微外弯，腹部不完整或不发育。Haeckel（1887, p. 1224）将原定属 *Lychnocanium* Ehrenberg（1847）分为三个亚属 Subgenus 1. *Lychnocanella* Haeckel、Subgenus 2. *Lychnocanissa* Haeckel 和 Subgenus 3. *Lychnocanoma* Haeckel。然而，在没有原则说明的情况下，Foreman（1973, p. 437）将后一亚属视为属级 Genus *Lychnocanoma*，并沿用至今。这是放射虫分类历史上的一个典型混乱范例。实际上，*Lychnocanoma* 属应等同于 *Lychnocanium* 属。

381. 美灯犬虫 *Lychnocanoma bellum* (Clark and Campbell)

（图版 117，图 1–10）

Lychnocanium bellum Clark and Campbell, 1942, p. 72, pl. 9, figs. 35, 39.

该种特征与 *Lychnocanoma conica* (Clark and Campbell)较相似，作者在同时建立这两个种时关于头与胸的描述以及图版均未能清晰地给予相互区别，存在模糊不清之处，后人引用中也存在此问题。实际上，这些标本中有着两类不同特征。在此，为了更好区别和辨认，认为本种的三个基脚一般较短、不发育或折断，而后一种的基脚常稍长或发育完整。

地理分布 南极罗斯海，中古新世—全新世。

382. 圆锥灯犬虫 *Lychnocanoma conica* (Clark and Campbell)

（图版 117，图 11–14）

Lychnocanium conicum Clark and Campbell, 1942, p. 71, pl. 9, fig. 38.

Lychnocanoma conica (Clark and Campbell), Petrushevskaya, 1975, p. 583, pl. 12, figs. 2, 11–15；Abelmann, 1990, p. 697, pl. 6, fig. 8, pl. 7, figs. 1A–B；Takemura, 1992, p. 747, pl. 2, figs. 13–14.

Pterocanium korotnevi (Dogiel), Kling, 1973, p. 638, pl. 4, figs. 1–4, pl. 10, figs. 6–9.

Lychnocanoma sphaerothorax Weaver, 1976, p. 581, pl. 5, figs. 4–5.

Sethochytris babylonis (Clark and Campbell), Hollis, 1997, p. 65, pl. 5, fig. 32.
Lychnocanium cf. *conicum* (Clark and Campbell), Hollis, 1997, p. 64, pl. 5, figs. 40–41.

地理分布　南极罗斯海，晚中新世。

383. 秀灯犬虫（新种）*Lychnocanoma eleganta* sp. nov.

（图版 117，图 15–24）

Lychnocanium sp. aff. *L. grande* Campbell and Clark, Chen, 1975, p. 462, pl. 1, figs. 6–7；Petrushevskaya, 1975, pl. 12, fig. 4.
Lychnocanoma sp. B, Abelmann, 1990, p. 697, pl. 7, figs. 2A–B.

头近球形，壁孔很细，顶角角锥形，较短；胸呈半球形，略膨凸，壁孔类圆形，大小相近，规则或亚规则排列，有六角形框架，孔径约为孔间桁宽的 2–3 倍，胸口开阔，无附属物，口缘环较细；末端脚呈三棱角锥状，沿壳壁方向斜下生长，较直，或在末端略内敛与微外弯，脚长与胸长近等；壳体的壳壁稍薄，壳表光滑，无棘刺。

标本测量：头直径 35–50μm，胸长 85–100μm、宽 115–130μm，顶角长 15–50μm，基脚长 85–110μm。

模式标本：DSDP274-12-3 2（图版 117，图 15–16），来自南极罗斯海的 DSDP Leg 28 航次 Site 274 孔的 274-12-3 岩心样品中，保存在中国科学院南海海洋研究所沉积标本室。

地理分布　南极罗斯海，晚始新世—晚中新世。

该新种与 *Lychnocanoma grande* (Campbell and Clark)的主要区别在于后者的壳壁较厚，三个基脚非常粗壮，且长度明显大于胸壳的长度。新种在罗斯海出现于较老的地层中。

384. 大灯犬虫 *Lychnocanoma grande* (Campbell and Clark) group

（图版 118，图 1–10）

Lychnocanium grande Campbell and Clark, 1944a, p. 42, pl. 6, figs. 3, 4, 6；Petrushevskaya and Kozlova, 1972, p. 553, pl. 29, fig. 6；Kling, 1973, p. 637, pl. 10, figs. 10–14；Lazarus, 1990, p. 717, pl. 7, fig. 9.
Lychnocanoma sp. cf. *grande* Campbell and Clark, Kling, 1973, pl. 4, figs. 9–10.
Lychnocanium grande Campbell and Clark group, Petrushevskaya, 1975, p. 583, pl. 12, figs. 5–6.

地理分布　南极罗斯海，早上新世。

385. 大灯犬虫皱亚种 *Lychnocanoma grande rugosum* (Riedel)

（图版 118，图 11–12）

Lychnocanium grande rugosum Riedel and Campbell, 1952, p. 6, pl. 1, fig. 1; Hays, 1965, p. 175, pl. 3, fig. 5; Weaver, 1976, p. 581, pl. 2, fig. 10, pl. 9, fig. 5.

地理分布　南极罗斯海，出现于中渐新世。

386. 大孔灯犬虫（新种）*Lychnocanoma macropora* sp. nov.

（图版 118，图 13–14）

个体稍大，钟罩形；头部不规则，一半是凸起的小球形，另一半为低矮丘形，壁孔较少，大小相近，不规则分布，顶角短小或不发育；胸呈长钟罩形，壁孔圆形或椭圆形，较大，一般在上部稍小，往下渐大，亚规则排列，具六角形框架，孔径为孔间桁宽的

3–4 倍，横向有 4–5 孔，无胸肋，阔开口近截平，无口缘环；在口端发育有三个斜下的笔直基脚，为圆柱形，末端缩尖，基脚长度小于胸长；壳壁稍厚，在孔间桁各节点上有锥形凸起，使壳表粗糙，但无棘刺。

标本测量：头长 45μm、宽 55–60μm，胸长 150–160μm、宽 125–150μm，基脚长 115–125μm。

模式标本：DSDP274-25-5 1（图版 118，图 13–14），来自南极罗斯海的 DSDP Leg 28 航次 Site 274 孔的 274-25-5 岩心样品中，保存在中国科学院南海海洋研究所沉积标本室。

地理分布 南极罗斯海，出现于中渐新世。

该新种以头呈脑丘状、胸壳较长不膨凸和壁孔较大而少等特征区别于其他种类。

花篮虫科 Family Anthocyrtidae Haeckel, 1887

壳两节，被一横缢分成头胸二部，具 4–9 条或更多的放射肋。

筛锥虫属 Genus *Sethopyramis* Haeckel, 1881

角锥形的胸壁上有很多直或微弯的放射肋，织网结构简单，无头角。

387. 方筛锥虫 *Sethopyramis quadrata* Haeckel

（图版 119，图 1–23）

Sethopyramis quadrata Haeckel, 1887, p. 1254, pl. 54, fig. 2；谭智源、宿星慧，1982，173 页，图版 16，图 6–7；陈木宏、谭智源，1996，212 页，图版 29，图 2–3，图版 50，图 9；陈木宏等，2017，199 页，图版 71，图 1。

Bathropyramis woondringi Campbell and Clark, Kling, 1973, p. 635, pl. 2, figs. 20–23, pl. 9, figs. 4–7.

Bathropyramis quadrata Haeckel, van de Paverd, 1995, p. 252, pl. 75, fig. 15 (not 16, 17).

Bathropyramis magnifica (Clark and Campbell), Hollis, 2002, p. 303, pl. 6, fig. 6 (only).

放射肋挺直，角锥形壳，具方形孔，无次生网，壳表光滑无刺。

地理分布 南极罗斯海，早古新世—中渐新世。

388. 筛锥虫（未定种 1）*Sethopyramis* sp. 1

（图版 120，图 1–5）

壳的上部呈角锥状，下部呈圆筒状；下部出现个别次生放射肋，有些水平桁不连续或交错，与 *Sethopyramis quadrata* Haeckel 有明显区别。

地理分布 南极罗斯海，晚始新世。

裹锥虫属 Genus *Peripyramis* Haeckel, 1881

具双层细长的锥形壳，有一些放射桁。格孔壳被外部的蛛网状或海绵状物所包围。

389. 围裹锥虫 *Peripyramis circumtexta* Haeckel

（图版 120，图 6–14）

Peripyramis circumtexta Haeckel, 1887, p. 1162, pl. 54, fig. 5; Riedel, 1958, p. 231, pl. 2, figs. 8–9; Benson, 1966, pp. 426–427, pl. 29, fig. 4; Petrushevskaya, 1968, pp. 111–112, figs. 64I–II, 65I–II; Kling, 1973, p. 637, pl. 2, figs. 15–19; Weaver, 1976, p. 582, pl. 1, fig. 5；Weaver, 1983, p. 678, pl. 1, fig. 1.

头球形，很小，壳角锥形；壳孔方形，渐增大，无次生桁；蛛网状外壳破损或消失，仅留各节点凸起的末端痕迹。

地理分布　南极罗斯海，中渐新世—全新世。

梯锥虫属 Genus *Bathropyramis* Haeckel, 1881

壳简单，呈细长或宽阔的角锥体，头小或退化消失，有一些放射桁，网孔简单，方形，较大。

390. 分枝梯锥虫 *Bathropyramis ramosa* Haeckel

（图版 121，图 1–21）

Bathropyramis ramosa Haeckel, 1887, p. 1161, pl. 54, fig. 4.
Bathropyramis quadrata (Haeckel), van de Paverd, 1995, p. 252, pl. 75, figs. 16–17 (not 15).

罗斯海标本特征与该种较为接近，但该种在各节点上的凸起末端有分叉，而罗斯海标本的分叉不明显，而且在胸下部的壳孔上还可见一些次生桁。

地理分布　南极罗斯海，中始新世—中渐新世。

391. 扩口梯锥虫 *Bathropyramis tenthorium* (Haeckel)

（图版 122，图 1–2）

Litharachnium tenthorium Haeckel, 1862, p. 281, pl. 4, figs. 7–10；Haeckel, 1887, p. 1163；Jørgensen, 1905, p. 138, pl. 16, fig. 91.

壳体略呈喇叭形，有一些不连续的次级肋将壳孔分隔成更小的方形孔，壳表光滑，无顶角。

地理分布　南极罗斯海，出现于晚始新世。

392. 光面梯锥虫 *Bathropyramis trapezoides* Haeckel

（图版 122，图 3–4）

Bathropyramis trapezoides Haeckel, 1887, p. 1160, pl. 54, fig. 3.

放射肋弯曲，胸下部外扩，壳呈喇叭状。方形孔，无次生网，壳表光滑。

地理分布　南极罗斯海，早渐新世。

盾虫属 Genus *Aspis* Nishimura, 1992

壳二或三节，头部表面粗糙，有散开的管或孔，常具一顶角，胸圆柱形或纺锤形（腹常缺失）。

393. 小瓶盾虫（新种）*Aspis ampullarus* sp. nov.

（图版 122，图 5–6）

个体很小，呈酒瓶形；头球形，部分陷入胸顶，有一个粗壮的角锥状顶角，顶角基宽与头径近等；胸的上部呈圆锥形，中-下部为圆筒形，壳壁稍厚，上部壁孔类圆形，较小，下部壁孔椭圆形，稍大，亚规则排列，横向有 5–6 孔，孔径是孔间桁宽的 1–2 倍，孔间桁较宽；口端截平，壳壁稍厚，壳表略粗糙，无棘刺。

标本测量：壳长 105μm，头直径 25μm，胸宽 65–75μm，顶角长 35–40μm，顶角基宽 20μm。

模式标本：DSDP274-40-1 1（图版 122，图 5–6），来自南极罗斯海的 DSDP Leg 28 航次 Site 274 孔的 274-40-1 岩心样品中，保存在中国科学院南海海洋研究所沉积标本室。

地理分布 南极罗斯海，中古新世。

该种以个体小，瓶形，顶角粗，壁孔少为典型特征，区别于其他种类。

394. 枝角盾虫（新种）*Aspis cladocerus* sp. nov.

（图版 122，图 7–8）

壳近圆柱形，上部略收窄；头近圆球形，被包覆在胸壳内，顶角很小，为圆锥形，有 3–4 个侧角，从边缘斜上生长，呈枝形，有侧刺，或可延长连接两个枝状侧角；胸壳的下部完全圆筒形，上部渐缩窄，无缩缢和内隔环，壳顶圆弧形；壁孔类圆形和椭圆形，孔数较少，自上而下渐增大，大小不等，分布无规律，孔径约为孔间桁宽的 1–2 倍，孔间桁较宽；壳壁较厚，表面粗糙，有一些小棘刺，口端截平。

标本测量：壳长 165μm，头宽 30μm，胸宽 50–80μm，顶角长 15μm，侧角长 30–40μm。

模式标本：DSDP274-23-5 1（图版 122，图 7–8），来自南极罗斯海的 DSDP Leg 28 航次 Site 274 孔的 274-23-5 岩心样品中，保存在中国科学院南海海洋研究所沉积标本室。

地理分布 南极罗斯海，早渐新世。

该新种与 *Aspis murus* Nishimura 的主要区别在于后者的头上仅有一个顶角，无侧角，壁孔较多，大小相近，排列较规则，具六角形框架，孔间桁较细，壳表光滑无棘刺。

395. 墙盾虫 *Aspis murus* Nishimura

（图版 122，图 9–20）

Aspis murus Nishimura, 1992, p. 358, pl. 10, figs. 4, 5, 9, pl. 13, fig. 12；Hollis, 2002, p. 303, pl. 5, figs. 11–14.

地理分布 南极罗斯海，晚始新世—早渐新世。

格锥虫属 Genus *Cinclopyramis* Haeckel, 1881

壳简单，具复网。

396. 格锥虫（未定种 1）*Cinclopyramis* sp. 1

（图版 122，图 21–22）

Cinclopyramis sp.，陈木宏等，2017，201–202 页，图版 72，图 4–7。

壳呈喇叭形，下部迅速扩大，表面光滑无刺；头球形，光滑，无壁孔，头角很小（可能已折断）；胸部的半侧上有 8–9 条放射肋，水平环连续，在壳的中下部发育有次级放射肋，网孔呈近正方形，排列规则整齐；壳的下部或口缘处常不同程度破损。

筛笼虫科 Family Sethocyrtida Haeckel, 1887

由Sethocorida和Sethocapsida(Haeckel, 1881, pp. 439, 433)合并组成，属于Dicyrtoidea类，具二节壳，被一横缩缢分隔为头和胸，无放射状辅枝（侧翼和脚），口部开放或有筛板封闭。

筛盔虫属 Genus *Sethocorys* Haeckel, 1881

胸呈卵形或亚圆柱形，口部收缩，具环状的口缘，头上有一顶角。

397. 奥地苏筛盔虫 *Sethocorys odysseus* Haeckel

（图版 112，图 6–18）

Sethocorys odysseus Haeckel, 1887, p. 1302, pl. 62, fig. 10；Petrushevskaya and Kozlova, 1972, pl. 22, fig. 16；谭智源、宿星慧，1982，175 页，图版 XVII，图 14；陈木宏、谭智源，1996，215 页，图版 30，图 5–6。

Carpocanistrum brevispina Vinassa de Regny, 1900, pl. 2, fig. 23.

Cystophormis brevispina (Vinassa de Regny) group, Petrushevskaya, 1975, p. 588, pl. 13, figs. 3–7, pl. 44, figs. 1–2.

Sethocorys bussonii Carnevale, 1908, p. 31, pl. 4, fig. 17.

Cryptocarpium bussonii (Carnevale) gr., Hollis, 1997, p. 66, pl. 6, figs. 26–27.

Dictyocephalus bergontianus Carnevale, 1908, p. 32, pl. 4, fig. 20.

Sethocorys sp., Hays, 1965, p. 177, pl. 3, fig. 8.

Carpocanistrum spp., Riedel and Sanfilippo, 1971(in part), p. 1596, pl. 3D, figs. 4, 8, 9；Abelmann, 1990, p. 695, pl. 5, fig. 13；Takemura, 1992, p. 744, pl. 3, figs. 5–6；O'Connor, 1993, p. 57, fig. 14.

Carpocanarium sp. O, Petrushevskaya, 1973, p. 1044, p. 8, pl. 3, figs. 10–11.

Plannapus mauricei O'Connor, 1999, p. 8, pl. 1, figs. 11–14, pl. 5, figs. 12–15；O'Connor, 2000, p. 208, pl. 1, figs. 11–14.

地理分布 南极罗斯海，晚始新世。

冠明虫属 Genus *Lophophaena* Ehrenberg, 1847, emend. Petrushevskaya, 1971

壳近圆柱形，胸呈卵形或亚圆柱形，口端截平或缩缢；头角数量不定。

398. 头冠明虫（新种）*Lophophaena cephalota* sp. nov.

（图版 112，图 19–24）

壳体稍大，呈头靶形，头宽是胸宽的 1/3–1/2；头近圆球形，格孔壳或孔间桁叠交形成壳壁，壁孔较小，无顶角，有少量棘突；胸呈圆筒形，肩部圆弧状，壳壁较厚，壁孔大而少，类圆形或椭圆形，大小相近或不等，不规则排列，似有六角形框架，孔间桁相对较细，无侧翼；口部开放，截平，无口缘环；胸壳表基本平滑，可见个别小棘突。

标本测量：头长 55–65μm、宽 55–70μm，胸长 70–100μm、宽 85–100μm，口宽 75–85μm。

模式标本：DSDP274-23-3 2（图版 112，图 20–21），来自南极罗斯海的 DSDP Leg 28 航次 Site 274 孔的 274-23-3 岩心样品中，保存在中国科学院南海海洋研究所沉积标本室。

地理分布 南极罗斯海，中渐新世—中中新世。

该新种的形态特征与 *Lophophaena cylindrica* (Cleve)有些相似，主要区别是后者的壳体偏瘦，头型拉长，有颈部，胸宽常与头宽接近，而且壳壁稍薄，壁孔较小，有侧翼。

399. 圆头冠明虫（新种）*Lophophaena cyclocapita* sp. nov.

（图版 112，图 25–40）

个体稍大，壳近圆筒形或钟罩形，壁稍厚，头与胸间的缩缢较深或有颈部；头较大，正常圆球形，壁孔类圆形，大小相近，规则或亚规则排列，有六角形框架，有的标本壁孔与孔间桁模糊；胸壳呈圆筒形，或向口端略扩大，壁孔类圆形，大小相近或不等，亚规则或不规则排列，具六角形框架，有的孔间桁较宽、较薄，壳表光滑或有小棘凸略显粗糙，但无针刺；无顶针和侧翼，胸口开放。

标本测量：壳长 170–215μm，头直径 85–115μm，胸长 85–125μm、宽 125–160μm。

模式标本：DSDP274-33-1 1（图版 112，图 25–26），来自南极罗斯海的 DSDP Leg 28 航次 Site 274 孔的 274-33-1 岩心样品中，保存在中国科学院南海海洋研究所沉积标本室。

地理分布 南极罗斯海，晚始新世—中渐新世。

该新种以头胸间分隔、有颈部、壳表无刺、无顶针和无侧翼等特征区别于 *Psilomelissa tricuspidata* Popofsky（1908, p. 284, pl. 32, fig. 9, pl. 33, fig. 8）、*Sethoconus tabulatus* (Ehrenberg)（Cleve, 1899, pl. 4, fig. 2）、*Antarctissa longa* (Popofsky)（Chen, 1975, p. 457, pl. 17, figs. 6–8）和 *Antarctissa strelkovi* Petrushevskaya（Lazarus, 1990, p. 713, pl. 3, figs. 13–15）。

400. 圆筒冠明虫 *Lophophaena cylindrica* (Cleve)

（图版 113，图 1–14）

Lophophaena cylindrica (Cleve), Petrushevskaya, 1971d, p. 117, figs. 57V, 61IV–VI.
Dictyocephalus cylindricus Cleve, 1900, p. 154；Cleve, 1901, p. 20, pl. 4, fig. 10.
Dictyocephalus sp., Cleve, 1899, pl. 2, fig. 1.
Lophophaenoma sp., Petrushevskaya, 1975, pl. 9, fig. 20.

壳体较小，头稍长；壁薄，孔较小而多，类圆形，无规律分布。

地理分布 南极罗斯海，晚始新世—中渐新世。

401. 橄榄冠明虫（新种）*Lophophaena olivacea* sp. nov.

（图版 113，图 15–17）

壳近橄榄形，两节壳之间无缩缢；头呈半球形，简单，稍厚，壁孔较小，类圆形，大小相近，具六角形框架，孔径与孔间桁宽近等；胸呈长腰鼓形，中部膨凸，口端略收缩，口宽约为壳宽的一半，无口缘，壁孔类圆形，多数孔的大小相近，个别较小，规则或亚规则排列，胸壳单面约有 7–8 纵排孔；在整个头部及胸上部的表面上覆盖有一层较薄的海绵组织，壳表无棘刺；无顶角和侧翼。

标本测量：头长 60–65μm、宽 90–95μm，胸长 200–210μm、宽 175–180μm，口宽 75–85μm。

模式标本：DSDP274-24-3 2（图版 113，图 15–17），来自南极罗斯海的 DSDP Leg 28 航次 Site 274 孔的 274-24-3 岩心样品中，保存在中国科学院南海海洋研究所沉积标本室。

地理分布 南极罗斯海，出现于中渐新世。

壳体形态与结构特征与 *Lophophaena spongiosa* (Petrushevskaya)相似，但在胸的中部

稍膨凸，下部略收缩；主要区别是此新种的整个壳壁为格孔状，壁孔类圆形，大小相近，规则或亚规则分布，具六角形或多角形框架；无顶角与侧翼。

402. 球冠明虫？（新种）*Lophophaena*? *orbicularis* sp. nov.

（图版 113，图 18–29）

Lophophaena ? *capito* Ehrenberg group, Petrushevskaya and Kozlova, 1972, p. 535, pl. 33, figs. 20–23.

个体较小，头部近球形，有数个很小的锥凸顶角，接近胸处收缩，壁孔较小，类圆形或椭圆形，孔间桁较宽，呈近板片状，表面光滑；胸壳常欠发育或很短，不同标本有一定的结构组成变化；无侧翼和基脚。

标本测量：头长 70–125μm、宽 90–125μm，胸长 0–30μm，口宽 40–75μm。

模式标本：DSDP274-31-2 2（图版 113，图 20–21），来自南极罗斯海的 DSDP Leg 28 航次 Site 274 孔的 274-31-2 岩心样品中，保存在中国科学院南海海洋研究所沉积标本室。

地理分布　南极罗斯海，早始新世—晚更新世。

该新种与 *Lophophaena capito* Ehrenberg 的主要区别是后者明显具有胸壳，略扩展，胸上有实心侧翼。该种的胸壳常欠发育，壳壁格孔状，表面光滑，无明显棘刺，壳体基本特征与 *Lophophaena* 和 *Lithomelissa* 等属均有一定区别。

403. 海绵冠明虫 *Lophophaena spongiosa* (Petrushevskaya)

（图版 114，图 1–19）

Botryometra (?) *spongiosa* Petrushevskaya, 1975, p. 590, pl. 13, fig. 11.

罗斯海标本的顶角较短或发育不明显，部分壳体近似橄榄形，胸的下部逐渐缩小，口端较窄，无侧翼。

地理分布　南极罗斯海，中始新世—中渐新世。

404. 冠明虫（未定种 1）*Lophophaena* sp. 1

（图版 115，图 1–2）

壳近圆柱形，壳壁有不规则状交叉桁框，壁孔较大，近圆形，似有六角形框架，各节点上生出许多棘突，表面粗糙。

地理分布　南极罗斯海，全新世。

筛圆锥虫属 Genus *Sethoconus* Haeckel, 1881

具圆锥形或钟形、逐渐拓宽的胸部和阔开口。头具一角或多角。

405. 杜氏筛圆锥虫？*Sethoconus*? *dogieli* Petrushevskaya

（图版 123，图 1–6）

Sethoconus (?) *dogieli* Petrushevskaya, 1968, pp. 94–95, figs. 53I–II；Chen, 1975, p. 462, pl. 19, fig. 4；Weaver, 1976, p. 582, pl. 1, fig. 8.

地理分布 南极罗斯海，早上新世—全新世。

406. 板筛圆锥虫 *Sethoconus tabulata* (Ehrenberg)

（图版 123，图 7–11）

Sethoconus tabulatus (Ehrenberg), Haeckel, 1887, p. 1293；Cleve, 1899, p. 33, pl. 4, fig. 2；Bjørklund and Kruglikova, 2003, pl. 6, fig. 10；Cortese et al., 2003, p. 69 (not figured)；陈木宏等，2017，203–204 页，图版 73，图 1–10。
Sethoconus? *tabulates* Petrushevskaya, 1967, pp. 94–96, figs. 54I–V；Cortese and Abelmann, 2002, pl. 2, figs. 14–15.

地理分布 南极罗斯海，早上新世—全新世。

407. 筛圆锥虫（未定种 1）*Sethoconus* sp. 1

（图版 123，图 12–13）

头较大，圆球形，光滑；胸壁孔上部较小，中部与下部很大，圆形或六角形，数量少，纵向仅有 4–5 排，具六角形框架，节点上有小角锥形凸起，无顶针。

地理分布 南极罗斯海，出现于早上新世。

花笼虫属 Genus *Anthocyrtium* Haeckel, 1887

二节壳，无胸肋，口缘上有一些末端脚，头具一顶角。

408. 拜洛花笼虫 *Anthocyrtium byronense* Clark

（图版 123，图 14–15）

Anthocyrtium byronense Clark, 1947, p. 73, pl. 7, figs. 1–4, 7.

个体较小，头半球形，胸钟罩形，口端略收缩，壁孔近圆形，具六角形框架，纵向规则排列，壳表光滑无刺，顶角和末端脚均不发育。罗斯海标本特征与 Clark（1947）的标本很相似，均呈现顶角和末端脚较细小或不甚发育特征。

地理分布 南极罗斯海，出现于早渐新世。

409. 五脚花笼虫（新种）*Anthocyrtium quinquepedium* sp. nov.

（图版 123，图 16–17）

壳呈钟罩形，个体很小；头近球形，部分被胸壁包围，较薄，一个顶角三棱柱形，较短，头与胸之间的缩缢较浅；胸呈钟罩形，壳壁稍厚，自上而下渐扩大，口部开放，在口缘处向下生出 5–6 个末端脚，均呈角锥状，较短；整个壳体的壁孔很小，数量少，类圆形，分布不规律，孔间桁较宽，表面光滑。

标本测量：壳长 85μm，头直径 25μm，胸长 70μm、宽 65μm，顶角长 15μm，末端脚长 15–25μm。

模式标本：DSDP274-32-1 2（图版 123，图 16–17），来自南极罗斯海的 DSDP Leg 28 航次 Site 274 孔的 274-32-1 岩心样品中，保存在中国科学院南海海洋研究所沉积标本室。

地理分布 南极罗斯海，出现于晚始新世。

该种特征与 *Anthocyrtium reticulatum* Haeckel（1887, p. 1274）较接近，主要区别是后

者的顶角较长，有侧齿，末端脚较多（12 个以上），壁孔为不规则多角形，表面有棘刺。

盔蜂虫属 Genus *Corythomelissa* Campbell, 1951

圆锥形壳，有三个侧脚和一个顶角，壳表不光滑。

410. 钩盔蜂虫 *Corythomelissa adunca* (Sanfilippo and Riedel)

（图版 123，图 18–25）

Spongomelissa adunca Sanfilippo and Riedel, 1973, p. 529, pl. 19, figs. 3–4, pl. 34, figs. 1–6.
Corythomelissa adunca (Sanfilippo and Riedel), Strong et al., 1995, p. 208, fig. 9B；Hollis, 2002, p. 295, pl. 3, fig. 14.

地理分布　南极罗斯海，渐新世。

411. 头盔蜂虫（新种）*Corythomelissa corysta* sp. nov.

（图版 124，图 1–2）

Corythomelissa sp., Petrushevskaya, 1975, p. 590, pl. 11, fig. 11.

壳呈钢盔形，壳壁稍厚；头较大，近半球形，顶角较小或不甚发育，角锥状，壁孔类圆形，大小略有差异，一般孔径较小，不规则分布，孔间桁较粗，或有些叠交，近似海绵结构；头腔内的中心主桁向上形成顶角，侧桁斜下延伸形成胸肋和侧脚；胸较短，扩展开阔，或发育不明显，壁孔亚规则分布，横向有 2–3 排，三个角锥状侧脚斜下生长（已折断），侧角上有穿孔；壳体孔间桁上有棘突，表面粗糙。

标本测量：头长 125μm、宽 125μm，胸长 45μm、宽 145μm，侧脚长>40μm。

模式标本：DSDP274-29-1 1（图版 124，图 1–2），来自南极罗斯海的 DSDP Leg 28 航次 Site 274 孔的 274-29-1 岩心样品中，保存在中国科学院南海海洋研究所沉积标本室。

地理分布　南极罗斯海，出现于古新世—早渐新世。

该新种特征与 *Spongomelissa adunca* Sanfilippo and Riedel 较为接近，主要区别在于后者的头较小，有分叶（2–3），顶角粗大，头与胸的缩缢较深，胸壳较宽，发育完整。我们的标本特征与 *Corythomelissa* sp.（Petrushevskaya, 1975, p. 590, pl. 11, fig. 11）较为相似，该未定种出现于南极海区的古新世地层。

412. 粗糙盔蜂虫 *Corythomelissa horrida* Petrushevskaya

（图版 124，图 3–5）

Corythomelissa horrida Petrushevskaya, 1975, p. 590, pl. 11, figs. 14–15, pl. 21, fig. 9；Takemura, 1992, p. 744, pl. 3, fig. 14.
Pseudodictyophimus gracilipes (Bailey), Hollis, 1997, p. 53, pl. 3, fig. 27.

地理分布　南极罗斯海，早渐新世。

413. 盔蜂虫（未定种 1）*Corythomelissa* sp. 1

（图版 124，图 6–8）

Clathromitra ? sp., Petrushevskaya, 1975, pl. 21, fig. 6.

地理分布 南极罗斯海，早渐新世。

角笼虫属 Genus *Ceratocyrtis* Bütschli, 1882

头稍大，头角一个或更多，领隔清楚，胸圆锥形或钟罩形，渐膨大，平滑，阔开口。

414. 扩角笼虫 *Ceratocyrtis amplus* (Popofsky) group

（图版 124，图 9–38）

Helotholus amplus Popofsky, 1908, p. 283, pl. 34, fig. 3；Petrushevskaya, 1975, p. 590, pl. 11, figs. 3–6, 13, pl. 19, fig. 2, pl. 44, fig. 4.
Lithomelissa sp. C, Chen, 1975, p. 458, pl. 11, figs. 3–4.
Ceratocyrtis sp., Petrushevskaya, 1975, pp. 590–591, pl. 11, fig. 12, pl. 18, figs. 1–3, pl. 19, fig. 1.

地理分布 南极罗斯海，中始新世—晚中新世。

415. 筒角笼虫（新种）*Ceratocyrtis cylindris* sp. nov.

（图版 125，图 1–4）

壳呈钟罩形或圆筒形，壳壁较厚；头球形，部分陷入胸内，缩缢较深，有 3–4 个较短的细柱状顶角，壁孔类圆形；胸壳近圆筒形，肩部平直，或有斜倾，中下部无收缩或扩大，壁孔类圆形或椭圆形，大小不等，亚规则或不规则排列，孔间桁较细；壳表有许多棘刺或棘片状凸起，表面较为粗糙；口端截平，呈不规则状，无口缘环或齿。

标本测量：壳长 135–145μm、宽 100–110μm，头直径 45–50μm。

模式标本：DSDP274-10-5 1（图版 125，图 1–2），来自南极罗斯海的 DSDP Leg 28 航次 Site 274 孔的 274-10-5 岩心样品中，保存在中国科学院南海海洋研究所沉积标本室。

地理分布 南极罗斯海，早上新世。

该未定种特征与 *Ceratocyrtis manumi* Goll and Bjørklund（1989, p. 730, pl. 5, figs. 21–23）较相似，主要区别是后者胸壳较宽（胖），肩部下斜，壳中部略缩窄，下部明显增大，最大胸宽在口端，而新种的胸宽上下基本不变。

416. 不规则角笼虫（新种）*Ceratocyrtis irregularis* sp. nov.

（图版 125，图 5–39）

壳呈不规则钟罩形，个体较小，壳壁较厚；头部脑丘状，分室 3–4 个或更多，堆积或排列为丘形、压扁长条形或不规则形，有的位于胸顶，且与胸有较深的缩缢，但多数压扁形的脑部较宽，几乎包覆胸顶，甚至延伸至胸壳上部的外围，致头壳与胸壳的界线不清；头角 1–3 个，较短小，常侧向生长；胸为钟罩形或圆筒形，壁孔类圆形或不规则形，大小不等，不规则排列，孔间桁在不同标本上粗细有差异，胸口开放，近截平，口端常有孔间桁形成的小齿；壳体表面光滑，无棘刺。

标本测量：壳长 105–150μm、宽 95–115μm，头长 25–50μm、宽 45–95μm。

模式标本：DSDP274-26-5 2（图版 125，图 5–6），来自南极罗斯海的 DSDP Leg 28 航次 Site 274 孔的 274-26-5 岩心样品中，保存在中国科学院南海海洋研究所沉积标本室。

地理分布 南极罗斯海，晚始新世—中渐新世。

该新种以其发育不规则的多室叶形态特征区别于其他各种。

417. 阔角笼虫 *Ceratocyrtis manumi* Goll and Bjørklund

（图版 125，图 40–41）

Gen. and sp. indet., Bjørklund, 1976, p. 1124, pl. 24, figs. 1–2.
Ceratocyrtis manumi Goll and Bjørklund, 1989, p. 730, pl. 5, figs. 21–23.

地理分布　南极罗斯海，晚始新世。

418. 雄角笼虫 *Ceratocyrtis mashae* Bjørklund

（图版 126，图 1–31）

Ceratocyrtis mashae Bjørklund, 1976, p. 1125, pl. 17, figs. 1–5；Abelmann, 1990, p. 694, pl. 4, figs. 15A–C.
Sethoconus sp., Chen, 1975, p. 462, pl. 10, figs. 5–6；Weaver, 1976, p. 582, pl. 7, fig. 5.
Ceratocyrtis stoermeri Goll and Bjørklund, 1989, p. 731, pl. 5, figs. 5–9.
Ceratocyrtis sp., Abelmann, 1990, p. 694, pl. 4, fig. 14；Lazarus, 1992, p. 798, pl. 7, figs. 10–17.

地理分布　南极罗斯海，晚始新世—晚渐新世。

419. 围玛角笼虫（新种）*Ceratocyrtis perimashae* sp. nov.

（图版 127，图 1–9）

壳呈钟罩形，壳壁较厚；头圆球形，较小，下部陷入胸内，缩缢较深，壁孔圆形，很小；胸呈钟罩形或圆锥形，自上而下渐扩大，口端截平，壁孔类圆形或椭圆形，大小不一，不规则排列，具六角形或多角形框架，孔径约为孔间桁宽的 1–2 倍，孔间桁较宽；壳体表面在各节点上有较长的棘刺状突起，刺的末端有侧分枝，呈细丝状相互连接，在壳表外形成一个蛛网状的次生组织，组成了一个较薄的外围壳；表面粗糙。

标本测量：壳长 160–245μm、宽 130–175μm，头直径 30–50μm，内壳与外壳间距 10–25μm。

模式标本：DSDP274-32-1 2（图版 127，图 1–2），来自南极罗斯海的 DSDP Leg 28 航次 Site 274 孔的 274-32-1 岩心样品中，保存在中国科学院南海海洋研究所沉积标本室。

地理分布　南极罗斯海，晚始新世—晚渐新世。

该新种特征与 *Ceratocyrtis mashae* Bjørklund 较为相似，主要区别在于前者具有一个外围壳，而后者胸壳孔间桁节点上的棘突起为角锥形，无末端侧刺或蛛网状的次生组织与外围壳，孔间桁一般较细。

420. 强壮角笼虫 *Ceratocyrtis robustus* Bjørklund

（图版 127，图 10–13）

Ceratocyrtis robustus Bjørklund, 1976, p. 1125, pl. 17, figs. 6–10.

地理分布　南极罗斯海，晚始新世—早渐新世。

421. 角笼虫（未定种 1）*Ceratocyrtis* sp. 1

（图版 127，图 14–15）

地理分布 南极罗斯海，出现于早渐新世。

格头虫属 Genus *Dictyocephalus* Ehrenberg, 1860

壳圆锥形或钟罩形，胸口截平或有环状物，无顶角。

422. 南方格头虫 *Dictyocephalus australis* Haeckel

（图版 127，图 16–17）

Dictyocephalus australis Haeckel, 1887, pp. 1306–1307, pl. 62, fig. 1.
Tricolocapsa papillosa (Ehrenberg) group, Petrushevskaya, 1975, p. 588, pl. 13, fig. 13.

无顶角，无侧翼；头半球形，胸亚柱状，双型孔，不规则排列。
地理分布 南极罗斯海，出现于晚上新世。

423. 厚格头虫 *Dictyocephalus crassus* Carnevale

（图版 127，图 18–20）

Dictyocephalus crassus Carnevale, 1908, p. 32, pl. 4, fig. 21.
Cryptocarpium bussonii (Carnevale) group, Hollis, 1997, p. 66, pl. 6, figs. 26–27.
Cystophormis sp., Petrushevskaya, 1975, p. 27, fig. 9.

头半球形，无顶角；胸近圆球形，较宽，壳孔较小，类圆形，亚规则排列；口部缩窄，仅为胸宽的约 1/3，口缘很短。
地理分布 南极罗斯海，中渐新世。

424. 乳格头虫 *Dictyocephalus papillosus* (Ehrenberg)

（图版 127，图 21–31）

Eueyrtidium papillosus Ehrenberg, 1872, S. 310, Taf. 7, Fig. 10.
Carpocanium calycothes Stöhr, 1880, S. 96, Taf. 3, Fig. 8.
Dictyocephalus papillosus (Ehrenberg), Haeckel, 1887, p. 1307；Riedel, 1958, p. 236, pl. 3, fig. 10, textfig. 8；陈木宏、谭智源，1996，216 页，图版 30，图 11–12；陈木宏等，2017，206 页，图版 74，图 13–22。
Dictyocryphalus papillosus (Ehrenberg), Nigrini, 1967, p. 63, pl. 6, fig. 6；Renz, 1976, p. 139, pl. 6, fig. 9.
Dictyocephalus? *papillosus* (Ehrenberg), Petrushevskaya, 1967, pp. 112–113, figs. 66I–III.
Carpocanarium papillosum (Ehrenberg) group, Nigrini and Moore, 1979, p. N27, pl. 21, fig. 3.

地理分布 南极罗斯海，中渐新世—早上新世。

425. 塔格头虫（新种）*Dictyocephalus turritus* sp. nov.

（图版 128，图 1–15）

个体较小，壳呈圆锥形或尖塔形，壁厚中等或稍厚；头呈脑丘形，有一个稍大的球形室，部分陷入胸内，缩缢不明显，无顶针；胸壳为圆塔形，自上而下逐渐加宽，最大宽度在口端，壁孔类圆形，大小相近，规则或亚规则斜向排列，孔径与孔间桁近等宽，胸口截平开阔，锯齿状或有一个较细口缘环；壳表光滑。

标本测量：壳长 105–120μm、宽 75–120μm，头长 25–30μm、宽 30–50μm。

模式标本：DSDP274-22-3 2（图版 128，图 1–2），来自南极罗斯海的 DSDP Leg 28 航次 Site 274 孔的 274-22-3 岩心样品中，保存在中国科学院南海海洋研究所沉积标本室。

地理分布　南极罗斯海，晚始新世—中渐新世。

该新种与 *Dictyocephalus variabilus* sp. nov.较相似，主要区别在于后者的最大壳宽在中部，口端缩窄，有明显的口缘环。

426. 变形格头虫（新种）*Dictyocephalus variabilus* sp. nov.

（图版 128，图 16–51）

Phormostichoartus fistula Nigrini, Hollis, 1997, pl. 4, figs. 5–7.
Theocampe sp., Ling, 1991b, p. 320, pl. 1, fig. 9.

壳分两节，近瓶形，下部略缩窄，壳体形态与壁孔特征有一定变化；头亚球形或半球形，部分陷入胸内，胸的上部近 1/3 处为最大壳宽，往口部渐收缩，口截平；壁厚中等，壁孔较小，较多，类圆形，大小相近，规则或亚规则排列，在最大壳宽处单面约有 7–9 孔；壳表平滑，口端常有透明的缘环。

标本测量：壳长 125–170μm、宽 70–105μm，头直径 25–45μm。

模式标本：DSDP274-25-1 1（图版 128，图 16–17），来自南极罗斯海的 DSDP Leg 28 航次 Site 274 孔的 274-25-1 岩心样品中，保存在中国科学院南海海洋研究所沉积标本室。

地理分布　南极罗斯海，中始新世—晚渐新世。

罗斯海的标本特征与 Hollis（1997, pl. 4, figs. 5–7）基本相同，仅有两节壳，虽然壳体外形常有变化，但在第二节壳内无任何内隔环，此类标本与 *Phormostichoartus fistula* Nigrini（1977, p. 253, pl. 1, figs. 11–13）的主要区别在于后者壳体有 3–4 节，具明显的内隔环，头侧有一耳管，壳体形态较规则，壁孔稍大而少，分布有规律。

427. 格头虫（未定种 1）*Dictyocephalus* sp. 1

（图版 129，图 1–2）

头半球形，无顶角；胸近圆球形，口端收缩，截平；壁孔类圆形，分布无规律，有六角形框架。该未定种特征与 *Cryptocarpium bussonii* (Carnevale) gr.（Hollis, 1997, p. 66, pl. 6, figs. 26–27）较接近，主要区别是后者有一个较厚实的口缘环，而该未定种开口处简单，无口缘等。

地理分布　南极罗斯海，出现于早渐新世。

428. 格头虫（未定种 2）*Dictyocephalus* sp. 2

（图版 129，图 3–4）

头圆球形，较大；胸短腰鼓形，腰长约为头径的 1.5 倍；壳壁较厚，壁孔类圆形，大小不等，分布无规律；口部略收缩，似有一透明口缘环。该未定种特征与 *Dictyocephalus amphora* Haeckel（1887, p. 1305, pl. 62, fig. 4）的主要区别是后者壁孔规则排列，自上而下略增大；与 *Theocampe amphora* (Haeckel)（Hollis, 1997, p. 56, pl. 4, figs. 38–39）较为接近，但后者似乎壳分头、胸、腹三节，壁孔大小近等，横向规则排列。

地理分布 南极罗斯海，出现于中渐新世。

果篮虫属 Genus *Carpocanium* Ehrenberg, 1847

头部藏在胸腔顶部，无顶角，无胸肋，口缘有一些末端脚。

429. 海绵果篮虫（新种）*Carpocanium spongiforma* sp. nov.

（图版 129，图 5–12）

壳近椭球形，外表无缩缢，壳壁较厚，似海绵状；头部脑丘状，较大，完全被包覆于胸壳内；胸壳为规则椭球形，壁孔较小而多，类圆形，孔径与孔间桁宽近等，分布无规律，较均匀散布，孔间桁上有一些凸起，使壳表粗糙；口端缩窄，截平，口径为壳宽的 1/3–1/2，口缘上有小齿，似为末端脚的雏形。

标本测量：壳长 140–175μm、宽 90–135μm，头长 38–60μm、宽 60–85μm，口宽 40–50μm。

模式标本：DSDP274-21-1 2（图版 129，图 5–6），来自南极罗斯海的 DSDP Leg 28 航次 Site 274 孔的 274-21-1 岩心样品中，保存在中国科学院南海海洋研究所沉积标本室。

地理分布 南极罗斯海，晚始新世—中渐新世。

该新种以具有似海绵状的壳壁组织和较大的头部形态为基本特征，区别于其他种类。

430. 果篮虫（未定种 1）*Carpocanium* sp. 1

（图版 129，图 13–20）

壳亚球形，个体较小，壁厚；头呈脑丘状位于胸腔顶部，胸壳亚球形或类椭球形，中下部较宽，标本的口部发育不完整；壁孔类圆形，大小不一，分布无规律，孔径是孔间桁宽的 1–2 倍，壳表光滑。

地理分布 南极罗斯海，晚始新世—中渐新世。

果圆球虫属（新属） Genus *Carpoglobatus* gen. nov.

壳球形，分三节，被内隔环所分隔，无缩缢，有一个口管。

模式种：*Carpoglobatus tubulus* sp. nov.

431. 管果圆球虫（新种）*Carpoglobatus tubulus* sp. nov.

（图版 129，图 21–22）

个体较大，椭球形，三节壳间无缩缢，内隔环弯曲或倾斜；头部结构简单，近半球形，壳壁较薄，壁孔类圆形或椭圆形，大小不一，孔间桁较宽；胸壳与腹壳完全融为一体，两者间的内隔环侧斜，壳壁较厚，壁孔类圆形，大小相近，亚规则排列，孔径是孔间桁宽的 2–3 倍；口部明显缩窄，约为壳宽的 1/3，口缘向外延伸为一管状物，其长度是宽度的 3/4，管壁较薄，透明状，有一些小孔，管口不平整，无口缘齿。

标本测量：壳长 230μm、宽 175μm，头长 55μm、宽 115μm，管口长 45μm、宽 55μm。

模式标本：DSDP274-12-5 1（图版 129，图 21–22），来自南极罗斯海的 DSDP Leg 28 航次 Site 274 孔的 274-12-5 岩心样品中，保存在中国科学院南海海洋研究所沉积标本室。

地理分布　南极罗斯海，出现于中中新世。

该种与 *Carpocanium* 属各个种类的主要区别是后者有口缘，但无口管，有末端脚，脑丘状的头部位于胸腔内。

足篮虫科 Family Podocyrtidae Haeckel, 1887

壳分三节，由两个横缢分为头、胸、腹，具三个辐射状骨突。

里曼虫属 Genus *Lipmanella* Loeblich and Tappan, 1961

具一顶角，胸壳上有三个格孔状的侧翼，无放射肋延伸至腹壁，无末端脚。

432. 始弹里曼虫 *Lipmanella archipilium* (Petrushevskaya)

（图版 129，图 23–46）

Dictyophimus (?) *archipilium* Petrushevskaya, 1975, p. 583, pl. 25, figs. 1–2.
Anthocyrtella ? *kruegeri* (Popofsky) group, Petrushevskaya, 1975, p. 587, pl. 25, figs. 9–10.

地理分布　南极罗斯海，早—晚始新世。

433. 网角里曼虫 *Lipmanella dictyoceras* (Haeckel)

（图版 130，图 1–4）

Lithornithium dictyoceras Haeckel, 1860, p. 840.
Lipmanella dictyoceras (Haeckel), Kling, 1973, p. 636, pl. 4, figs. 24–26.

个体较小，顶针小；胸翼实心或格孔状，较细小；壁孔小而多，分布不规则。该种的形态结构特征与 *Dictyophimus* aff. *borisenkoi* Nishimura（Hollis, 2002, p. 312, pl. 9, figs. 1–2）有些接近，但后者壳体较为粗大，顶角和侧翼也很粗壮，壁孔较大，具六角形框架，规则排列。

地理分布　南极罗斯海，早渐新世。

434. 里曼虫（未定种 1）*Lipmanella* sp. 1

（图版 130，图 5–6）

壳体较小，头近半球形，头胸间的缩缢较深，胸钟罩形，腹不完整或为胸的延伸部分；未见顶角，在胸侧上有 5–6 个具窗孔的侧翼，较短；壳壁孔类圆形或椭圆形，大小不一，分布不规则，孔径约为孔间桁宽的 1–4 倍；口部开放，不收缩，无口缘。该类标本与 *Lipmanella* 属（仅具三个侧翼）有明显区别，可能是一新属。

地理分布　南极罗斯海，出现于晚始新世。

翼篮虫属 Genus *Pterocanium* Ehrenberg, 1847

具三个辐射状肋，肋由胸部延至腹部并延长变成三个格孔状足。

435. 多门翼篮虫 *Pterocanium polypylum* Popofsky

（图版 130，图 7–18）

Pterocanium polypylum Popofsky, 1913, pp. 388–390, text-figs. 101–103.

地理分布 南极罗斯海，早始新世—中渐新世。

436. 翼篮虫（未定种 1）*Pterocanium* sp. 1

（图版 130，图 19–20）

地理分布 南极罗斯海，出现于中上新世。

假网杯虫属 Genus *Pseudodictyophimus* Petrushevskaya, 1971

壳体第一节较小，第二节明显增大并常在其底部封闭，具一个顶针和三个侧脚，壁孔大小不等，上部小，下部大，排列不规则。

437. 头盔假网杯虫 *Pseudodictyophimus galeatus* Caulet

（图版 131，图 1–15）

Pseudodictyophimus galeatus Caulet, 1991, p. 534, pl. 2, figs. 9–10.
Lithomelissa challengerae Chen, Hollis, 1997, p. 51, pl. 3, fig. 21.
Ceratocyrtis sp., Hollis, 2002, pl. 3, fig. 17 (only).

地理分布 南极罗斯海，晚始新世—早上新世。

438. 大针假网杯虫（新种）*Pseudodictyophimus gigantospinus* sp. nov.

（图版 132，图 1–2）

个体较大，两节壳，倒圆筒形，每节的顶部近水平形，与侧面近垂直交汇，两壳间呈阶梯状扩大；第一节头的下部略阔宽，壁孔类圆形，大小差异较大，上部壁孔细小，下部壁孔较大，孔间桁较细；胸壳陡然增大，肩部水平，往下略扩大，壁孔椭圆形或不规则形，孔径为孔间桁宽的 1–5 倍，壁孔相差悬殊，孔间桁较细，口开阔，无口缘；具一个顶角、一个侧角和三个末端脚，均呈三棱角锥状，较为粗长；孔间桁上有一些棘片状突起，壳表粗糙。

标本测量：壳长 180μm、宽 175μm，头长 85μm、宽 140μm，顶针长 165μm，侧针长 145μm，基脚长 65–120μm。

模式标本：DSDP274-10-1 1（图版 132，图 1–2），来自南极罗斯海的 DSDP Leg 28 航次 Site 274 孔的 274-10-1 岩心样品中，保存在中国科学院南海海洋研究所沉积标本室。

地理分布 南极罗斯海，出现于早上新世。

该新种特征与 *Pseudodictyophimus gracilipes* (Bailey)较接近，主要区别是后者的头部

较小，顶针较短，无侧针，壁孔较多，壳表光滑。

439. 单薄假网杯虫 *Pseudodictyophimus gracilipes* (Bailey)

（图版 132，图 3–24）

Dictyophimus gracilipes Bailey, 1856, p. 4, pl. 1, fig. 8；Cleve, 1899, p. 29, pl. 2, fig. 2；Jørgensen, 1900, pl. 5, fig. 26；Benson, 1966, pl. 25, figs. 4–6；Petrushevskaya, 1967, figs. 38–39；Abelmann, 1992b, p. 380, pl. 4, figs. 1–2.

Pseudodictyophimus gracilipes (Bailey), Petrushevskaya, p. 91, pls. 47–49；Bjørklund, 1976, p. 1124, pl. 9, figs. 1–5, pl. 11, figs. 6–7；Boltovskoy and Riedel, 1987, pl. 4, fig. 25；Takahashi, 1991, p. 116, pl. 37, figs. 12–13；van de Paverd, 1995, pl. 68, figs. 3–6；Bjørklund et al., 1998, pl. 2, figs. 7–8.

Dictyophimus spp., Abelmann, 1992b, p. 380, pl. 4, figs. 3, 5 (only).

该种的壳体特征变化较大。

地理分布　南极罗斯海，晚始新世—全新世。

440. 单薄假网杯虫（亲近种）*Pseudodictyophimus* sp. aff. *P. gracilipes* (Bailey)

（图版 133，图 1–4）

Pseudodictyophimus sp. aff. *P. gracilipes* (Bailey), Bjørklund, 1976, p. 1124, pl. 16, figs. 1–5.

地理分布　南极罗斯海，早上新世。

441. 纤细假网杯虫（新种）*Pseudodictyophimus tenellus* sp. nov.

（图版 133，图 5–6）

个体较小，头呈半球形，胸为钟罩形，缩缢较深，顶角很短小，细柱状；整个壳体壁较薄，壁孔较大，圆形、椭圆形或近方形，大小差异较大，分布不规则，孔间桁较细，孔径是孔间桁宽的 2–6 倍，壁孔较稀少；三个末端脚较细小，呈细柱状，末端缩尖；胸口不平整，残缺状；壳表有一些棘刺。

标本测量：壳长 110μm、宽 90μm，头长 30μm、宽 45μm，顶针长 10μm，基脚长 30–50μm。

模式标本：DSDP274-7-1 1（图版 133，图 5–6），来自南极罗斯海的 DSDP Leg 28 航次 Site 274 孔的 274-7-1 岩心样品中，保存在中国科学院南海海洋研究所沉积标本室。

地理分布　南极罗斯海，出现于晚上新世。

该新种特征与 *Pseudodictyophimus gracilipes* (Bailey)接近，但两者间的区别较为明显，主要是后者壳壁稍厚，壁孔较多，大小相近，胸壳在侧肋间常有膨凸，末端脚与顶角相对粗大，呈三棱角锥状。

辫篓虫科 Family Phormocyrtidae Haeckel, 1887

三节壳由两个横缩环分为头、胸、腹。具 4–9 个或更多的辐射肋，自头基部的领缩环长出，沿着胸部和腹部下伸，常发育为末端脚。

圆蜂虫属 Genus *Cycladophora* Ehrenberg, 1847, emend. Lombari and Lazarus, 1988

以 *Cycladophora davisiana* Ehrenberg 为典型种（模式种），圆锥形壳，一般呈二节壳，

偶见三节；亚球形头较小，格孔状胸壳渐扩大，无末端脚和侧翼或肋。如有腹节，常呈现为不规则的附属物或具隔环的齿与盖板。

442. 双角圆蜂虫 *Cycladophora bicornis* (Popofsky) group

（图版 133，图 7–16）

Pterocorys bicornis Popofsky, 1908, p. 228, pl. 34, figs. 7–8.

Theocalyptra bicornis (Popofsky), Riedel, 1958, p. 240, pl. 4, fig. 4；Petrushevskaya, 1967, p. 126, pl. 71, figs. 2–7；Petrushevskaya and Kozlova, 1972, p. 540, pl. 33, figs. 11–12.

Clathroyclas bicornis Hays, 1965, p. 179, pl. 3, fig. 3；Petrushevskaya, 1975, p. 586, pl. 15, fig. 25, pl. 23, fig. 3.

Clathrocyclas bicornis (Popofsky) group, Petrushevskaya, 1975, p. 586, pl. 15, fig. 26, pl. 23, fig. 2.

Cycladophora bicornis (Popofsky) Lombari and Lazarus, 1988, pp. 106–114, pl. 4, figs. 1–12, pl. 5, figs. 1–12；Takemura, 1992, p. 745, pl. 2, fig. 15.

Cycladophora bicornis amphora Lombari and Lazarus, 1988, p. 110, pl. 4, figs. 6–12.

Cycladophora pliocenica Lazarus, 1990, p. 715, pl. 4, figs. 6–7.

Cycladophora bicornis (Popofsky)，陈木宏等，2017，213–214 页，图版 76，图 14–16。

地理分布 南极罗斯海，中中新世—全新世。

443. 锥圆蜂虫 *Cycladophora conica* Lombari and Lazarus

（图版 133，图 17–22）

Cycladophora conica Lombari and Lazarus, 1988, p. 105, pl. 3, figs. 1–16；Abelmann, 1990, p. 697, pl. 7, fig. 12；Takemura, 1992, p. 745, pl. 2, figs. 16–17.

地理分布 南极罗斯海，早渐新世—晚更新世。

444. 戴维斯圆蜂虫 *Cycladophora davisiana* Ehrenberg group

（图版 133，图 23–33）

Cycladophora? *davisiana* Ehrenberg, 1862, p. 297；Ehrenberg, 1873, pl. 2, fig. 11.

Pterocodon davisianus Ehrenberg, 1862, pp. 300–301；Ehrenberg, 1873, pl. 2, fig. 10.

Pterocanium davisianus Haeckel, 1862, p. 332.

Eucyrtidium davisianum Haeckel, 1862, pp. 328–329.

Stichopilium davisianum (Ehrenberg), Cleve, 1899, p. 33, pl. 4, fig. 6.

Theocalyptra davisiana (Ehrenberg), Riedel, 1958, p. 239, pl. 4, figs. 2–3, text-fig. 10；Benson, 1966, pl. 29, figs. 14–16；Nigrini and Moore, 1979, p. N57, pl. 24, figs. 2a–b；陈木宏、谭智源，1996，223 页，图版 33，图 1–2，图版 53，图 9。

Cycladophora cf. *humerus* Petrushevskaya, 1975, p. 586, pl. 15, figs. 22–23(only), pl. 43, figs. 1–2；Lombari and Lazarus, 1988, p. 123, pl. 9, figs. 1–6；Hollis, 1997, p. 59, pl. 5, fig. 21.

Cycladophora davisiana Ehrenberg, Petrushevskaya, 1967, pp. 120–122, pl. 69, figs. 1–7；Weaver, 1983, p. 675, pl. 1, fig. 3；Abelmann and Gowing, 1997, p. 22; Bjørklund et al., 1998, pl. 2, fig. 6; Boltovskoy, 1998, fig. 15.131; Itaki and Ikehara, 2004, pl. I, figs. 18–20.

Cycladophora davisiana Ehrenberg group，陈木宏等，2017，215 页，图版 77，图 16–28。

地理分布 南极罗斯海，早渐新世—全新世。

445. 戴维斯圆蜂虫似角亚种 *Cycladophora davisiana cornutoides* (Petrushevskaya)

（图版 134，图 1–22）

Theocalyptra davisiana (Ehrenberg) var. *cornutoides*, Petrushevskaya, 1967, p. 124, figs. 70 I–III；Takahashi and Honjo, 1981, p. 153, pl. 9, fig. 18；Takahashi, 1991, p. 123, pl. 41, figs. 12–16.

Diplocyclas sp. aff. *D. bicorona* Haeckel group, Petrushevskaya, 1975, p. 587, pl. 15, figs. 8–10, pl. 24, figs. 1–3.

地理分布 南极罗斯海，晚始新世—晚中新世。

446. 假异圆蜂虫 *Cycladophora pseudoadvena* (Kozlova)

（图版 134，图 23–28）

Ceratocyrtis pseudoadvena Kozlova, 1999, p. 114, pl. 12, fig. 6.

Cycladophora aff. *Ceratocyrtis pseudoadvena* Kozlova, Hollis, 2002, p. 309. pl. 7, figs. 1–6.

地理分布 南极罗斯海，晚始新世。

447. 棉胸圆蜂虫 *Cycladophora spongothorax* (Chen)

（图版 134，图 29–31）

Theocalyptra bicornis spongothorax Chen, 1975, p. 462, pl. 12, figs. 1–3；Weaver, 1976a, p. 582, pl. 2, figs. 1–4, pl. 6, figs. 2–4；Weaver, 1983, p. 678, pl. 5, fig. 7.

Cycladophora spongothorax (Chen), Lombari and Lazarus, 1988, p. 122, pl. 9, figs. 7–12；Lazarus, 1990, p. 715, pl. 4, figs. 1–3；Vigour and Lazarus, 2002, p. 5, pl. P3, figs. 1–5.

地理分布 南极罗斯海，出现于中中新世。

448. 长钟圆蜂虫（新种）*Cycladophora tinocampanula* sp. nov.

（图版 135，图 1–5）

壳分两节，头圆球形，较小，有一顶角和一个侧角，头角较小，细柱状，头下部略陷入胸壳；胸的顶部较窄，至中部逐渐增大，之后壳宽不变，胸壳的下部呈圆筒状；头部壁孔细小稀疏，胸的顶部壁孔较小而密集，逐渐变大，中部和下部壁孔类圆形，大小近等，横向排列较为规则，具六角形孔间桁，较细，孔径约为孔间桁宽的 3–4 倍；胸口开放，截平；在孔间桁各节点上有一些很小的棘突，壳表略显粗糙。

标本测量：壳长 155–175μm、宽 100–120μm，头直径 25–30μm，头角长 12–20μm。

模式标本：DSDP274-6-3 2（图版 135，图 1–2），来自南极罗斯海的 DSDP Leg 28 航次 Site 274 孔的 274-6-3 岩心样品中，保存在中国科学院南海海洋研究所沉积标本室。

地理分布 南极罗斯海，早—晚上新世。

该新种的壳体形态与 *Lophocyrtis golli* Chen 较相似，主要区别是后者壳分三节，有一个加宽的腹节壳，而且胸壳的壁孔类型与腹壳不同。

冠笼虫属 Genus *Lophocyrtis* Haeckel, 1887

头具两个顶角，或为一个分叉的顶角；腹部常呈圆柱形，阔开口，口端截平。

449. 戈氏冠笼虫 *Lophocyrtis golli* Chen

（图版 135，图 6–29）

Lophocyrtis golli Chen, 1975, p. 461, pl. 12, figs. 4–5；Weaver, 1976, p. 581, pl. 4, figs. 9–10, pl. 9, figs. 8–9.

Cycladophora pliocenica (Hays), Lombari and Lazarus, 1988, p. 104；Lazarus, 1990, p. 715, pl. 4, figs. 6–7；Vigour and Lazarus, 2002, pl. P3, fig. 7.

地理分布 南极罗斯海，早中新世—早上新世。

窗袍虫属 Genus *Clathrocyclas* Haeckel, 1881, emend. Foreman, 1968

具一简单花冠形末端足，环绕在张大的口部。腹部张大，截角圆锥状或盘状。壳壁无肋。

450. 君窗袍虫 *Clathrocyclas alcmenae* Haeckel

（图版 136，图 1–34）

Clathrocyclas alcmenae Haeckel, 1887, p. 1388, pl. 49, fig. 6.
Diplocyclas sp. aff. *D. bicorona* Haeckel group, Petrushevskaya, 1975, p. 587, pl. 15, figs. 8–10 (only).
Cycladophora cosma cosma Lombari and Lazarus, 1988, p. 104, pl. 1, figs. 1–6.
Cycladophora aff. *cosma* Lombari and Lazarus, Hollis, 2002, p. 307, pl. 6, figs. 13–17.

头半球形，有两个角锥状头角；胸呈喇叭形，向口端渐增大，阔开口，最大胸宽在口端，近等或略大于胸长，胸壳壁孔六角形，较大，孔间桁细；腹部为简单花冠，常破损不完整或消失。

地理分布 南极罗斯海，中始新世—中渐新世。

451. 车窗袍虫（相似种）*Clathrocyclas* cf. *C. danaes* Haeckel

（图版 136，图 35–36）

cf. *Clathrocyclas danaes* Haeckel, 1887, p. 1388, pl. 59, figs. 13–14；陈木宏、谭智源，1996，221 页，图版 32，图 3–4。
Euceryphalus gegenbauri Haeckel, Petrushevskaya, 1971d, p. 222, figs. 103I–III.

罗斯海标本的个体非常小。

地理分布 南极罗斯海，全新世。

宽口斯科虫属 Genus *Eurystomoskevos* Caulet, 1991

两节壳，或为三节壳（第二节壳内偶见一隔环）。头有一长而坚实的顶角，胸壁上有外伸的骨针。

452. 匹楚宽口斯科虫 *Eurystomoskevos petrushevskaae* Caulet

（图版 137，图 1–28）

Diplocyclas sp. A, Petrushevskaya and Kozlova, 1972, p. 541, pl. 33, figs. 14–16; Petrushevskaya, 1975, p. 587, pl. 24, fig. 4; Chen, 1975, p. 460, pl. 7, figs. 4–5; Takemura, 1992, p. 746, pl. 3, fig. 16.
Eurystomoskevos petrushevskaae Caulet, 1991, p. 536, pl. 3, figs. 14–15; O'Connor, 1993, p. 74, pl. 8, figs. 12–14; Crouch and Hollis, 1996, p. 26; Hollis et al., 1997, p. 62, pl. 5, figs. 22–23; Hollis, 2002, p. 310, pl. 7, figs. 7–10.
Eurystomoskevos cauleti Hollis, 1997, p. 61, pl. 5, figs. 4–5.

地理分布 南极罗斯海，晚始新世—早上新世。

453. 宽口斯科虫（未定种 1）*Eurystomoskevos* sp. 1

（图版 137，图 29–35）

壳呈长钟罩形，顶针较短或不太发育；壳上部的壁孔类圆形，下部呈长椭圆形，大小不一，不规则分布，孔间桁较宽；胸外侧有个别实心骨针。该未定种可能是

Eurystomoskevos petrushevskaae Caulet 的早期类型。

地理分布　南极罗斯海，晚始新世—中渐新世。

丽篮虫属 Genus *Lamprocyrtis* Kling, 1973

头近圆柱形，具三棱角锥状顶角，壳的早期为三节，晚期为二节，阔开口。

454. 射光丽篮虫 *Lamprocyrtis aegles* (Ehrenberg)

（图版 138，图 1–4）

Podocyrtis aegles Ehrenberg, 1854, pl. 35, fig. 18.

Lamprocyclas aegles (Ehrenberg), Petrushevskaya, 1971b, p. 201, pl. 116, figs. 1–2；Petrushevskaya and Kozlova, 1972, p. 544, pl. 36, fig. 13.

Calocyclas semipolita robusta Clark and Campbell, 1942, p. 84, pl. 8, fig. 21.

Theocotyle robusta (Clark and Campbell), Petrushevskaya, 1975, p. 580, pl. 8, fig. 9, pl. 22, fig. 1；Caulet, 1991, p. 539, pl. 4, fig. 14.

Lamprocyrtis (?) cf. *hannai* (Campbell and Clark), Abelmann, 1992a, p. 777, pl. 3, figs. 5–6.

地理分布　南极罗斯海，中渐新世。

455. 丽篮虫（未定种 1）*Lamprocyrtis* sp. 1

（图版 138，图 5–6）

头近圆柱形，有一个三棱角锥状顶角；胸呈葫芦形，与腹之间缩缢较深；腹不完整，呈喇叭状阔开口。

地理分布　南极罗斯海，出现于早中新世。

神篓虫科 Family Theocyrtidae Haeckel, 1887, emend. Nigrini, 1967

壳三节，由两个横缢环分为头、胸、腹，无放射肋。

冈瓦纳虫属 Genus *Gondwanaria* Petrushevskaya, 1975

头球形，有一头颈，胸呈冲天炉形，往下略扩大，胸壳上有或无侧翼，腹部发育程度不定。

456. 狄弗朗冈瓦纳虫 *Gondwanaria deflandrei* Petrushevskaya

（图版 138，图 7–8）

Gondwanaria deflandrei Petrushevskaya, 1975, p. 584, pl. 9, figs. 8–9.

Calocyclas sp. A, Takemura, 1992, p. 745, pl. 1, figs. 3–4.

胸壳上无侧翼，胸宽与腹宽近等或略小于腹宽，腹部开阔或发育形态有变化，壳壁孔较细，类圆形，规则或亚规则排列。

地理分布　南极罗斯海，出现于晚渐新世。

457. 日本冈瓦纳虫 *Gondwanaria japonica* (Nakaseko) group

（图版 138，图 9–14）

Sethocyrtis japonica Nakaseko, 1963, p. 176, pl. 1, figs. 10a–b；Nakaseko and Sugano, 1973, pl. 3, fig. 2.

Gondwanaria japonica (Nakaseko) group, Petrushevskaya, 1975, p. 584, pl. 8, fig. 15, pl. 9, figs. 2–7, pl. 12, fig. 1, pl. 21, figs. 4–5.

个体略小，壁孔稍大，有的标本具六角形框架或胸上部有侧翼。

地理分布 南极罗斯海，晚古新世—晚始新世。

458. 米露冈瓦纳虫 *Gondwanaria milowi* (Riedel and Sanfilippo) group

（图版 138，图 15–20；图版 139，图 1–15）

Cyclampterium (?) *milowi* Riedel and Sanfilippo, 1971, p. 1953, pl. 3B, fig. 3, pl. 7, figs. 8–9；Riedel and Sanfilippo, 1978, p. 67, pl. 4, fig. 14；Chen, 1975, p. 460, pl. 2, figs. 4–5；Abelmann, 1990, p. 696, pl. 7, fig. 8；Takemura, 1992, pp. 745–746, pl. 5, figs. 1–3.

Cyclampterium (?) *longiventer* Chen, 1975, pp. 459–460, pl. 10, fig. 7.

Thyrsocyrtis sp., Petrushevskaya, 1975, p. 580, pl. 8, fig. 10.

Lophocyrtis (*Paralampterium*) *longiventer* (Chen), Sanfilippo, 1990, p. 309, pl. 3, figs. 1–5；Strong et al., 1995, p. 208, fig. 11；Crouch and Hollis, 1996, p. 26；Hollis, 1997, p. 63, pl. 6, figs. 13–17；Hollis et al., 2020, pl. 16, figs. 19–20.

Calocyclas asperum (Ehrenberg), Caulet, 1991, p. 537, pl. 4, fig. 8.

头球形，较小；胸圆锥形，较短；腹近圆筒形，较长，壁孔类圆形，纵向排列，具六角形框架，孔径约为孔间桁宽的 2–3 倍；无放射肋，阔开口；壳体较粗实，一般壳壁较厚。*Cyclampterium* 是 *Cycladophoras* 属的一个亚属名称（Haeckel, p. 1379），该亚属的腹部具有多条放射肋，与此类标本特征不符，归入 *Gondwanaria* 属较合适。

地理分布 南极罗斯海，中始新世—中渐新世。

459. 翼冈瓦纳虫（新种）*Gondwanaria pteroforma* sp. nov.

（图版 140，图 1–12）

Gondwanaria japonica (Nakaseko) group, Petrushevskaya, 1975, pl. 21, figs. 4–5.

壳仅二节，头较大，半球形，头宽约为胸宽的一半，有一个瘦角锥形顶角，顶角长略小于头长；胸近胖圆筒状，上、下部等宽或下部略缩窄，在胸上部有三个较粗壮的三棱角锥状侧翼，侧翼基部较宽，末端缩尖，阔开口，无口缘；壳壁孔类圆形或椭圆形，亚规则或不规则分布，孔间桁有一些凸起，壳表显粗糙。

标本测量：壳长 125–180μm、宽 120–150μm，头长 50–75μm、宽 45–85μm，顶角长 25–50μm，侧翼长 45–75μm，侧翼基宽 25–40μm。

模式标本：DSDP274-29-1 2（图版 140，图 1–2），来自南极罗斯海的 DSDP Leg 28 航次 Site 274 孔的 274-29-1 岩心样品中，保存在中国科学院南海海洋研究所沉积标本室。

地理分布 南极罗斯海，晚始新世。

该新种特征与 *Sethocyrtis japonica* Nakaseko 区别较大，后者个体较小，胸近圆球形，口端缩窄，顶角不发育，一般无侧翼；新种与 Petrushevskaya（1975, pl. 21, figs. 4–5）的标本较相似，Petrushevskaya（1975）将此类标本归入并称其为非典型的 *Gondwanaria japonica* (Nakaseko)类型，显然，两者之间存在较大的特征差异。

460. 隶董冈瓦纳虫 *Gondwanaria redondoensis* (Campbell and Clark)

（图版 140，图 13–14）

Theocyrtis redondoensis Campbell and Clark, 1944a, p. 49, pl. 7, fig. 4；Casey, 1972, pl. 2, fig. 3.
Theocorys redondoensis (Campbell and Clark), Kling, 1973, p. 638, pl. 11, figs. 26–28；Chen, 1975, p. 505, pl. 20, figs. 2–3；Weaver, 1976, p. 582, pl. 2, fig. 11.
Gondwanaria deflandrei Petrushevskaya, Abelmann, 1990, p. 697, pl. 7, fig. 7.

胸壳上无侧翼，胸宽明显大于腹宽，胸与腹之间有一较深的勒缢，腹部呈圆筒形，口端截平，无口缘；壳壁孔较细，类圆形，亚规则排列。

地理分布　南极罗斯海，出现于早上新世。

461. 壮实冈瓦纳虫 *Gondwanaria robusta* (Abelmann)

（图版 140，图 15–22）

Cyrtocapsella robusta Abelmann, 1990, p. 696, pl. 5, figs. 10–11；Caulet, 1991, p. 538, fig. 4；Takemura, 1992, p. 746, pl. 1, figs. 5–6.

地理分布　南极罗斯海，早—晚始新世。

462. 半滑冈瓦纳虫 *Gondwanaria semipolita* (Clark and Campbell)

（图版 141，图 1–20）

Calocyclas semipolita Clark and Campbell, 1942, p. 83, pl. 8, figs. 12, 14, 17–19, 21–23.
Calocyclas semipolita (?) Clark and Campbell, Chen, 1975, p. 459, pl. 6, figs. 3–6.
Theocotyle robusta Petrushevskaya, 1975, p. 580, pl. 8, fig. 9, pl. 22, fig. 1.
Calocyclas cf. *semipolita* Clark and Campbell, Abelmann, 1990, p. 697, pl. 7, fig. 4.
Calocyclas semipolita Clark and Campbell group, Caulet 1991, p. 537, figs. 4: 1–4 (only).
Calocyclas asperum (Ehrenberg), Caulet, 1991, p. 537, pl. 4, fig. 8.
Aphetocyrtis rossi Sanfilippo and Caulet, 1998, p. 18, pl. 2, figs. 8, 9, 12, 13, pl. 7, figs. 1–9.
Pterocodon? *lex* Sanfilippo and Riedel, Hollis, 2002, p. 311, pl. 8, figs. 11, 12a–b.
Aphetocyrtis rossi ? Sanfilippo and Caulet, Hollis et al., 2020, pl. 15, figs. 8–11.

该种特征与 *Gondwanaria milowi* (Riedel and Sanfilippo) group 较接近，主要区别是前者的壳壁孔较小，类圆形，不规则分布，孔间桁一般稍宽。实际上这两个种存在着一些过渡类型标本，难以完全区别，可能合并为一个种群更加合理。

地理分布　南极罗斯海，晚始新世—中渐新世。

463. 半滑冈瓦纳虫（相似种）*Gondwanaria* cf. *semipolita* (Clark and Campbell)

（图版 142，图 1–18）

Calocyclas cf. *semipolita* Clark and Campbell, Abelmann, 1990, p. 697, pl. 7, fig. 4；Takemura, 1992, p. 745, pl. 4, figs. 5–6.

壳分三节，壁厚中等，壁孔六角形，排列较规则，孔间桁较细，有六角形框架。

地理分布　南极罗斯海，晚始新世—中渐新世。

464. 瘦冈瓦纳虫（新种）*Gondwanaria tenuoria* sp. nov.

（图版 142，图 19–28）

Lophophaena (?) sp., Petrushevskaya, 1975, pl. 9, fig. 13.

Gen. and sp. indet., Chen, 1975, pl. 19, fig. 5.

个体较瘦小，近葫芦形，壳分三节，壳壁较薄；头圆球形，头腔内的中心骨针发育为细柱状的顶针，壁孔较细；胸近圆锥形，下部略收缩，与胸壳间的缩缢较深，壁孔较小，类圆形，规则排列；腹呈圆筒形或有一定变化，向口端略变宽，壁孔类圆形，大小相近，比胸壁孔大，一般排列规则，具六角形框架，孔径是孔间桁宽的 2–3 倍，孔间桁较细；壳壁常有一些角锥形棘刺，口开阔，不平整。

标本测量：壳长 145–160μm、宽 75–90μm，头直径 25–35μm，胸长 50–60μm、宽 65–70μm，顶针长 5–15μm。

模式标本：DSDP274-23-1 1（图版 142，图 19–20），来自南极罗斯海的 DSDP Leg 28 航次 Site 274 孔的 274-23-1 岩心样品中，保存在中国科学院南海海洋研究所沉积标本室。

地理分布 南极罗斯海，早–中渐新世。

该新种特征与 *Gondwanaria* cf. *semipolita* (Clark and Campbell)较相似，主要区别是后者个体较大，壳壁稍厚，无顶针，壳表光滑，无棘刺。罗斯海的标本与 Gen. and sp. indet.（Chen, 1975, pl. 19, fig. 5）很相似，但后者口缘上有若干末端脚，可能是该种发育更加完善的结果。

465. 冈瓦纳虫（未定种 1）*Gondwanaria* sp. 1

（图版 142，图 29）

头球形，较小；胸圆锥形，较短；腹近圆筒形，较长，在下部近开口处有一浅缩缢；壁孔类圆形，大小近等，纵向规则排列，有六角形框架；壳表光滑。

地理分布 南极罗斯海，出现于早上新世。

吊筐虫属 Genus *Artophormis* Haeckel, 1881

壳卵形或纺锤形，壳壁内有一些侧肋，延伸为末端脚，口收缩。

466. 简单吊筐虫（相似种）*Artophormis* sp. cf. *A. gracilis* Riedel

（图版 143，图 1–13）

Artophormis gracilis Riedel, Sanfilippo and Riedel, 1973, p. 219, pl. 4, fig. 4；Abelmann, 1990, p. 697, pl. 7, fig. 5.

Artophormis sp. cf. *A. gracilis* Riedel, Sanfilippo and Riedel, 1973, p. 219, pl. 4, figs. 1–3; Sanfilippo and Riedel, 1974, p. 219, pl. 2, fig. 2.

Theocorys anaclasta Riedel and Sanfilippo, Sanfilippo et al., 1985, p. 683, figs. 24: 1a–b.

壳呈塔状或葫芦形，头小；胸分三段，自上而下迅速扩大，三段的壳宽比例约为 1∶2∶3，壁孔自上而下增大，为六角形或多角形，孔间桁较细；壳表有一些棘刺，表面粗糙；无侧肋和末端脚，口开阔。罗斯海的标本特征与 Sanfilippo 和 Riedel（1973, 1974）、Abelmann（1990）的标本基本相似，均无侧肋和末端脚等，但与原种 *Artophormis gracilis* Riedel（1959, p. 300, pl. 2, figs. 12–13）特征差别较大，后者的末节缩窄，壁孔大小不等且分布规律，具有末端脚。因此，罗斯海标本与 Sanfilippo 和 Riedel（1973, 1974）、Abelmann（1990）标本应属另外一个属种更加合理。

我们的标本特征也与 *Cymaetron sinolampas* Caulet（1991, p. 536, pl. 4, figs. 10–12）较相似，但后者胸壳增大不明显，壳孔类圆形，孔间桁较宽，而且壳表光滑无刺。该种类归入 *Cymaetron* 属（Caulet, 1991）[两节壳，呈蜿蜒的轮廓；头小，有一顶角；胸壳上有 2–3 个收缩将之分为上、中、下三段或四段（无内隔环），口部阔开]较合适。

有两个标本（30-2 1、30-6-2）的壳体形态特征与 *Lophocorys polyacantha* Popofsky（1913, p. 400, text-fig. 122）完全一致，头呈半球形，胸仅二节且均无头角、侧翼和末端脚。

此类标本有一定的变化范围，又存在相互关联或相似之处的过渡类型，对于鉴定种类尚存质疑。在属征上，与 *Artophormis* 和 *Lophocorys* 属基本不符。

地理分布　南极罗斯海，晚始新世—中渐新世。

467. 吊筐虫（未定种 1）*Artophormis* sp. 1

（图版 143，图 14–15）

地理分布　南极罗斯海，出现于早渐新世。

468. 吊筐虫（未定种 2）*Artophormis* sp. 2

（图版 143，图 16–17）

地理分布　南极罗斯海，出现于晚始新世。

毛虫科 Family Podocampidae Haeckel, 1887

壳呈环状，由三个或更多的横缢将壳体分为四节或五节以上，有三个辐射状肋骨。

篮袋虫属 Genus *Cyrtopera* Haeckel, 1881

具三个格孔状的侧翼，或为三纵排的格孔翼，头部有一角。

469. 小壶篮袋虫 *Cyrtopera laguncula* Haeckel

（图版 143，图 18–23）

Cyrtopera laguncula Haeckel, 1887, p. 1451, pl. 75, fig. 10；Benson, 1966, p. 510, pl. 35, figs. 3–4；Chen, 1975, p. 460, pl. 18, fig. 9；Renz, 1976, p. 120, pl. 4, fig. 7；Abelmann, 1992a, p. 378, pl. 5, fig. 18；陈木宏、谭智源，1996，226 页，图版 34，图 4，图版 54，图 7；陈木宏等，2017，221–222 页，图版 79，图 10–12，15–21。

Stichopera pectinata Haeckel group, Kling, 1973, p. 638, pl. 3, figs. 25–27, pl. 10, figs. 1–5.

地理分布　南极罗斯海，晚上新世。

470. 巨篮袋虫（新种）*Cyrtopera magnifica* sp. nov.

（图版 144，图 1–2）

壳长圆锥形或尖塔形，个体较大，至少有六节，各节高度相近，宽度逐渐增大；头部很小，圆球形，完全被顶角的基部包围；顶角较长，基部较宽，呈圆柱形或长角锥形，末端缩尖；胸和腹共有五节以上，各节间的缩缢较深，内隔环很细，壳壁孔均为类圆形，大小相近，亚规则排列，自上而下孔径略增大；壳口开阔，末节壳缺失；壳表无棘刺，较平滑，侧翼缺失。

标本测量：壳长>300μm、宽>170μm，头直径 20μm，胸长 50–60μm、宽 65–70μm，顶角长 115μm。

模式标本：DSDP274-33-3 2（图版 144，图 1–2），来自南极罗斯海的 DSDP Leg 28 航次 Site 274 孔的 274-33-3 岩心样品中，保存在中国科学院南海海洋研究所沉积标本室。

地理分布 南极罗斯海，出现于晚始新世。

该新种基本特征与 *Cyrtopera laguncula* Haeckel 很相似，主要区别是后者的壳体相对较小，头部外露，顶角较为短小，壁孔六角形，排列规则，有六角形框架，孔间桁细。

石毛虫科 Family Lithocampidae Haeckel, 1887

壳多节环形，由三个以上的横缢分隔为四个以上的环形节，无放射肋。

石螺旋虫属 Genus *Lithostrobus* Bütschli, 1882

壳圆锥形，向扩开口端逐渐膨胀。头部有一角。

471. 长石螺旋虫（相似种）*Lithostrobus* cf. *longus* Grigorjeva

（图版 144，图 3–15）

Lithostrobus cf. *longus* Grigorjeva, Hollis, 2002, p. 310, pl. 8, figs. 1–3.

地理分布 南极罗斯海，晚始新世。

窄旋虫属 Genus *Artostrobus* Haeckel, 1887

壳呈圆筒状，上部圆形，下部截平，头部有一角。

472. 环窄旋虫 *Artostrobus annulatus* (Bailey)

（图版 144，图 16–33）

Cornutella annulata Bailey, 1856, p. 3, pl. 1, figs. 5a–b.

Artostrobus annulatus (Bailey), Haeckel, 1887, p. 1481；Cleve, 1899, p. 27, pl. 1, fig. 6；Petrushevskaya, 1967, p. 98, figs. 56I–V；Petrushevskaya, 1975, p. 578, pl. 10, figs. 4–5；Renz, 1976, p. 117, pl. 4, fig. 5；Takahashi and Honjo, 1981, p. 154, pl. 10, fig. 8；Abelmann, 1992a, p. 378, pl. 5, fig. 16；Hollis, 1997, p. 57, pl. 5, figs. 1–2；陈木宏等，2017，226 页，图版 81，图 10–12。

Cycladophora annulata (Bailey), van de Paverd, 1995, p. 234, pl. 76, figs. 29–30.

地理分布 南极罗斯海，中中新世—全新世。

473. 米斯窄旋虫 *Artostrobus missilis* (O'Connor)

（图版 145，图 1–30）

Artostrobus sp., Petrushevskaya, 1975, pl. 10, fig. 1.

Cyrtocalpis sp. aff. *C. operosa* Tan Sin Hok, Foreman, 1978, p. 746, pl. 5, fig. 6；Ling, 1991, p. 320, pl. 1, fig. 10.

Artostrobus sp., Ling and Lazarus, 1990, p. 355, pl. 3, fig. 14.

Siphocampe missilis O'Connor, 1994, p. 340, pl. 1, figs. 7, 9–12, pl. 3, figs. 8–12.

地理分布　南极罗斯海，早古新世—晚始新世。

474. 多节窄旋虫（新种）*Artostrobus multiartus* sp. nov.

（图版 145，图 31–48）

Lithomitra sp., Bjørklund, 1976, p. 1124, pl. 15, figs. 26, 28 (only), pl. 23, figs. 15–21.

个体较小，圆筒形或中部略膨大，壳壁稍厚；头部脑丘状，近半球形，透明，无壁孔；胸较短，扁腰鼓形或渐扩大，胸与腹壳之间的隔环清楚，勒缢较深；腹壳较长，圆筒形，分三节以上，各腹节间有内隔环，但壳表无勒缢；壳壁孔类圆形，大小相近，横向规则排列，表面光滑；口开放，截平，无口缘环。

标本测量：壳长 70–95μm、宽 30–40μm，头直径 12–20μm。

模式标本：DSDP274-42-2 1（图版 145，图 31–32），来自南极罗斯海的 DSDP Leg 28 航次 Site 274 孔的 274-42-2 岩心样品中，保存在中国科学院南海海洋研究所沉积标本室。

地理分布　南极罗斯海，早古新世—晚始新世。

该新种特征与*Artostrobus pachyderma* (Ehrenberg)接近，主要区别是后者的壳壁很厚，最大的壁厚位于壳体中部，且连续包覆胸壳和腹壳，外表平滑无勒缢；新种与 *Artostrobus missilis* O'Connor 的主要区别是后者的胸与腹较难区分，两节之间无明显的勒缢和隔环。

475. 厚壁窄旋虫 *Artostrobus pachyderma* (Ehrenberg) group

（图版 146，图 1–20）

Eucyrtidium pachyderma Ehrenberg, 1873, p. 231；Ehrenberg, 1876, pl. 11, fig. 21.
Theocampe sp. aff. *T. minuta* (Clark and Cam.), Petrushevskaya, 1975, pl. 26, fig. 3.
Siphocampe pachyderma (Ehrenberg), Caulet, 1991, p. 539, pl. 3, fig. 12.
Archicorys sp., Ling, 1991, pp. 318–319, pl. 1, fig. 11.
Plannapus ? *aitai* O'Connor, 2000, pp. 199–200, pl. 2, figs. 16–21, pl. 3, figs. 9–18.

地理分布　南极罗斯海，早古新世—中渐新世。

476. 平板窄旋虫? *Artostrobus*? *pretabulatus* Petrushevskaya

（图版 146，图 21–26）

Artostrobus (?) *pretabulatus* Petrushevskaya, 1975, p. 580, pl. 10, figs. 2–3.

地理分布　南极罗斯海，晚上新世—全新世。

477. 窄旋虫（未定种 1）*Artostrobus* sp. 1

（图版 146，图 27–28）

Theocampid multisegmented, Petrushevskaya, 1975, p. 578, pl. 10, fig. 6.

腹壳不完整或变异。

地理分布　南极罗斯海，出现于中渐新世。

细篮虫属 Genus *Eucyrtidium* Ehrenberg, 1847, emend. Nigrini, 1967

具卵形或纺锤形壳，壳口缢缩，但不延长成管。头具一实心的角。

478. 环节细篮虫 *Eucyrtidium annulatum* (Popofsky)

（图版 146，图 29–36）

Stichopilium annulatum Popofsky, 1913, pp. 403–404, pl. 37, figs. 2–3.
Lithostrobus hexastichus Benson, 1966, pp. 506–508, pl. 34, figs. 13–14.
Eucyrtidium sp., Petrushevskaya, 1971d, figs. 99I–II.
Eucyrtidium? *hexastichum* group Benson, 1983, p. 503, pl. 9, fig. 10.
Eucyrtidium annulatum (Popofsky)，陈木宏等，2017，226 页，图版 81，图 13–22。

地理分布 南极罗斯海，晚始新世—早渐新世，全新世。

479. 歪细篮虫（新种）*Eucyrtidium anomurum* sp. nov.

（图版 146，图 37–42）

Eucyrtidium sp. A, Petrushevskaya, 1975, p. 581, pl. 14, figs. 21–22.

个体较大，有 5–8 节壳或更多，壳体发育不正常，有的呈扭曲状，壳室之间的隔环不规律弯曲或接近螺旋状；头球形，很小，腔内有一主骨针或延伸为很小的顶针；胸壳圆锥形，宽度比腹宽小，缢缩较深；腹壳多节，各节的形态不同，宽度有变化，缢缩较深；壁孔类圆形，大小相近，纵向亚规则排列或不规则分布；壳表略显粗糙。

标本测量：壳长 180–300μm、宽 105–120μm，头直径 20–30μm，胸长 35–50μm、宽 75–95μm，顶针长 5–20μm。

模式标本：DSDP274-33-3 1（图版 146，图 39–40），来自南极罗斯海的 DSDP Leg 28 航次 Site 274 孔的 274-33-3 岩心样品中，保存在中国科学院南海海洋研究所沉积标本室。

地理分布 南极罗斯海，晚始新世—早渐新世。

该新种特征与 *Eucyrtidium antiquum* Caulet 较接近，主要区别是后者的壳形规则不扭曲，腹部各节的形态与大小基本一致，内隔环均呈水平状，相互平行。

480. 丽转细篮虫 *Eucyrtidium calvertense* Martin

（图版 147，图 1–12）

Eucyrtidium calvertense Martin, 1904, p. 450, pl. 130, fig. 5；Hays, 1965, p. 181, pl. III, fig. 4；Hays, 1970, p. 213, pl. 1, fig. 6；Kling, 1973, p. 636, pl. 4, figs. 16, 18, 19, pl. 11, figs. 1–5；Chen, 1975, p. 460, pl. 15, fig. 9；Weaver, 1983, p. 675, pl. 2, fig. 9, pl. 4, fig. 7；Vigour and Lazarus, 2002, pl. P2, figs. 1–4；陈木宏等，2017，227 页，图版 81，图 23–24。

该种壳体特征与 *Buryella tetradica* Foreman（1973, p. 433, pl. 8, figs. 4–5, pl. 9, figs. 13–14）较接近，主要区别是前者壳体各节长度相近，而后者的第一腹节较长，最宽，壳壁最厚，第二腹节渐缩窄，且壳壁变薄。

地理分布 南极罗斯海，中中新世—早上新世。

481. 芊口细篮虫 *Eucyrtidium cienkowskii* Haeckel group

（图版 147，图 13–20）

Eucyrtidium cienkowskii Haeckel, 1887, p. 1493, pl. 80, fig. 9.
Eucyrtidium cienkowskii Haeckel group, Sanfilippo et al., 1973, p. 221, pl. 5, figs. 7–11；Chen, 1975, p. 460, pl. 15, fig. 7；Weaver, 1976a, p. 581, pl. 4, figs. 3–5, pl. 8, figs. 7–9；Weaver, 1983, p. 675, pl. 5, fig. 6；Abelmann, 1990, p. 696, pl. 6, figs. 3A–D.
Stichopodium (?) sp., Petrushevskaya, 1975, pl. 22, fig. 4.

地理分布　南极罗斯海，晚始新世—早上新世。

482. 冰细篮虫（新种）*Eucyrtidium gelidium* sp. nov.

（图版 147，图 21–24）

壳呈胖子弹形，壳壁较厚；头球形，很小，有一很小的顶角或突起；胸圆锥形，腹 2–3 节，圆筒形，各节之间缩缢不明显，但有较细的内隔环；壳壁稍厚，壁孔类圆形或椭圆形，大小不匀，分布无规律，孔间桁较宽；壳表似有一些锥形或三角形片状小凸起，还附着一些枝状杂物；口部开放，不收缩，口缘不平整。

标本测量：壳长 140–150μm、宽 110–125μm，头直径 15–20μm，胸长 30–55μm、宽 50–75μm，顶针长 5–15μm。

模式标本：DSDP274-25-1 2（图版 147，图 21–22），来自南极罗斯海的 DSDP Leg 28 航次 Site 274 孔的 274-25-1 岩心样品中，保存在中国科学院南海海洋研究所沉积标本室。

地理分布　南极罗斯海，中渐新世。

该新种壳体特征与 *Eucyrtidium annulatum* (Popofsky)较相似，但后者的壳壁较薄，壁孔较小，大小相同，排列规则，壳表光滑。

483. 颗粒细篮虫 *Eucyrtidium granulatum* (Petrushevskaya) group

（图版 147，图 25–28；图版 148，图 1–24）

Lithocampe sp. A, Dumitrica, 1973, p. 789, pl. 10, fig. 3, pl. 11, fig. 3.
Lithocampe ? *granulata* Petrushevskaya, 1977, p. 18, pl. 3, figs. a, b, v.
Buryella cf. *clinata* Foreman, Blome, 1992, p. 644, pl. 1, fig. 3.
Buryella sp. A, Blome, 1992, p. 644, pl. 1, figs. 1–2.
Stichomitra granulata (Petrushevskaya), Hollis, 1993, p. 321, pl. 1, figs. 10–11.
Buryella granulata (Petrushevskaya), Hollis, 1997, p. 80, pl. 21, figs. 1–5；Hollis, 2002, p. 300, pl. 4, fig. 10；Hollis et al., 1997, p. 57, pl. 5, fig. 6；Strong et al., 1995, p. 209, fig. 8J；O'Connor, 2001, p. 7, pl. 1, figs. 1a–5, pl. 3, figs. 1–4.
Eucyrtidium antiquum Caulet, 1991, pp. 535–536, pl. 4, figs. 1–2；Hollis, 1997, p. 60, pl. 5, figs. 30–31；Hollis et al., 2020, pl. 10, fig. 1.
Eucyrtidium sp., Chen, 1975, p. 461, pl. 7, figs. 6–8.
Stichocorys peregrina (Riedel), Weaver, 1983, p. 678, pl. 6, figs. 3, 9.
Eucyrtidium calvertense Martin group, Abelmann, 1990, p. 696, pl. 6, figs. 4, 5A–C.
Eucyrtidium cheni Takemura, 1992, p. 746, pl. 4, figs. 1–4.

该种形态变化较大，壳体有 5–10 节，上部圆锥形，下部一般圆柱形，各节高度相近；口部一般不缩窄，有时缩窄；不同个体的壳宽不同，或整个较窄，或稍宽，或在中部较宽；壳壁一般较厚，表面显粗糙；顶角或顶针有的发育不明显。Petrushevskaya（1973, 1977）首先报道和建立了这一种类，并展示口部缩窄与不缩窄两种类型个体，后人的鉴定中出现了上述的各种相似特征，仅 Hollis（1993）、Hollis 等（2020）等的标本图示就呈现完全不同的壳体特征，其他多位作者的标本也属类似情况，因此，该种的鉴定引用存在一定的混乱或困难。Hollis 等（2020）认为该种是澳大利亚古新世—始新世的地层种。由于罗斯海的 Site 274 岩心中的大量标本特征包含了该种的基本特征，呈现了一定的变化范围与相互的过渡类型，从中难以划分这类标本之间的区别界线，因此将之归为一个种群较为合理。

地理分布 南极罗斯海，中始新世—晚渐新世。

484. 小孔细篮虫 *Eucyrtidium punctatum* (Ehrenberg) group

（图版 149，图 1–16）

Lithocampe punctata Ehrenberg, 1844, p. 84.
Eucyrtidium punctatum (Ehrenberg), Ehrenberg, 1858, p. 43；Ehrenberg, 1854, pl. 22, fig. 24；Sanfilippo et al., 1973, p. 221, pl. 5, figs. 15–16.
Eucyrtidium punctatum (Ehrenberg) group, Chen, 1975, pp. 460–461, pl. 15, fig. 8.
Eucyrtidium inflatum Kling, 1973, p. 636, pl. 11, fig. 7, pl. 15, figs. 7–10.
Stichopodium sp. aff. *Eucyrtidium matuyamai* Hays, Petrushevskaya and Kozlova, 1972, p. 549, pl. 26, figs. 15–16 (only).
Stichopodium inflatum (Kling) group, Petrushevskaya, 1975, p. 581, pl. 26, figs. 7–8.
Eucyrtidium sp. aff. *E. inflatum* Kling, Weaver, 1976, p. 581, pl. 2, fig. 12.
Eucyrtidium pseudoinflatum Weaver, 1983, pp. 675–676, pl. 5, figs. 8–9；Lazarus, 1990, p. 716, pl. 6, figs. 12–14.

地理分布 南极罗斯海，中中新世—早上新世。

上述异名表所列文献的标本特征基本一致，均被归为 *Eucyrtidium punctatum* (Ehrenberg) group 较合适。其中包括 Chen（1975）的 *Eucyrtidium punctatum* (Ehrenberg) group 应属此种（第一腹壳明显膨大），其与 Sanfilippo 等（1973, p. 221, pl. 5, figs. 15–16）的 *Eucyrtidium punctatum* (Ehrenberg) group 有一定区别，后者腹壳近圆筒形，近口端收缩不明显（与 *Eucyrtidium cienkowskii* Haeckel group 较接近）；与 Abelmann（1990, p. 696, pl. 6, figs. 6A–B）的 *Eucyrtidium punctatum* (Ehrenberg) group 也存在差异，后者壳体中部略增大，壁厚，口端缩窄。因此，该种 *Eucyrtidium punctatum* (Ehrenberg) group 存在着一定特征变化，罗斯海标本中呈现出其不同类型的标本。

485. 洁雅细篮虫 *Eucyrtidium lepidosa* (Kozlova)

（图版 149，图 17–31）

Lithocampe (?) *lepidosa* Kozlova, Petrushevskaya, 1971d, p. 174, figs. 90I–II.

头小，亚球形或半球形，有一细小的顶针；壳体呈圆锥形，各节壳渐增大，壳壁稍厚，壁孔类圆形，大小略有差异，排列不规则。该种壳壁较厚，孔间桁有尖突，壳壁粗糙，而罗斯海的标本壳壁稍薄和平滑。

地理分布 南极罗斯海，中始新世—中渐新世。

486. 细篮虫（未定种 1）*Eucyrtidium* sp. 1

（图版 150，图 1–2）

壳串珠状，壳表光滑无刺。头部破损，胸圆锥形，腹部有两节，腰鼓形，隔环有歪曲，第一腹节壁孔圆形，排列规则，第二腹节壁孔大小不等，排列不规则，第三腹节开始已破缺。

地理分布　南极罗斯海，出现于中渐新世。

石帽虫属 Genus *Lithomitra* Bütschli, 1881

壳呈圆筒形，头圆形，腹口平截，头无角。

487. 线石帽虫 *Lithomitra lineata* (Ehrenberg) group

（图版 150，图 3–28）

Lithocampe lineata Ehrenberg, 1838, S. 130 (partim)；Ehrenberg, 1854, Taf. 22, Fig. 26, Taf. 36, Fig. 16.

Lithomitra lineata (Ehrenberg), Haeckel, 1887, p. 1484；Cleve, 1899, pl. 2, fig. 7；Schröder, 1914, p. 137, fig. 113；Bjørklund, 1976, p. 1124, pl. 11, fig. 16；Cortese et al., 2003, pl. 4, figs. 23–26.

Lithomitra lineata (Ehrenberg) group, Riedel and Sanfilippo, 1971, p. 1600, pl. II, figs. 1–11, pl. 21, figs. 14–16, pl. 3E, fig. 14；Petrushevskaya and Bjørklund, 1974, fig. 3；Bjørklund et al., 2014, p. 88, pl. 9, figs. 12–14；谭智源、陈木宏，1999，351 页，图 5-294；陈木宏等，2017，229–230 页，图版 82，图 14–20。

Siphocampe arachnea (Ehrenberg), Riedel, 1958, p. 242, pl. 4, figs. 7–8；Abelmann, 1990, p. 698, pl. 8, fig. 4B (only)；Abelmann, 1992b, p. 382, pl. 5, fig. 15.

Siphocampe imbricata (Ehrenberg), Caulet, 1991, p. 539, pl. 3, fig. 13.

Siphocampe nodosaria (Haeckel), Takemura, 1992, p. 743, pl. 3, fig. 15.

Tricolocampe lineata (Ehrenberg), van de Paverd, 1995, p. 253, pl. 76, figs. 13–18.

Eucyrtidium lineatum Ehrenberg, Suzuki et al., 2009, pl. 44, figs. 5a–b, 6a–b.

该种与 *Siphocampe arachnea* (Ehrenberg)很相似，壳表均有一些纵纹，主要区别在于前者各腹壳的形态有一定变化，每节腹壳仅有一排壳孔；而后者腹部一般有 5–7 个圆弧状缢缩，各腹壳发育较规则，每节腹壳上有 2–4 排壳孔，头部有一斜生的顶管。

地理分布　南极罗斯海，早始新世—全新世。

管毛虫属 Genus *Siphocampe* Haeckel, 1881

壳呈卵形或纺锤形。壳口缢缩，但不延长成管状。头具一斜生开口的顶管。

488. 高奇剑管毛虫 *Siphocampe altamiraensis* (Campbell and Clark)

（图版 150，图 29–33）

Lithomitra altamiraensis Campbell and Clark, 1944a, p. 53, pl. 7, fig. 9.

Siphocampe altamontensis (Campbell and Clark), Foreman, 1968, p. 53, pl. 6, figs. 14a–b；Ling, 1991, p. 320, pl. 1, fig. 12.

Foreman（1968）和 Ling（1991）鉴定种类 *Siphocampe altamontensis* (Campbell and Clark) 所引用的文献 *Tricolocampe* (*Tricolocamptra*) *altamontensis* Campbell and Clark（1944a, p. 33, pl. 7, figs. 24, 26）有误，经查该文献中不存在这个种，所标记处为其他种类，可能他们都未能细查 Campbell 和 Clark（1944）的原始文献。罗斯海的标本与 Ling（1991）的鉴定标本较为相似，与 Campbell 和 Clark（1944a, p. 53, pl. 7, fig. 9）的种类特征接近。

地理分布　南极罗斯海，早始新世。

489. 隶杂管毛虫 *Siphocampe elizabethae* (Clark and Campbell)

（图版 151，图 1–4）

Lithomitra elizabethae Clark and Campbell, 1942, p. 92, pl. 9, fig. 18.
Lithomitrella elizabethae (Clark and Campbell), Caulet, 1986, p. 853.
Siphocampe elizabethae (Clark and Campbell), Nigrini, 1977, p. 256, pl. 3, fig. 6.
Siphocampe ? *elizabethae* Nigrini, Hollis et al., 1997, pp. 55–56, pl. 4, figs. 21–26.

地理分布 南极罗斯海，晚始新世。

490. 中突管毛虫（新种）*Siphocampe mesinflatis* sp. nov.

（图版 151，图 5–13）

壳呈圆筒形，仅四节壳；头较大，脑丘状，半球形，顶角很小，有一侧管；头与胸的缩缢很浅，胸近圆筒形，往下略扩大；第一腹节膨大，呈腰鼓形，壳缘圆弧形，为整个壳体的最宽处；第二腹节圆筒形，与胸等宽，壳缘直线形；胸与腹各节高度近等，内隔环清楚，外缩缢较深；壳体的壁孔均很小，六角形或类圆形，横向规则排列，每一节壳体上的壁孔横向有 7–10 个；壳口收缩不明显，截平，无口缘环。

标本测量：壳长 105–130μm、宽 55–75μm，头长 20–25μm、宽 40–45μm。

模式标本：DSDP274-1-3 1（图版 151，图 5–6），来自南极罗斯海的 DSDP Leg 28 航次 Site 274 孔的 274-1-3 岩心样品中，保存在中国科学院南海海洋研究所沉积标本室。

地理分布 南极罗斯海，中中新世—全新世。

该种特征与 *Siphocampe corbula* (Harting)较接近，但后者壳体近纺锤形，壳节高度不等，缩缢较浅，口部收缩，似有口缘环。

491. 方管毛虫 *Siphocampe quadrata* (Petrushevskaya and Kozlova)

（图版 151，图 14–15）

Lithamphora sacculifera quadrata Petrushevskaya and Kozlova, 1972, p. 539, pl. 30, figs. 4–6.
Lithomitra docilis Foreman, 1973, p. 431, pl. 8, figs. 20–22, pl. 9, figs. 3–5；Blome, 1992, p. 645, pl. 1, fig. 4；Nishimura, 1992, p. 329, fig. 11, pl. 13, fig. 19.
Siphocampe? *quadrata* (Petrushevskaya and Kozlova), Nigrini, 1977, p. 257, pl. 3, fig. 12；Caulet, 1991, p. 539；Takemura, 1992, p. 743, pl. 7, fig. 7.

地理分布 南极罗斯海，出现于晚始新世。

石毛虫属 Genus *Lithocampe* Ehrenberg, 1838

壳呈卵形或纺锤形。壳口缢缩，但不延长成管状，头无角也无管。

492. 微缚石毛虫 *Lithocampe subligata* Stöhr group

（图版 151，图 16–33）

Lithocampe subligata Stöhr, 1880. p. 102, pl. 4, fig. 1；Petrushevskaya and Kozlova, 1972, p. 546, pl. 25, figs. 7–10；Petrushevskaya, 1973, pl. 3, fig. 4；Petrushevskaya, 1975, p. 581, pl. 14, figs. 6–9, 12.

地理分布 南极罗斯海，早始新世—中渐新世。

旋篮虫属 Genus *Spirocyrtis* Haeckel, 1881, emend.

壳具螺旋状缢，或各节间缩缢呈斜向等非规则状，头有一角。

493. 丽旋篮虫（新种）*Spirocyrtis bellulis* sp. nov.

（图版 152，图 1–4）

Spirocyrtis scalaris Haeckel, Petrushevskaya, 1971d, p. 211, figs. 126i, vii.

个体较小，仅三节壳；头呈脑丘状，近圆球形，顶角和侧管均很小；胸呈圆锥形，自上而下渐扩大，与腹壳间的勒缢较深；腹壳为圆筒形，腹壳顶部比胸壳底部明显增大，腹壳长约为胸壳长的 1.3 倍，口部开放，无口缘环；壳壁孔类圆形或六角形，大小相近，排列较规则，有六角形框架，胸与腹各有 5–7 排孔；壳表平滑。

标本测量：壳长 85–95μm，头直径 15μm，胸长 25–30μm、宽 30–38μm，腹长 40–50μm、宽 50–55μm。

模式标本：DSDP274-23-3 2（图版 152，图 1–2），来自南极罗斯海的 DSDP Leg 28 航次 Site 274 孔的 274-23-3 岩心样品中，保存在中国科学院南海海洋研究所沉积标本室。

地理分布　南极罗斯海，早–中渐新世。

该新种特征壳形与 *Spirocyrtis scalaris* Haeckel 有些接近，头部结构相似，均具头角和侧管，壳节的宽度渐加大，呈梯状，主要区别是此新种的壳体勒缢不螺旋，呈水平状。

494. 圆梯旋篮虫?（亲近种）*Spirocyrtis*? aff. *gyroscalaris* Nigrini

（图版 152，图 5–8）

Spirocyrtis? aff. *gyroscalaris* Nigrini, Hollis, 2002, p. 301, pl. 5, figs. 7a–c.

个体较大，圆锥形，头角较小，有约 7–8 节壳，各壳节的边缘呈圆弧形，勒缢较深，每节壳上有 2–5 排类圆形孔，从第四节壳开始发育有六条放射肋，在口缘延伸为末端脚。此种特征与 *Spirocyrtis gyroscalaris* Nigrini（1977, p. 258, pl. 2, figs. 10, 11）较接近，但后者无放射肋和末端脚。

地理分布　南极罗斯海，晚始新世。

495. 矩形旋篮虫（新种）*Spirocyrtis rectangulis* sp. nov.

（图版 152，图 9–11）

个体较宽，壳分三节，壳壁较薄；头较大，脑丘状，不规则形，下部陷入胸壳内，头侧有一小角；胸的顶部为圆锥形，中下部近圆筒形，在中间段膨凸，底部侧面与加宽的腹壳顶部垂直接触呈 90°角，内隔环清楚；腹壳横向迅速增宽，呈规矩的圆筒形，顶面水平状，与侧面相互垂直，口端截平，不收缩；壁孔均为六角形，大小相近，排列规则，胸与腹横向上各有 8 排以上的壁孔，有六角形框架，孔间桁较细，孔径是孔间桁宽的 2–3 倍；壳表光滑无刺。

标本测量：壳长 155μm，头长 30μm（部分陷入胸内）、宽 38μm，胸长 75μm、宽 100μm

（中点），腹长 65μm，腹宽 130μm。

模式标本：DSDP274-30-6 2（图版 152，图 9–11），来自南极罗斯海的 DSDP Leg 28 航次 Site 274 孔的 274-30-6 岩心样品中，保存在中国科学院南海海洋研究所沉积标本室。

地理分布 南极罗斯海，出现于晚始新世。

该新种特征与 *Spirocyrtis scalaris* Haeckel 接近，主要区别是后者呈塔形，自上而下渐扩大，螺旋形的壳节较多。罗斯海标本壳形与（Petrushevskaya, 1971d, p. 211, figs. 126i–iv）较相似。

496. 梯盘旋篮虫 *Spirocyrtis scalaris* Haeckel

（图版 152，图 12–21）

Spirocyrtis scalaris Haeckel, 1887, p. 1509, pl. 76, fig. 14；Popofsky, 1913, S. 406, Text-figs. 128–130；Nigrini, 1967, pp. 88–90, pl. 8, fig. 7, pl. 9, fig. 4；陈木宏、谭智源，1996，229 页，图版 35，图 2–3，图版 54，图 13，16；陈木宏等，2017，233 页，图版 84，图 3–8。

地理分布 南极罗斯海，晚始新世—早渐新世。

陀螺虫科 Family Artostrobiidae Riedel, 1967, emend. Foreman, 1973

壳多节，具头角和侧管。

陀螺虫属 Genus *Artostrobium* Haeckel, 1887

壳的各节有若干横排小孔，排数可变。

497. 耳陀螺虫 *Artostrobium auritum* (Ehrenberg) group

（图版 153，图 1–8）

Lithocampe aurita Ehrenberg, 1844a, S. 84.

Lithocampe australe Ehrenberg, 1844b, p. 187.

Eucyrtidium auritum Ehrenberg, 1854, Taf. 22, Fig. 25.

Artostrobus auritus Haeckel, 1887, p. 1482.

Siphocampium cf. *seriatus* Haeckel, Benson, 1966, pp. 521–523, pl. 35, figs. 12–13.

Artostrobium auritum (Ehrenberg) group, Riedel and Sanfilippo, 1971, p. 1599, pl. 1H, figs. 5–8；Kling, 1973, p. 639, pl. 5, figs. 27–30, pl. 12, figs. 24–27；谭智源、宿星慧，1982，182 页，图版 20，图 11；陈木宏、谭智源，1996，232 页，图版 35，图 21–23；陈木宏等，2017，234–235 页，图版 84，图 12–17。

Botryostrobus auritus australis (Ehrenberg) group, Nigrini, 1977, pp. 246–248, pl. 1, figs. 2–5；Abelmann, 1992a, p. 378, pl. 5, figs. 1–12.

地理分布 南极罗斯海，晚中新世—全新世。

498. 糙角陀螺虫 *Artostrobium rhinoceros* Sanfilippo and Riedel

（图版 153，图 9–16）

Artostrobium rhinoceros Sanfilippo and Riedel, 1974, p. 1000, pl. 4, figs. 8–9.

Spirocyrtis n. sp. A, Hollis, 1997, p. 56, pl. 4, figs. 33–35.

该种特征与 *Stichocorys seriata*（Jørgensen, 1905, p. 140, pl. 18, figs. 102–104）较为接近，主要区别是前者各节壳宽变化不大，而后者第二节开始明显加宽，最大壳宽在第三、四节，第五节略缩窄较长，呈圆筒状。

地理分布　南极罗斯海，中渐新世—晚中新世。

499. 浪陀螺虫 *Artostrobium undulatum* (Popofsky)

（图版 153，图 17–21）

Artopilium undulatum Popofsky, 1913, p. 405, pl. 36, figs. 4–5.
Peromelissa undulatum (Popofsky), van de Paverd, 1995, p. 224, pl. 71, figs. 12–13.

地理分布　南极罗斯海，晚始新世。

旋葡萄虫属 Genus *Botryostrobus* Haeckel, 1887

壳圆锥形，具直的轴线，各壳节长度不等。头部叶状，有一些不规则的缩缢。

500. 布拉旋葡萄虫 *Botryostrobus bramlettei* (Campbell and Clark)

（图版 153，图 22–32）

Lithomitra (*Lithomitrissa*) *bramlettei* Campbell and Clark, 1944a, p. 53, pl. 7, figs. 10–14.
Artostrobium rhinoceros Sanfilippo and Riedel, 1974, p. 1000, pl. 4, figs. 8–11.
Botryostrobus euporus (Ehrenberg) group, Petrushevskaya, 1975, p. 585, pl. 10, figs. 22–24.
Stichocorys seriata (Jørgensen), Bjørklund, 1976, p. 1124, pl. 10, figs. 7–12.
Botryostrobus bramlettei (Campbell and Clark), Nigrini, 1977, p. 248, pl. 1, figs. 7–8; Alexandrovich, 1989, pl. 3, fig. 5; 陈木宏等，2017，235–236 页，图版 84，图 25–28。
Botryostrobus bramlettei pretumidulus Caulet, 1979, p. 129, pl. 1, fig. 5.

地理分布　南极罗斯海，晚中新世—全新世。

501. 佐得旋葡萄虫 *Botryostrobus joides* Petrushevskaya

（图版 154，图 1–16）

Botryostrobus joides Petrushevskaya, 1975, p. 585, pl. 10, fig. 37.

地理分布　南极罗斯海，晚始新世—中中新世。

筐列虫属 Genus *Phormostichoartus* Campbell, 1951, emend. Nigrini, 1977

壳分四节，圆柱形，口端略收缩，口缘清楚，垂直管在胸壳上，无顶角。

502. 管筐列虫 *Phormostichoartus fistula* Nigrini

（图版 154，图 17–28）

Phormostichoartus fistula Nigrini, 1977, p. 253, pl. 1, figs. 11–13.

地理分布　南极罗斯海，晚始新世。

参考文献

Abelmann A. 1990. Oligocene to Middle Miocene radiolarian stratigraphy of southern high latitudes from Leg 113, Sites 689 and 690, Maud Rise. In: Barker P F, Kennett J P, Connell S O (eds). Proceedings of the Ocean Drilling Program, Scientific Results, Volume 113. Ocean Drilling Program, College Station, TX. 675–708, pls. 1–8

Abelmann A. 1992a. Early to Middle Miocene radiolarian stratigraphy of the Kerguelen Plateau, Leg 120. In: Wise S W Jr, Schlich R, Palmer A A (eds). Proceedings of the Ocean Drilling Program, Scientific Results, Volume 120. College Station, TX: Ocean Drilling Program. 757–783

Abelmann A. 1992b. Radiolarian taxa from Southern Ocean sediment traps (Atlantic sector). Polar Biology, 12(3–4): 373–385

Abelmann A. 1992c. Radiolarian flux in Antarctic waters (Drake Passage, Powell Basin Bransfield Strait). Polar Biology, 12(3–4): 357–372

Abelmann A, Gowing M M. 1997. Spatial distribution pattern of living polycystine radiolarian taxa-baseline study for paleoenvironmental reconstructions in the Southern Ocean (Atlantic sector). Marine Micropaleontology, 30(1): 3–28

Abelmann A, Nimmergut A. 2005. Radiolarians in the Sea of Okhotsk and their ecological implication for paleoenvironmental reconstructions. Deep Sea Research Part II: Topical Studies in Oceanography, 52(16): 2302–2331

Abelmann A, Gersonde R, Spiess V. 1990. Pliocene-Pleistocene paleoceanography in the Weddell Sea—siliceous microfossil evidence. In: Bleil U, Thiede J (eds). Geological History of the Polar Oceans: Arctic Versus Antarctic. Dordrecht: Kluwer. 729–759

Ainley D G, Jacobs S S. 1981. Sea-bird affinities for ocean and ice boundaries in the Antarctic. Deep Sea Research Part A. Oceanographic Research Papers, 28(10): 1173–1185

Alexandrovich J M. 1989. Radiolarian biostratigraphy of ODP Leg 111, Site 677, eastern equatorial Pacific, late Miocene through Pleistocene. Proceedings of the Ocean Drilling Program, Scientific Results, 111: 245–262

Bailey J W. 1856. Notice of miocroscopic forms found in the soundings of the Sea of Kamtschacka. American Journal of Science and Arts, 2nd Series, 22: 1–6

Bandy O L, Casey R E, Wright R C. 1971. Late Neogene planktonic zonation, magnetic reversals, and radiometric dates, Antarctic to the tropics. Antarctic Oceanology I, 15: 1–26. doi: 10.1029/AR015p0001.

Benson R N. 1966. Recent radiolaria from the Gulf of California. Ph. D. Thesis. Twin Cities: University of Minnesota, Minneapolis. 1–577

Benson R N. 1972. Radiolaria: Leg 12 of Deep Sea Drilling Project. In: Laughton A S, Berggren W A (eds). Initial Reports of the Deep Sea Drilling Project, Volume 12. Washington: US Government Printing Office. 1085–1113

Benson R N. 1983. Quaternary Radiolarians from the Mouth of the Gulf of California, Leg 65 of the Deep Sea Drilling Project. In: Lewis B T R, Robinson P et al. (eds). Initial Reports of the Deep Sea Drilling Project, Volume 65. Washington: US Government Printing Office. 491–523

Bjørklund K R. 1973. Radiolarians from the surface sediment in Lindåspollene, western Norway. Sarsia, 53(1): 71–75

Bjørklund K R. 1974. The seasonal occurrence and depth zonation of radiolarians in Korsfjorden, Western Norway. Sarsia, 56(1): 13–42

Bjørklund K R. 1976. Radiolaria from the Norwegian Sea, Leg 38 of the Deep Sea Drilling Project. In: Talwani M, Udintsev G et al. (eds). Initial Reports of the Deep Sea Drilling Project, Volume 38. Washington: US Government Printing Office. 1101–1168

Bjørklund K R, Goll R M. 1979. Internal skeletal structures of *Collosphaera* and *Trisolenia*: a case of repetitive evolution in the Collosphaeridae (Radiolaria). Journal of Paleontology, 53(6): 1293–1326

Bjørklund K R, Kruglikova S B. 2003. Polycystine radiolarians in surface sediments in the Arctic Ocean basins and marginal seas.

Marine Micropaleontology, 49(3): 231–273

Bjørklund K R, Swanberg N. 1987. The distribution of two morphotypes of the radiolarian *Amphimelissa setosa* Cleve (Nassellarida): a result of environmental variability? Sarsia, 72: 245–254

Bjørklund K R, Cortese G, Swanberg N, Schrader H J. 1998. Radiolarian faunal provinces in surface sediments of the Greenland, Iceland and Norwegian (GIN) Seas. Marine Micropaleontology, 35(1): 105–140

Bjørklund K R, Itaki T, Dolven J K. 2014. Per Theodor Cleve: a short résumé and his radiolarian results from the Swedish Expedition to Spitsbergen in 1898. Journal of Micropalaeontology, 33(1): 59–93

Blome C D. 1992. Radiolarians from Leg 122, Exmouth and Wombat Plateaus, Indian Ocean. In: von Rad U, Haq B, et al. (eds). Proc ODP Sci Results, 122: 633–652

Blueford J R. 1982. Miocene actinommid radiolaria from the equatorial Pacific. Micropalaeontology, 28(2): 189–213

Blueford J R. 1983. Distribution of Quaternary Radiolaria in the Navarin Basin geologic province, Bering Sea. Deep Sea Research Part A. Oceanographic Research Papers, 30(7): 763–781

Blueford J. 1988. Radiolarian biostratigraphy of siliceous Eocene deposits in central California. Micropaleontology, 34(3): 236–258

Boltovskoy D. 1998. Classification and distribution of South Atlantic Recent polycystine Radiolaria. Palaeontologia Electronica, 1(2): 1–116

Boltovskoy D, Jankilevich S S. 1985. Radiolarian distribution in east equatorial Pacific plankton. Oceanol Acta, 8: 101–123

Boltovskoy D, Riedel W R. 1980. Polycystine radiolaria from the southwestern Atlantic Ocean plankton. Revista Espanola de Micropaleontologia, 12(1): 99–146

Boltovskoy D, Riedel W R. 1987. Polycystine Radiolaria of the California Current region: seasonal and geographic patterns. Mar Micropaleontol, 12: 65–104

Brandt K. 1885. Die Kolonibildenden Radiolarien des Golfes von Neapel. und der angrenzenden Meeresabschnitte. Monogr Fauna und Flora d. Golfes V. Neapel, 13: 1–275

Brandt K. 1905. Zur Systematik der Kolonibebildenden Radiolarien. Zool Jahrg Suppl, 8: 311–352

Bütschli O. 1882. Beiträge zur Kenntnis der Radiolarienskelette, insbesondere der der Cyrtida. Zeitschr f wiss Zool, 36: 485–541

Campbell A S. 1954. Radiolaria. In: Moore R C (ed). Treatise on Invertebrate Paleontology. Part D, Protista 3. Lawrence: University Press of Kansas. D11–D163

Campbell A S, Clark B L. 1944a. Miocene radiolarian faunas from southern California. Geological Society of America Special Paper, 51: 1–76

Campbell A S, Clark B L. 1944b. Radiolarian from the Upper Cretaceous of middle California. Geological Society of America Special Paper, 57: 1–61

Carnevale P. 1908. Radiolarie Silicoflagellati di Bergonzano (Beggio Emilia). Reale Istituto Veneto di Scienze Lettreed Arti, Memorie, 28(3): 1–46

Casey R E. 1971a. Distribution of polycystine radiolaria in the oceans in relation to physical and chemical conditions. In: Funnell B, Riedel W R (eds). The Micropalaeontology of Oceans. Cambridge: Cambridge University Press. 151–159

Casey R E. 1971b. Radiolarians as indicators of past and present water-masses. In: Funnell B, Riedel W R (eds). The Micropalaeontology of Oceans. Cambridge: Cambridge University Press. 331–337

Casey R E. 1972. Neogene radiolarian biostratigraphy and paleotemperatures: southern California, the experimental Mohole, Antarctic core E14–8. Palaeogeography, Palaeoclimatology, Palaeoecology, 12(1–2): 115–130

Caulet J P. 1979. Les dépots à Radiolaires d'age pliocène supérieur à pléistocène dans l'Océan Indien central: nouvelle zonation biostratigraphique. In: Recherches Océanographiques dans l'Océan Indien. Paris 20–22 Juin 1977. Mémoires du Muséum National d'Histoire Naturelle, Paris, série C, 43: 119–141

Caulet J P. 1986. Radiolarians from the southwest Pacific. In: Kennett J P, von der Borch C C, et al. (eds). Initial Reports of the Deep Sea Drilling Project, Volume 90 (Part 2). Washington: US Government Printing Office. 835–861

Caulet J P. 1991. Radiolarians from the Kerguelen Plateau, Leg 119. In: Barron J, Larsen B, et al. (eds). Proceedings of the Ocean Drilling Program, Scientific Results, Volume 119. College Station, TX: Ocean Drilling Program. 513–546

Caulet J P, Nigrini C. 1988. The genus *Pterocorys* (Radiolaria) from the tropical late Neogene of the Indian and Pacific Oceans.

Micropaleontology, 34(3): 217–235

Chen M H, Tan Z Y. 1989. Description of a new genus and 12 new species of radiolaria in sediments from the South China Sea. Tropical Oceanology, 8(1): 1–9 (in Chinese with English abstract) [陈木宏, 谭智源. 1989. 南海沉积物中放射虫 1 新属 12 新种. 热带海洋, 8(1): 1–9]

Chen M H, Tan Z Y. 1996. Radiolaria in the Sediments of Central and Northern South China Sea. Beijing: Science Press. 1–271 (in Chinese) [陈木宏, 谭智源. 1996. 南海中、北部沉积物中的放射虫. 北京: 科学出版社. 1–271]

Chen M H, Zhang Q, Zhang L L, Zarikian C A, Wang R J. 2014. Stratigraphic distribution of the radiolarian *Spongodiscus biconcavus* Haeckel at IODP Site U1340 in the Bering Sea and its paleoceanographic significance. Palaeoworld, 23(2014): 90–104

Chen M H, Zhang Q, Zhang L L, Liu L. 2017. Radiolaria in the Sediemnts from the Northwest Pacific and Its Marginal Seas. Beijing: Sceince Press. 1–279 (in Chinese with English abstract) [陈木宏, 张强, 张兰兰, 刘玲. 2017. 西北太平洋及其边缘海沉积物中的放射虫. 北京: 科学出版社. 1–279]

Chen P H. 1974. Some new Tertiary Radiolaria from Antarctic deep-sea sediments. Micropaleontology, 20(4): 480–492

Chen P H. 1975. Antarctic Radiolaria. In: Hayes D E, Frakes L A, et al. (eds). Initial Reports of the Deep Sea Drilling Project, Volume 28. Washington: US Government Printing Office. 437–513

Cienkowski L. 1871. Ueber Schwarmerbildung bei Radiolarien. Arch f Mikrosk Anat, 7: 371–381

Clark B L. 1947. Eocene radiolarian faunas, Mt. Diablo, California. Geological Society of America Special Paper, 87: 1–112

Clark B L, Campbell A S. 1942. Eocene radiolarian faunas from the Mount Diablo area, California. Geological Society of America, Special Paper, 39: 1–106

Clark B L, Campbell A S. 1945. Radiolaria from the Kreyenhagen Formation near Los Banos, California. Geological Society of America, Memoir, 10: 1–62

Cleve P T. 1899. Plankton collected by the Swedish Expedition to Spitzbergen in 1898. Küngliga Svenska Vetenskaps Akademiens Handlingar, 32(3): 1–51

Cleve P T. 1900. Notes on some Atlantic planktonic organisms. Kgl Svenska Ventensk Akad Handl, 34(1): 1–22

Cleve P T. 1901. Plankton from the Indian Ocean and the Malay Archipelago. Kgl Svenska Ventensk Akad Handl, 35(5): 1–58

Cortese G, Abelmann A. 2002. Radiolarian-based paleotemperatures during the last 160 kyr at ODP Site 1089 (Southern Ocean, Atlantic Sector). Palaeogeography, Palaeoclimatology, Palaeoecology, 182(3): 259–286

Cortese G, Bjørklund K R. 1997. The morphometric variation of *Actinomma boreale* (Radiolaria) in Atlantic boreal waters. Marine Micropaleontology, 29(3): 271–282

Cortese G, Bjørklund K R. 1998a. Morphometry and taxonomy of *Hexacontium* species from western Norwegian fjords. Micropaleontology, 44(2): 161–172

Cortese G, Bjørklund K R. 1998b. The taxonomy of boreal Atlantic Ocean Actinommida (Radiolaria). Micropaleontology, 44(2): 149–160

Cortese G, Bjørklund K R, Dolven J K. 2003. Polycystine radiolarians in the Greenland-Iceland-Norwegian (GIN) Seas: species and assemblage destribution. Sarsia, 88(1): 65–88

Cramer B S, Toggweiler J R, Wright J D, Katz M E, Miller K G. 2009. Ocean overturning since the Late Cretaceous: inferences from a new benthic foraminiferal isotope compilation. Paleoceanography, 24: PA4216, doi: 10.1029/2008PA001683

Crouch E M, Hollis C J. 1996. Paleogene palynomorph and radiolarian biostratigraphy of DSDP Leg 29, Sites 280 and 281, South Tasman Rise. Lower Hutt, Institute of Geological and Nuclear Sciences, Science Report, 96/19: 1–46

Dogiel V A, Reschetnjak V V. 1952. Material on Radiolarians of the northwestern part of the Pacific Ocean. Investigation of the Far East Seas of the USSR, 3: 5–36

Dolven J K. 1998. Late Pleistocene to late Holocene biostratigraphy and paleotemperatures in the SE Norwegian Sea, based on Polycystine Radiolarians. Master's Degree Thesis. Oslo (Norway): University of Oslo

Dreyer F. 1889. Die Pylombildungen in vergleichend-anatomischer und entwicklungsgeschichtlicher Beziehung bei Radiolarien und bei Protisten uberhaupt, nebst System und Beschreibung neuer und der bis jetzt bekannten pylomatischen Spumellarien. Jenaische Zeitschrift für Naturwissenschaft, Jena, 23(n. ser. 16): 1–138

Dreyer F. 1890. Die Tripoli von Caltanisetta (Steinbruch Gessolungo) auf Sizilien. Jenaische Zeitschrift Naturwiss, 24: 471–548

Dreyer F. 1913. Die polycystinen der plankton Expedition, Ergebn. Plankton Expedition Humboldt Stiftung, 3(L, d and e): 1–104

Dumitrica P. 1968. Consideratii micropaleontologice asupra orizontului argilos cu radiolari din tortonianul regiunii Carpatice. Studii si Cercetari de Geologie, Geofizca Geografie, Bucharest, Serie Geologie, 13: 227–241

Dumitrica P. 1973. Cretaceous and Quaternary Radiolaria in deep sea sediments from the Northwest Atlantic Ocean and Mediterranean Sea. In: Ryan W B F, Hsü K J, et al. (eds). Initial Reports of the Deep Sea Drilling Project, Volume 13. Washington: US Government Printing Office. 829–901

Dumitrica P. 1988. New families and subfamilies of Pyloniacea (Radiolaria). Revue de Micropaléntologie, 31(3): 178–195

Dumitrică P. 2014. On the status of the radiolarian genera *Lonchosphaera* Popofsky, 1908 and Arachnostylus Hollande and Enjumet, 1960. Acta Palaeontologica Romaniae, 9(2): 59–66

Dzinoridze R N, Jousé A P, Koroleva-Golikova G S. 1978. Diatom and radiolarian cenozoic stratigraphy, Norwegian Basin；DSDP Leg 38. In: Talwani M, Udintsev G, et al. (eds). Initial Reports of the Deep Sea Drilling Project, Volume 38. Washington: US Government Printing Office. 289–427

Dzhinoridze R N, Zhuze A P, Ignatova G V, Kozlova G E, Koltun V M, Golikova G S, Nagaeva G S, Likina T G, Petrushevskaya M G. 1979. The history of the microplankton of the Norwegian Sea (on the deep sea drilling materials). Issledovaniya Fauny Morei, 23(31): 3–190

Ehrenberg C G. 1838. Polycystna (*Lithocampe*, *Coenutella*, *Haliomma*) in Upper die Bildung der Kreidefelsen und des Kredemergels durch unsichtbare Organismen. Konigliche Preussische Akademie der Wissenschaften zu Berlin, Abhandlungen, 1838: 1–117

Ehrenberg C G. 1844. Uber 2 neue Lager von Gebirgsmassen aus Infusorien als Meeres-Absatz in Nord-Amerika und eine Vergleichung derselben mit den organischen Kreide-Gebilden in Europa und Afrika. Konigliche Preussische Akademie der Wissenschaften zu Berlin, Bericht, Jahre, 1844: 57–97

Ehrenberg C G. 1847. Beobachtungen über die mikroskopischen kieselschaligen Polycystinen als mächtige Gebirgsmasse von Barbados und über das Verhältnis der aus mehr als 300 neuen Arten bestehenden ganz eigentümlichen Formengruppe jener Felsmasse zu den Iebenden Thieren und zur Kreidebildung. Konigliche Preussische Akademie der Wissenschaften zu Berlin, Bericht, Jahre, 1847: 40–60

Ehrenberg C G. 1854. Mikrogeology. Das Erden und Felsen Schaffende Wirken des unsichtbar Kleinen selbstandigen Lebens auf der Erde. Berlin: Verhandlungen der Königl. Preufs. Akademie der Wissenschaften zu Berlin, Bericht, 1854: 1–490

Ehrenberg C G. 1858. Kurze Characteristik der 9 neuen Genera und der 105 neuen Species des ägäischen Meeres und des Tiefgrundes des Mittel-Meeres. Berlin: Königliche Preussische Akademie der Wissenschaften zu Berlin, Monatsberichte, 1858: 10–40

Ehrenberg C G. 1860. Ueber den Tiefgründe des stillen Oceans zwischen Californien und den Sandwich-Inseln. Berlin, Monatsber: K. Preuss. Akad. d. Wiss. 819–833

Ehrenberg C G. 1862. Über die Tiefgrund-Verhältnisse des Oceans am Eingange der Davisstrasse und bei Island. Königlichen Preufs. Berlin: Akademie der Wissenschaften zu Berlin, Monatsberichte, 1861: 275–315

Ehrenberg C G. 1872. Mikrogeologischen Studien über das kleinste Leben der Meeres-Tiefgrunde aller Zonen und dessen geologischen Einfluss. Berlin: Königliche Preussische Akademie der Wissenschaften zu Berlin, Monatsberichte, 1873: 131–399

Ehrenberg C G. 1873. Mikrogeologische studien als Zusammenfassung seiner Beobachtungen des Kleinsten Lebens der Meeres Tiefgrunde aller Zonen und dessen geologischen Einfluss. Berlin: Königliche Preussische Akademie der Wissenschaften zu Berlin, Monatsberichte, 1872: 265–322

Ehrenberg C G. 1874. Mikrogeologische studien als Zusammenfassung seiner Beobachtungen des Kleinsten Lebens der Meeres Tiefgrunde aller Zonen und dessen geologischen Einfluss. Berlin: Königliche Preussische Akademie der Wissenschaften zu Berlin, Monatsberichte. 213–263

Ehrenberg C G. 1876. Fortsetzung der mikrogeologischen Studien als Gesammt-Übersicht der mikroskopischen Paläontologie gleichartig analysirter Gebirgsarten der Erde, mit specieller Rücksicht auf den Polycystinen-Mergel von Barbados. Berlin: Königliche Preussische Akademie der Wissenschaften zu Berlin, Abhandlungen, 1875: 1–226

Flower B P, Kennett J P. 1994. The middle Miocene climatic transition: East Antarctic ice sheet development, deep ocean circulation and global carbon cycling. Palaeogeography, Palaeoclimatology, Palaeoecology, 108: 537–555

Foreman H P. 1968. Upper Maestrichtian Radiolaria of California. Special Papers in Palaeontology, 3: 1–82

Foreman H P. 1973. Radiolaria of Leg 10 with systematics and ranges for the families Amphipyndacidae, Artostrobiidae and Theoperidae. In: Worzel J L, Bryant W, et al. (eds). Initial Reports of the Deep Sea Drilling Project, Volume 10. Washington: US Government Printing Office. 407–474

Foreman H P. 1975. Radiolaria from the North Pacific, Deep Sea Drilling Project, Leg 32. In: Larson R L, Moberly R, et al. (eds). Initial Reports of the Deep Sea Drilling Project, Volume 32. Washington: US Government Printing Office. 579–676

Foreman H P. 1978. Mesozoic radiolaria in the Atlantic ocean off the northwest coast of Africa, Deep Sea Drilling Project, Leg 41. In: Larson R L, Moberley R, et al. (eds). Initial Reports of the Deep Sea Drilling Project, Volume 32. Washington: U.S. Government Printing Office. 739–761

Goll R M. 1968. Classification and phylogeny of Cenozoic Trissocyclidae (Radiolaria) in the Pacific and Caribbean Basins, Part 1. Journal of Paleontology, 42(6): 1409–1432

Goll R M. 1969. Classification and phylogeny of Cenozoic Trissocyclidae (Radiolaria) in the Pacific and Caribbean Basins, Part 2. Journal of Paleontology, 43(2): 322–339

Goll R M. 1972. Section on Radiolaria for synthesis chapter, Leg 9. In: Hays J D, Harry E C, et al. (eds). Initial Reports of the Deep Sea Drilling Project, Volume 9. Washington: US Government Printing Office. 947–1058

Goll R M. 1976. Morphological intergradation between modern population of *Lophosphaerid* and *Phormospyris* (Trissocyclidae, Radiolaria). Micropaleontology, 22(4): 379–418

Goll R M. 1978. Five trissocyclid radiolaria from site 338. In: Talwani M, Udintsev G, et al. (eds). Initial Reports of the Deep Sea Drilling Project, Volume 38. Washington: US Government Printing Office. 177–191

Goll R M. 1979. The Neogene evolution of *Zygocircus*, *Neosemantis* and *Callimitra*: their bearing on Nassellarian classification: a revision of the Plagiacanthoidea. Micropaleontology, 25(4): 365–396

Goll R M, Bjørklund K R. 1971. Radiolaria in surface sediments of the South Atlantic. Micropaleontology, 17(4): 434–454

Goll R M, Bjørklund K R. 1974. Radiolaria in surface sediments of the South Atlantic Ocean. Micropaleontology, 20(1): 38–75

Goll R M, Bjørklund K R. 1989. A new radiolarian biostratigraphy for the Neogene of the Norwegian Sea: ODP Leg 104. In: Rldholm O, Thiede J, Taylor E, et al. (eds). Proceedings of the Ocean Drilling Program, Scientific Results, Volume 104. College Station, TX: Ocean Drilling Program. 697–737

Haeckel E. 1860. Fernere Abbildungen und Diagnosen neuer Gattungen und Arten von lebenden Radiolarien des Mittelmeeres. Königlichen Preufs. Akademie der Wissenschaften zu Berlin, Monatsherichte, 1860: 835–845

Haeckel E. 1862. Die Radiolarien (Rhizopoda Radiaria)-Eine Monographie. Berlin: Druck und Verlag von Georg Reimer, Monographie. 1–572

Haeckel E. 1879. Ueber die Phaeodarien, eine neue Gruppe kieselschaliger mariner Rhizopoden. Sitzungsberichte der Medizinisch-Naturwissenschaftlichen Gesellschaft Jena, 13: 151–157

Haeckel E. 1881. Radiolarien-Systems auf Grund von Studien der Challenger-Radiolarien. Jenaische Zeitschrift fur Naturwissenschaften, 15: 418–472

Haeckel E. 1887. Report on the Radiolaria collected by the H. M. S. Challenger during the Years 1873–1876. Report on the Scientific Results of the Voyage of the H. M. S. Challenger, Zoology, Volume 18. London: HerMajesty's Stationary Office. 1–1803

Haecker V. 1904. Bericht über die Tripyleen Ausbeute der Deutschen Tiefsee Expedition. Verhandl Deutsch Zool Ges, 14: 122–156

Haecker V. 1907. Altertumliche Spharellarien und Cyrtellarien aus grossen Meerestiefen. Archiv fur Protistenkunde, 10: 114–126

Haecker V. 1908a. Tiefsee-Radiolarien. Spezieller Teil. Die Tripyleen, Collodarien und Mikroradiolarien der Tiefsee. Deutsch Tiefsee Exped. auf dem Dempfer "Valdivia" 1898–1899, Wiss Ergebn, 14: 1–476

Haecker V. 1908b. Tiefsee Radiolarien Allg. 1. Form und Formbildung bei den Radiolarien. Wiss Ergebn Deutschen Tiefsee Expedition, 14: 477–706

Harting P. 1863. Bijdrage tot de kennis der mikroskopische faune en flora van de Banda Zee. Verhandelingen der Koninklijke Akademie van Wetenschappen, Amsterdam, 10(1): 2–54

Hayes D E, Davey F J. 1975. A geophysical study of the Ross Sea, Antarctica. In: Kulm L D, von Huene R, et al. Initial Reports of the Deep Sea Drilling Project, Volume 34. Washington: U.S. Government Printing Office. 887–907. https://doi.org/10.2973/dsdp.proc.28.134.1975

Hayes D E, Frakes L A, Barrett P J, Chen P H, Kaneps A G, Kemp E M, McCollum D W, Piper D J W, Wall R E, Webb P N. 1974. Site 274. In: Kulm L D, von Huene R, et al. (eds). Initial Reports of the Deep Sea Drilling Project, Volume 10. Washington: U.S. Government Printing Office. 369–433.

Hays J D. 1965. Radiolaria and late Tertiary and Quaternary history of Antarctic seas. In: Llano G A (ed). Biology of the Antarctic Seas II. Antarct Res Ser, 5: 125–184

Hays J D. 1970. Stratigraphy and evolutionary trends of Radiolaria in North Pacific deep sea sediments. In: Hays J D (ed). Geological Investigations of the North Pacific. Geological Society of America Memoirs, 126: 185–218

Hays J D, Opdyke N D. 1967. Antarctic radiolaria, magnetic reversals, and climatic change. Science, 158(3804): 1001–1011

Hertwig R. 1879. Der Organismus der Radiolarien. Jenaische Denkschr, 2: 129–277

Heusser L E, Morley J J. 1996. Pliocene climate of Japan and environs between 4.8 and 2.8Ma: a joint pollen and marine faunal study. Marine Micropaleontology, 27(1): 85–106

Hilmers C. 1906. Zur kenntnis der Collosphaeriden. Doctoral Dissertation, Kgl. Christain Albrecht Univ Kiel. 1–95

Holdsworth B K, Nell P A R. 1992. Mesozoic radiolarian faunas from the Antarctic Peninsula: age, tectonic and palaeoceanographic significance. Journal of the Geological Society London, 149: 1003–1020

Hollande A, Enjumet M. 1960. Cytologie, évolution et systématique des Sphaeroidés (Radiolaires). Mus Natl Hist Nat, Paris, Arch, 7: 1–134

Hollis C J. 1993. Latest Cretaceous to Late Paleocene radiolarian biostratigraphy: a new zonation from the New Zealand region. Marine Micropaleontology, 21: 295–327

Hollis C J. 1997 Cretaceous-Paleocene Radiolaria from eastern Marlborough, New Zealand. Lower Hutt Institute of Geological and Nuclear Sciences Monograph, 17: 1–152

Hollis C J. 2002. Biostratigraphy and paleoceanographic significance of Paleocene radiolarians from offshore eastern New Zealand. Marine Micropaleontology, 46: 265–316

Hollis C J. 2006. Radiolarian faunal turnover through the Paleocene-Eocene transition, Mead Stream, New Zealand. Eclogae Geol Helv, 99 (2006) Supplement 1: 79–99

Hollis C J, Waghorn D B, Strong C P, Crouch E M. 1997. Integrated Paleogene biostratigraphy of DSDP Site 277 (Leg 29): foraminifera, calcareous nannofossils, radiolaria, and palynomorphs. Institute of Geological & Nuclear Sciences Science Report, 97(7): 1–73

Hollis C J, Pascher K M, Sanfilippo A, Nishimura A, Kamikuri S, Shepherd C L. 2020. An Austral radiolarian biozonation for the Paleogene. Stratigraphy, 17(4): 213–278

Ikenoue T, Okazaki Y, Takahashi K, Sakamoto T. 2016. Bering Sea radiolarian biostratigraphy and paleoceanography at IODP Site U1341 during the last four million years. Deep Sea Research Part II: Tropical Studies in Oceanography, 61: 17–49

Itaki T. 2003. Depth-related radiolarian assemblage in the water-column and surface sediments of the Japan Sea. Marine Micropaleontology, 47(3): 253–270

Itaki T, Ikehara K. 2004. Middle to late Holocene changes of the Okhotsk Sea intermediate water and their relation to atmospheric circulation. Geophysical Research Letters, 31(24): L24309. doi: 10.1029/ 2004GL021384

Itaki T, Takahashi K. 1995. Preliminary results on radiolarian fluxes in the central subarctic Pacific and Bering Sea. Proceedings Hokkaido Tokai University Science and Engineering, 7: 37–47 (in Japanese with English Abstract)

Itaki T, Uchida M, Kim S, Shin H S, Tada R, Khim B K. 2009. Late Pleistocene stratigraphy and palaeoceanographic implications in northern Bering Sea slope sediments: evidence from the radiolarian species *Cycladophora davisiana*. Journal of Quaternary Science, 24(8): 856–865

Jacobs S S, Giulivi C F. 2010. Large multidecadal salinity trends near the Pacific-Antarctic continental margin. Journal of Climate, 23(17): 4508–4524

Jacobs S S, Giulivi C F, Mele P A. 2002. Freshening of the Ross Sea During the Late 20th Century. Science, 297: 386–389

Johnson D A, Nigrini C. 1980. Radiolarian biogeography in surface sediments of the western Indian Ocean. Marine Micropaleontology, 5 (2): 111–152

Jørgensen E. 1900. Protophyten und Protozöen in Plankton aus der norwegischen Westküste. Bergens Museums Aarbog [1899], 6: 51–112

Jørgensen E. 1905. The Protist plankton and the diatoms in bottom samples. VII. Radiolaria. In: Nordgaard O (ed). Hydrographical and Biological Investigations in Norwegian Fiords. Bergen: Bergens Museum. 114–142

Kamikuri S. 2010. New late Neogene radiolarian species from the middle to high latitudes of the North Pacific. Revue de Micropaléontologie, 53 (2010): 85–106

Kamikuri S, Nishi H, Motoyama I, Saito S. 2004. Middle Miocene to Pleistocene radiolarian biostratigraphy in the Northwest Pacific Ocean, ODP Leg 186. The Island Arc, 13: 191–226

Kamikuri S, Nishi H, Motoyama I. 2007. Effects of late Neogene climatic cooling on North Pacific radiolarian assemblages and oceanographic conditions. Palaeogeography, Palaeoclimatology, Palaeoecology, 249(3): 370–392

Kamikuri S, Motoyama I, Nishimura A. 2008. Radiolarian assemblages in surface sediments along longitude 175°E in the Pacific Ocean. Marine Micropaleontology, 69: 151–172

Kamikuri S, Moore T C, Lyle M, Ogane K, Suzuki N. 2013. Early and Middle Eocene radiolarian assemblages in the eastern equatorial Pacific Ocean (IODP Leg 320 Site U1331): faunal changes and implications for paleoceanography. Marine Micropaleontology, 98: 1–13

Keany J. 1979. Early Pliocene radiolarian taxonomy and biostratigraphy in the Antarctic Region. Micropaleontology, 25(1): 50–74

Keany J, Kennett J P. 1972. Pliocene-early Pleistocene palaeoclimatic history recorded in Antarctic-Subantarctic deep-sea cores. Deep-Sea Research and Oceanography Abstracts, 19: 529–548

Kennett J P. 1977. Cenozoic evolution of Antarctic glaciation, the Circum-Antarctic Ocean, and their impact on global paleoceanography. Journal of Geophysical Research, 82: 3843–3860

Kling S A. 1971. Radiolaria. In: Fischer A G, Heezen B C, et al. (eds). Initial Reports of the Deep Sea Drilling Project, Volume 6. Washington: US Government Printing Office. 1069–1117

Kling S A. 1973. Radiolaria from the eastern North Pacific, Deep Sea Drilling Project, Leg 18. In: Musich L F, Weser O E, et al. (eds). Initial Reports of the Deep Sea Drilling Project, Volume 18. Washington: US Government Printing Office. 617–671

Kling S A. 1976. Relation of radiolarian distributions and subsurface hydrography in the North Pacific. Deep Sea Research and Oceanographic Abstracts, 23(11): 1043–1058

Kling S A. 1977. Local and regional imprints on radiolarian assemblages from California coastal basin sediments. Marine Micropaleontology, 2: 207–221

Kling S A. 1979. Vertical distribution of polycystine radiolarians in the central North Pacific. Marine Micropaleontology, 4: 295–318

Kling S A, Boltovskoy D. 1995. Radiolarian vertical distribution patterns across the southern California Current. Deep Sea Research Part I: Oceanographic Research Papers, 42(2): 191–231

Kozlova G E. 1999. Paleogene boreal radiolarians from Russia. Practical manual on microfauna, Vol. 9. St. Petersburg: All-Russ. Pet. Res. Explor. Inst. (VNIGRI). 1–323

Kozlova G E, Gorbovetz A N. 1966. Radiolaria of the upper Cretaceous and Upper Eocene deposits of the west Siberian Lowland. Tr Vses Nauch-Issled Geo. Neft Inst, (16): 1–271

Kruglikova S B, Bjorklund K R, Zas'ko D N. 2007. Distribution of Polycystina (Euradiolaria) in the Bottom Sediments and Plankton of the Arctic Ocean and Marginal Arctic Seas. Doklady Biological Sciences, 415: 284–287

Kunitomo Y, Sarashima I, Iijima M, Endo K, Sashida K. 2006. Molecular phylogeny of acantharean and polycystine radiolarians based on ribosomal DNA sequences, and some comparison with data from the fossil record. European Journal of Protistology, 43(2): 143–153

Lazarus D B. 1990. Middle Miocene to Recent radiolarians from the Weddell Sea, Antarctica, ODP Leg 113. In: Barker P F, Kennett J P, et al. (eds). Proceedings of the Ocean Drilling Program, Scientific Results, Volume 113. College Station, TX: Ocean Drilling Program. 709–727

Lazarus D. 1992. Antarctic Neogene radiolarians from the Kerguelen Plateau, Legs 119 and 120. In: Wise S W Jr, Schlich R, et al.

Proc ODP, Sci Results, 120. College Station, TX: Ocean Drilling Program. 785–809

Lazarus D B, Pallant A. 1989. Oligocene and Neogene radiolarians from the Labrador Sea, ODP Leg 105. In: Srivastava S P, Arthur M, Clement B, et al. (eds). Proceedings of the Ocean Drilling Program, Scientific Results, Volume 105. College Station, TX: Ocean Drilling Program. 349–380

Lazarus D B, Faust K, Popova-Goll I. 2005. New species of prunoid radiolarians from the Antarctic Neogene. Journal of Micropaleontology, 24(2): 97–121

Ling H Y. 1972. Polycystine radiolaria from surface sediments of the South China Sea and adjacent seas of Taiwan. Acta Oceanographica Taiwanica, 2: 159–178

Ling H Y. 1973. Radiolaria: Leg 19 of the Deep Sea Drilling Project. In: Creager J S, Scholl D W, et al. (eds). Initial Reports of the Deep Sea Drilling Project, Volume 19. Washington: US Government Printing Office. 777–797

Ling H Y. 1975. Radiolaria: Leg 31 of the Deep Sea Drilling Project. In: Karig D E, Ingle J C Jr, et al. (eds). Initial Reports of the Deep Sea Drilling Project, Volume 31. Washington: US Government Printing Office. 703–761

Ling H Y. 1980. Radiolarians from the Emperor Seamounts of the Northwest Pacific, Leg 55 of the Deep Sea Drilling Project. In: Jackson E D, Koizumi I, et al. (eds). Initial Reports of the Deep Sea Drilling Project, Volume 55. Washington: US Government Printing Office. 365-373

Ling H Y. 1991. Cretaceous (Maestrichtian) radiolarians: Leg 114. In: Ciesielski P F, Kristoffersen Y, et al. Proc ODP, Sci Results, 114. College Station, TX: Ocean Drilling Program. 317–324

Ling H Y, Lazarus D B. 1990. Cretaceous radiolaria from the Weddell Sea: Leg 113 of the Ocean Drilling Program. In: Barker P F, Kennett J P, et al., Proc ODP, Sci Results, 113. College Station, TX: Ocean Drilling Program. 353–363

Ling H Y, Stadum C J, Welch M L. 1971. Polycystine radiolaria from Bering Sea surface sediments. In: Farinacci A (ed). Proceeding of the Second Planktonic Conference, Roma, 2: 705–729

Liu L, Zhang Q, Chen M H, Zhang L L, Xiang R. 2016. Radiolarian biogeography in surface sediments of the Northwest Pacific marginal seas. Science China (Earth Sciences), 60(3): 517–530

Loeblich A R, Tappan H. 1961. Remarks on the systematics of the Sarcodina (Protozoa), renamed homonyms and new and validated genera. Proceedings of the Biological Society of Washington, 74: 213–214

Lombari G, Lazarus D B. 1988. Neogene cycladophorid radiolarians from the North Atlantic, Antarctic, and North Pacific deep-sea sediments. Micropaleontology, 34(2): 97–135

Martin G C. 1904. Radiolaria. In: Clark W B, Eastman C R, Glenn L C, Bagg R M, Bassler R S, Boyer C S, Case E C, Hollick C A (eds). Systematic Paleontology of the Miocene Deposits of Maryland. Baltimore: Johns Hopkins University Press. 447–459

Mast H. 1910. Die Astrosphaeriden, Wissenschaftliche Ergebnisse der Deutschen Tiefsee Expedition auf dem Dampfer “Valdivia” (1898–1899). Jena Germany: Gustav Fischer. 123–190

Matsuzaki K M, Nishi H, Suzuki N, Takashima R, Kawate Y, Sakai T. 2014. Middle to Late Pleistocene radiolarian biostratigraphy in the water mixed region of the Kuroshio and Oyashio currents, northeastern margin of Japan (JAMSTEC Hole 902-C9001C). Journal of Micropalaeontology, 33(2): 205–222

Matsuzaki K M, Suzuki N, Nishi H. 2015. Middle to Upper Pleistocene polycystine radiolarians from Hole 902-C9001C, northwestern Pacific. Paleontological Research, 19 (s1): 1–77

Matul A G. 2011. The recent and Quaternary distribution of the Radiolarian species *Cycladophora davisiana*: a biostratigraphic and paleoceanographic tool. Oceanology, 51(2): 335–346

Matul A, Abelmann A. 2001. Quaternary water structure of the Sea of Okhotsk based on radiolarian data. Doklady Earth Sciences, 381(8): 1005–1007

Matul A, Abelmann A. 2005. Pleistocene and Holocene distribution of the radiolarian *Amphimelissa setosa* Cleve in the North Pacific and North Atlantic: evidence for water mass movement. Deep Sea Research Part II: Topical Studies in Oceanography, 52(16): 2351–2364

Matul A, Abelmann A, Tiedemann R, Kaiser A, Nürnberg D. 2002. Late Quaternary polycystine radiolarian datum events in the Sea of Okhotsk. Geo-Marine Letters, 22(1): 25–32

McKay R M, de Santis L, Kulhanek D K, the Expedition 374 Scientists. 2018. Expedition 374 Preliminary Report: Ross Sea West

Antarctic Ice Sheet History. International Ocean Discovery Program, 374. https://doi.org/10.14379/iodp.pr.374.2018

Molina-Cruz A. 1977. Radiolarian assemblages and their relationship to the oceanography of the subtropical southeastern Pacific. Marine Micropaleontology, 2: 315–352

Molina-Cruz A. 1982. Radiolarians in the Gulf of California: Deep Sea Drilling Project Leg 64. In: Curray J R, Moore D G, et al., Init Repts DSDP, 64. Washington: U.S. Govt Printing Office. 983–1001

Molina-Cruz A. 1991. Holocene palaeo-oceanography of the northern Iceland Sea, indicated by Radiolaria and sponge spicules. Journal of Quaternary Science, 6(4): 303–312

Morley J J. 1980. Analysis of the abundance variations of the subspecies of *Cycladophora davisiana*. Marine Micropaleontology, 5: 205–214

Morley J J. 1985. Radiolarians from the Northwest Pacific, Deep Sea Drilling Project Leg 86. In: Heath G R, Burckle L H, et al. (eds). Initial Reports of the Deep Sea Drilling Project, Volume 86. Washington: US Government Printing Office. 399–422

Morley J J, Hays J D. 1979. *Cycladophora davisiana*: a stratigraphic tool for Pleistocence North Atlantic and interhemispheric correlation. Earth Planetary Science Letter, 44(3): 383–389

Morley J J, Hays J D. 1983. Oceanographic conditions associated with high abundances of the radiolarian *Cycladophora davisiana*. Earth and Planetary Science Letters, 66: 63–72

Morley J J, Nigrini C. 1995. Miocene to Pleistocene radiolarian biostratigraphy of North Pacific Sites 881, 884, 885, 886 and 887. In: Rea D K, Basov I A, Scholl D W, Allan J F (eds). Proceedings of the Ocean Drilling Program, Scientific Results, Volume 145. College Station, TX: Ocean Drilling Program. 55–91

Morley J J, Robinson S W. 1986. Improved method for correlating late Pleistocene/Holocene records from the Bering Sea: application of a biosiliceous/geochemical stratigraphy. Deep Sea Research Part A, Oceanographic Research Papers, 33(9): 1203–1211

Morley J J, Hays J D, Robertson J H. 1982. Stratigraphic framework for the late Pleistocene in the northwest Pacific Ocean. Deep Sea Research Part A, Oceanographic Research Papers, 29(12): 1485–1499

Morley J J, Tiase V L, Ashby M M, Kashgarian M. 1995. A high-resolution stratigraphy for Pleistocene sediments from North Pacific Sites 881, 883, and 887 based on abundance variations of the radiolarian *Cycladophora davisiana*. In: Rea D K, Basov I A, Scholl D W, Allen J F (eds). Proceedings of the Ocean Drilling Program, Scientific Results, Volume145. College Station, TX: Ocean Drilling Program. 133–140

Motoyama I. 1996. Late Neogene radiolarian biostratigraphy in the subarctic Northwest Pacific. Micropaleontology, 42(3): 221–262

Motoyama I. 1997. Origin and evolution of *Cycladophora davisiana* Ehrenberg (Radiolarian) in DSDP Site 192, Northwest Pacific. Marine Micropaleontology, 30(1–3): 45–63

Motoyama I, Maruyama T. 1998. Neogene diatom and radiolarian biochronology for the middle-to-high latitudes of the Northwest Pacific region: calibration to the Cande and Kent's geomagnetic polarity time scales (CK 92 and CK 95). Journal of the Geological Society of Japan, 104: 171–183 (in Japanese with English abstract)

Motoyama I, Nishimura A. 2005. Distribution of radiolarians in North Pacific surface sediments along the 175°E meridian. Paleontological Research, 9(2): 95–117

Müller J. 1855. Uber Sphaerozoum und Thalassicolla. Berlin: Verhandlungen der Königl. Preufs. Akademie der Wissenschaften zu Berlin, Bericht, 1855: 229–253

Müller J. 1858. Über die Thalassicollen, Polycystinen und Acanthemettren des Mittelmeeres. Berlin: Koniglichen Akademie der Wissenschaften zu Berlin, Abhandlungen, 1958: 1–54

Mullineaux L, Smith M J. 1986. Radiolarians as paleoceanographic indicators in the Miocene Monterey Formation, Upper Newport Bay, California. Micropaleontology, 32(1): 48–71

Nakaseko K. 1955. Miocene radiolarian fossil assemblage from the southern Tojama Prefecture in Japan. Science Reports, College of General Education, Osaka University, 4: 65–127

Nakaseko K. 1959. On superfamily Liosphaericae (Radiolaria) from sediments in the sea near Antarctica. Part 1. On Radiolaria from sediments in the sea near Antarctica. Special Publications from the SetoMarine Biological Laboratory, 1–13

Nakaseko K. 1963. Neogene Crytoidea (Radiolaria) from the Isozaki Formation in Ibaraki Prefecture, Japan. Science Reports,

College of General Education, Osaka University, 12(2): 165–198

Nakaseko K. 1964. Liosphaeridae and Collosphaeridae (radiolarian) from the sediment of the Japan Trench. Science Reports, College of General Education, Osaka University, 13(1): 39–57

Nakaseko K. 1971. On some species of the Genus *Thecosphaera* from the Neogene formations, Japan. Science Reports, College of General Education, Osaka University, 20(2): 59–66

Nakaseko K, Nishimura A. 1971. A new species of *Actinomma* from the Neogene formation, Japan. Science Reports, College of General Education, Osaka University, 20(2): 67–71

Nakaseko K, Nishimura A. 1982. Radiolaria from the bottom sediments of the Bellingshausen Basin in the Antarctic Sea. Report of the Technology Research Center, JNOC, 16: 91–244.

Nakaseko K, Sugano K. 1973. Neogene radiolarian zonation in Japan. The Memoirs of the Geological Society of Japan, 8: 23–33

Nigrini C. 1967. Radiolaria in pelagic sediments from the Indian and Atlantic Oceans. Bulletin of the Scripps Institution of Oceanography, University of California, 11: 1–125

Nigrini C A. 1968. Radiolaria from eastern tropical Pacific sediments. Micropaleontology, 14 (1): 51–63

Nigrini C A. 1970. Radiolarian assemblages in the North Pacific and their application to a study of Quaternary sediments in core V20–130. Geological Society of America Memoirs, 126: 139–183.

Nigrini C A. 1971. Radiolarian zones in the Quaternary of the equatorial Pacific Ocean. In: Funnel B M, Riedel W R (eds). The Micropaleontology of Oceans. Cambridge: Cambridge University Press. 443–461

Nigrini C A. 1977. Tropical Cenozoic Artostrobiidae [Radiolaria]. Micropaleontology, 23(3): 241–269

Nigrini C A. 1991. Composition and biostratigraphy of radiolarian assemblages from an area of upwelling (northwestern Arabian Sea, Leg 117). In: Prell W L, Niitsuma N, et al. (eds). Proceedings of the Ocean Drilling Program, Scientific Results, Volume 117. College Station, TX: Ocean Drilling Program. 89–126

Nigrini C A, Lombari G. 1984. A guide to Miocene Radiolaria. Cushman Foundation for Foraminiferal Research, Special Publication, 22: 1–320

Nigrini C A, Moore T C. 1979. A guide to modern radiolarian. Cushman Foundation for Foraminiferal Research, Special Publication, 16: 1–260

Nigrini C, Sanfilippo A. 2001. Cenozoic Radiolarian Stratigraphy for Low and Middle Latitudes with Descriptions of Biomarkers and Stratigraphically Useful Species. ODP Technical Note, 27. College Station, TX: Ocean Drilling Program. 1–486.

Nishimura A. 1992. Paleocene radiolarian biostratigraphy in the northwest Atlantic at Site 384, Leg 43, of the Deep Sea Drilling Project. Miceopaleontology, 38(4): 317–362

Nishimura A. 2003. The skeletal structure of *Prunopyle antarctica* Dreyer (Radiolaria) in sediment samples from the Antarctic Ocean. Micropaleontology, 49(2): 197–200

Nishimura A, Yamauchi M. 1984. Radiolarians from the Nankai Trough in the Northwest Pacific. News of Osaka Micropaleontologists, Special Volume 6: 1–148

O'Connor B. 1993. Radiolaria from the Mahurangi limestone, Northland, New Zealand. Unpublished M Sc thesis. Auckland: University of Auckland

O'Connor B. 1994. Seven new radiolarian species from the Oligocene of New Zealand. Micropaleontology, 40(4): 337–350

O'Connor B. 1997a. New Radioalaria from the Oligocene and early Miocene of Northland, New Zealand. Micropaleontology, 43(1): 63–100

O'Connor B. 1997b. Lower Miocene Radiolaria from Te Kopua Point, Kaipara Harbour, New Zealand. Micropaleontology, 43(2): 101–128

O'Connor B. 1999. Radiolaria from the Late Eocene Oamaru Diatomite, South Island, New Zealand. Micropaleontology, 45(1):1–55

O'Connor B. 2000. Stratigraphic and geographic distribution of Eocene–Miocene Radiolaria from the southwest Pacific. Micropaleontology, 46(3): 189–228

O'Connor B M. 2001. *Buryella* (Radiolaria, Artostrobiidae) from DSDP Site 208 and ODP Site 1121. Micropaleontology, 47: 1–22

Odette V S, Margarita M S M, Giglio S. 2008. Polycystina Radiolaria (Protozoa: Nassellaria and Spumellaria) sedimented in the center-south zone of Chile (36°–43°S). Gayana, 72(1): 79–93

Okazaki Y, Takahashi K, Yoshitani H, Nakatsuka T, Ikehara M, Wakatsuchi M. 2003a. Radiolarians under the seasonally sea-ice covered conditions in the Okhotsk Sea: flux and their implications for paleoceanography. Marine Micropaleontology, 49(3): 195–230

Okazaki Y, Takahashi K, Nakatsuka T, Honda M C. 2003b. The production scheme of *Cycladophora davisiana* (Radiolaria) in the Okhotsk Sea and the northwestern North Pacific: implication for the paleoceanographic conditions during the glacials in the high latitude oceans. Geophysical Research Letters, 30(18): 1–5, doi: 10.1029/2003GL018070

Okazaki Y, Takahashi K, Itaki T, Kawasaki Y. 2004. Comparison of radiolarian vertical distributions in the Okhotsk Sea near Kuril Islands and the northwestern North Pacific off Hokkaido Island. Marine Micropaleontology, 51(3–4): 257–284

Okazaki Y, Takahashi K, Asahi H, Katsuki K, Hori J, Yasuda H, Tokuyama H. 2005a. Productivity changes in the Bering Sea during the late Quaternary. Deep Sea Research Part II: Topical Studies in Oceanography, 52(16): 2150–2162

Okazaki Y, Takahashib K, Onoderab J, Honda M C. 2005b. Temporal and spatial flux changes of radiolarians in the northwestern Pacific Ocean during 1997–2000. Deep Sea Research Part II: Topical Studies in Oceanography, 52(16): 2240–2274

Okazaki Y, Takahashi K, Katsuki K, Ono A, Hori J, Sakamoto T, Uchida M, Shibata Y, Ikehara M, Aoki K. 2005c. Late Quaternary paleoceanographic changes in the southwestern Okhotsk Sea: evidence from geochemical, radiolarian, and diatom records. Deep Sea Research Part II: Topical Studies in Oceanography, 52(16): 2332–2350

Okazaki Y, Seki O, Nakatsuka T, Sakamoto T, Ikehara M, Takahashi K. 2006. *Cycladophora davisiana* (Radiolaria) in the Okhotsk Sea: a key for reconstructing glacial ocean conditions. Journal of Oceanography, 62(5): 639–648

Okazaki Y, Takahashi K, Asahi H. 2008. Temporal fluxes of radiolarians along the W-E transect in the central and western equatorial Pacific, 1999–2002. Micropaleontology, 54(1): 71–86

Orsi A H, Smethie Jr W M, Bullister J L. 2002. On the total input of Antarctic waters to the deep ocean: a preliminary estimate from chlorofluorocarbon measurements. Journal of Geophysical Research, 107(C8): 3122, doi: 10.1029/2001JC000976

Pascher K M, Hollis C J, Bohaty S M. 2015. Expansion and diversification of high-latitude radiolarian assemblages in the late Eocene linked to a cooling event in the southwest Pacific. Climate of the Past Discussions, 11(12): 1599–1620

Perner J. 1892. O Radiolariich z Ceskeho Utvaru Kridoveho. Praze: Kralovske Ceske Spolecnosti Nauk, Rozpravy, 1890–1891: 255–269

Petrushevskaya M G. 1962. The importance of skeleton growth in Radiolaria for their systematics. Zoological Journal, 41(3): 331–341

Petrushevskaya M G. 1964. On homologies in the elements of the inner skeleton of some Nassellaria. Zoologicheskii Zhurnal, 43(8): 1121–1128

Petrushevskaya M G. 1965. Osobennosti konstruktsii skeleta radiolyarii Botryoidae (otr. Nassellaria). Tr Zool Inst, Leningrad, 35: 79–118

Petrushevskaya M G. 1967. Radiolaria of orders Spumellaria and Nassellaria of the Antarctic region. In: Andriyashev A P, Ushakov P V (eds). Studies of Marine Fauna, Biological Reports of the Soviet Antarctic Expedition (1955–1958), Volume 3. Leningrad: Academy of Sciences of the USSR. 1–186

Petrushevskaya M G. 1968. Gomologii v skeletakh radiolyarii Nassellaria. 2. Osnovnye skeletnye dugi slozhnoustroennykh tsefalisov Cyrtoidae i Botryoidae. Zool Zhurn, 47: 1766–1776

Petrushevskaya M G. 1969. Radiolyarii Spumellaria i Nassellaria v donnykh osadkakh kak indikatory gydrologycheskikh uslovii [Spumellarian and Nassellarian radiolarians in bottom sedimentsas indicators of hydrological conditions]. In: Jouse A P (ed). Osnovnye Problemy Micropaleontologii i Organogennovo Osadkonakopleniya v Okeanakh i Moryakh [Basic Problems of Micropaleontology and the Accumulation of Organogenic Sediments in Oceans and Seas]. Moscow: Nauka. 127–150

Petrushevskaya M G. 1971a. Spumellarian and Nassellarian radiolariain Plankton and bottom sediments of the central Pacific. In: Funnell B M, Riedel W R (eds). The Micropaleontology of Oceans. Cambridge: Cambridge University Press. 309–317

Petrushevskaya M G. 1971b. On the natural system of polycystine radiolaria (Class Sarcodina). Proceedings of the II Planktonic Conference, Roma. 981–992

Petrushevskaya M G. 1971c. Radiolaria in the plankton and recent sediments from the Indian Ocean and Antarctic. In: Funnel B M, Riedel W R (eds). The Micropalaeontology of Oceans. Cambridge: Cambridge University Press. 319–329

Petrushevskaya M G. 1971d. Radiolarii Nassellarida v Planktone Mirovogo Okeana. Radiolarii Microvogo Okeana poMaterialam Sovetskikh Ekspeditsii, Issled. Fauni Morei. Leningrad: Nauka. 5–294

Petrushevskaya M G. 1973. Radiolarii v donnych otiojeniach ujnogo poloucharia. Okeanologia, 13(6): 1041–1051

Petrushevskaya M G. 1975. Cenozoic radiolarians of the Antarctic, Leg 29, DSDP. Cenozoic Radiolarians of the Antarctic, Leg 29, DSDP. In: Kennett J P, Houtz R E, et al. (eds). Initial Reports of the Deep Sea Drilling Project, Volume 29. Washington: US Government Printing Office. 541–675

Petrushevskaya M G. 1976. Bottom sediments of the Indian Ocean and Antarctic: radiolarian stratigraphy. Journal of the Marine Biological Association of India, 18(3): 626–631

Petrushevskaya M G. 1977. New species of the radiolarians of the order Nassellaria. In: Exploration of the Fauna of the Seas - New Species and Genera of Marine Invertebrates. Acad Sci USSR Zool Inst Leningrad, 21: 10–19 (in Russian)

Petrushevskaya M G. 1979. New Variants of the System of Polycystina. Leningrad: Akad Nauk SSSR. 103–118

Petrushevskaya M G. 1986. Evolution of the *Antarctissa* group. Marine Micropaleontology, 11: 185–195

Petrushevskaya M G, Bjørklund K R. 1974. Radiolarians in Holocene sediments of the Norwegian-Greenland Seas. Sarsia, 57(1): 33–46

Petrushevskaya M G, Kozlova G E. 1972. Radiolaria: Leg 14, Deep Sea Drilling Project, Initial Reports of the Deep Sea Drilling Project, Volume 14. Washington: US Government Printing Office. 495–648

Popofsky A. 1908. Die Radiolarien der Antarktis (mit Ausnahme der Tripyleen). Deuts. Südpolar Exped. (1901–1903), Zool Vol. 2, 10: 183–305

Popofsky A. 1912. Die Sphaerellarien des Warmwassergebietes. Deutsche Südpolar Expedition, 1901–1903, 13: 73–159

Popofsky A. 1913. Die Nasselarien des Warmwassergebietes. Deutschen Südpolar Expedition, 1901–1903, 14: 217–416

Popofsky A. 1917. Die collosphaeriden der Deutschen Südpolar Expedition 1901–1903. Aarit Nachrag zu den Spumellarien und der Nassellarien. Deutschen Südpolar Expedition, 16: 236–278

Principi P. 1909. Contributo allo studio dei Radiolari Miocenici Italiani [Contribution to the study of the Miocene Radiolaria of Italy]. Bollettino della Societa Geologica Italiana, 28: 1–22

Purkey S G, Johnson G C. 2010. Warming of global abyssal and deep southern ocean waters between the 1990s and 2000s: contributions to global heat and sea level rise budgets. Journal of Climate, 23: 6336–6351

Renaudie J. 2012. A synthesis of Antarctic Neogene radiolarians: taxonomy, macroevolution and biostratigraphy. Dissertation zur Erlangung des akademischen Grades doctor rerum naturalium (Dr. rer. nat.) im Fach Biologie, eingereicht an derMathematisch-Naturwissenschaftlichen Fakultät I der Humboldt-Universität zu Berlin. 1–300

Renz G W. 1974. Radiolaria from Leg 27 of the Deep Sea Drilling Poject. In: Veevers J J, Heirtzler J R, et al. (eds). Initial Reports of the Deep Sea Drilling Project, Volume 27. Washington: US Government Printing Office. 769–841

Renz G W. 1976. The distribution and ecology of Radiolaria in the Central Pacific plankton and surface sediments. Bulletin of the Scripps Institution of Oceanography, University of California, 22: 1–267

Riedel W R. 1953. Mesozoic and late Tertiary radiolaria of Rotti. Journal of Paleontology, 27(6): 805–813

Riedel W R. 1957. Radiolaria: a preliminary stratigraphy. In: Pettersson H (ed). Reports of the Swedish Deep Sea Expedition, Volume 6(3). Goteborg: Elanders Boktryckeri Aktiebolag. 59–96

Riedel W R. 1958. Radiolaria in Antarctic sediments, B.A.N.Z. Antarctic Research Expedition Reports, Series B, 6(10): 217–255

Riedel W R. 1959. Siliceous organic remains in pelagic sediments. In: Iseland H A (ed). Silica in Sediments. Special Publication, No. 7. Tulsa: Society of Economic Paleontologists and Mineralogists. 80–91

Riedel W R. 1967. Class Actinopoda. Protozoa. In: Harland W B, Holland C H, House M R, Hughes N F, Reynolds A B, Rudwick M J S, Satterthwaite G E, Tarlo L B H, Willey E C (eds). The Fossil Record. London: Geological Society of London. 291–298

Riedel W R. 1971. Systematic classification of polycystine radiolaria. In: Funnell B M, Riedel W R (eds). The Micropaleontology of Oceans. Cambridge: Cambridge University Press. 649–661

Riedel W R, Campbell A S. 1952. A new Eocene radiolarian genus. Journal of Paleontology, 26: 667–669

Riedel W R, Sanfilippo A. 1970. Radiolaria, Leg 4, Deep Sea Drilling Project. In: Bader R G, et al. (eds). Initial Reports of the Deep Sea Drilling Project, Volume 4. Washington: US Government Printing Office. 503–575

Riedel W R, Sanfilippo A. 1971. Cenozoic Radiolaria from the western tropical Pacific, Leg 7. In: Winterer E L, Riedel W R, et al. (eds). Initial Reports of the Deep Sea Drilling Project, 7 (Part 2). Washington: US Government Printing Office. 1529–1672

Riedel W R, Sanfilippo A. 1973. Cenozoic Radiolaria from the Caribbean, Deep Sea Drilling Project, Leg 15. In: Edgar N T, Saunders J B, et al. (eds). Initial Reports of the Deep Sea Drilling Project, Volume 15. Washington: US Government Printing Office. 705–751

Riedel W R, Sanfilippo A. 1977. Cainozoic Radiolaria. In: Ramsay A T S (ed). Oceanic Micropalaeontology. New York: Academic Press. 847–912

Riedel W R, Sanfilippo A. 1978. Stratigraphy and evolution of tropical Cenozoic radiolarians. Micropaleontology, 24(1): 61–96

Riedel W R, Sanfilippo A, Cita M B. 1974. Radiolarians from the stratotype Zanclean (Lower Pliocens, Sicily). Rivista Italiana Paleontogiae Stratigrafia, 80(4): 699–734

Robertson J H. 1975. Glacial to interglacial oceanographic changes in the north-west Pacific, including a continuous record of the last 400, 000 years. Ph. D. Thesis. New York: Columbia University

Sakai T. 1980. Radiolarians from Sites 434, 435, and 436, Northwest Pacific, Leg 56, Deep Sea Drilling Project. In: Langseth M, Hakuyu O, et al. (eds). Initial Reports of the Deep Sea Drilling Project, Volumes 56–57, Part 2. Washington: US Government Printing Office. 695–733

Sanfilippo A. 1990. Origin of the Subgenera *Cyclampterium*, *Paralampterium* and *Sciadiopeplus* from *Lophocyrtis* (*Lophocyrtis*) (Radiolaria, Theoperidae). Marine Micropaleontology, 15: 287–312

Sanfilippo A, Caulet J P. 1998. Taxonomy and evolution of Paleogene Antarctic and tropical Lophocyrtid radiolarians. Micropaleontology, 44: 1–43

Sanfilippo A, Nigrini C. 1998. Code numbers for Cenozoic low latitude radiolarian biostratigraphic zones and GPTS conversion tables. Marine Micropaleontology, 33(1998): 109–156

Sanfilippo A, Riedel W R. 1970. Post-Eocene "closed" theoperid radiolarians. Micropaleontology, 16(4): 446–462

Sanfilippo A, Riedel W R. 1973. Cenozoic Radiolaria (exlusive of Theoperids, Artostrobiida and Amphipyndacids) from the Gulf of Mexico, DSDP Leg. 10. In: Worzel J L, Bryant W, et al. (eds). Initial Reports of the Deep Sea Drilling Project, Volume 10. Washington: US Government Printing Office. 475–612

Sanfilippo A, Riedel W R. 1974. Radiolaria from the weat-central Indian Ocean and Gulf of Aden. In: Fischer R L, Bunce E T, et al. (eds). Initial Reports of the Deep Sea Drilling Project, Volume 24. Washington: US Government Printing Office. 997–1035

Sanfilippo A, Burckle L H, Martini E, Riedel W R. 1973. Radiolarians, diatoms, silicoflagellates and calcareous nannofossils in the Mediterranean Neogene. Micropaleontology, 19(2): 209–234

Sanfilippo A, Westberg-Smith M J, Riedel W R. 1985. Cenozoic Radiolaria. In: Bolli H M, Saunders J B, Perch-Nieisen K (eds). Plankton Stratigraphy. 631–712

Schröder O. 1909a. Die nordischen Spumellarien. Teil II. Unterlegion Sphaerellaria. Nordisches Plankton, 7(11): 1–66

Schröder O. 1909b. Die nordischen Nassellarien. Nordisches Plankton, 7(11): 67–146

Schröder O. 1913. Die Tripyleen Radiolarien (Phaeodarien) der Deutschen Südpolar-Expedition 1901–1903. Deutsche Südpolar-Expedition, 14: 133–215

Schröder O. 1914. Die nordischen Nassellarien. Nordisches Plankton (1929), 7 (11): 67–146

Schröder-Ritzrau A. 1995. Aktuopaläontologische Untersuchung zu Verbreitung und Vertikalfluss von Radiolarien sowie ihre räumliche und zeitliche Entwicklung im Europäischen Nordmeer. Berichte aus dem Sonderforschungsbereich 313, Universität zu Kiel, 52: 1–99

Shevenell A E, Kennett J P. 2004. Paleoceanographic change during the Middle Miocene climate revolution: an Antarctic stable isotope perspective. Geophysical Monograph Series, 151: 235–252

Shilov V V. 1995. Miocene–Pleistocene radiolarians from Leg 145, North Pacific. In: Rea D K, Basov I A, Scholl D W, Allan J F (eds). Proceedings of the Ocean Drilling Program, Scientific Result, 145: 93–116

Stöhr E. 1880. Die Radiolarienfauna der Tripoli von Grotte, Provinz Girgenti in Sicilien [The radiolarian fauna of the Tripoli of Grotte, Girenti Province, Sicily]. Palaeontographica, 26 (series 3, Volume 2): 71–124

Strelkov A A, Reshetnyak B B. 1962. Colonial Radiolaria-Spumellaria in the southern Hainan Islands of the South China Sea.

Studia Marina Sinica, 1: 121–139 (in Chinese) [斯特列尔科夫 A. A., 列雪特尼阿克 B. B. 1962. 中国南海海南岛南端地区的群体放射虫类——泡沫放射虫. 海洋科学集刊, 1: 121–139]

Strelkov A A, Reshetnyak V V. 1971. Colonial spumellarian radiolarians of the World Ocean. In: Strelkov A A (ed). Radiolarians of the Ocean-Reports on the Soviet Expeditions, Explorations of the Fauna of the Seas, Academy of Sciences of the U. S. S. R., 9(7): 295–369 (in Russian, Translated to English by W R Riedel)

Strong C P, Holls C J, Wilson G J. 1995. Foreminiferal, radiolarian and dinoflagellate biostratigraphy of Late Cretaceous to Middle Eocene pelagic sediments (Muzzle Group), Mead Stream, Marlborough, New Zealand. New Zealand Journal of Geology and Geophysics, 38: 171–212

Suzuki N, Ogane K, Aita Y, Sakai T, Lazarus D. 2009. Reexamination of Ehrenberg's Neogene radiolarian collections and its impact on taxonomic stability. In: Tanimura Y, Aita Y (eds). Joint Haeckel and Ehrenberg Project: Reexamination of the Haeckel and Ehrenberg Microfossil Collections as a Historical and Scientific Legacy. Tokyo: National Museum of Nature and Science Monographs, 40: 87–96

Takahashi K. 1991. Radiolaria: flux, ecology, and taxonomy in the Pacific and Atlantic. In: Honjo S (ed). Ocean Biocoenosis, Series No. 3. Massachusetts: Woods Hole Oceanographic Institution Press. 1–303

Takahashi K, Honjo S. 1981. Vertical flux of Radiolaria: a taxon-quantitative sediment trap study from the western tropical Atlantic. Micropaleontology, 27 (2): 140–190

Takahashi O, Mayama S, Matsuoka A. 2003. Host-symbiont associations of polycystine Radiolaria: epifluorescence microscopic observation of living Radiolaria. Marine Micropaleontology, 49(3): 187–194

Takemura A. 1992. Radiolarian paleogene biostratigraphy in the southern Indian Ocean, Leg 1201. Proceedings of the Ocean Drilling Program, Scientific Results, 120: 735–756

Takemura A, Ling H Y. 1997. Eocene and Oligocene radiolarian biostratigraphy from the Southern Ocean: correlation of ODP Leg 114 (Atlantic Ocean) and 120 (Indian Ocean). Marine Micropaleontology, 30: 97–116

Tan Z Y. 1993. Spumellaria in the Xisha Islands. Studia Marina Sinica, 34: 181–226 (in Chinese) [谭智源. 1993. 西沙群岛的泡沫放射虫. 海洋科学集刊, 34: 181–226]

Tan Z Y. 1998. Fauna Sinica, Phylum Protozoa, Class Sacodina, Order Acantharia, Order Spumellaria. Beijing: Science Press. 1–315 (in Chinese with English abstract) [谭智源. 1998. 中国动物志, 原生动物门肉足虫纲, 等辐骨虫目, 泡沫虫目. 北京: 科学出版社. 1–315]

Tan Z Y, Chen M H. 1999. Chinese Offshore Radiolarians. Beijing: Science Press. 1–404 (in Chinese) [谭智源, 陈木宏. 1999. 中国近海的放射虫. 北京: 科学出版社. 1–404]

Tan Z Y, Su X H. 1982. Radiolaria in the continental shelf sediments of the East Sea. Studia Marina Sinica, 19: 129–216 (in Chinese) [谭智源, 宿星慧. 1982. 东海大陆架沉积物中的放射虫. 海洋科学集刊, 19: 129–216]

Tan Z Y, Tchang Z R. 1976. Radiolarian study II in the East Sea, Spumellaria, Nassellaria, Acantharia, and Taxopodia. Studia Marina Sinica, 11: 217–314 (in Chinese) [谭智源, 张作人. 1976. 东海放射虫的研究II, 泡沫虫目、罩笼虫目、稀孔虫目和棒矛虫目. 海洋科学集刊, 11: 217–314]

van de Paverd P J. 1995. Recent Polycystine Radiolaria from the Snellius-II Expedition. Ph. D. thesis. Oslo (Norway): Center for Marine Earth Science (the Netherlands) and Paleontological Museum in Oslo.

Vigour R, Lazarus D. 2002. Biostratigraphy of late Miocene–early Pliocene radiolarians from ODP Leg 183 Site 1138. In: Frey F A, Coffin M F, Wallace P J, Quilty P G (eds). Proceedings of the Ocean Drilling Program, Scientific Results, 183: 1–17

Vinassa de Regny P E. 1900. Radiolari Miocenici Italiani. [Miocene Radiolaria from Italy]. Memorie della R. Accademia delle Scienze dell'Istituto di Bologna, Serie 5, 8: 565–595

Wang R J, Chen R H. 2005. *Cycladophora davisiana* (Radiolarian) in the Bering Sea during the late Quaternary: a stratigraphic tool and proxy of the glacial Subarctic Pacific Intermediate Water. Science in China Ser. D Earth Sciences, 48(10): 1698–1707

Weaver F M. 1976. Antarctic radiolaria from the southeast Pacific Basin, DSDP, Leg 35. In: Hollister C D, Craddock C, et al. Init Repts DSDP, 35. Washington: U.S. Govt Printing Office. 569–603

Weaver F M. 1983. Cenozoic radiolarians from the southwest Atlantic, Falkland Plateau region, Deep Sea Drilling Project Leg 71. In: Ludwig W J, Krasheninnikov V A, et al. Init Repts DSDP, 71 (Pt. 2). Washington: U.S. Govt Printing Office. 667–686

Whitworth T, Orsi A H, Kim S J, Nowlin W D, Locarnini R A. 1998. Water masses and mixing near the Antarctic slope front. Antarctic Research Series, 75: 1–27

Zachos J, Pagani M, Sloan L, Thomas E, Billups K. 2001. Trends, rhythms, and aberrations in global climate 65Ma to present. Science, 292: 686–693

Zhang L L, Suzuki N, Nakamura Y., Tuji A. 2018. Modern shallow water radiolarians with photosynthetic microbiota in the western North Pacific. Marine Micropaleontology, 139: 1–27.

Zhang Q, Chen M H, Zhang L L, Wang R J, Xiang R, Hu W F. 2014. Radiolarian Biostratigraphy in the Southern Bering Sea since Pliocene. Science China Earth Sciences, 57(4): 682–692

Zhang Q, Chen M H, Zhang L L, Su X, Xiang R. 2016. Changes and influencing factors in biogenic opal export productivity in the Bering Sea over the last 4.3Ma: evidence from the records at IODP Site U1340. Journal of Geophysical Research: Oceans, 121A: 5789–5804, doi: 10.1002/2016JC011750

种名索引

Cenozoic Radiolarians from the Ross Sea, Antarctic

Summary

This work aims to analyse the radiolarian diversity, species composition, systematics, assemblage, biostratigraphy and evolution in relation to environmental changes in the Antarctic during the Cenozoic.

The Antarctic is a polar region that includes the ice shelves, island territories and sea waters located south of 66°33′39″S. To date, limited research has been conducted on radiolarians in this specific area, with only a few papers discussing radiolarian species found in coring samples from three sites drilled by the Deep Sea Drilling Project (DSDP). Chen (1974, 1975) described 60 species (including 15 new species and 11 unnamed species) from Site 274 in the northwestern Ross Sea. Weaver (1976) reported 27 species from Site 324 on the Antarctic continental rise, and both Lazarus (1990) and Ling and Lazarus (1990) reported 19 species and 34 species, respectively, from Site 693 in the northeastern Weddell Sea. Some species overlap among these studies, resulting in a total of 113 radiolarian species (including 34 unnamed species) reported by previous researchers in the Antarctic region, which actually represents only a small fraction of radiolarian species in the Antarctic. Additionally, no comprehensive discussion of radiolarian taxonomy in this region has been conducted up to now.

In this study, we used 89 coring samples from DSDP Site 274, located at 68°59.81′S and 173°25.64′E with a water depth of 3305m in the Ross Sea, for detailed species identification and systematics analysis. We also utilized 98 samples from DSDP Site 266, located at 56°24.13′S and 110°06.70′E with a water depth of 4173m in the southern Indian Ocean for comparison. In total, we identified 502 radiolarian species (including 68 unnamed species) belonging to 36 family and 133 genera, among which 4 new genus and 126 new species were established. All of these species have been discussed, and descriptions, synonyms, taxonomy and references for them, along with 154 plates illustrating the 502 radiolarian species, are included in this book.

All samples are provided by Gulf Coast Repository of International Ocean Discovery Program. We sincerely thank Dr. Phil Rumford of Texas A&M University, USA, and Dr. Lallan Gupta of IODP KCC Curator for their assistance with sampling. We would also like to express our gratitude to the IODP-China Office, and Professor Xuefa Shi from the First Institute of Oceanography, Ministry of Natural Resources for their support to this work. This research was funded by National Natural Science Foundation of China (Grant Nos. 42176080, 42076073), the Taishan Scholar Foundation of Shandong Province (Grant No. tspd20181216), and the development fund of South China Sea Institute of Oceanology of the Chinese Academy of Sciences (Grant Nos. SCSIO2023PT10, SCSIO202201).

I. Cenozoic radiolarian biostratigraphy in the Ross Sea, Antarctic

Up to now, DSDP Leg 28 Site 274 is the only location that preserves the comparatively complete Cenozoic sedimentary record in the Antarctic Sea. This site thus plays a vital role in the analysis and establishment of Cenozoic radiolarian biostratigraphy in the Antarctic region, as well as in providing essential data for the comparison and discussion of comprehensive stratigraphy in high-latitude regions.

Previous studies on Antarctic radiolarians have been conducted, but they primarily focused on late Neogene periods and neglect Paleogene stratigraphy due to sampling restriction (Hays, 1965; Chen, 1974, 1975; Abelmann, 1992a; Lazarus, 1992). However, a recent study by Hollis et al. (2020) analyzed Paleogene radiolarian stratigraphy in the mid-high latitude areas of the southwest Pacific and southeast Indian Ocean. This study established a framework of the stratigraphic chronology, providing valuable information on fossil events and ages for radiolarian stratigraphic division and correlation in the Southern Ocean and its adjacent areas. Despite this, due to differences in ecological environment and geographic feature, the characteristics of radiolarian assemblage and evolution in the Cenozoic Ross Sea differ significantly from those in southwest Pacific and southeast Indian Ocean, probably resulting in different types of radiolarian fauna in various geological periods. Therefore, it is challenging to apply the fossil zones that were established in the mid-high latitude areas in the South Ocean to stratigraphic divisions in the Antarctic regions. Our objective here is to provide evidence of radiolarian zonation fossils, including some new species and a few common species mentioned in Hollis et al. (2020), for the discussion of radiolarian stratigraphy in the Ross Sea, Antarctic.

This work mainly focuses on the analysis of the division, determination, and ages of Cenozoic radiolarian biostratigraphic boundaries in the Ross Sea, based on the distribution and evolution of radiolarian characteristics species and events preserved in core samples of Site 274 (Fig. 5). By combining the presence of cold water species in the stratum, which was used to assess the formation process of Antarctic ice sheet and its correlation with global climatic cooling during the Oligocene, we have preliminarily established a Cenozoic radiolarian biostratigraphic framework (Fig. 6) and revealed the process of environmental evolution in Antarctic.

1. Holocene/Pleistocene boundary (at hole depth of 6.75m, age of 12ka)

It is defined by the last appearance datums (LAD) of *Solenosphaera bitubula* sp. nov., *Zygocircus productus* (Hertwig), *Antarctissa clausa* (Popofsky) and *Cycladophora conica*

Lombari and Lazarus.

2. Middle Pleistocene/early Pleistocene boundary (at hole depth of 16.25m, age of 0.9Ma)

It is assumed by the first appearance datum (FAD) of *Acanthosphaera marginata* Popofsky, but lack other fossil information in the Ross Sea.

3. Pleistocene/Pliocene boundary (at hole depth of 38.8m, age of 2.4Ma)

It is determined by the LAD of *Prunopyle tetrapila* Hays group, rather than *Helotholus vema* Hays that might go down to the Holocene. Stratigraphic distributions of these two species in the Ross sea are different with those in the Southern Ocean or other areas.

4. Late Pliocene/middle Pliocene boundary (at hole depth of 76.7m, age of 3.6Ma)

It is ascertained by the LAD of *Spongoplegma antarcticum* Haeckel and *Eucyrtidium calvertense* Martin.

5. Middle Pliocene/early Pliocene boundary (at hole depth of 88.9m, age of 4.2Ma)

It is ascertained by FAD of *Antarctissa strelkovi* Petrushevskaya and *Helotholus vema* Hays. *Helotholus vema* Hays was reported as FAD at 4.2Ma by Lazarus (1992). Meanwhile, this boundary is also confirmed by the LAD of some other species: *Spongodiscus inconcavus* sp. nov., *Larcopyle adelstoma* (Kozlova and Gobovets), *Larcopyle titan* (Campbell and Clark) group, *Antarctissa pachyoma* sp. nov., *Velicucullus oddgurneri* Bjørklund and *Pseudodictyophimus galeatus* Caulet.

6. Pliocene/Miocene boundary (at hole depth of 98.7m, age of 5.33Ma)

It is marked by the FAD of *Helotholus vema* Hays and *Lychnocanoma grande* (Campbell and Clark) group, and the LAD of *Stylosphaera coronata laevis* Ehrenberg, *Larcopyle polyacantha* (Campbell and Clark) and *Antarctissa conradae* Chen. Perhaps, it is needed to be confirmed by other stratigraphic tools for its age.

7. Late Miocene/middle Miocene boundary

No radiolarian event might be found by our examination for samples of Site 274.

However, a middle Miocene climatic optimum (MMCO) happened at ~17–15Ma, while ice sheet retreated and a lot of meltwater flowed into the sea, and immediately followed by an interval of Antarctic ice growth and cooling, termed the middle Miocene climate transition (MMCT, during 14.2–13.8Ma; Kennett, 1977; Flower and Kennett, 1994; Zachos et al., 2001; Shevenell et al., 2004; Holbourn et al., 2005; Shevenell et al., 2008; Cramer et al., 2009). It was a turbulent period in the middle Miocene for Antarctic ice sheet growth and global climatic change, which affected sedimentary condition in the Ross Sea and caused probably lack of sedimentary strata. There was also no information of the cooling event from this work. Instead, some warm radiolarian species, *Tetrapyle* spp., appeared in strata of middle-late Miocene, showing mainly a warm period and a mixed ecologic environment of warm-cold waters.

8. Middle Miocene/early Miocene boundary (at hole depth of 111.3m, age of 17.3Ma)

Here some FAD species are *Eucyrtidium punctatum* (Ehrenberg) group, *Stylosphaera coronata laevis* Ehrenberg, *Desmospyris spongiosa* Hays, *Botryopera cylindrita* sp. nov., *Cycladophora bicornis* (Popofsky) group, *Artostrobus annulatus* (Bailey), *Eucyrtidium calvertense* Martin and *Siphocampe mesinflatis* sp. nov. Abelmann (1992a) and Lazarus (1992) recognised the FAD age of *Eucyrtidium punctatum* (Ehrenberg) group at 17.3Ma in the Kerguelen Plateau site. Meanwhile, species of *Actinomma boreale* Cleve and *Larcopyle hayesi* (Chen) group are basicly disappeared (LAD) by this boundary.

9. Miocene/Oligocene boundary (at hole depth of 143.25m, age of 23Ma)

It is marked by the FAD of *Antarctissa denticulata* (Ehrenberg) group with its large number individuals. The LAD species are *Thecosphaera zittelii* Dreyer and *Stylosphaera minor* Clark and Campbell.

10. Late Oligocene/middle Oligocene boundary (at hole depth of 181m, age of 27.7–27.8Ma)

It is ascertained by the LAD species of *Cenosphaera solantarctica* sp. nov., *Cenellipsis monikae* (Petrushevskaya), *Stylatractus gracilus* sp. nov., *Perichlamydium limbatum* Ehrenberg, *Spirema lentellipsis* Haeckel, *Amphimelissa setosa* (Cleve), *Lithomelissa sphaerocephalis* Chen, *Lophophaena cylindrica* (Cleve), *Dictyocephalus variabilus* sp. nov. and *Eucyrtidium granulatum* (Petrushevskaya) group.

Lithomelissa sphaerocephalis Chen was last appearing at 27.8Ma in the southern Ocean (Hollis et al., 2020). *Eucyrtidium granulatum* (Petrushevskaya) group here is synonymous with *Eucyrtidium antiquum* (Hollis, 1997, 2020), which appeared last at 27.8Ma (Hollis et al., 2020).

11. Middle Oligocene/early Oligocene boundary (at hole depth of 203m, age of 28.6Ma)

By this boundary, the FAD species are *Dictyocephalus crassus* Carnevale, and some species of LAD are *Cenosphaera favosa* Haeckel, *Haliomma stylota* sp. nov., *Actinomma hexactis* Stöhr, *Lophophaena spongiosa* (Petrushevskaya), *Dictyocephalus turritus* sp. nov. and *Gondwanaria milowi* (Riedel and Sanfilippo) group. Species *Gondwanaria milowi* (Riedel and Sanfilippo) group here is synonymous with *Lophocyrtis* (*Paralampterium*) *longiventer* (Hollis et al., 2020).

A little below this boundary, cold species *Spongotrochus glacialis* Popofsky start to occur in stratum.

12. Oligocene/Eocene boundary (at hole depth of 266.75m, age of 33.9Ma)

It is marked by the FAD of *Cycladophora conica* Lombari and Lazarus, *Cycladophora davisiana* Ehrenberg group and *Gondwanaria tenuoria* sp. nov. It is also the LAD of some species: *Velicucullus spongiformus* sp. nov., *Gondwanaria pteroforma* sp. nov., *Thecosphaera entocuba* Chen et al., *Artostrobium undulatum* (Popofsky), *Solenosphaera monotubulosa* (Hilmers) group, *Amphiplecta siphona* sp. nov., *Lithostrobus* cf. *longus* Grigorjeva, *Pylonium neoparatum* sp. nov., *Actinomma holtedahli* Bjørklund, *Lithomelissa cratospina* sp. nov., *Lipmanella archipilium* (Petrushevskaya), *Spongurus cauleti* Goll and Bjørklund, *Stylodictya circularis* (Clark and Campbell), *Triosphaera bulloidis* sp. nov., *Sethocorys odysseus* Haeckel, *Artostrobus multiartus* sp. nov., reflecting the drastic changes of ecologic environment and radiolarian fauna on this boundary.

13. Late Eocene/middle Eocene boundary (at hole depth of 314m, age of 35.9–36.8Ma)

Many new types first appear on this boundary. The FAD species were *Cenosphaera favosa* Haeckel, *Actinomma impolita* sp. nov., *Pylonium neoparatum* sp. nov., *Triosphaera bulloidis* sp. nov., *Lithomelissa cratospina* sp. nov., *Lithomelissa ministyla* sp. nov., *Lophophaena cylindrica* (Cleve), *Amphiplecta siphona* sp. nov., *Ceratocyrtis irregularis* sp. nov., *Ceratocyrtis perimashae* sp. nov., *Gondwanaria semipolita* (Clark and Campbell) and *Phormostichoartus fistula* Nigrini. The LAD species are *Spirema vetula* sp. nov. and *Siphocampe quadrata* (Petrushevskaya and Kozlova).

Gondwanaria semipolita (Clark and Campbell) is synonymous with *Aphetocyrtis rossi*, which age of FAD is considered at 36.8Ma (Hollis et al., 2020). Age of *Siphocampe quadrata* LAD was given at 35.9Ma by Hollis et al. (2020).

14. Middle Eocene/early Eocene boundary (at hole depth of 320m, age of 48.6Ma)

Due to fewer or short of information from radioalarian fossil, it is difficult to determine this boundary certainly. The FAD species is only *Eucyrtidium granulatum* (Petrushevskaya).

15. Eocene/Paleocene boundary (at hole depth of >333.25m, age of 56Ma)

It is marked by the FAD species, including some new species. They are *Cenosphaera solantarctica* sp. nov., *Carposphaera globosa* Clark and Campbell, *Thecosphaera akitaensis* Nakaseko, *Thecosphaera reticularis* sp. nov., *Solenosphaera monotubulosa* (Hilmers) group, *Actinomma plasticum* Goll and Bjørklund, *Cenellipsis monikae* (Petrushevskaya), *Stylatractus gracilus* sp. nov., *Spongurus cauleti* Goll and Bjørklund, *Spongocore cylindricus* Haeckel, *Stylodictya heliospira* Haeckel, *Larcopyle frakesi* (Chen) group, *Staurodictya medusa* Haeckel,

Larcopyle polyacantha (Campbell and Clark), *Spirema lentellipsis* Haeckel, *Lithelius rotalarius* sp. nov., *Amphimelissa setosa* (Cleve), *Lithomelissa thoracites* Haeckel, *Lipmanella archipilium* (Petrushevskaya), *Pterocanium polypylum* Popofsky, *Gondwanaria robusta* (Abelmann), *Lithomitra lineata* (Ehrenberg) group and *Lithocampe subligata* Stöhr group.

Foraminiferal species *Schenckiella* cf. *levis* of early Eocene was found from the sediment (Hayes et al., 1974).

16. Late Paleocene/early Paleocene boundary (at hole depth of 361.8m, age of 58.7Ma)

It is speculated from the FAD of *Hexalonche anaximensis* Haeckel, *Actinomma henningsmoeni* Goll and Bjørklund, *Astrophacus perplexus* (Clark and Campbell), *Larcopyle ovata* (Kozlova and Gorbovetz) and *Desmospyris haysi* (Chen).

17. Paleocene/Cretaceous boundary (at hole depth of 402.77–390.27m, age of 65.5Ma)

It is marked by the FAD species of many new and local typies. These species are *Cenosphaera archantarctica* sp. nov., *Haliomma microlaris* sp. nov., *Stylodictya circularis* (Clark and Campbell), *Larcopyle adelstoma* (Kozlova and Gobovets), *Larcopyle hayesi* (Chen) group, *Pylozonium saxitalum* sp. nov., *Prunopyle quadrata* sp. nov., *Spirema vetula* sp. nov., *Streblacantha circumyexta* (Jørgensen) and *Sethopyramis quadrata* Haeckel.

Simultaneously, some very small and vague specimens of possible Cretaceous types are disapeared, such as *Prunopyle archaeota* sp. nov. and *Pylozonium saxitalum* sp. nov. They appear in bottom sediment (402.77m depth) overlaid directly on the base of basalt. The chronology of the bottom sediment at Site 274 has been determined to be late Cretaceous based on the occurrences of foraminiferal species *Globotruncana* and *Rugoglobigerina* (Hayes et al., 1974, p. 375).

II. Description of new species

Order SPUMELLARIA Ehrenberg, 1875

Suborder SPHAEROIDEA Haeckel, 1887

Family Liosphaeridae Haeckel, 1887, emend.

Genus *Cenosphaera* Ehrenberg, 1854

1. *Cenosphaera archantarctica* sp. nov.

(Pl. 1, Figs. 1–9)

Single spherical lattice-shell, smaller or larger, with slight thick walled, smooth or rough surface, covered with short thin spine at some specimen. Pores roundish, similar or different size, subregular or irregular arrangement, with hexagonal frame, pore diameter 1–2 times as broad as the bars, 6–9 pores on the half equator, which not clear on surface of some specimens, for filling of sediments, from the old stratum at bottom of Site 274 and no appearance in newer stratum.

Measurements: Diameter of the shell 106–210μm, pores 6–18μm.

Holotype: DSDP274-42-2 1 deposited in the South China Sea Institute, CAS, from sample of DSDP 274 in the Ross Sea, pictured in pl. 1, figs. 4–5.

Distribution: Paleocene in Ross Sea, Antarctic.

This new species is similar to *Cenosphaera vesparia* Haeckel, but the latter has same pore size, regularly arranged, thinner bars and wall, and is distributing in newer stratum.

7. *Cenosphaera miniantarctica* sp. nov.

(Pl. 3, Figs. 10–17)

Cenosphaera sp. D, Lazarus and Pallant, 1989, p. 365, pl. 7, fig. 1.

Small shell, with slight thick walled, smooth surface; pores roundish, with hexagonal frames, regularly arranged, 5–6 pores on the half equator; pores 3–4 times as broad as the thin bars. This is a very special group of small shell, which characteristics is similar to *Cenosphaera* sp. D (Lazarus and Pallant, 1989, p. 365, pl. 7, fig. 1), although the later has some rough surface and more than 8–9 pores on the half equator, they have basically the same features. So, we regard them as a same species.

Measurements: Diameter of the shell 50–95μm.

Holotype: DSDP274-2-3 1 deposited in the South China Sea Institute, CAS, from sample of DSDP 274 in the Ross Sea, pictured in pl. 3, fig. 15.

Distribution: Holocene, and rare in late Eocene in Ross Sea, Antarctic.

This new species is distinguished from other known species by its very small or tiny individuals (shells) with less pores regularly arranged.

9. *Cenosphaera pseudocoela* sp. nov.

(Pl. 3, Figs. 29–32)

Shell moderate walled, with roundish, polygonal or elliptical pores in different sizes, irregularly arranged, diameter of pores is generally larger than the thin bars; many spines arosefrom shell joint points, furcated at the end, and interconnected each other forming a thin pseudo-outer shell, seems to have a double shell of this species, some spines even protrude outer shell as conical or cylindrical radiative spines; rough surface.

Measurements: Diameter of the lattice shell 170–180μm, outer shell 195–205μm, length of spines 15–30μm.

Holotype: DSDP274-23-5 1 deposited in the South China Sea Institute, CAS, from sample of DSDP 274 in the Ross Sea, pictured in pl. 3, figs. 29–30.

Distribution: Late Eocene–late Oligocene in Ross Sea, Antarctic.

Character of this new species is close or similar to *Cenosphaera solantarctica* sp. nov., distinction between them is the latter with nearly same pore size, hexagonal frame and regularly arranged, no spine on the flat surface.

11. *Cenosphaera solantarctica* sp. nov.

(Pl. 4, Figs. 7–21)

Cenosphaera sp., Chen, 1975, p. 453, pl. 7, figs. 1–2.
Cenosphaera sp., Keany, 1979, pl. 5, fig. 1.

Shell variable in size, spherical or exceptionally ellipsoidal; thick walled, with quasi-circular pores in similar size and hexagonal frame, regularly or sub-regularly arranged, pores 2–3 times as broad as the bars, 14–18 pores on the half equator (less pores in smaller individuals); thorny surface, many spines arosefrom frame corners or joint points of bars as long pyramid, with same length and furcated at the end, and these divarications may connect each other forming a thorny pseudo-peripheral thick shell.

Measurements: Diameter of the lattice shells 125–340μm, diameter of pores 9–15μm, length of spines 15–25μm.

Holotype: DSDP274-29-3 2 deposited in the South China Sea Institute, CAS, from sample of DSDP 274 in the Ross Sea, pictured in pl. 4, figs. 7–8.

Distribution: Early Eocene–Oligocene in Ross Sea, Antarctic.

This new species is somewhat similar to *Cenosphaera favosa* Haeckel, the key distinction is the latter with conical spines, sharp end, not ramify, and relatively shorter.

12. *Cenosphaera spongiformis* sp. nov.

(Pl. 5, Figs. 1–2)

Single lattice shell enveloped by thick texture of flocculence. Lattice shell walled as medium thickness, with roundish, elliptical or polygonal pores, in different sizes, 2–3 times as broad as the bars; some spines on shell surface developed as irregular pergola for supporting surrounding flocculence, which thickness is about 1/6–1/4 times of lattice shell diameter, several spines even reach out texture of flocculence. This is a very special shell structure of *Cenosphaera*.

Measurements: Diameter of the lattice shells 168–190μm, outer diameter of all the shell 210–260μm, length of spines 7–18μm.

Holotype: DSDP274-34-1 2 deposited in the South China Sea Institute, CAS, from sample of DSDP 274 in the Ross Sea, pictured in pl. 5, figs. 1–2.

Distribution: Early Eocene in Ross Sea, Antarctic.

This new species has a single lattice shell surrounded by texture of flocculence, forming a thick outer shell of flocculence or sponge, which character is distinctly known from other species.

15. *Cenosphaera xiphacantha* sp. nov.

(Pl. 5, Figs. 26–27)

Single spheroidal shell, large, thick walled, with roundish pores and hexagonal frames, similar size regularly arranged, pores 2–3 times as broad as bars, small raised cones at each joint point on surface, especially has a strong single pyramidal spine developed from inner side of the wall to outward, which bladed at first half section and pyramid at the latter half section, and sharp at end, like a dagger; surface slight coarse, no other spine.

Measurements: Diameter of the lattice shells 310–325μm, length of spines 80–90μm.

Holotype: DSDP274-31-2 1 deposited in the South China Sea Institute, CAS, from sample of DSDP 274-31-2 in the Ross Sea, pictured in pl. 5, figs. 26–27.

Distribution: Late Eocene–early Oligocene in Ross Sea, Antarctic.

This new species is similar to *Cenosphaera vesparia* Haeckel, but the latter there is no any spine, also no raised cones, with smooth surface.

Genus *Carposphaera* Haeckel, 1881

21. *Carposphaera anomala* sp. nov.

(Pl. 6, Figs. 16–21)

Individual small. Cortical shell ellipsoidal or round trilateral, medullary shell spheroidal, the rate of two shell 2:1 or 3:1; pores of cortical shell small and many, round or oval, in different sizes, irregularly arranged, with polygonal bar frames, 8–15 pores across the half equator; medullar shell near spherosome, inside of medullar shell have several radial beams join at centrally; outside of cortical shell surrounded by a thin and fragile wrap (usually not incomplete), supported by many radial thin beams; rough surface, no radial spine.

Measurements: Diameter of cortical shells 100–120μm, medullary shell 30–60μm.

Holotype: DSDP274-34-5 2 deposited in the South China Sea Institute, CAS, from sample of DSDP274-34-5 in the Ross Sea, pictured in pl. 6, figs. 20–21.

Distribution: Middle Eocene in Ross Sea, Antarctic.

This new species is similar to *Carposphaera magnaporulosa* Clark and Campbell, main distinction between them is the latter has spheroidal cortical shell, with larger and less pores, and the former cortical shell is ellipsoidal or round trilateral.

24. *Carposphaera sterrmona* sp. nov.

(Pl. 7, Figs. 28–29)

Cortical shell spheroidal or near cuboidal, thick walled, with slight large quasi-circular pores, in different sizes, 6–7 pores across half equator, irregularly arranged, with hexagonal or irregular frames; medullar shell very small, spherical, only 1/5–1/4 size of cortical shell, with small round pores; many radial beams connecting medullar shell and cortical shell, with some branch at early or medium beams, which also connected to the cortical shell and together with the main radial beams formed outer radial spines; there are 50–60 dumpy radial spines on shell surface, spines pyramid and triangular at their bases, with length of 1/4 shell diameter; rough surface.

Measurements: Diameter of cortical shells 150–170μm, medullary shell 25–30μm, spine length 25–40μm.

Holotype: DSDP274-28-1 2 deposited in the South China Sea Institute, CAS, from sample of DSDP274-28-1 in the Ross Sea, pictured in pl. 7, figs. 28–29.

Distribution: Early Oligocene in Ross Sea, Antarctic.

This new species is similar to *Carposphaera magnaporulosa* Clark and Campbell, main distinction between them is the latter has smaller individual, with smooth surface and no dumpy spine.

Genus *Eccentrisphaera* gen. nov.

Three or four lattice shells, with a cortical shell and 2–3 medullar shells, or 2 cortical shells and 1–2 medullar shells, all medullar shells have a same center, which is different with the center of cortical shell.

Type species: *Eccentrisphaera bimedullaris* sp. nov.

27. *Eccentrisphaera biderma* sp. nov.

(Pl. 8, Figs. 1–4)

Two cortical shells and 1–2 medullar shells; outer cortical shell thin walled, with round pores, hexagonal frames, regularly arranged, 12–14 pores across half equater, surface smooth or clavula; inner cortical shell about 2/3 size of outer cortical shell, distance between these two shells less than 1/3 diameter of inner cortical shell, and inner cortical shell with circular or quasi-circular pores, regularly or irregularly arranged, 9–10 pores across half equater; two medullar shells are not situated at the centre of cortical shells, outer medullar shell growth stick close to inner cortical shell, inner medullar shell very small and situated at the centre of outer medullar shell; size rate of four shells as 1 : 3 : 7 : 10.

Measurements: Diameter of outer cortical shells 185–210μm, inner cortical shell 130–145μm, outer medullary shell 62–75μm, inner medullar shell 22μm.

Holotype: DSDP274-12-5 1a deposited in the South China Sea Institute, CAS, from sample of DSDP274-12-5 in the Ross Sea, pictured in pl. 8, figs. 1–2.

Distribution: Late Miocene in Ross Sea, Antarctic.

This new species is distinguished from other species by two medullar shells being not situated at center of two cortical shells, outer medullar shell growing closely stick to inner cortical shell, smooth surface or rare clavula.

28. *Eccentrisphaera bimedullaris* sp. nov.

(Pl. 8, Figs. 5–20)

Actinomma cocles Renaudie and Lazarus, Renaudie, 2012, p. 36, pl. 4, fig. 6 (only).

Individual somewhat small with one cortical shell and two medullar shells. Cortical shell thin or medium walled, with roundish pores, in different or near similar sizes, regularly or irregularly arranged, hexagonal frames, thin bars of between pores, diameters of pores 4–5 times as broad as the bars, 6–9 pores across half equator, which generally less pores on thinner shell and more pores on thicker shell; two medullar shells not located at center of cortical shell, outer medullar shell near or closer to one side of cortical shell, inner medullar shell at center of outer medullar shell; diameter of outer medullar shell about 1/3–1/2 size of cortical shell and 3 times of inner shell; rare or some smaller cylindrical spines on shell surface.

Measurements: Diameter of cortical shells 106–143μm, pores of cortical shell 13–25μm, length of spine 10–30μm.

Holotype: DSDP274-6-1 1 deposited in the South China Sea Institute, CAS, from sample of DSDP274-6-1 in the Ross Sea, pictured in pl. 8, figs. 8–9.

Distribution: Pliocene to present in Ross Sea, Antarctic.

This new species has two medullar shells with a same shell center of them, but not located at the center of cortical shell, near or closer to one side of cortical shell, which features are obviously distinguished from other species.

30. *Eccentrisphaera porolaris* sp. nov.

(Pl. 8, Figs. 25–26)

Two cortical shells and one medullar shell that is placed sticking to the wall of inner cortical shell, size rate of three shells as 1:2.5:4; outer cortical shell spheroidal or near ellipsoidal, one front opening with a short peristoma of transparency, this opening as broad as 1/3–1/4 diameter of shell, pores of cortical shell quasi-circular, with similar or different sizes, sub-regularly distributed, 15–17 pores across half equater; inner cortical shell spheroidal, with roundish pores, in differet sizes, apparently hexagonal frames, sub-regularly distributed, 9–11 pores across half equater; medullar shell situated not at the centre of cortical shell and sticked to the wall of inner cortical shell; short thorny on surface, but no radial spine.

Measurements: Diameter of outer cortical shell 200–230μm, inner cortical shell 115–120μm, medullar shell 50–55μm.

Holotype: DSDP274-12-5 1b deposited in the South China Sea Institute, CAS, from sample of DSDP274-12-5 in the Ross Sea, pictured in pl. 8, figs. 25–26.

Distribution: Late Miocene in Ross Sea, Antarctic.

This new species is different from other species by its ellipsoidal outer cortical shell, with a large front opening. All other similar species are enclosing spheroidal shells, and no opening.

31. *Eccentrisphaera trimedullaris* sp. nov.

(Pl. 8, Figs. 27–29)

Large individual; one cortical shell thick walled, many quasi-circular pores in similar sizes, sub-regular distributed, with hexagonal frames, diameter of pores 3–4 times as broad as bars, 15–17 pores across half equater; three medullar shells situated not at the centre of cortical shell, out medullar shell stick to a side of cortical shell, two other inner medullar shells in centre of outer medullar shell; size rate of four shells (three medullar shells and one cortical shell) as 1 : 3 : 9 : 18; shell surface somewhat coarse, with some convex thorns and several small radial spines.

Measurements: Diameter of cortical shell about 250μm, pores of cortical shell 10–15μm, length of spine 8–15μm.

Holotype: DSDP274-7-1 1 deposited in the South China Sea Institute, CAS, from sample of DSDP274-7-1 in the Ross Sea, pictured in pl. 8, figs. 27–29.

Distribution: Late Pliocene in Ross Sea, Antarctic.

This new species is different from other species by with three medullar shells and one cortical shell, three medullar shells have a same center which is different to the cortical shell, outer medullar shell stick to inner wall of cortical shell.

Genus *Liosphaera* Haeckel, 1881

32. *Liosphaera carpolaria* sp. nov.

(Pl. 8, Figs. 30–31)

Double shells, no medullar shell; outer shell nearly ellipsoidal, slight thin walled, larger pores, polygonal, different sizes, irregularly arranged, 7–9 pores across half equater, with thin bars, diameter of pores 3–7 times as broad as bars, surface arised small bulges at join points; inner shell spheroidal, pores circular of similar size, with hexagonal frames, sub-regularly arranged, many radial beams connecting inner shell and outer shell; surface rough, no radial spine.

Measurements: Diameter of outer shell 155–170μm, inner shell 100–106μm, distance between two shell walls 25–35μm.

Holotype: DSDP274-10-5 2 deposited in the South China Sea Institute, CAS, from sample of DSDP274-10-5 in the Ross Sea, pictured in pl. 8, figs. 30–31.

Distribution: Early Pliocene in Ross Sea, Antarctic.

This new species is similar to *Liosphaera hexagonia* Haeckel (1887, pl. 76, pl. 20, fig. 3), main distinction is the latter with more pores on both two shells, hexagonal, about 20 pores across half equator, regularly arranged, smaller space between two shells, bars of outer shell thin filiform, and smooth surface.

Genus *Thecosphaera* Haeckel, 1881

38. *Thecosphaera minutapora* sp. nov.

(Pl.10, Figs. 5–12)

One cortical shell and two medullar shells; cortical shell thick walled, pored quasi-circular, very small, numerous, similar size, 12–14 pores across half equator, regularly or sub-regularly arranged, pores diameter 1–2 times as broad as bars, without or with slight hexagonal frames; two medullar shells very small, connected with cortical shell by several radial beams, rate of three shells 1 : 3 : 12; surface generally smooth, only individual specimen with very short pyramidal spines.

Measurements: Diameter of cortical shell 170–210μm, outer medullar shell 45–60μm, inner medullar shell 15–20μm.

Holotype: DSDP274-30-6 1 deposited in the South China Sea Institute, CAS, from sample of DSDP274-30-6 in the Ross Sea, pictured in pl. 10, figs. 10–11.

Distribution: Late Eocene in Ross Sea, Antarctic.

This new species is somewhat similar to *Actinomma henningsmoeni* Goll and Bjørklund (1989, pl. 2, figs. 10–15), but the latter with a cortical shell of deformation, rough surface, slight larger medullar shell of 1/3 cortical shell size. This new species is distinguished from other species of genus *Thecosphaera* by its smaller medullar shells, only less 1/4 of specimen, and more smaller pores.

40. *Thecosphaera ovata* sp. nov.

(Pl. 10, Figs. 21–24)

One cortical shell and two medullar shells; cortical shell ovate or ellipsoidal, pores quasi-circular or polygonal, similar sizes, irregularly arranged, with or without hexagonal frames, pores diameters 2–4 times as broad as bars, surface smooth or coarse, some prominences arose from join points of shells with hexagonal frames; outer medullar shell spheroidal or ellipsoidal, relatively large, thick walled; inner medullar shell spheroidal, smaller; rate of diameters of two medullar shells with short axis of cortical shell as 1 : 2.5 : 4, many radial beams connecting three shells, no radial spine on surface.

Measurements: Diameter of cortical shell long axis 190–200μm, cortical shell short axis 150–155μm, outer medullar shell 100–125μm, inner medullar shell 45–55μm.

Holotype: DSDP274-12-3 1 deposited in the South China Sea Institute, CAS, from sample of DSDP274-12-3 in the Ross Sea, pictured in pl. 10, figs. 23–24.

Distribution: Late Miocene in Ross Sea, Antarctic.

This new species is distinguished from other species of *Thecosphaera* by it being a typical ellipsoidal cortical shell, which is enclosed.

41. *Thecosphaera pachycortica* sp. nov.

(Pl. 10, Figs. 25–32)

One cortical shell and two medullar shells; cortical shell thick walled, pores quasi-circular, 9–11 pores across half equator, with hexagonal frames, bars between pores relatively thinner and higher, bar width about 1/4–1/3 diameter of pores, with larger distance between inner wall and outer wall of cortical shell, seemly radial beams connecting the inner and outer walls, forming approximately a cortical shell of two layers; outer medullar shell calabash-shaped, its neck elongated to stick with cortical shell, diameter of outer medullar shell short axis about 1/2 inner diameter of cortical shell; inner medullar shell spheroidal, small, about 1/3 size of outer medullar shell; surface generally smooth, with a number of dumpy cone shape spines.

Measurements: Diameter of cortical shell outer wall 140–170μm, cortical shell inner wall 100–113μm, outer medullar shell 44–50μm, inner medullar shell 18μm, length of spines 6–18μm.

Holotype: DSDP274-23-5 1 deposited in the South China Sea Institute, CAS, from sample of DSDP274-23-5 in the Ross Sea, pictured in pl. 10, figs. 25–26.

Distribution: Late Paleocene-early Oligocene in Ross Sea, Antarctic.

This new species is similar to *Thecosphaera sanfilippoae* Blueford, but the latter with thin walled and outer medullar shell not calabash-shaped.

43. *Thecosphaera reticularis* sp. nov.

(Pl. 11, Figs. 5–19)

Individual sizes variable, some specimens with larger shells; cortical shell thick walled, pores quasi-circular, similar sizes, sub-regularly arranged, 12–18 pores across half equator, hexagonal frames. Prominences on the surface of the cortical shell generally possess branches at the extremities, which are interconnected to create a loosely reticular structure. In some cases, the prominences arising from bar joints of the cortical shell can further develop into spines that have branches at their ends and interconnect with each other, thereby forming an outer shell with a spongy structure. Medullar shells small, size rate of two medullar shells with cortical shell (inner diameter) about 1∶3∶9, outer medullar shell with small pores of roundness, approximately regularly arranged, 8–10 pores across half equator; shell surface rough, with a few of short radial spines.

Measurements: Diameter of cortical shell 160–280μm, total shell (including layer of outer net) 180–375μm, outer medullar shell 56–105μm.

Holotype: DSDP274-36-1 2 deposited in the South China Sea Institute, CAS, from sample of DSDP274-36-1 in the Ross Sea, pictured in pl. 11, figs. 15–17.

Distribution: Early Eocene–early Oligocene in Ross Sea, Antarctic.

This new species is similar to *Thecosphaera zittelii* Dreyer, key distinction between them is the latter without spine branches and nets, or spongy textures on outer surface of cortical shell.

Genus *Cromyosphaera* Haeckel, 1881

49. *Cromyosphaera asperata* sp. nov.

(Pl. 13, Figs. 6–7)

Shell spheroidal, with two cortical shells and two medullar shells; diameter of outer cortical near 2 times as inner cortical shell, rate of four shells about as 1 : 3 : 8 : 15, medullar shells small; outer cortical shell composed of irregular coarse bars, like spongy construction, thick walled, with surface of calthrops; wall composition of inner cortical shell and outer medullar shell similar to outer cortical shell, but thin walled, the spongy likeness appeared also in some areas of inside shell, exceptionally, inner medullar shell latticed with small pores regularly arranged; these four shells connected by 17–22 thick radial beams, no protrude out of shell; surface rough, without radial spine.

Measurements: Diameter of outer cortical shell 400μm, inner cortical shell 220μm, outer medullar shell 90μm, inner medullar shell 30μm.

Holotype: DSDP274-25-1 2 deposited in the South China Sea Institute, CAS, from sample of DSDP274-25-1 in the Ross Sea, pictured in pl. 13, figs. 6–7.

Distribution: Early Oligocene in Ross Sea, Antarctic.

This new species is similar to *Cromyosphaera quadruplex* Haeckel, key distinction between them is the latter with lattice of all shells, spheroidal or hexagonal pores, hexagonal frames, without spongy construction, and smooth surface.

Genus *Plegmosphaera* Haeckel, 1882

50. *Plegmosphaera asperula* sp. nov.

(Pl. 13, Figs. 8–9; Pl. 14, Figs. 1–14)

Plegmosphaera sp., Chen et al., 2017, p. 93, pl. 6, figs. 4–6.

Large individuals, a single shell, near spheroidal, irregularly spherical or ellipsoidal, a great emptiness inside shell; thick walled, formed by similar spongy parenchyma of loose and irregular bar intersections, no nodes rise on some plane of shell; pores in various irregular forms, with different shapes and sizes, dispersedly distributed, slippy for inner edge of pores; bars general solid, with some different sizes, three-dimensional growth at some areas; rough surface, or with concave-convex surface, no radial spine.

Measurements: Diameter of shell 250–460μm.

Holotype: DSDP274-25-3 2 deposited in the South China Sea Institute, CAS, from sample of DSDP274-25-3 in the Ross Sea, pictured in pl. 13, figs. 8–9.

Distribution: Early Eocene–early Oligocene in Ross Sea, Antarctic.

This new species is distinguished from *Plegmosphaera coelopila* Haeckel (1887, p. 88) by the latter being thin bars and smooth surface, but this new species has often a shell of irregular shape, thick bars, rough surface.

51. *Plegmosphaera globula* sp. nov.

(Pl. 15, Figs. 1–14)

Single shell, spheroidal or elliptical, thick walled, with construction of composite sponge, pores quasi-circular or irregular, different sizes, numberous, bars irregularly interlaced; shell generally enclosed, but a roundish fenestrule may exist for individual specimen at unsteady position; one or more small shells commonly placed at different part inside shell (not at centre), which is suspected as smaller shells of devoured or symbiotic diatoms; smooth surface, or slightly coarse, no radial spine and raised.

Measurements: Diameter of spheroidal shell 190–270μm, long axis of ellipsoidal shell 210–350μm, short axis of ellipsoidal shell 150–290μm.

Holotype: DSDP274-26-1 2 deposited in the South China Sea Institute, CAS, from sample of DSDP274-26-1 in the Ross Sea, pictured in pl. 15, figs. 13–14.

Distribution: Early Eocene–early Oligocene in Ross Sea, Antarctic.

This new species is distinguished from *Plegmosphaera petrushevia* sp. nov. by the latter with all ellipsoidal shell, thin walled, thinner bars, smaller pores, some miniature ridges dividing surface as a number of plots, in different sizes.

52. *Plegmosphaera petrushevia* sp. nov.

(Pl. 15, Figs. 15–20)

Plegmosphaera monikae Petrushevskaya, 1975, p. 655, pl. 30, figs. 1–7.

Cenosphaera sp. A, Abelmann, 1992a, pl. 1, fig. 6.

Single shell, large, nearly ellipsoidal or ovate, slightly irregular shape, thin walled, constituted as desmachyme, approximately spongy; pores very small, near roundish, with similar sizes, irregularly distributed, divided by some linear miniature ridges, forming a number of plots of different sizes, number of pores variable for each small plot, fine bars, about 30–40 pores across half equator; one or more small and round shells usually appeared at different positions inside shell (not at centre), which is suspected as symbiotic diatoms or devoured small diatoms; a small fenestra of roundness always appeared at one end of shell; very smooth surface, no radial spine.

Measurements: Diameter of shell 240–375μm.

Holotype: DSDP274-27-1 2 deposited in the South China Sea Institute, CAS, from sample of DSDP274-27-1 in the Ross Sea, pictured in pl. 15, figs. 15–16.

Distribution: Middle Oligocene in Ross Sea, Antarctic.

This new species is distinguished from *Plegmosphaera coelopila* Haeckel (1887, p. 88) by the latter with thick wall, large pores and thin bars, diameter of pores 5–10 times as width of bars. Character of our specimen is very similar to *Plegmosphaera monikae* Petrushevskaya (1975), they should be a same species. However, while established this new species, Petrushevskaya (1975) could not give any description for it, only provided pictures and name of *Plegmosphaera monikae* Petrushevskaya, which can not be take as a valid name of new species, because of the inconformity of naming norm.

53. *Plegmosphaera spiculata* sp. nov.

(Pl. 16, Figs. 1–19)

Single spherical shell, spheroidal, elliptical or long ovate, thick walled, which appeared as spongy, araneose or latticed; pores quasi-circular or irregular, very small, irregularly arranged, bars slight broader; there is always a openings at one end of shell, its peristoma with thick conic teeth, on surfaces of some shells with some dumpy spines of cone or anomaly; 1–2 specific smaller shells, perhaps diatom, of rotundity or plates often irregularly posited inside shell; rough surface.

Measurements: Diameter of shell long axis 205–430μm, short axis 130–340μm; length of teeth and spines 12–30μm, basal width of teeth and spines 10–30μm.

Holotype: DSDP274-33-5 1 deposited in the South China Sea Institute, CAS, from sample of DSDP274-33-5 in the Ross Sea, pictured in pl. 16, figs. 14–15.

Distribution: Late Eocene–middle Oligocene in Ross Sea, Antarctic.

This new species is similar to *Plegmosphaera globula* sp. nov., their distinction is the latter without shell opening, peristomal tooth and no surface spine.

Genus *Spongoplegma* Haeckel, 1881

56. *Spongoplegma quadratum* sp. nov.

(Pl. 17, Figs. 19–21)

Small shell, nearly quadrate; outer shell composed of loose sponges or clutter meshes, which wraps medullar shell, pores with irregular shapes, sizes and distributions, loose construction, all bars as similar widths; medullar shell of simple spherosome, with quasi-circular pores, thin bars, diameter of medullar shell about 1/3–2/5 of outer shell; rough surface, without radial spine.

Measurements: Diameter of outer shell 118–150μm, medullar shell 40–45μm.

Holotype: DSDP274-1-5 1 deposited in the South China Sea Institute, CAS, from sample of DSDP274-1-5 in the Ross Sea, pictured in pl. 17, figs. 19–21.

Distribution: Holocene in Ross Sea, Antarctic.

Shell structural feature of this new species is similar to *Spongoplegma antarcticum*

Haeckel, differences of them are the latter with larger shell and spherosome.

Genus *Acrosphaera* Haeckel, 1882

63. *Acrosphaera acuconica* sp. nov.

(Pl. 19, Figs. 1–4)

Hexastyliidae gen. sp. indet., Petrushevskaya, 1975, p. 567, pl. 17, fig. 1.

Single spheroidal shell, median size, slight thick walled, vacant inside shell; many coniform bulges arised from bar joint points on shell surface, forming radial spines of sharp cone; pores subround or irregular, in different sizes, irregularly arranged, no hexagonal frame, 9–11 pores across half equator, thick bars, diameter of pores about 1–5 times as width of bars.

Measurements: Diameter of shell 200–215μm, pores 12–36μm, length of spines 18–25μm.

Holotype: DSDP274-25-3 2 deposited in the South China Sea Institute, CAS, from sample of DSDP274-25-3 in the Ross Sea, pictured in pl. 19, figs. 1–2.

Distribution: Late Eocene–early Oligocene in Ross Sea, Antarctic.

Shell structural feature of this new species is similar to *Haliomma acanthophora* Popofsky (Takahashi 1991, p. 68, pl. 9, fig. 3), but the latter with a small medullar shell at the center, which has a form of loose latticed and some filiform bars connecting to cortical shell (not appeared in our specimen); this new species is also similar to *Elatomma penicillus* Haeckel (Takahashi, 1991, p. 69, pl. 9, fig. 10), but the latter has a medullar shell.

Genus *Siphonosphaera* Müller, 1858

71. *Siphonosphaera mixituba* sp. nov.

(Pl. 21, Figs. 5–6)

Single spheroidal shell, large individual, thick walled; shell surface with many pores, spheroidal, elliptical and irregular, in different sizes and irregular distributions, 12–16 pores across half equator; small short tubules elongated from walls of every pore, slant flattened, sharp at one side of the end, all these arised tubules of different shapes make a periphery of mussy embossments; all walls including tubules and shell as transparent, no other radial spine on surface.

Measurements: Diameter of shell 200–215μm, pores (width of tubule) 10–25μm, length of tubules 15–25μm.

Holotype: DSDP274-30-4 2 deposited in the South China Sea Institute, CAS, from sample of DSDP274-30-4 in the Ross Sea, pictured in pl. 21, figs. 5–6.

Distribution: Late Eocene in Ross Sea, Antarctic.

This new species is similar to *Siphonosphaera cyathina* Haeckel (1887, p. 105, pl. 6, fig. 10), but the latter with thinner wall, shorter tubule and regular cut at the end, periphery tidy. This new species is partly different with *Acrosphaera spinosa* (Haeckel) by the latter with conical spines, but no tubule.

Genus *Solenosphaera* Haeckel, 1887

72. *Solenosphaera bitubula* sp. nov.

(Pl. 21, Figs. 7–16)

Small shell, with two openings at two opposite ends, extended to form as short tube-like mouths, width of tubes (or mouths) about half diameter of shell, flat margin of mouth; thick walled for every part of the shell, transparent, dispersed some quasi-circular pores in different sizes, and irregularly distributed; smooth surface of shell.

Measurements: Diameter of spheroidal shell 80–110μm, length of shell 120–150μm, width of tubal mouth 50–60μm, length of tubule 18–30μm.

Holotype: DSDP274-1-5 2 deposited in the South China Sea Institute, CAS, from sample of DSDP274-1-5 in the Ross Sea, pictured in pl. 21, figs. 11–12.

Distribution: Late Pliocene–late Pleistocene in Ross Sea, Antarctic.

This new species is characterized by its two opposite tubes of shell to distinguish from *Siphonosphaera monotubulosa* Hilmers (1906, p. 82, pl. 1, fig. 5) and other species.

Family Hexastylidae Haeckel, 1881, emend. Petrushevskaya, 1975
Genus *Lonchosphaera* Popofsky, 1908, emend. Dumitrică, 2014

77. *Lonchosphaera scabrata* sp. nov.

(Pl. 22, Figs. 1–11)

Lonchosphaera sp. C, Petrushevskaya, 1975, p. 567, pl. 17, figs. 11–15.

Cortical shell thick walled, pores quasi-circular or elliptical, different sizes, irregularly arranged, 13–16 pores across half equator, diameter of pores 1–4 times as broad as bars; inner shell indistinct, with very thin radial beams disarrayed intracavity and connected to cortical shell; shell surface developed 14–20 strong radial spines, three prismatic shape, length of spines about half diameter of cortical shell; especially, at various joints of bars forming many complementary spines, with shape of sharp corner cones or prismatic pillars of furcated ends, similar sizes and lengths; very rough surface.

Measurements: Diameter of cortical shell 200–225μm, length of radial spines 40–105μm, complementary spines 18–35μm.

Holotype: DSDP274-26-3 1 deposited in the South China Sea Institute, CAS, from sample of DSDP274-26-3 in the Ross Sea, pictured in pl. 22, figs. 5–6.

Distribution: Middle Eocene–early Oligocene in Ross Sea, Antarctic.

This new species is similar to *Lonchosphaera spicata* Popofsky, but the latter with thin walled, undeveloped complementary spines, and comparatively smooth on shell surface.

Family Cubosphaeridae Haeckel, 1881, emend. Campbell, 1954

Genus *Hexalonche* Haeckel, 1881

82. *Hexalonche gelidis* sp. nov.

(Pl. 23, Figs. 28–29)

Two-layer shells, small space between two shells, inner shell large, diameter of inner shell about 3/4 of outer shell; thin walled, smaller pores of outer shell, quasi-circular, similar sizes, irregularly arranged, 14–16 pores across half equator, diameter of pores about 1–3 times as broad as bars; larger pores of inner shell, nearly hexagonal, regularly distributed, only 4–5 pores across half equator, thin bars, diameter of pores 5–6 times as broad as bars; six radial beams connecting inner and outer shells, stretched out of outer shell, forming six radial spines, which are long and thin cone or aciculiform, extreme sharp, length of spines nearly half diameter of outer shell; smooth surface.

Measurements: Diameter of cortical shell 180μm, inner shell 130μm, length of spines 80μm.

Holotype: DSDP274-2-3 2 deposited in the South China Sea Institute, CAS, from sample of DSDP274-2-3 in the Ross Sea, pictured in pl. 23, figs. 28–29.

Distribution: Late Pleistocene in Ross Sea, Antarctic.

This new species is classified into Genus *Hexalonche*, because of it has six mutual symmetrical radial spines. It is distinguished from other species by its very big medullar (inner) shell with specially large pores (hexagonal).

Genus *Hexacontium* Haeckel, 1881

91. *Hexacontium parallelum* sp. nov.

(Pl. 26, Figs. 1–6)

Hexacontid juveniles van de Paverd, 1995, pl. 32, figs. 3, 6.

Cortical shell incomplete, generally only retain side branches or side spines of radial spines, and their ramification as part of cortical shell, which is easy fragile and difficult to be retained; two medullar shells unbroken, smaller, outer medullar shell middle thick walled, with quasi-circular pores, similar sizes, irregularly distributed, width of bars similar to diameter of pores, 6–8 pores across half equator, connected with inner medullar shell by several thin radial beams; inner medullar shell very small, clathrate or reticular; rate of three shells 1 : 3 : 6–9; six radial beams of triangular prism developed from outer medullar shell, extended outside as 6 strong triangular pyramidal radial spines, terminal sharp, smooth surface, without complementary spine.

Measurements: Diameter of cortical shell 115–180μm, outer medullar shell 60–75μm, inner medullar shell 25–30μm, length of spine 30–60μm.

Holotype: DSDP274-21-1 2 deposited in the South China Sea Institute, CAS, from sample of DSDP274-21-1 in the Ross Sea, pictured in pl. 26, figs. 5–6.

Distribution: Middle Oligocene–early Pliocene in Ross Sea, Antarctic.

This new species is similar to *Actinomma boreale* Cleve (1899, pl. 1, fig. 5a), but the latter has more than 7 radial spines, and a completed cortical shell; this new species is also similar to specimen of *Hexacontid juveniles* (van de Paverd, 1995, pl. 32, figs. 3, 6), which was taken as juvenile of unknown species. Based on overall consideration, we think that this new species and *Hexacontid juveniles* are a same species of *Hexacontium*.

92. *Hexacontium quadrangulum* sp. nov.

(Pl. 26, Figs. 7–11)

One cortical shell and two medullar shells, rate of three shells 1:3:8; cortical shell middle thick walled, pores quasi-circular, similar sizes, sub-regularly arranged, 10–13 pores across half equator, width of bars varied not obviously, no hexagonal frame, diameter of pores 2–4 times as broad as bars; outer medullar shell showed as a cube, pores small, nearly circular, similar sizes, sub-regularly arranged, 7–8 pores on each side; inner medullar shell spheroidal, very small, connected to outer medullar shell by several radial beams; six main radial beams developed from the apex angle of quadrate outer medullar shell and connected to cortical shell, triangular prism, extended to outside forming as six radial spines of orthogonality, these spines triangular pyramidal shape, relatively short, length of spines less than half diameter of cortical shell; smooth surface, no complementary spine or other protrude.

Measurements: Diameter of cortical shell 180–235μm, outer medullar shell 75–90μm, inner medullar shell 30–40μm, length of spine 45–60μm.

Holotype: DSDP274-32-1 1 deposited in the South China Sea Institute, CAS, from sample of DSDP274-32-1 in the Ross Sea, pictured in pl. 26, figs. 9–11.

Distribution: Late Eocene–middle Oligocene in Ross Sea, Antarctic.

This new species is similar to *Joergensenium 233pollo* Kamikuri (2010, pl. 3, figs. 1–3, 5, 6), they have similar outer medullar shell of cube, but the latter has smaller individuals and radial beams not protrude outside, no six radial spines. This new species is differenced from *Hexacontium senticetum* Tan and Su by the latter with spheroidal medullar shells, and from *Hexacontium quadratum* Tan by the latter with cortical shell of near cube and medullar shells of spherosome.

Family Astrosphaeridae Haeckel, 1881

Genus *Acanthosphaera* Ehrenberg, 1858

100. *Acanthosphaera polageota* sp. nov.

(Pl. 27, Figs. 19–22)

Acanthosphaera sp., Hays, 1965, p. 169, pl. 2, fig. 8.

Single spheroidal shell, thickness of wall variable; pores circular or sub-circular, different sizes, with polygonal or hexagonal frames, diameter of pores 2–3 times as broad as bars, 10–15 pores across half equator, mostly about 10 pores; 6–9 radial main spines of similar length, triangular pyramid, length of main spines about 1/4 of shell diameter , main spines of some specimens even longer than half diameter of shell, usually irregularly or pairwise distributed on surface of shell; complementary thin spines or pyramid protrudes grown from bar nodes of some specimens, rough surface.

Measurements: Diameter of cortical shell 115–125μm, length of spines 30–40μm.

Holotype: DSDP274-16-1 1 deposited in the South China Sea Institute, CAS, from sample of DSDP274-16-1 in the Ross Sea, pictured in pl. 27, figs. 19–20.

Distribution: Middle Paleocene–late Oligocene in Ross Sea, Antarctic.

This new species is distinguished from other species by its smaller pores and much more spines, irregularly distributed.

Genus *Heliosphaera* Haeckel, 1862

104. *Heliosphaera miniopora* sp. nov.

(Pl. 28, Figs. 3–6)

Lattice shell slight thick walled, rough surface; pores smaller, similar sizes, quasi-circular, sub-regularly arranged, 7–9 pores across half equator, width of bars nearly equal to diameter of pores, with hexagonal frames; 12–18 long and strong radial spines, prismatic shape of three slices, very thick, lateral margin of spine smooth and without side spine, spine length greater than diameter of shell, spine width invariant from its base to terminal end, no sharp; some small bulges raised from nodes of surface, or forming small complementary spines.

Measurements: Diameter of shell 105–120μm, length of spines 95–130μm.

Holotype: DSDP274-16-1 1 deposited in the South China Sea Institute, CAS, from sample of DSDP274-16-1 in the Ross Sea, pictured in pl. 28, figs. 3–4.

Distribution: Late Oligocene in Ross Sea, Antarctic.

This new species is distinguished from *Heliosphaera dentata* (Mast) and *Heliosphaera macrohexagonaria* Tan by its smaller and more pores, slight thicker wall, no side teeth and smooth lateral margin of spine.

Genus *Haliomma* Ehrenberg, 1844

107. *Haliomma asteranota* sp. nov.

(Pl. 28, Figs. 19–22)

Spumellarian gen. et sp. indet #1 and #2, Lazarus and Pallant, 1989, p. 367, pl. 7, figs. 25–28.

Small shell, clathrate cortical shell polygonal or sub-spherical, pores quasi-circular, irregularly arranged, diameter of pores 1–3 times as broad as bars, 6–8 pores across half equator; medullar shell irregular form, made up of disorder grids, scattered, about half size of cortical shell; with 8–14 radial spines, grown from corners of polygonal cortical shell, relatively thick and strong, triangular pyramid, terminal sharp, length of spines near diameter of cortical shell; no complementary spine, smooth surface.

Measurements: Diameter of cortical shell 60–75μm, medullar shell 30–45μm, length of spines 50–65μm.

Holotype: DSDP274-24-3 1 deposited in the South China Sea Institute, CAS, from sample of DSDP274-24-3 in the Ross Sea, pictured in pl. 28, figs. 21–22.

Distribution: Early Oligocene in Ross Sea, Antarctic; Oligocene in seas of Greenlandand north Atlantic Ocean.

This new species is somewhat similar to *Haliomma microlaris* sp. nov., main distinction is the latter with small cylinder radial spines, relatively shorter, terminal flattened, and smaller medullar shell. Features of Ross Sea specimens is basically consistent with Spumellarian gen. et sp. indet (Lazarus and Pallant, 1989, p. 367, pl. 7, figs. 25–28).

108. *Haliomma microlaris* sp. nov.

(Pl. 28, Figs. 23–32; Pl. 29, Figs. 1–2)

Shell very small, spheroidal, rate of two shells near 3∶1; pores of cortical shell quasi-circular or hexagonal, similar sizes, regularly arranged or not, diameter of pores 3–4 times of bar width, 5–7 pores across half equator; medullar shell polygonal grid, formed by connections of several thin beams; some small radial spines of thin cylinder grown from nodes of shell surface, no terminal sharp, length of spines about 1/5–1/3 proportion of shell diameter.

Measurements: Diameter of cortical shell 56–81μm, medullar shell 19–30μm, length of spines 13–40μm.

Holotype: DSDP274-40-1 1 deposited in the South China Sea Institute, CAS, from sample of DSDP274-40-1 in the Ross Sea, pictured in pl. 28, figs. 27–28.

Distribution: Early–late Paleocene in Ross Sea, Antarctic.

This new species is distinguished from other known species by characters of its small shell, medullar shell of reticulation and small radial spines of thin cylinder. It may be a special type of Antarctic environment.

110. *Haliomma stylota* sp. nov.

(Pl. 29, Figs. 5–20)

One cortical shell and one medullar shell; cortical shell latticed, middle thick walled, with circular or quasi-circular pores, small, sub-regularly arranged, 9–11 pores across half equator, slightly hexagonal frame, diameter of pores 2–3 times as broad as bars; medullar shell spheroidal or pear-shaped, about 1/3 size of cortical shell, less pores; near 6–12 redial beams connecting medullar and cortical shells, and extended outside to form strong radial spines of prismatic shape with three slices, no terminal sharp, relatively longer, with side spines or teeth; generally smooth surface, no by-spine.

Measurements: Diameter of cortical shell 120–140μm, medullar shell 50–60μm, length of spine 65–135μm.

Holotype: DSDP274-28-3 1 deposited in the South China Sea Institute, CAS, from sample of DSDP274-28-3 in the Ross Sea, pictured in pl. 29, figs. 11–12.

Distribution: Middle Eocene–middle Oligocene in Ross Sea, Antarctic.

This new species is distinguished from *Stylosphaera megaxyphos tetraxyphos* Clark and Campbell (1942, pl. 6, figs. 1, 8) and *Actinomma* sp. (ancestor to *A. golownini* ?) (Lazarus, 1992, pl. 4, figs. 1, 4, 9, 13) by the latter without side spines or teeth on their radial spines, which are shaped of triangular pyramid.

Genus *Haliommetta* Haeckel, 1887, emend. Petrushevskaya, 1972

113. *Haliommetta hadraspina* sp. nov.

(Pl. 29, Figs. 27–32)

One cortical shell and two medullar shells, rate of three shells 1 : 3 : 6; cortical shell thick walled, pores quasi-circular, with similar or different sizes, hexagonal frames, irregularly or sub-regularly distributed, 6–9 pores across half equator; outer medullar shell near spheroidal, obscure, inner medullar shell spheroidal, very small, with fine-pores regularly arranged; generally rough surface, thorns on various nodes, and 6–9 strong radial spines, presented as triangular prism, very short, wider at the base, no by-spine.

Measurements: Diameter of cortical shell 150–165μm, outer medullar shell 65–100μm, inner medullar shell 25–38μm, length of spine 15–25μm.

Holotype: DSDP274-30-4 2 deposited in the South China Sea Institute, CAS, from sample of DSDP274-30-4 in the Ross Sea, pictured in pl. 29, figs. 31–32.

Distribution: Eocene in Ross Sea, Antarctic.

This new species is distinguished from *Haliommetta rossina* sp. nov. by the latter with smaller shell and cortical shell slight thin walled, thin bars, and longer radial spines, which are triangular pyramid, some by-spines on surface.

115. *Haliommetta rossina* sp. nov.

(Pl. 30, Figs. 12–15)

One cortical shell and two medullar shells, rate of three shells 1:3:8; cortical shell middle thick walled, pores quasi-circular or ellipsoidal, similar sizes, sub-regularly arranged, with hexagonal frames, diameter of pores 3–4 times as broad as bars, 7–8 pores across half equator; two medullar shells spheroidal, both lattice shells, outer medullar shell with quasi-circular pores, similar sizes, hexagonal frames, some thin radial beams between inner and outer medullar shells, connecting two medullar shells, but no correlating with cortical shell; other thick radial beams from outer medullar to cortical shell, and extending as 10–16 main radial spines, as long triangular pyramid, terminal sharp, stingless on side margin; shell surface rough, with some small by-spines.

Measurements: Diameter of cortical shell 125–138μm, outer medullar shell 60–70μm, inner medullar shell 25–30μm, length of spines 60–105μm.

Holotype: DSDP274-30-6 2 deposited in the South China Sea Institute, CAS, from sample of DSDP274-30-6 in the Ross Sea, pictured in pl. 30, figs. 14–15.

Distribution: Eocene in Ross Sea, Antarctic.

This new species is distinguished from *Haliomma stylota* sp. nov. by the former with two medullar shells, less pores, rough surface, short by-spines, and the latter with main spines of three slices prismatic shape and side spines, instead of cone-shaped.

Genus *Actinomma* Haeckel, 1862, emend. Nigrini, 1967

119. *Actinomma dumitricanis* sp. nov.

(Pl. 31, Figs. 3–15)

Usually three latticed shells, similar distance among them, rate of three shells 1:3:5; outer cortical shell thick walled, quasi-circular pores, small, sub-regularly or irregularly distributed, bars relatively wider, equal or greater than diameter of pores, 14–20 pores across half equator; inner cortical shell and medullar shell slight thin walled, three shells with similar pore features, many radial beams connected three shells; shell surface smooth or thorny, sometimes short conical radial spines presented at individual specimens; a circular opening typically develops on the cortical shell, with about 1/4 of the cortical shell's diameter.

Measurements: Diameter of cortical shell 170–250μm, inner cortical shell 110–140μm, medullar shell 40–55μm, length of spine 15–40μm, width of opening 40–60μm.

Holotype: DSDP274-24-5 1 deposited in the South China Sea Institute, CAS, from sample of DSDP274-24-5 in the Ross Sea, pictured in pl. 31, figs. 8–10.

Distribution: Late Paleocene–late Pliocene in Ross Sea, Antarctic.

This new species is similar to *Actinomma holtedahli* Bjørklund (1976, p. 1121, pl. 20, figs. 8, 9), but the latter with spongy medullar shell, relatively smaller, very closed between two cortical shells, and thinner walled for outer cortical shell than inner cortical shell; features of *Actinomma* cf. *holtedahli* Bjørklund (Abelmann, 1990, p. 690, pl. 1, figs. 4A–B) and *Actinomma kerguelenensis* Caulet (Caulet, 1991, p. 531, pl. 1, figs. 1–2) seem also to be similar to this new species, but they both with thin or thick long radial spines of triangular cone and shorter distance between two cortical shell.

125. *Actinomma impolita* sp. nov.

(Pl. 32, Figs. 14–25)

Large individuals, rate of three shell sizes 1∶3∶7.5; cortical shell thick walled, pores quasi-circular, ellipsoidal or irregular, different sizes, irregularly arranged, with hexagonal frames, different widths of bars, diameter of pores 1–5 times as broad as bars, 12–16 pores across half equator; outer medullar shell medium thick walled, pores quasi-circular, diameter of pores 2–3 times as broad as bars, 7–9 pores across half equator; inner medullar shell very small, clathrate; about 12–20 radial beams connecting medullar and cortical shells, and extended outside of shell forming main radial spines, which presented as triangular pyramid, thick and short, length of spines about 1/4 size of cortical shell diameter; rough surface, many bulges at nodes of bars, which may grow as short conical or thorny complementary spines.

Measurements: Diameter of cortical shell 220–310μm, outer medullar shell 106–125μm, inner medullar shell 40–50μm, length of spines 40–110μm.

Holotype: DSDP274-33-5 2 deposited in the South China Sea Institute, CAS, from sample of DSDP274-33-5 in the Ross Sea, pictured in pl. 32, figs. 18–19.

Distribution: Middle Eocene–middle Oligocene in Ross Sea, Antarctic.

This new species is similar to *Actinomma laminata* sp. nov., but the latter with thin walled and smooth surface, no thorny or complementary spines.

127. *Actinomma laminata* sp. nov.

(Pl. 33, Figs. 9–20)

Large individuals, rate of three shell sizes 1∶2.2∶4.5; cortical shell thin walled, quasi-circular pores, similar sizes, sub-regularly arranged, with hexagonal or polygonal frames, thin bars, diameter of pores 3–6 times as broad as bars, 12–14 pores across half equator; outer medullar shell middle thick walled, quasi-circular pores, diameter of pores 2–3 times as broad as bars, 7–9 pores across half equator; inner medullar shell very small, clathrate structure; about 10–16 radial beams connecting medullar and cortical shells, extended outside to form radial spines of triangular pyramid, with different lengths for various specimens, generally less than half diameter of cortical shell, and that the longest may near diameter; smooth surface, no by-spine.

Measurements: Diameter of cortical shell 210–250μm, outer medullar shell 110–130μm, inner medullar shell 50–60μm, length of spines 50–220μm.

Holotype: DSDP274-23-5 2 deposited in the South China Sea Institute, CAS, from sample of DSDP274-23-5 in the Ross Sea, pictured in pl. 33, figs. 9–10.

Distribution: Middle Eocene–middle Oligocene in Ross Sea, Antarctic.

This new species is similar to *Actinomma leptoderma* (Jørgensen), but the latter with a smaller shell, less pores and shorter radial spines.

130. *Actinomma magicula* sp. nov.

(Pl. 35, Figs. 1–13)

Two latticed shells, spheroidal or ellipsoidal, medium thick walled, smaller distance between two shells, rate of their sizes 3:4; outer shell with quasi-circular or irregular pores, irregularly arranged, diameter of pores 2–4 times as broad as bars, thinner bars, 7–9 pores across half equator; inner shell with quasi-circular pores, similar sizes, sub-regularly arranged, diameter of pores 2–3 times as broad as bars, 6–7 pores across half equator; inner cavity with often some looser irregular reticulate structures, filling partly or all the inner cavity; surface smooth or some rough, individually with very short conical spines.

Measurements: Diameter of outer shell 105–180μm, inner shell 75–120μm, length of spines 10–20μm.

Holotype: DSDP274-24-5 1 deposited in the South China Sea Institute, CAS, from sample of DSDP274-24-5 in the Ross Sea, pictured in pl. 35, figs. 1–3.

Distribution: Middle Paleocene–middle Pleistocene in Ross Sea, Antarctic.

Character of this new species is some similar to *Actinomma dumitricanis* sp. nov., but the latter with three shells, larger individual, outer shell thick walled, more pores, very small. This new species has obvious different shell sizes for different specimens, the smaller size in the older stratum.

135. *Actinomma solidula* sp. nov.

(Pl. 36, Figs. 29–32)

Amphisphaera kina Hollis, 1993, p. 318, pl. 1, figs. 3–4; Hollis, 2002, p. 277, pl. 1, figs. 6a–b.

Two latticed shells, relatively small individual; outer shell thick walled, with circular pores and hexagonal frames, diameter of pores 2–3 times as broad as bars, pores of similar sizes, regularly arranged, 7–9 pores across half equator; inner shell small, with 1/4 size of out shell, pores circular or hexagonal, very thin bars, 5–7 pores across half equator; smooth surface, conical bulge on each node, 8–12 radial spines, very short, pyramid, no by-spine.

Measurements: Diameter of outer shell 120–125μm, inner shell 35–45μm, length of spine 12–25μm, width of spine base 10–13μm.

Holotype: DSDP274-22-3 2 deposited in the South China Sea Institute, CAS, from sample of DSDP274-22-3 in the Ross Sea, pictured in pl. 36, figs. 29–30.

Distribution: Middle Eocene–middle Oligocene in Ross Sea, Antarctic.

Specimen of this new species looks some similar to *Amphisphaera kina* Hollis (1993, p. 318, pl. 1, figs. 3–4; Hollis, 2002, p. 277, pl. 1, figs. 6a–b), but the latter with radial spines assembled around two poles of shell, outer shell slight thinner walled, but relatively larger inner shell. Actually, specimens of Hollis (1993) and Hollis (2002) are discrepant. Comparing to the former (Hollis, 1993), the latter has cortical shell thinner walled, longer spines (different lengths), and two medullar shells, which is queried for them to be a same species? In the meantime, specimen of Hollis et al. (2020, pl. 1, figs. 1a–b) have only one short pole spine, thick walled and unrevealed inner shells, all these features are basically different to Hollis (1993) and Hollis (2002). Genus *Amphisphaera* is defined as 3 or 4 latticed concentric spherical shells, two pole spines of same size, similar shape, equal or not equal lengths. Therefore, affiliation of *Amphisphaera kina* Hollis is still unclear. This new species is distinguished from *Actinomma plasticum* Goll and Bjørklund by the latter having 10 radial spines, by-spines, 4 or 3 shells, outer shell thin walled, polygonal pores in similar sizes, regularly arranged and thin bars. So, discrepancy between this new species and *Actinomma plasticum* Goll and Bjørklund is obvious.

Suborder PRUNOIDEA Haeckel, 1883

Family Stylosphaeridae Haeckel, 1881

Genus *Ellipsoxiphus* Dunikowski, 1882, emend. Haeckel, 1887

152. *Ellipsoxiphus leptodermicus* sp. nov.

(Pl. 43, Figs. 1–6)

Single ellipsoidal shell, thin walled, circular pores of similar sizes, no mesh, regularly arranged, 6–7 pores across half equator, diameter of pores 4–6 times as broad as bars, thin bars between pores, with slight hexagonal frames; two pole spines of similar sizes, quite fine, long conical or cylindrical (terminal sharp), great change in lengths of spines for different specimens, length of spines generally less than short axis, or longer than long axis of shell in some specimens; smooth shell surface, no by-spine or clear thorn.

Measurements: Shell long axis 100–180μm, short axis 85–160μm, length of pole spines 50–120μm.

Holotype: DSDP274-1-3 1 deposited in the South China Sea Institute, CAS, from sample of DSDP274-1-3 in the Ross Sea, pictured in pl. 43, figs. 3–4.

Distribution: Early Pliocene–Holocene in Ross Sea, Antarctic.

Feature of this new species is relatively closer to *Xiphostylus alcedo* Haeckel (1887, p. 127, pl. 13, fig. 4), main differences are the latter with thick walled, deep hexagonal frames, two pole spines of different sizes or lengths, triangular pyramid, thick and strong.

153. *Ellipsoxiphus rotundalus* sp. nov.

(Pl. 43, Figs. 7–9)

Single spherical shell, spheroidal, very thick walled, circular pores, deep pores and high bars, with hexagonal frames, regularly arranged, 11–13 pores across half equator, thin bar width about 1/4–1/3 of pore diameter; form of internal face of wall fitted with surface of shell, larger interval between them forming a pseudo double-shell; two pole spines of different lengths, one very long and strong, prismatic shape of three slices, no lateral spine; smooth surface, no by-spine and thorn.

Measurements: Shell external diameter 210μm, inner diameter 160μm, shell thickness 25μm, length of pole spines 45–160μm.

Holotype: DSDP274-36-1 2 deposited in the South China Sea Institute, CAS, from sample of DSDP274-36-1 in the Ross Sea, pictured in pl. 43, figs. 7–9.

Distribution: Early Eocene in Ross Sea, Antarctic.

This new species is somewhat similar to *Xiphostylus alcedo* Haeckel (1887, p. 127, pl. 13, fig. 4), but the latter with ellipsoidal shell, less pores, two pole spines of triangular pyramid.

154. *Ellipsoxiphus echinatus* sp. nov.

(Pl. 43, Figs. 10–11)

Single ellipsoidal shell, comparably thin walled, quasi-circular pores, 7–8 pores across half equator, sub-regularly arranged, large pores and thin bars, diameter of pores 4–6 times as broad as bars; rough surface, thorny at every node, or formed as by-spines; two pole spines near needle, of different lengths, early stage of pole spines as triangular shape, relatively thick, last stage as thin coniform, rapid sharp; some pyramidal by-spines developed from nodes of shell surface.

Measurements: Shell long axis 110–120μm, short axis 105–110μm, length of pole spines 50–60μm.

Holotype: DSDP274-1-5 2 deposited in the South China Sea Institute, CAS, from sample of DSDP274-1-5 in the Ross Sea, pictured in pl. 43, figs. 10–11.

Distribution: Holocene in Ross Sea, Antarctic.

This new species is distinguished from *Ellipsoxiphus leptodermicus* sp. nov. by the latter with regularly arranged pores, smooth surface, no thorny or by-spine, thin and long pole spines, entirely conical or cylindrical, non-deformation.

Genus *Druppatractus* Haeckel, 1887

163. *Druppatractus carpocus* sp. nov.

(Pl. 46, Figs. 1–4)

Two shells nearly spheroidal, rate of shell sizes 3∶5, both with larger pores, small interval between two shells; outer shell medium thick walled, pores quasi-circular, similar sizes, regularly arranged, with hexagonal frames, thin bars, diameter of pores 4–5 times as broad as bars, 6–7 pores across half equator; inner shell large, pores hexagonal or quasi-circular for different specimens, regularly or irregularly arranged, similar sizes or different, 5–7 pores across half equator; many radial beams connecting two shells, one of them extended outside forming a thin pole spines, slenderly conical or cylindrical, slightly terminal sharp, another pole spines undeveloped, different lengths of spines for different specimens; smooth surface, no by-spines.

Measurements: Diameter outer shell 110–120μm, inner shell 70–75μm, length of pole spines 25–80μm.

Holotype: DSDP274-10-3 1 deposited in the South China Sea Institute, CAS, from sample of DSDP274-10-3 in the Ross Sea, pictured in pl. 46, figs. 3–4.

Distribution: Early Pliocene in Ross Sea, Antarctic.

This new species is distinguished from *Druppatractus hastatus* Blueford by the latter with smaller inner shell, smaller and more pores, 11–13 pores across half equator, longer spines, length of pole spine greater than diameter of shell.

166. *Druppatractus pyriformus* sp. nov.

(Pl. 46, Figs. 12–14)

Small individual, size rate of two shells 1∶2; outer shell ellipsoidal, slightly thick walled, pores round, similar sizes, with hexagonal frames, regularly arranged, diameter of pores 3–4 times as broad as bars, 6–7 pores across half equator; inner shell nearly pyriform, with two terminals shrinked; rough surface, thorny at nodes; two acicular pole spines with different degree of development, no lateral edge.

Measurements: Shell diameter of long axis 85–95μm, short axis 80–90μm, length of pole spines 20–80μm.

Holotype: DSDP274-11-3 1 deposited in the South China Sea Institute, CAS, from sample of DSDP274-11-3 in the Ross Sea, pictured in pl. 46, figs. 12–14.

Distribution: Late Miocene in Ross Sea, Antarctic.

This new species is similar to *Druppatractus irregularis* Popofsky, but the latter with cortical shell thin walled, pores sub-regularly arranged, no hexagonal frame, and medullar shell of more pores, round, regularly arranged, very smooth surface.

Genus *Stylatractus* Haeckel, 1887

168. *Stylatractus gracilus* sp. nov.

(Pl. 46, Figs. 17–23; Pl. 47, Figs. 1–19)

Amphisphaera sp., Chen, 1975, p. 453, pl. 6, figs. 1–2.
Axoprunum bispiculum (Popofsky), Takemura, 1992, p. 741, pl. 1, figs. 1–2; Chen, 1975, p. 454, pl. 21, figs. 1–2; Hollis, 2002, pl. 1, figs. 7–9.
Axoprunum liostylum (Ehrenberg) group, Petrushevskaya, 1975, p. 571, pl. 2, fig. 22.
Stylacontarium sp. aff. *S. bispiculum*, Kling, 1973, p. 634, pl. 6, figs. 19–23, pl. 14, figs. 5–8.
Stylacontarium bispiculum Popofsky, Kling, 1973, pl. 15, figs. 11–14; Benson, 1983, p. 508, pl. 1, figs. 7–10.
Axoprunum pierinae (Clark and Campbell), Hollis et al., 1997, p. 44, pl. 1, figs. 9–10, 14 (only).

Shell nearly spheroidal or ellipsoidal, one cortical shell and two medullar shells, rate of three shell sizes 1:2:6; cortical shell generally thick walled, some slightly thin walled, pores quasi-circular or oval, similar or different sizes, sub-regularly or irregularly arranged, with hexagonal frames, generally 10–12 pores across half equator, exceptional only 8 pores for specimens with great different sizes of pores, diameter of pores 2–3 times as broad as bars; two medullar shells very small, nearly spheroidal, with smaller quasi-circular pores; two pole spines of similar shape, relatively long, thin conical or aciculiform; smooth surface, or some tiny spinous process on nodes of bars presented as slightly rough surface. Thickness and pores of cortical shell variable, but they appeared basically in a same stratigraphy.

Measurements: Shell diameter of long axis 110–150μm, short axis 95–145μm, length of pole spines 60–110μm.

Holotype: DSDP274-20-5 1 deposited in the South China Sea Institute, CAS, from sample of DSDP274-20-5 in the Ross Sea, pictured in pl. 47, figs. 1–2.

Distribution: Early Eocene–late Oligocene in Ross Sea, Antarctic.

This new species is obviously distinguished from *Stylacontarium bispiculum* Popofsky (1912, p. 91, pl. 2, fig. 2) by the latter with thin walled, large pore, diameter of pores 3–4 times as broad as bars, no thorny on nodes and smooth surface. It may be mistaken that Kling (1973), Chen (1975) and Takemura (1992) considered specimens with features like this new species as *Stylacontarium bispiculum*, they should be two different species. This new species is similar to *Stylosphaera liostylus* Ehrenberg (1873, p. 259; Ehrenberg, 1876, pl. 25, fig. 7), but the latter with regularly arranged pores, many conical calthrops on surface, terminal sharp, and pole spines conical, thick and short.

Genus *Xiphatractus* Haeckel, 1887

172. *Xiphatractus longostylus* sp. nov.

(Pl. 49, Figs. 7–10)

Shell ellipsoidal or nearly spheroidal, one cortical shell and two medullar shells, size rate of three shells 1:2.5:5; cortical shell medium thick walled, with quasi-circular pores and hexagonal frames, similar sizes, regularly or sub-regularly arranged, 6–7 pores across half equator, diameter of pores 3 times as broad as bars, bulge of pyramid on each node, terminal sharp; outer medullar shell ellipsoidal, inner medullar shell spheroidal, both medullar shells latticed, with slightly thin walled, pores quasi-circular and less; two pole spines longer, lengths near diameter of cortical shell, similar sizes and shapes, triangular prism and wider at spine base, thin cylindrical for rest part, slightly terminal sharp; rough surface.

Measurements: Diameter of cortical shell long axis 120–140μm, short axis 110–125μm; outer medullar shell long axis 70–75μm, short axis 55–60μm, inner medullar shell 25μm, length of pole spines 60–110μm.

Holotype: DSDP274-10-5 1 deposited in the South China Sea Institute, CAS, from sample of DSDP274-10-5 in the Ross Sea, pictured in pl. 49, figs. 7–8.

Distribution: Late Miocene in Ross Sea, Antarctic.

This new species is similar to *Xiphatractus spumeus* Dumitrica, distinction between them by the latter with thicker walled, shorter pole spines, triangular pyramid, which was reported by Dumitrica (1973) appearing in the North Atlantic Ocean, showing these two species on behalf of different regional environment of biogeography. Specimens of this new species is also some similar to *Xiphatractus* sp. B (Takahashi, 1991), but the latter with shorter pole spines that is basically the same of *Xiphatractus spumeus* Dumitrica.

Genus *Amphisphaera* Haeckel, 1881, emend. Petrushevskaya, 1975

181. *Amphisphaera sperlita* sp. nov.

(Pl. 50, Figs. 24–25)

Three latticed shells; outer shell ellipsoidal, thin walled, smooth, larger circular pores, 5–6 pores across half equator, with hexagonal frames, thin bars, diameter of pores 4–5 times as broad as bars; middle shell ellipsoidal, similar shape with outer shell, slightly thicker walled than outer shell, large pores; inner shell nearly spheroidal, unstable form, large pores; two pole spines relatively long, thin rodlike, terminal sharp; smooth surface, no by-spines.

Measurements: Diameter of cortical shell long axis 140μm, short axis 120μm; middle shell long axis 95μm, short axis 80μm, inner shell 50μm, length of pole spines 70–110μm.

Holotype: DSDP274-10-3 2 deposited in the South China Sea Institute, CAS, from sample of DSDP274-10-3 in the Ross Sea, pictured in pl. 50, figs. 24–25.

Distribution: Early Pliocene in Ross Sea, Antarctic.

Feature of this new species is close to *Lithatractus santaennae* Campbell and Clark (1944b, p. 19, pl. 2, figs. 20–22; Petrushevskaya, 1975, p. 570, pl. 2, fig. 21), but the latter with thicker walled, shorter pole spines of triangular pyramid.

Suborder DISCOIDEA Haeckel, 1862

Family Porodiscidae Haeckel, 1881

Subfamily Trematodiscidae Haeckel, 1862

Genus *Perichlamydium* Ehrenberg, 1847

198. *Perichlamydium irregularmus* sp. nov.

(Pl. 56, Figs. 12–19)

Shell of spongy disk roundish, nearly quadrate or irregular shape, often variable; less disk rings, only 2–4 rings mainly concentrated at central area, with interrupted rings or mussy structure in peripheral area, irregularly; a extra heavy sideband around disk margin, which make sharpness of border and smoothness, sideband often discontinuous and not enclosed completely, sometimes spongy disk with partly deficiency or incomplete; generally no radial spine on the disk edge, only exceptional specimen with tiny spines at where of incomplete edge.

Measurements: Diameter of disk 140–300μm, breadth of sideband 6–12μm.

Holotype: DSDP274-23-1 2 deposited in the South China Sea Institute, CAS, from sample of DSDP274-23-1 in the Ross Sea, pictured in pl. 56, figs. 14–15.

Distribution: Early-middle Oligocene in Ross Sea, Antarctic.

This new species is similar to *Perichlamydium limbatum* Ehrenberg, but the latter with more disk rings, regularly arranged, and no extra heavy sideband around disk margin.

Subfamily Stylodictyinae Haeckel, 1881

Genus *Stylodictya* Ehrenberg, 1847

207. *Stylodictya variata* sp. nov.

(Pl. 60, Figs. 9–12)

Large individual, spongy disk circular, with loose spongy tissue, larger interval of disk rings, 4–5 rings on disk; central chamber circular, gradually increasing interval of rings outward, the first and second rings ellipsoidal, the third and fourth rings quasi-circular or irregular, rings of periphery usually incomplete or discontinuous, with pores of different sizes and ragged, maldistribution of spongy tissue; radial beam on spongy disk undeveloped, spongy disk with irregular margin, some small spines, or breakage.

Measurements: Diameter of disk 150–240μm.

Holotype: DSDP274-25-5 1 deposited in the South China Sea Institute, CAS, from sample of DSDP274-25-5 in the Ross Sea, pictured in pl. 60, figs. 11–12.

Distribution: Early-middle Oligocene in Ross Sea, Antarctic.

This new species is some similar to *Stylodictya heliospira* Haeckel, main differences are the latter with more regular rings, spiral arranged, smaller interval of rings, obvious radial beams, uniform formation of spongy tissue, compacted, no spine at periphery of disk.

Family Spongodiscidae Haeckel, 1881

Genus *Spongodiscus* Ehrenberg, 1854

217. *Spongodiscus inconcavus* sp. nov.

(Pl. 64, Figs. 1–6)

Spongy disk roundish, relatively larger, central area very thin walled, usually fragile or broken, forming a hole; from the central area outward increased gradually thickness of spongy disk, generally a double-faced concave area at the central, and slight thinning at margin of disk; form of disk just the reverse of *Spongodiscus biconcavus* Haeckel of thickening at central area; a gap of V shape may be appeared at peripheral disk, or absent, without radial spine and radial beam.

Measurements: Diameter of disk 205–340μm.

Holotype: DSDP274-10-3 1 deposited in the South China Sea Institute, CAS, from sample of DSDP274-10-3 in the Ross Sea, pictured in pl. 64, fig. 2.

Distribution: Middle Oligocene–early Pliocene in Ross Sea, Antarctic.

This new species is distinguished from *Spongodiscus biconcavus* Haeckel by the latter with thickening at the center of spongy disk, shell structure of these two species is just the reverse.

218. *Spongodiscus perizonatus* sp. nov.

(Pl. 64, Figs. 7–12)

Large spongy disk, irregular or quasi-circular shape, disk margin of forniciform or wavy, enclosed by an incrassation or overstriking enclosure wall, forming an obvious band, with a clear outline; spongy tissue rather thick and dense at central area, and slightly thinner and looser outward; without radial spine and radial beam, smooth surface.

Measurements: Diameter of disk 220–275μm.

Holotype: DSDP274-23-1 1 deposited in the South China Sea Institute, CAS, from sample of DSDP274-23-1 in the Ross Sea, pictured in pl. 64, figs. 9–10.

Distribution: Late Eocene–middle Oligocene in Ross Sea, Antarctic.

This new species is distinguished from other species by character of its disk margin with an enclosure wall of incrassation or overstriking, clear outline, spongy disk of irregular shape.

219. *Spongodiscus planarius* sp. nov.

(Pl. 65, Figs. 1–2)

Spongy disk roundish, relatively larger, slightly thin disk, even surface, no incrassation area, basically the same of spongy texture for entire disk, homogeneous structure; pores of disk distinct, quasi-circular, similar sizes, uniform distribution, diameter of pores 1–2 times as broad of bars, all quietly fine; without structures of chamber and ring, no radial beam, no feature of concave-convex for disk surface; individual small spine as short and thin at disk margin.

Measurements: Diameter of disk 290–310μm, lenth of spines 9–13μm.

Holotype: DSDP274-23-1 1 deposited in the South China Sea Institute, CAS, from sample of DSDP274-23-1 in the Ross Sea, pictured in pl. 65, figs. 1–2.

Distribution: Middle Oligocene in Ross Sea, Antarctic.

This new species is distinguished from *Spongodiscus biconcavus* Haeckel by the latter with a central incrassation area, convex and concave disk, spongy texture variable, rather than uniform.

220. *Spongodiscus radialinus* sp. nov.

(Pl. 65, Figs. 3–6)

Spongy disk roundish, both surfaces flat, no feature of convex or concave, spongy texture relatively denser at central disk, gradually looser forward periphery, without structure of central chamber and ring; 35–40 radial beams developed outward from the center, distribution of radial beams as similar interval, between every two beams only one row of pores or superposed, increasing pore size along with beam interval increased outward; no appurtenance surround disk, disk edge of slight unevenness; no radial spine.

Measurements: Diameter of disk 165–215μm.

Holotype: DSDP274-1-5 1 deposited in the South China Sea Institute, CAS, from sample of DSDP274-1-5 in the Ross Sea, pictured in pl. 65, fig. 5.

Distribution: Holocene in Ross Sea, Antarctic.

This new species is distinguished from *Spongodiscus trachodes* (Renz) by the latter with a ring of central chamber, no radial beam, spongy texture of entire disk relatively denser, smaller, and uniform distribution.

Suborder LARCOIDEA Haeckel, 1887

Family Pyloniidae Haeckel, 1881

Subfamily Diplozonaria Haeckel, 1887, emend. Tan and Chen, 1990

Genus *Pylonium* Haeckel, 1881

250. *Pylonium palaeonatum* sp. nov.

(Pl. 79, Figs. 32–33)

Shell of ellipsoidal lentoid, thick walled, pores smaller, quasi-circular, bars between pores slightly thick and solid; outer medullar shell with trizonal girdle system, large ellipsoidal box, lenth nearly equal to transverse girdle of cortical shell (width of cortical shell), both walls cling each other; inner medullar shell small, connected with outer medullar shell by many radial beams, about 1/3 size of outer medullar shell; cortical lateral girdle enclosed at both ends, forming four gates of semilunar shape, several thick beam branchs in gates connecting outer medullar shell and cortical shell; shell surface without radial spine, but small conical raised.

Measurements: Cortical shell length 140μm, width 110μm; outer medullar shell length 100μm, width 75μm, inner medullar shell length 45μm, width 30μm.

Holotype: DSDP274-40-1 1 deposited in the South China Sea Institute, CAS, from sample of DSDP274-40-1 in the Ross Sea, pictured in pl. 79, figs. 32–33.

Distribution: Late Paleocene in Ross Sea, Antarctic.

This new species is relatively similar to *Pylonium quadricorne* Haeckel (1887, p. 655, pl. 9, fig. 14), but the latter with many spines on shell surface, thin walled and large pore, gate of kidney shape. Obviously, this new species is a type of older stratum and *Pylonium quadricorne* Haeckel of modern type.

251. *Pylonium neoparatum* sp. nov.

(Pl. 79, Figs. 34–39)

Cortical lateral girdle opened at both ends, forming broad apertures, opening about 3/4 width of shell, pores quasi-circular, small, regularly distributed; cotical shell and outer medullar shell connected by many thick radial beams, medullar shells with trizonal system, relatively smaller, ellipsoidal or near roundish; size of outer medullar shell only half width of cortical shell (length of transverse girdle), size of inner medullar shell about 1/3 of outer medullar shell; smooth surface, without spine.

Measurements: Cortical shell length 110–145μm, width 95–115μm; outer medullar shell length 55–65μm, width 50–55μm, inner medullar shell length 20–25μm, width 15–20μm.

Holotype: DSDP274-30-4 1 deposited in the South China Sea Institute, CAS, from sample of DSDP274-30-4 in the Ross Sea, pictured in pl. 79, figs. 34–35.

Distribution: Late Eocene in Ross Sea, Antarctic.

This new species is some similar to *Pylonium palaeonatum* sp. nov., main distinction for them is the former with smaller medullar shells, larger space between cortical and medullar shells, opened at two ends of cortical shell, but the latter enclosed cortical shell and part shells of cortical and medullar shells attached each other. On the other hand, this new species appears above stratigraphic position of *Pylonium palaeonatum* sp. nov., with a newer chronology, meaning development from the latter species.

Genus *Pylozonium* Haeckel, 1887

252. *Pylozonium saxitalum* sp. nov.

(Pl. 79, Figs. 40–53)

Shapes of three shells similar, ellipsoidal lentoid, connected by some radial beams; size rate of three shells 1 : 3 : 8, slightly thick walled; outer shell with quasi-circular pores, similar sizes, regularly distributed, multi-gates distinct; inner shells very small, clear outline, but uncertain inner structure; rough surface, or small thorny, no radial spine. This species only occurs in stratum of early Cenozoic.

Measurements: Cortical shell length 120–140μm, width 105–125μm; outer medullar shell length 50–75μm, width 40–55μm.

Holotype: DSDP274-42-2 1 deposited in the South China Sea Institute, CAS, from sample of DSDP274-42-2 in the Ross Sea, pictured in pl. 79, figs. 44–45.

Distribution: Early Paleocene in Ross Sea, Antarctic.

This new species is basically similar to *Pylozonium novemcinctum* Haeckel (1887, p. 659), but the latter with obviously larger individuals and clear four gates. Comparing to *Pylozonium* sp. 1, shell structure of this species is more completely and clearly, probably being early type of *Pylozonium* sp. 1, or developed from *Pylozonium* sp. 1.

Family Actinommidae Haeckel, 1862, sensu Riedel, 1967

Genus *Prunopyle* Dreyer, 1889

255. *Prunopyle archaeota* sp. nov.

(Pl. 80, Figs. 17–33)

Individuals small, shell of some irregular shape, nearly ellipsoidal or spheroidal, thick walled; one cortical shell smooth surface or some thorny, with a open aperture, contractive or broad; two medullar shells spheroidal or ellipsoidal, or presented as disorder interior, with obscure structure, forms of medullar shells unclear, seeming to be imcompletely developed. It may be a initial type of Genus *Prunopyle*, distributing at the bottom of Cenozoic stratigraphy.

Measurements: Cortical shell length 65–100μm, width 55–85μm; outer medullar shell length 30–50μm, width 30–45μm, aperture width of cortical shell 25–65μm.

Holotype: DSDP274-43-3 1 deposited in the South China Sea Institute, CAS, from sample of DSDP274-43-3 in the Ross Sea, pictured in pl. 80, figs. 21–22.

Distribution: Early Paleocene in Ross Sea, Antarctic.

This new species is distinguished from other species by its small individual, unclear interior structure and form. It is only appeared at the bottom of Site 274, with evidently stratigraphic significance, should be a new species of early Paleogene in the Ross Sea.

256. *Prunopyle minuta* sp. nov.

(Pl. 80, Figs. 34–57)

Prunopyle sp., O'Connor, 1997b, p. 116, pl. 3, fig. 9.

Individuals small, ellipsoidal or ovoid, shell 3–4 layers, general one or two cortical shell and two medullary shells, sometimes a incomplete layer cover on one end of cortical shell; cortical shell with small quasi-circular pores, a open mouth (aperture) at one end of the cortical shell, and a few small spines surrounding peristoma; outer medullar shell generally ellipsoidal or long ovoid, exceptional irregular shape, about 1/2–3/5 size of cortical shell; inner medullar shell spheroidal or long ellipsoidal, very small; shell surface smooth, no spine.

Measurements: Cortical shell length 80–120μm, width 60–75μm; outer medullar shell length 60–70μm, width 35–50μm, diameter of inner medullar shell 20–30μm, aperture width of cortical shell 15–30μm.

Holotype: DSDP274-32-1 1 deposited in the South China Sea Institute, CAS, from sample of DSDP274-32-1 in the Ross Sea, pictured in pl. 80, figs. 40–41.

Distribution: Late Paleocene–late Pliocene in Ross Sea, Antarctic.

This new species is basically small individuals, distinguished from *Prunopyle tetrapila* Hays by the latter with relatively larger shell, only one cortical shell without cover layer, two medullar shells both spheroidal, and located at centre of cortical shell.

257. *Prunopyle opeocila* sp. nov.

(Pl. 81, Figs. 1–4)

Shell of three layers; outer shell ellipsoidal or slightly irregular, pores quasi-circular, with hexagonal frames, diameter of pores nearly equaling to width of bars, smooth surface, no radial spine, pole hole (open mouth) relatively larger, diameter of mouth about 1/2 of shell, no teeth or spine on peristoma; two inner shells (medullar shells) both spheroidal, not at centre of outer shell, or closed to the mouth, with structure of slight irregulation; radial beams developed among three shells discontinuously, rate of three shells about 1∶2.5∶4.

Measurements: Outer shell length 140–200μm, width 125–180μm; diameter of outer medullar shell 90–110μm, inner medullar shell 40–50μm, width of open mouth 65–70μm.

Holotype: DSDP274-10-1 2 deposited in the South China Sea Institute, CAS, from sample of DSDP274-10-1 in the Ross Sea, pictured in pl. 81, figs. 3–4.

Distribution: Early Pliocene in Ross Sea, Antarctic.

This new species is similar to *Prunopyle tetrapila* Hays, but the latter with two inner shells located at centre of cortical shell, regular structure, smaller pole hole, generally mouth width less than 1/3 diameter of cortical shell, and with teeth or spines on peristoma.

258. ***Prunopyle quadrata*** **sp. nov.**

(Pl. 81, Figs. 5–32)

Individuals small, with shell of 4–5 layers, outer cortical shell cylindrical or box shape, lateral view of rectangle, rectilinear figure for two sides, individual specimens of spheroidicity, thick walled, quasi-circular pores, smaller and denser, solid bars, wide open at one end of shell, usually some small teeth at peristoma, no ring of peristoma; shape of inner cortical shell ellipsoidal, similar to outer cortical shell, both closed to each other; 2–3 medullar shells, outer medullar shell general spheroidicity, inner medullar shells spheroidal or ellipsoidal, rate of four shells about 1 : 3 : 5 : 6, variable for different specimens; smooth surface, no spine.

Measurements: Outer cortical shell length 115–150μm, width 80–105μm; outer medullar shell length 75–85μm, width 70–80μm, width of open mouth 40–75μm.

Holotype: DSDP274-24-1 2 deposited in the South China Sea Institute, CAS, from sample of DSDP274-24-1 in the Ross Sea, pictured in pl. 81, figs. 9–10.

Distribution: Early Paleocene–middle Oligocene in Ross Sea, Antarctic.

This is a special species found in older stratum of Ross Sea, Antarctic. This new species is some similar to *Prunopyle minuta* sp. nov., but the latter with smaller individuals, ellipsoidal shell, arc shape at two ends, less layers; and this new species is generally with nearly right angle joins for its both ends and two sides.

Family Tholonidae Haeckel, 1887

Genus *Triosphaera* gen. nov.

Cortical shells composited of three global chambers, they have intersection angle of 120° mutually in a plane, with deep constrictions between chambers; medullar shell very small at the interior centre of cortical shell, spherical or irregular.

Type species: *Triosphaera bulloidis* sp. nov.

262. ***Triosphaera bulloidis*** **sp. nov.**

(Pl. 83, Figs. 7–10)

Individuals small, with shape of round triangle; cortical shell composed of three spheroidal chambers of similar sizes, they adjoined together, arranged at a plane with 120° intersection angles each other, constricted between chambers, and enclosed entire shell; some beams (twigs) developed from shell centre to inner walls of each chamber, supporting the walls; pores of wall quasi-circular or oval, irregularly distributed, diameter of pores 1–2 times as broad as the bars, with hexagonal or polygonal frames; thick walled, smooth surface, individual short conical spine appeared at adjoining place of chambers; medullar shell with simple or irregular structures.

Measurements: Diameter of shell 75–90μm, width of chamber 50–65μm.

Holotype: DSDP274-31-2 1 deposited in the South China Sea Institute, CAS, from sample of DSDP274-31-2 in the Ross Sea, pictured in pl. 83, figs. 9–10.

Distribution: Late Eocene in Ross Sea, Antarctic.

This new species has similar morphological structure of shell with *Globigerina* or *Globigerinoides* of planktonic foraminifera, but the latter with shell of calcium carbonate, more chambers helically arranged, and an aperture at the last chamber, no pore and frame on the wall.

Family Litheliidae Haeckel, 1862

Genus *Spirema* Haeckel, 1881

266. *Spirema vetula* sp. nov.

(Pl. 85, Figs. 1–10)

Shell of spherosome, interior with asymmetrical helical or irregular structure, developed from a round initial chamber at shell center, to outside forming a completely enclosed outer shell, thick walled, with many small quasi-circular pores, irregularly distributed, bars with hexagonal frames or connected each other disorderly; initial chamber very small; smooth surface, without spine.

Measurements: Diameter of outer shell 95–160μm, medullar shell (initial shell) 10–20μm.

Holotype: DSDP274-42-2 1 deposited in the South China Sea Institute, CAS, from sample of DSDP274-42-2 in the Ross Sea, pictured in pl. 85, figs. 5–6.

Distribution: Early Paleocene–late Eocene in Ross Sea, Antarctic.

This new species is similar to *Spirema melonia* Haeckel, but the latter with shell of ellipsoidal or irregular shape, front opening on surface, and unclosed outer shell.

Genus *Lithelius* Haeckel, 1862

269. *Lithelius coelomalis* sp. nov.

(Pl. 85, Figs. 43–46)

Shell nearly spheroidal; initial chamber very large, spheroidal, diameter of initial chamber about 2/3–3/5 of whole shell, interior cavity empty, wall of first chamber clathrate, with quasi-circular pores, similar sizes, sub-regularly or irregularly arranged, diameter of pores 2–3 times as bar width, with hexagonal or polygonal frames, 10–14 pores across half equator; only 1–2 rotary shells of periphery, small distance between rotary shells, nearly uniformly-spaced, peripheral wall latticed, with similar pores, but disorderly overlapping distribution; rough surface, many small spinous, no radial spine.

Measurements: Shell length 140–165μm, width 120–145μm, diameter of initial shell 90–100μm, interval of rotary shell 13–15μm.

Holotype: DSDP274-26-1 2 deposited in the South China Sea Institute, CAS, from sample of DSDP274-26-1 in the Ross Sea, pictured in pl. 85, figs. 45–46.

Distribution: Middle Oligocene in Ross Sea, Antarctic.

This new species is distinguished from other species of Genus *Lithelius* by its very large initial chamber and less rotary shells.

274. *Lithelius regularis* sp. nov.

(Pl. 86, Figs. 33–38)

Individuals median size, ellipsoidal or near box shape, shell short axis about 1/2–2/3 of long axis; initial chamber very small, spherosome; volution of 3–4 layers, seemed single rotary, similar interval between every two layers, connected by many radial beams, well distributed; wall latticed, small and dense pores, irregularly distributed bars; rough surface, with some short small calthrops, no radial spine.

Measurements: Shell long axis 135–170μm, short axis 100–130μm, interval of rotary layers 12–18μm.

Holotype: DSDP274-25-3 1 deposited in the South China Sea Institute, CAS, from sample of DSDP274-25-3 in the Ross Sea, pictured in pl. 86, figs. 33–34.

Distribution: Early Eocene–middle Oligocene in Ross Sea, Antarctic.

This new species is similar to *Lithelius nautiloides* Popofsky, distinction between them is the latter with nearly spheroidal shell, increasing layer interval outward, and some radial spines generally.

275. *Lithelius rotalarius* sp. nov.

(Pl. 87, Figs. 1–9)

Lithelius sp. E, Petrushevskaya, 1975, p. 573, pl. 3, fig. 2.

Large individuals, spheroidal shell, thick walled, enclosed periphery; with double helix structure, 3–4 volutions, increasing interval gradually outward; medullar (initial) shell roundish, situated at center of shell, with about 1/5 size of all the shell; wall latticed, quasi-circular pores, similar sizes, disorderly distributed, relatively denser, diameter of pores about 1–3 times as broad as bars; rough surface, thorny bulges on bar joints, some small spines of slender cylindrical shape or thick conical shape on surface.

Measurements: Diameter of shell 205–240μm, length of spine 20–45μm.

Holotype: DSDP274-36-1 2 deposited in the South China Sea Institute, CAS, from sample of DSDP274-36-1 in the Ross Sea, pictured in pl. 87, figs. 7–9.

Distribution: Early Eocene–middle Oligocene in Ross Sea, Antarctic.

Feature of this new species is relatively near to *Lithelius nautiloides* Popofsky, distinction between them is the latter with unclosed shell on periphery, one aperture at the last shell, surface smooth or rough, with some radial spines generally.

Features of specimens from Ross Sea is basically the same with *Lithelius* sp. E (Petrushevskaya, 1975, p. 573, pl. 3, fig. 2), which was found in subantarctic stratum of Eocene.

Family Phorticidae Haeckel, 1881

Genus *Carpodrupa* gen. nov.

Shell is of ellipsoidal or the box shape, with 2–3 layers of homocentric and non-spiral structure, similar shape of each layer and similar distance between laters, no opening of shell.

Type species: *Carpodrupa rosseala* sp. nov.

285. *Carpodrupa rosseala* sp. nov.

(Pl. 90, Figs. 1–8)

Amphitholus (?) sp. A, Lazarus and Pallant, 1989, p. 367, pl. 8, figs. 4, 10.

Shell ellipsoidal or box shape, relatively small, enclosed external, about 3 layers of concentric structure, not spiral, similar intervals of shell layers, or slightly increased, with similar shapes for each shell; medullar shell very small, spheroidal, presented as one or couple shells stacked; other outer shells with latticed pores, slightly thin walled, pores of multi-type, smaller sizes, bars disordered, forming superposed meshes; rough surface, with small calthrops, no radial spine, no front opening.

Measurements: Shell long axis 125–145μm, short axis 100–115μm, interval of shell layers 10–19μm.

Holotype: DSDP274-25-1 1 deposited in the South China Sea Institute, CAS, from sample of DSDP274-25-1 in the Ross Sea, pictured in pl. 90, figs. 1–2.

Distribution: Late Eocene–middle Oligocene in Ross Sea, Antarctic.

This new species seems similar to *Lithelius regularis* sp. nov., distinction between them is the latter with shells of helical structure and an opening at the last shell.

Order NASSELLARIA Ehrenberg, 1875

Su.border SPYROIDEA Ehrenberg, 1847, emend. Petrushevskaya, 1971c

Family Zygospyridae Haeckel, 1887

Genus *Liriospyris* Haeckel, 1881, emend. Goll, 1968

289. *Liriospyris glabra* sp. nov.

(Pl. 91, Figs. 11–26)

Liriospyris sp., Chen et al., 2017, p. 169, pl. 54, figs. 9–10.

Lophospyris pentagona (Ehrenberg) *hyperborea* (Jørgensen) emend. Gol, Takahashi, 1991, p. 103, pl. 29, fig. 8 (only).

Lophospyris stabilis stabilis (Goll), van de Paverd, 1995, p. 210, pl. 63, fig. 20 (only).

Small individuals, shell of the bilocular cephalis long ellipsoidal or reniform, with a obvious sagittal constriction, which separated two cephalis chambers as near spherical respectively, medium thick walled, both lateral ends of circular arc form; pores quasi-circular or hexagonal, relatively small, similar or different sizes, irregularly arranged, unsymmetrical distribution on both cephalis shells, slightly wider bars, seemed to be with hexagonal frames, but no large pores and three-bladed intervening bars beside the sagittal and basal rings; vertex spine and basal feet very small or nearly disappeared; smooth surface, only individual spines. All pore shapes, bar widths, wall thickness and individual sizes of this species are variable for different specimens.

Measurements: Shell (cephalis) width 70–135μm, height 55–115μm.

Holotype: DSDP274-12-3 1 deposited in the South China Sea Institute, CAS, from sample of DSDP274-12-3 in the Ross Sea, pictured in pl. 91, figs. 13–14.

Distribution: Late Paleocene–Holocene in Ross Sea, Antarctic.

Distinctions of this new species with *Ceratospyris pentagona* Ehrenberg and *Lophospyris stabilis stabilis* (Goll) are mainly by the latters with large pores on both cephalis chambers, less number of pores, pentagonal, some spines on surface, and longer basal feet; also is different with *Lophospyris pentagona* (Ehrenberg) *hyperborea* (Jørgensen) by the latter with relatively larger pores beside sagittal ring and smaller toward laterally, usually thorny and longer basal feet.

Family Tholospyridae Haeckel, 1887

Genus *Tholospyris* Haeckel, 1881

292. *Tholospyris peltella* sp. nov.

(Pl. 91, Figs. 39–54)

Trissocyclid sp., Abelmann, 1992b, pl. 3, fig. 10.
Trissospyris sp. A, Bjørklund, 1976, pl. 18, figs. 7–11.

Individuals very small, bilocular cephalis kidney-shaped or lung-shaped, with deep sagittal constriction; pores quasi-circular or irregular, different sizes, indefinite numbers, irregularly arranged, 1–2 large pores at two sides of D ring for some specimens, different width of bars in different shells; apical spine very short, three basal feet often aggregated together at their ends and connected each other, forming a small enclosed construction as semi-circular or triangular under the middle base; shell surface smooth, no spine or thorn.

Measurements: Shell (cephalis) width 65–85μm, height 45–55μm.

Holotype: DSDP274-33-3 2 deposited in the South China Sea Institute, CAS, from sample of DSDP274-33-3 in the Ross Sea, pictured in pl. 91, figs. 39–40.

Distribution: Middle Paleocene–late Eocene in Ross Sea, Antarctic.

Feature of this new species is relatively similar to *Tholospyris antarctica* (Haecker), main distinction between them is the latter with larger individuals, basal feet downward or extroverted, not aggregated together at their ends, surface slightly rough and thorny.

Genus *Desmospyris* Haeckel, 1881

293. *Desmospyris coryatis* sp. nov.

(Pl. 91, Figs. 55–56)

Individuals very small, cephalis bilocular separated by incomplete sagittal ring inside, without sagittal stricture presented as a long helmet shape; cephalis wall with structural composition of 2–3 layers of vine bars, pores mutual overlay, quasi-circular, different sizes, thin bars, irregularly cross-linking, various joints grown many crests or thorns on shell surface; the front opening under cephalis nearly flattened, with a border, only one basal foot imperfectly grown at the middle place, short, with lateral spines, 2–3 lateral branches from lower middle part of basal foot connecting to cephalis underneath; rough surface, no apical spine.

Measurements: Shell (cephalis) width 55–65μm, height 65–70μm, length of basal foot 18–25μm.

Holotype: DSDP274-1-3 1 deposited in the South China Sea Institute, CAS, from sample of DSDP274-1-3 in the Ross Sea, pictured in pl. 91, figs. 55–56.

Distribution: Holocene in Ross Sea, Antarctic.

Shell form and size of this new species are somewhat similar to *Desmospyris multichyparis* sp. nov., main distinction of them is the latter with latticed wall, five basal feet of no lateral spine, these feet are separated and no connection.

296. *Desmospyris multichyparis* sp. nov.

(Pl. 92, Figs. 25–26)

Individuals very small, with only one cephalis chamber, elongated campanulate, enlarged gradually from top to down, no constriction at open mouth; pores of cephalis quasi-circular or polygonal, different sizes, irregularly distributed, thin bars, growth of some disorder, obvious node rises on shell surface, coniform; opening mouth flatted, no peristoma, developed five basal feet of thin cylinder-shaped, slightly terminal introversion and no connection among these feet; rough surface, several coniform bulges on the top of shell.

Measurements: Shell (cephalis) width 50–55μm, height 55–65μm, length of basal foot 20–25μm.

Holotype: DSDP274-1-3 1 deposited in the South China Sea Institute, CAS, from sample of DSDP274-1-3 in the Ross Sea, pictured in pl. 92, figs. 25–26.

Distribution: Holocene in Ross Sea, Antarctic.

Distinction of this new species with *Desmospyris trichyparis* sp. nov. is the latter with bilocular cephalis, slightly larger individuals, three basal feet connected by latticed wall, and smooth surface.

299. *Desmospyris trichyparis* sp. nov.

(Pl. 93, Figs. 21–23)

Trissocyclid sp. A, Abelmann, 1990, p. 695, pl. 5, figs. 5A–C.
Trissocyclidae gen. et sp. indet., Hollis, 2002, p. 293, pl. 3, fig. 3 (only).

Individuals medium size, slightly thick walled; bilocular cephalis, separated by sagittal ring of D shape, pores circular or sub-circular, similar sizes, slightly larger nearby sagittal ring, even distributed; three cone-shaped basal feet staunch, terminal sharp, downward growing, partly latticed structures between basal feet connected each other, forming incomplete lattice wall, with similar pores of cephalis, quasi-circular or elliptical, different sizes, less pores, irregularly distributed, wider bars; smooth surface, no spine.

Measurements: Shell (cephalis) width 105–115μm, height 70–80μm, length of basal foot 50–70μm.

Holotype: DSDP274-11-3 1 deposited in the South China Sea Institute, CAS, from sample of DSDP274-11-3 in the Ross Sea, pictured in pl. 93, fig. 21.

Distribution: Late Miocene in Ross Sea, Antarctic.

Shell features of this new species are relatively closed to *Desmospyris spongiosa* Hays, the main differences are the latter with disappeared basal feet, instead of feet by a completely barrel-shaped abdomen, opening mouth flatted, wall composed of spongy or meshed texture.

Suborder BOTRYODEA Haeckel, 1881

Family Cannobotryidae Haeckel, 1881, emend. Riedel, 1967

Genus *Amphimelissa* Jørgensen, 1905

300. *Amphimelissa quadrata* sp. nov.

(Pl. 93, Figs. 24–25)

Individuals slightly larger or medium size, presented as square block shape; cephalis large, separated by two sagittal constrictions of orthogonality as four nearly spherical chambers, with broad opening at underneath; thorax grown incompletely, only remnant wall very short and small, wide mouth; pores quasi-circular, different sizes, irregularly distributed, with hexagonal frames, diameter of pores 3–4 times as broad as bars; rough surface, some small calthrops. This species has a special cephalis of four chambers, and it is arranged into Genus *Amphimelissa* for the time being.

Measurements: Shell width 120–140μm, height 125–150μm.

Holotype: DSDP274-10-3 2 deposited in the South China Sea Institute, CAS, from sample of DSDP274-10-3 in the Ross Sea, pictured in pl. 93, figs. 24–25.

Distribution: Early Pliocene in Ross Sea, Antarctic.

Distinction of this new species from other species is mainly by its cephalis with four similar chambers separated by two orthogonal constrictions, and nearly quadrate shell shape.

Genus *Botryopera* Haeckel, 1887

304. *Botryopera cylindrita* sp. nov.

(Pl. 95, Figs. 5–10)

Antarctissa denticulata (?) (Ehrenberg) var. *faceata* Petrushevskaya, 1967, pp. 86–88, figs. 50II–III.
Lithomelissa sp. P, Petrushevskaya, 1973, p. 1044, fig. 3(12).

Shell of two sections, cylinder-shape, medium thick walled; primary chamber of cephalis relatively larger, nearly spheroidal or ellipsoidal, surrounded by several small head lobes, connecting directly to thorax (the second section); thorax completely long cylinder-shape, with straight or slight curve of lateral, opening mouth slight narrow, flatted, no peristoma ring and tooth, individual specimens with incomplete mouth; pores quasi-circular, very small, different sizes, diameter of pores similar to width bars, or smaller, irregularly arranged, rather even distributed; smooth surface, no spine.

Measurements: Primary chamber of cephalis length 30–60μm, width 35–55μm, thorax length 70–85μm, width 45–75μm.

Holotype: DSDP274-12-5 1 deposited in the South China Sea Institute, CAS, from sample of DSDP274-12-5 in the Ross Sea, pictured in pl. 95, figs. 7–8.

Distribution: Middle Miocene–late Pliocene in Ross Sea, Antarctic.

This new species is similar to *Antarctissa denticulata* (Ehrenberg), but the latter with shell of round triangle, shorter thorax, mouth of circular arc, appendants of thorns or spines on surface.

Suborder CYRTOIDEA Haeckel, 1862, emend. Petrushevskaya, 1971

Family Tripocalpidae Haeckel, 1887

Genus *Archipilium* Haeckel, 1881

314. *Archipilium vicinun* sp. nov.

(Pl. 96, Figs. 6–7)

Archipilium sp., Caulet, 1991, pl. 2, fig. 7.
Archipilium sp., O'Connor, 1997b, p. 111, pl. 3, fig. 17.

Individuals small, shell truncate-conical, cephalis with a small hemispherical arch, enlarged rapidly downward, widely open mouth, unfairness of peristoma, or fragmentary; pores quasi-circular or elliptical, with irregular distribution; pore diameter in different sizes, 1–4 times as broad as bars, bars width irregular; three lateral wings arising from the collar stricture, straightly divergent, or recurved, long cylindrical and tapering distally; some slightly introversion, lateral wings larger than shell length; shell surface smooth without any spine, no apical horn.

Measurements: Shell length 70–75μm, shell width 55–70μm, lateral ribs length 80–95μm.

Holotype: DSDP274-10-5 1 deposited in the South China Sea Institute, CAS, from sample of DSDP274-10-5 in the Ross Sea, pictured in pl. 96, figs. 6–7.

Distribution: Early Pliocene in Ross Sea, Antarctic; Oligocene to middle Micene in south Indian Ocean and New Zealand (Kaipara Harbour).

This new species is distinguished from other species of this Genus by its open mouth being unfairness, without peristoma or ring, pores of very different sizes and shapes, with a small apex of arch. Features of specimens from Ross Sea are very similar to that of Caulet (1991) and O'Connor (1997b).

Family Cyrtocalpidae Haeckel, 1887
Genus *Archicapsa* Haeckel, 1881

325. *Archicapsa olivaeformis* sp. nov.

(Pl. 98, Figs. 12–13)

Shell small, nearly oval, the broadest at middle of shell, gradually constricted toward two ends; slightly thick walled, with quasi-circular or elliptical pores, in different sizes, sub-regularly arranged, diameter of pores 1–2 times as broad as bars; cephalis hemispheric, no apex horn, no collar constriction between cephalis and thorax; no open mouth at the end of thorax that is enclosed, no sieve plate of pores; smooth surface, no spine.

Measurements: Shell length 110–115μm, width 75–80μm.

Holotype: DSDP274-34-1 2 deposited in the South China Sea Institute, CAS, from sample of DSDP274-34-1 in the Ross Sea, pictured in pl. 98, figs. 12–13.

Distribution: Late Eocene in Ross Sea, Antarctic.

This new species is similar to *Archicapsa quadriforis* Haeckel (1887, p. 1192), the distinction between them is the latter with shell of egg shape, relatively broader, shell length is only slightly larger than width, with sieve plate of four pores at mouth end, and rough surface.

Family Tripocyrtidae Haeckel, 1887, emend. Campbell, 1954
Subfamily Sethopilinae Haeckel, 1881, emend. Campbell, 1954
Genus *Dictyophimus* Ehrenberg, 1847

326. *Dictyophimus echinatus* sp. nov.

(Pl. 98, Figs. 14–29)

Ceratocyrtis sp., Petrushevskaya, 1975, p. 590, pl. 19, fig. 1.
Cyrtoidae gen. sp., Petrushevskaya, 1967, p. 91, fig. 52III.

Shell campanulate, gradually enlarged from top to bottom; cephalis small spherical, partly fallen into thorax, with very small pores, a apex horn of thin pyramid; thorax medium thick walled, pores quasi-circular or ellipsoidal, in different sizes, irregularly arranged, generally larger pores near opening, diameter of pores 1–4 times as broad as bars, thin bars; three main thoracal ribs developed, elongated as three basal feet of thin pyramid; thoracal mouth opened completely, other 3–5 long thin spines surround peristoma, their shapes and sizes somewhat similar to the basal feet; surface rough or smooth, with some spines of different sizes.

Measurements: Shell length 75–150μm, width 95–170μm, diameter of cephalis 40–50μm, length of apex horn 30–55μm, basal feet 30–95μm.

Holotype: DSDP274-20-5 1 deposited in the South China Sea Institute, CAS, from sample of DSDP274-20-5 in the Ross Sea, pictured in pl. 98, figs. 20–21.

Distribution: Late Eocene–Holocene in Ross Sea, Antarctic.

This new species is similar to *Dictyophimus mawsoni* Riedel (1958, p. 234, pl. 3, figs. 6–7), the distinction between them is the latter with three terminal feet of solid and very strong, and with a stouter and larger apex horn.

329. *Dictyophimus lubricephalus* sp. nov.

(Pl. 99, Figs. 13–14)

Dictyophimus sp. 4, Abelmann, 1992a, p. 380, pl. 4, fig. 6.

Individuals very small, shell nearly broad barrel-type, short, top circular arc, side vertical, composed mainly of cephalis structure, thorax unconspicuous, no collar stricture; shell thin walled, relatively larger pores, quasi-circular, with hexagonal frames, thinner bars, diameter of pores 2–5 times as broad as bars; apex spine undeveloped or very small, presented only as a small swelling; three solid lateral ribs developed from the central rod in thorax, extended and inclined outside, thin triangular pyramid, straight, terminal sharp; smooth surface, no spine.

Measurements: Shell length 50μm, width 75μm, length of solid lateral ribs 30–45μm.

Holotype: DSDP274-20-3 1 deposited in the South China Sea Institute, CAS, from sample of DSDP274-20-3 in the Ross Sea, pictured in pl. 99, figs. 13–14.

Distribution: Late Oligocene in Ross Sea, Antarctic.

This new species is similar to *Dictyophimus histricosus* Jørgensen, and can be distinguished by the latter possessing a conical or bell-shaped shell with a cephalis and apical horn, a delicate segment ring on the thorax, inwardly curved lateral wings, and remnants of the abdominal shell.

Genus *Lithomelissa* Ehrenberg, 1847

331. *Lithomelissa arrhencorna* sp. nov.

(Pl. 99, Figs. 17–20)

Individuals relatively small, nearly cylinder-shape with apex of arch, similar length and width for cephalis and thorax, thick walled; cephalis hemispherical, divided as 3–4 lobes of different sizes, a apex horn of triangular pyramid developed from the center of cephalis, thick and strong, terminal sharp; thorax of short cylinder, three lateral ribs small, thin pyramid, opening mouth, around peristoma with some teeth formed from bars; all the pores quasi-circular or elliptical, different sizes, irregularly distributed, similar sizes of bar width and pore diameter, wider bars, from joints arose small humps of pyramid, forming slightly not smooth surface, no spine.

Measurements: Shell length 105–140μm, width 90–120μm, length of cephalis 45–70μm, thorax 60–80μm, apex horn 50–80μm, lateral ribs 15–25μm.

Holotype: DSDP274-23-3 1 deposited in the South China Sea Institute, CAS, from sample of DSDP274-23-3 in the Ross Sea, pictured in pl. 99, figs. 17–18.

Distribution: Early Eocene–middle Oligocene in Ross Sea, Antarctic.

This new species is similar to *Lithomelissa ministyla* sp. nov., the distinction between them is the latter with thin walled, apex horn thin and small, with peristomal ring, no teeth, smooth surface.

332. *Lithomelissa callifera* sp. nov.

(Pl. 99, Figs. 21–25)

Lophophaenoma sp., Petrushevskaya, 1971d, p. 117, figs. 63III–IV.

Cephalis nearly spheroidal, interior structure of spines developed, one vertical spine growing upward as thin columnar or pyramid apex horn, and downward extended to the mouth of thorax, other three lateral spines forming sternal ribs; thorax near cylinder, sternal ribs may outspreading as thin spines, walls inflated slightly between sternal ribs, opening mouth broad or slightly constrictive, with different forms, seems to have three basal feet for some specimens; whole shell thin walled, relatively larger pores and thinner bars, irregularly arranged, with very smooth surface.

Measurements: Shell length 105–160μm, width 95–140μm, diameter of cephalis 50–80μm, length of thorax 55–95μm, apex horn 40–85μm, lateral ribs 25–40μm.

Holotype: DSDP274-10-3 1 deposited in the South China Sea Institute, CAS, from sample of DSDP274-10-3 in the Ross Sea, pictured in pl. 99, figs. 23–24.

Distribution: Early Pliocene in Ross Sea, Antarctic.

This new species is distinguished from other species by its developed spine structure of interior cephalis and their derivants, inflated thin wall between sternal ribs, and smooth surface.

335. *Lithomelissa cratospina* sp. nov.

(Pl. 100, Figs. 7–17)

Lithomelissa sphaerocephalis Chen, Hollis et al., 1997, p. 52, pl. 3, figs. 12–14.
Botryometra poljanskii Petrushevskaya, 1975, p. 588, pl. 26, fig. 13.

Shell of two sections, cephalis nearly spheroidal, thorax campanulate, deep collar constriction between cephalis and thorax, or with a neck, width of thorax slight greater than cephalis (diameter), thorax shorter, opening mouth, with incomplete margin, no peristoma; apex horn of triangular pyramid, sturdy and strong; pores quasi-circular, similar sizes, irregularly distributed, diameter of most pores equal to width of bars; bar joints with small clavulas, no spine, slightly rough surface.

Measurements: Shell length 105–170μm, width 110–130μm, diameter of cephalis 80–95μm, length of thorax 60–100μm, apex horn 65–75μm.

Holotype: DSDP274-10-3 2 deposited in the South China Sea Institute, CAS, from sample of DSDP274-10-3 in the Ross Sea, pictured in pl. 100, figs. 11–13.

Distribution: Late Eocene in Ross Sea, Antarctic.

This new species is similar to *Lithomelissa sphaerocephalis* Chen, main distinctions between them are the latter with unconspicuous pores of cephalis, smaller apex horn, and covering of spongy texture on surface.

339. *Lithomelissa globicapita* sp. nov.

(Pl. 102, Figs. 5–12)

Lithomelissa sp., Petrushevskaya, 1971d, p. 90, figs. 46I–V.

Individuals small, medium thick walled, surface smooth; cephalis large, nearly spheroidal, pores smaller, quasi-circular, different sizes, irregularly distributed, bars wider, apex spine undeveloped, or individual smaller spines existed laterally; thorax very short, incomplete, with similar width of cephalis, deep constriction between cephalis and thorax, with similar pores of cephalis, opening mouth, irregular.

Measurements: Shell length 110–130μm, cephalis length 90–100μm, width 80–100μm, thorax length 20–50μm, width 80–90μm.

Holotype: DSDP274-33-3 2 deposited in the South China Sea Institute, CAS, from sample of DSDP274-33-3 in the Ross Sea, pictured in pl. 102, figs. 11–12.

Distribution: Late Eocene in Ross Sea, Antarctic.

This new species is similar to *Lithomelissa thoracites* Haeckel, but the latter with features of wider thorax than cephalis, larger pores, or thinner bars, slightly rough surface. Features of specimens of this new species from Ross Sea are basically the same with that of

Petrushevskaya (1971, p. 90, figs. 46I–V).

341. *Lithomelissa hirundiforma* sp. nov.

(Pl. 102, Figs. 19–22)

Individuals relatively smaller, shell truncate-conical; cephalis hemispherical, small, unconspicuous collar constriction between cephalis and thorax, with a apical horn of cone; thorax coniform, or cylindrical for the lower part, shoulder declined, three thoracal ribs extended outside as small and short lateral wings of solid triangular pyramid, thoracal mouth opened, nearly flattened, with a peristoma of thin ring; whole shell slightly thick walled, pores quasi-circular or ellipsoidal, small, slightly increasing size near opening, irregularly distributed, bar joints with swellings; slightly rough surface, no spine.

Measurements: Shell length 105–125μm, width 85–100μm, cephalis width 40–50μm, thorax length 80–95μm, length of apical horn 15–25μm, lateral wings 40–50μm.

Holotype: DSDP274-12-3 2 deposited in the South China Sea Institute, CAS, from sample of DSDP274-12-3 in the Ross Sea, pictured in pl. 102, figs. 21–22.

Distribution: Late Miocene in Ross Sea, Antarctic.

This new species is similar to *Lithomelissa ehrenbergii* Bütschli, but the latter with larger shell and cephalis, cephalis nearly spheroidal, stronger apical horn and lateral wings, thorax cylindrical, without peristoma.

343. *Lithomelissa ministyla* sp. nov.

(Pl. 102, Figs. 27–35)

Shell of two segments, cephalis broad hemispherical, thorax short cylindrical, similar width of cephalis and thorax; where between cephalis and thorax with constriction on surface and lantern ring of inwall, internal skeletons developed, with one vertical spine and several horizonal spines, the vertical spine upward protruded wall outside, forming a apical horn, thin cylindrical and terminal sharp; similar pores for whole the shell, quasi-circular or elliptical, different sizes, irregularly distributed, with wider bars; smooth surface, or with complex structure formed by uprising of pore peritreme; usually a peristoma ring at thorax mouth, or incomplete.

Measurements: Shell length 95–145μm, cephalis length 30–65μm, width 65–95μm, thorax length 50–80μm, width 70–110μm.

Holotype: DSDP274-31-6 1 deposited in the South China Sea Institute, CAS, from sample of DSDP274-31-6 in the Ross Sea, pictured in pl. 102, figs. 32–33.

Distribution: Late Eocene–middle Oligocene in Ross Sea, Antarctic.

This new species is similar to *Lithomelissa poljanskii* (Petrushevskaya), but the latter with cephalis of spherosome, relatively higher, apical horn of triangular pyramid or thin pyramid, even pores, no peristoma ring.

344. *Lithomelissa pachyoma* sp. nov.

(Pl. 103, Figs. 1–4)

Lithomelissa sp., Petrushevskaya, 1971d, p. 89, figs. 45III–VI.

Individuals small, mainly shown as a cephalis, and a shorter or incomplete thorax, thick walled; cephalis round spheroidal, larger size, pores small and indistinct, different shapes and sizes, smaller inside and larger outside, irregularly distributed, bars broad, vertical spine of interior cephalis developed, forming a small apical horn of thin pillar or pyramid; thorax nearly cylindrical, length may less than or greater than radius of cephalis, width of thorax similar to, or less than diameter of cephalis, with a deep collar constriction between cephalis and thorax, thoracic pores quasi-circular, similar sizes, opening mouth nearly flatted; entire shell with small conic swellings on various bar joins, coarser, causing a very thick wall; rough surface.

Measurements: Shell length 100–115μm, cephalis length 65–75μm, width 65–80μm, thorax length 25–45μm, width 55–65μm.

Holotype: DSDP274-25-5 1 deposited in the South China Sea Institute, CAS, from sample of DSDP274-25-5 in the Ross Sea, pictured in pl. 103, figs. 1–2.

Distribution: Middle Oligocene–late Pleistocene in Ross Sea, Antarctic.

The cephalic spines and apical horn of this new species are similar to those of *Lithomelissa ministyla* sp. nov. The main differences between these two species are that the latter is characterized by a hemisphere cephalis, a thin wall, no obvious swellings at bar joins. The features of this new species in the Ross Sea are very similar to those of *Lithomelissa* sp. (Petrushevskaya, 1971d, p. 89, figs. 45III–VI).

349. *Lithomelissa stomaculeata* sp. nov.

(Pl. 104, Figs. 8–12)

Individuals slightly larger, medium thick walled; cephalis nearly round spherosome, larger pores, with a vertical horn of triangular pyramid and individual lateral spines, a deep collar constriction between cephalis and thorax; thorax campanulate, slightly expanded downward, the widest of shell at thoracic mouth, pores quasi-circular or ellipsoidal, of different sizes and shapes, irregularly distributed, thin bars; three thoracic ribs extended at middle part as three lateral wings, which has fenestratas in the basic or initial part of wings, the last parts of wings grown as thin pillar or pyramid; broad mouth, with many terminal pyramidal spines formed by bars underlaying, like long spines; smooth surface, or with some spines on surface.

Measurements: Shell length 150–180μm, width 125–145μm, diameter of cephalis 50–60μm, length of thorax 90–130μm, apical horn 20–40μm, lateral wings 30–70μm, terminal spines 20–50μm.

Holotype: DSDP274-10-3 2 deposited in the South China Sea Institute, CAS, from sample of DSDP274-10-3 in the Ross Sea, pictured in pl. 104, figs. 8–9.

Distribution: Early Pliocene in Ross Sea, Antarctic.

This new species is different to *Lithomelissa stigi* Bjørklund (1976) by the latter with smaller cephalis and thorax, undeveloped lateral wings, less spines at mouth and very short.

352. *Lithomelissa tryphera* sp. nov.

(Pl. 105, Figs. 25–33)

Individuals smaller, with slightly variable for morphological structure; shell campanulate, thin walled, pores irregularly distributed, surface or have secondary reticular tissue; cephalis larger, hemispherical, commonly transparent, pores of cephalis in different sizes and some obscured, unconspicuous collar constriction between cephalis and thorax, but with a inner lantern ring; thorax gradually expanded downward, pores of thorax quasi-circular, different sizes, irregularly distributed, with wider bars, smooth surface; a very small apophysis on cephalis, or of a apical horn of triangular pyramid, sometimes visible three solid lateral wings; opening mouth, no peristoma, often out of flatness.

Measurements: Shell length 115–160μm, width 125–145μm, cephalis length 50–60μm, width 80–110μm, length of apical horn 10–35μm, lateral wings 10–30μm.

Holotype: DSDP274-28-3 1 deposited in the South China Sea Institute, CAS, from sample of DSDP274-28-3 in the Ross Sea, pictured in pl. 105, figs. 27–28.

Distribution: Late Eocene–middle Oligocene in Ross Sea, Antarctic.

Morphological character of this new species is similar to *Lithomelissa tricornis* Chen, but the latter with obvious three apical horns on cephalis, deeper collar constriction between cephalis and thorax.

Genus *Antarctissa* Petrushevskaya, 1967

356. *Antarctissa cingocephalis* sp. nov.

(Pl. 106, Figs. 5–6)

Shell bullet-shape, surface wall consisted of thorax and its upside thicker wall extended continuously upward cladding, enclosing cephalis inside; upper part of cephalis hemispherical, underpart of cephalis constricted as a narrow long neck, both with similar length, width of cephalis two times of neck, and about half width of thorax, length of cephalis and neck nearly equaling to length of thorax; thorax cylindrical, rounding top, opening mouth flatted, without peristoma; pores of cephalis similar to thorax, quasi-circular or elliptical, irregularly distributed, with wider bars, of plate-like; all the shell covered by irregular lattice structure as oneness, no surface constriction, bars of out covering thicker and looser, with pores of different sizes, interior structure visible; smooth surface, no spine.

Measurements: Shell length 150μm, width 100μm, width of cephalis 55μm, neck 30μm, thorax length 80μm, width 90μm.

Holotype: DSDP274-16-1 1 deposited in the South China Sea Institute, CAS, from sample of DSDP274-16-1 in the Ross Sea, pictured in pl. 106, figs. 5–6.

Distribution: Early Miocene in Ross Sea, Antarctic.

This new species is distinguished from other species by with an enclosed outsourcing layer, covering all the cephalis and thorax inside.

363. *Antarctissa pachyoma* sp. nov.

(Pl. 109, Figs. 1–11)

Shell of two segments, bullet-shape, thicker walled; cephalis larger, nearly spheroidal, hemispherical or ellipsoidal, with similar width of thorax, a deep constriction between cephalis and thorax, but cladded by covering of thorax wall extended, their surface smoothly transited; shorter thorax, length of thorax less than its width and length of cephalis, mouth opening, generally flatted, without peristoma and tooth; pores of cephalis and thorax basically the same, quasi-circular, different sizes, irregularly distributed, diameter of pores similar to, or larger than width of bars; surface of whole the shell smooth, no spine, no apical horn and lateral wings.

Measurements: Shell length 125–175μm, width 100–130μm, diameter of cephalis 85–110μm, thorax length 30–60μm.

Holotype: DSDP274-10-3 1 deposited in the South China Sea Institute, CAS, from sample of DSDP274-10-3 in the Ross Sea, pictured in pl. 109, figs. 6–7.

Distribution: Early Oligocene–early Pliocene in Ross Sea, Antarctic.

This new species is similar to *Antarctissa antedenticulata* Chen, main distinctions of them are the later with smaller shell, not bullet-shape, thinner walled, a constriction on surface between cephalis and thorax, and with a thin peristoma at mouth.

364. *Antarctissa pterigyna* sp. nov.

(Pl. 109, Figs. 12–15)

Antarctissa (?) sp., Bjørklund, 1976, pl. 19, figs. 10–12.

Individuals slightly larger, thicker walled, no apical horn; cephalis spheroidal, larger, diameter of cephalis nearly equals to width of upper thorax, bars wider, some of them overlapped, pores blurring or quasi-circular, very small, with small conic swellings on bar joins; a septal collar or a deep collar constriction between cephalis and thorax; thorax relatively short, nearly cylindrical, narrow upper part, slightly increasing width downward, three trilateral wings of fenestra developed at thorax upside, which terminal solid, wall of thorax clathrate, pores quasi-circular or irregular, different sizes, irregularly arranged, with wider bars, partly irregular, smooth surface, the widest of shell at mouth, opening, incomplete and out of flatness; without spine on surface.

Measurements: Shell length 170–190μm, width 115–125μm, diameter cephalis 85–100μm, thorax length 80–115μm.

Holotype: DSDP274-32-3 2 deposited in the South China Sea Institute, CAS, from sample of DSDP274-32-3 in the Ross Sea, pictured in pl. 109, figs. 14–15.

Distribution: Late Eocene in Ross Sea, Antarctic.

This new species is similar to *Antarctissa antedenticulata* Chen, main distinctions of them are the later without lateral wings of thorax, opening mouth flatted, with partly sieve plate, smooth surface.

Genus *Spongomelissa* Haeckel, 1887

367. *Spongomelissa chaenothoraca* sp. nov.

(Pl. 109, Figs. 34–35)

Spongomelissa sp., Chen, 1975, p. 458, pl. 10, fig. 4; Bjørklund, 1976, p. 1124, pl. 22, figs. 6–9.
Lithomelissa sp., Hollis, 2002, p. 297, pl. 4, figs. 1–2.

Shell nearly campanulate, with one apical horn and three lateral wings; cephalis hemispherical, larger and wider, apical horn arose from central vertical spine inside cephalis, triangular pyramid; thorax cylindrical, very short, opening mouth, often broken or vestigital, three lateral wings at upper thorax, solid, with fenestrate at its proximal; whole shell thinner or slightly thick walled, with quasi-circular pores, similar or different sizes, irregularly distributed, thin bars; surface smooth, or with small spines, slightly different features of specimens from different sea areas.

Measurements: Shell length >125μm, width 115μm, cephalis length 70–75μm, width 95–110μm, thorax length incomplete, apical horn length 45–55μm.

Holotype: DSDP274-20-3 1 deposited in the South China Sea Institute, CAS, from sample of DSDP274-20-3 in the Ross Sea, pictured in pl. 109, figs. 34–35.

Distribution: Middle Oligocene in Ross Sea, Antarctic; late Paleocene–Eocene in Southwest Pacific; Oligocene in Norwegian Sea.

This new species is similar to *Spongomelissa cucumella* Sanfilippo and Riedel, but the latter with a smaller individuals, constricted mouth and longer lateral wings.

Genus *Velicucullus* Riedel and Campbell, 1952

371. *Velicucullus spongiformus* sp. nov.

(Pl. 112, Figs. 1–5)

Shell campanulate or shape of bamboo hat, thick wall constituted by spongy tissue; cephalis hemispherical or cylindrical, a window at the top, or with a lattice hole cover of hemisphere, without apical horn and spine; thorax broad conical, rapidly enlarged downward, wider opening mouth, some terminal solid spines at margin of mouth; no surface spine.

Measurements: Cephalis length 50–80μm, width 45–125μm, thorax length 50–100μm, width 90–260μm.

Holotype: DSDP274-29-1 2 deposited in the South China Sea Institute, CAS, from sample of DSDP274-29-1 in the Ross Sea, pictured in pl. 112, figs. 4–5.

Distribution: Late Paleocene–late Eocene in Ross Sea, Antarctic.

This new species is similar to *Velicucullus altus* Abelmann (1992a, p. 777, pl. 3, fig. 8), but it can be distinguished by the features of the latter, which are marked by a lattice wall in the middle-upper part, with a spongy framework along the lower periphery, and the absence of any peristoma spine.

Genus *Helotholus* Jørgensen, 1905

373. *Helotholus bulbosus* sp. nov.

(Pl. 115, Figs. 7–8)

Shell shape looks like lamp bulb, upside inflated, underpart narrowed; cephalis spheroidal, larger, no apex spine, inside spine structure developed, forming a porous skeleton at junction of cephalis and thorax, several slant bars protruding outside to form thin cylindrical lateral spines of cephalis, the central bar prolonged vertically downward, reached out mouth of thorax, developed as the only one basal feet; thorax cylinder-shape, very short, width of thorax about as half diameter of cephalis, breast height as 1/3 width, opening mouth, no sieve plate, out of flatness at margin; whole the shell slightly thin walled, pores quasi-circular, very small, numerous, irregularly distributed, bars wider, diameter of pores about equal to width of bars; shell surface slight unfairness, no general spine.

Measurements: Shell length 155–165μm, diameter of cephalis 125–135μm, thorax length 25–30μm, width 75–85μm, length of basal feet 40–45μm.

Holotype: DSDP274-23-5 1 deposited in the South China Sea Institute, CAS, from sample of DSDP274-23-5 in the Ross Sea, pictured in pl. 115, figs. 7–8.

Distribution: Middle Oligocene in Ross Sea, Antarctic.

This new species is distinguished from *Lophophaena*? *orbicularis* sp. nov. by the latter with undeveloped thorax, no spines on head, smooth surface.

377. *Helotholus vematella* sp. nov.

(Pl. 116, Figs. 19–26)

Individuals small; cephalis smaller, hemispherical, partly fallen into thorax, with deep constriction, 2–3 shorter apical and lateral spines of pyramid; thorax nearly cylindrical, slightly inflated at middle part, shrunken toward mouth, individual short conic spines on surface of thorax; one central bar of interior cephalis extended to the thoracic mouth, forming a horizontal cover plate of porous skeleton, mouth of incomplete open, nearly flattened; whole shell thin walled, with smaller pores, quasi-circular or elliptical, different sizes, irregularly distributed, diameter of pores equal or less width of bars, wider bars generally.

Measurements: Shell length 95–120μm, cephalis lenght 25–40μm, width 40–60μm, thorax length 55–75μm, width 95–115μm, mouth width 50–65μm.

Holotype: DSDP274-28-5 1 deposited in the South China Sea Institute, CAS, from sample of DSDP274-28-5 in the Ross Sea, pictured in pl. 116, figs. 19–20.

Distribution: Late Eocene–early Oligocene in Ross Sea, Antarctic.

This new species is similar to *Helotholus vema* Hays, but the latter with relatively larger individuals, thick walled, not shrunken at thoracic mouth, and distributed in the newer stratum.

Genus *Amphiplecta* Haeckel, 1881

379. *Amphiplecta siphona* sp. nov.

(Pl. 116, Figs. 29–38)

Shell two segments, both cylindrical; cephalis short cylinder-shape, top opening, nearly flattened, often incomplete, no apex spine; thorax broad cylinder-shape, quickly widened, shoulder arc-shaped, side of straightness, not expanded, thoracic mouth opening, incomplete margin, not flattened; whole shell slightly thick walled, clathrate, pores quasi-circular or irregular-shaped, different sizes, irregularly distributed, with different width of bars for different specimens; generally smooth surface, but some specimens with small conic spines at margins of pores, and causing slightly rough surface.

Measurements: Shell length 125–190μm, cephalis lenght 55–80μm, width 55–110μm, thorax length 65–125μm, width 125–150μm.

Holotype: DSDP274-32-1 1 deposited in the South China Sea Institute, CAS, from sample of DSDP274-32-1 in the Ross Sea, pictured in pl. 116, figs. 29–30.

Distribution: Late Eocene in Ross Sea, Antarctic.

This new species is similar to *Amphiplecta acrostoma* Haeckel (1887, p. 1223, pl. 97, fig. 10), but the latter with some upward spines at top wall of opening head, thorax coniform, extended thoracic mouth. It is distinguished from *Amphiplecta* sp. (Petrushevskaya, 1971d, p. 104, fig. 54) by the latter with a thorax of extensively conical disk.

Genus *Lychnocanoma* Haeckel, 1887, emend. Foreman, 1973

383. *Lychnocanoma eleganta* sp. nov.

(Pl. 117, Figs. 15–24)

Lychnocanium sp. aff. *L. grande* Campbell and Clark, Chen, 1975, p. 462, pl. 1, figs. 6–7; Petrushevskaya, 1975, pl. 12, fig. 4.
Lychnocanoma sp. B, Abelmann, 1990, p. 697, pl. 7, figs. 2A–B.

Cephalis nearly spherical, small, with tiny pores, apical horn of pyramid, short; thorax hemispherical, slightly swelled, pores quasi-circular, similar sizes, regularly or sub-regularly arranged, with hexagonal frames, diameter of pores 2–3 times as broad as bars, broad mouth, no appendant, peristoma ring thinner; three terminal feet triangular pyramid, down obliquely growing, straight, or slightly curved at the distal end, length of feet nearly equals to length of thorax; whole shell slightly thin walled, smooth surface, no spine.

Measurements: Diameter of cephalis 35–50μm, thorax length 85–100μm, width 115–130μm, length of apical horn 15–50μm, terminal feet 85–110μm.

Holotype: DSDP274-12-3 2 deposited in the South China Sea Institute, CAS, from sample of DSDP274-12-3 in the Ross Sea, pictured in pl. 117, figs. 15–16.

Distribution: Late Eocene–late Miocene in Ross Sea, Antarctic.

This new species is similar to *Lychnocanoma grande* (Campbell and Clark), but the latter with rather thick wall, three terminal feet very strong, obviously longer than thorax. This new species is generally appeared in older stratum in the Ross Sea.

386. *Lychnocanoma macropora* sp. nov.

(Pl. 118, Figs. 13–14)

Individuals slightly large, companulate; cephalis irregular, a half of prominent spherosome, another half of low dome, less pores, similar sizes, irregularly arranged, apical horn small or undeveloped; thorax long companulate, pores circular or elliptical, larger, generally smaller at upside, larger downward, sub-regularly arranged, with hexagonal frames, diameter of poers 3–4 times as broad as bars, 4–5 pores crosswise, without thoracic ribs, broad mouth opening, nearly flattened, no peristoma ring; three straight basal feet obliquely grown from the mouth downward, cylinder-shaped, terminal sharp, length of feet shorter than length of thorax; shell slightly thick walled, with small conical swelling on bar joints, rough surface, no spine.

Measurements: Cephalis length 45μm, width 55–60μm, thorax lenght 150–160μm, width 125–150μm, length of terminal feet 115–125μm.

Holotype: DSDP274-25-5 1 deposited in the South China Sea Institute, CAS, from sample of DSDP274-25-5 in the Ross Sea, pictured in pl. 118, figs. 13–14.

Distribution: Middle Oligocene in Ross Sea, Antarctic.

This new species is distinguished from other species by it with cephalis of more than one chambers, or of brain dome shape, thorax relatively longer and not inflated, pores larger and less.

Family Anthocyrtidae Haeckel, 1887

Genus *Aspis* Nishimura, 1992

393. *Aspis ampullarus* sp. nov.

(Pl. 122, Figs. 5–6)

Individuals very small, wine bottle shaped; cephalis spherical, partly falling into thoracic top, with a strong apical horn of pyramid, basic width of horn nearly equal to diameter of cephalis; thoracic upper part coniform, middle and lower parts cylindrical, relatively thick walled, pores of upside quasi-circular, smaller, pores below oval, slightly larger, sub-regularly arranged, 5–6 pores crosswise, diameter of pores 1–2 times as broad as bars, wider bars; mouth flattened, slightly rough surface, no spine.

Measurements: Shell length 105μm, diameter of cephalis 25μm, thorax width 65–75μm, length of apical horn 35–40μm, basic width of apical horn 20μm.

Holotype: DSDP274-40-1 1 deposited in the South China Sea Institute, CAS, from sample of DSDP274-40-1 in the Ross Sea, pictured in pl. 122, figs. 5–6.

Distribution: Middle Paleocene in Ross Sea, Antarctic.

This new species is distinguished from other species by its smaller individuals, shell of bottle-shaped, thick horn and less pores.

394. *Aspis cladocerus* sp. nov.

(Pl. 122, Figs. 7–8)

Shell nearly cylindrical, upper part slightly narrowed; cephalis spheroidal, wrapped in the thorax, apical horn very small, coniform, with 3–4 lateral horns, growing from margin and inclined top, as branch form, individual lateral spines extended to connect two lateral horns; thoracic lower part completely cylindrical, upper part gradually narrowed, without obvious constriction and interior lantern ring, with arc top of shell; pores quasi-circular and elliptical, less pores, different sizes, generally increasing sizes from top to bottom, irregularly distributed, diameter of pores 1–2 times as broad as bars, relatively wide bars of separate pores; thick walled, rough surface, with small calthrops, mouth transversal.

Measurements: Shell length 165μm, diameter of cephalis 30μm, thorax width 50–80μm, length of apical horn 15μm, lateral horns 30–40μm.

Holotype: DSDP274-23-5 1 deposited in the South China Sea Institute, CAS, from sample of DSDP274-23-5 in the Ross Sea, pictured in pl. 122, figs. 7–8.

Distribution: Early Oligocene in Ross Sea, Antarctic.

This new species is similar to *Aspis murus* Nishimura, their distinctions are the latter with only one apical horn, no lateral horn, more pores, similar sizes, regularly arranged, with hexagonal frames, thinner bars, smooth surface, no spine.

Family Sethocyrtida Haeckel, 1887

Genus *Lophophaena* Ehrenberg, 1847, emend. Petrushevskaya, 1971

398. *Lophophaena cephalota* sp. nov.

(Pl. 112, Figs. 19–24)

Individuals slightly smaller, shape of head target, width of cephalis 1/3–1/2 of thorax;

cephalis spheroidal, wall consisted of clathrate or bars overlapped, pores smaller, no apical horn, a few spinous process; thorax cylindrical, shoulders arc-shaped, thick walled, pores larger and less, quasi-circular or elliptical, similar or different sizes, irregularly arranged, seemly hexagonal frames, thinner bars, no lateral wings; opening mouth, flatted, no peristoma ring; generally smooth surface, with several spines.

Measurements: Cephalis length 55–65μm, width 55–70μm, thorax length 70–100μm, width 85–100μm, mouth width 75–85μm.

Holotype: DSDP274-23-3 2 deposited in the South China Sea Institute, CAS, from sample of DSDP274-23-3 in the Ross Sea, pictured in pl. 112, figs. 20–21.

Distribution: Middle Oligocene–middle Miocene in Ross Sea, Antarctic.

This new species is similar to *Lophophaena cylindrica* (Cleve), but the latter with thinnish shell, elongate cephalis, with a neck, similar width of cephalis and thorax, thinner wall, smaller pores, with lateral wings.

399. *Lophophaena cyclocapita* sp. nov.

(Pl. 112, Figs. 25–40)

Individuals larger, shell nearly cylindrical or campanulate, slightly thick walled, a deep constriction or neck between cephalis and thorax; cephalis larger, normally spherosome, pores quasi-circular, similar sizes, regularly or sub-regularly arranged, with hexagonal frames, some specimens obscured pores and bars; thorax cylindrical, or slightly expanded toward mouth, pores quasi-circular, similar or different sizes, sub-regularly or irregularly arranged, some specimens with hexagonal frames, and some specimens with wider and thinner bars, surface smooth or slightly rough for thorny, no spine; no apical horn and lateral wings, opening mouth.

Measurements: Shell length 170–215μm, diamter of cephalis 85–115μm, thorax length 85–125μm, width 125–160μm.

Holotype: DSDP274-33-1 1 deposited in the South China Sea Institute, CAS, from sample of DSDP274-33-1 in the Ross Sea, pictured in pl. 112, figs. 25–26.

Distribution: Late Eocene–middle Oligocene in Ross Sea, Antarctic.

This new species is distinguished by its separated cephalis and thorax, with a neck, no spine on surface, no apical horn and lateral wings from other species like *Psilomelissa tricuspidata* Popofsky (1908, p. 284, pl. 32, fig. 9, pl. 33, fig. 8), *Sethoconus tabulatus* (Ehrenberg) (Cleve, 1899, pl. 4, fig. 2), *Antarctissa longa* (Popofsky) (Chen, 1975, p. 457, pl. 17, figs. 6–8) and *Antarctissa strelkovi* Petrushevskaya (Lazarus, 1990, p. 713, pl. 3, figs. 13–15).

401. *Lophophaena olivacea* sp. nov.

(Pl. 113, Figs.15–17)

Shell olive-shaped, no constriction between two segments; cephalis hemispheroidal, simple, slightly thick walled, pores smaller, quasi-circular, similar sizes, with hexagonal frames, diameter of pores equaling to width of bars; thorax with shape of waist drum, expanded at middle part, shrinked toward mouth, width of mouth about half of shell, no peristoma, pores quasi-circular, most pores with similar sizes, several smaller, regularly or sub-regularly arranged, about 7–8 longitudinal rows of pores on one side of thorax; a thin layer of spongy tissue likely covering all the cephalis and upper thorax, no surface thorny, apical horn and lateral wings.

Measurements: Cephalis length 60–65μm, width 90–95μm, thorax length 200–210μm, width 175–180μm, mouth width 75–85μm.

Holotype: DSDP274-24-3 2 deposited in the South China Sea Institute, CAS, from sample of DSDP274-24-3 in the Ross Sea, pictured in pl. 113, figs. 15–17.

Distribution: Middle Oligocene in Ross Sea, Antarctic.

Shell shape and structure of this new species is similar to *Lophophaena spongiosa* (Petrushevskaya), but it is expanded at middle part and shrinked underpart; distinctions between them mainly are this new species with lattice shell, pores quasi-circular, similar sizes, nearly regular distributions, with hexagonal or polygonal frames, no apical horn and wings.

402. *Lophophaena*? *orbicularis* sp. nov.

(Pl. 113, Figs. 18–29)

Lophophaena ? *capito* Ehrenberg group, Petrushevskaya and Kozlova, 1972, p. 535, pl. 33, figs. 20–23.

Individuals small, cephalis nearly spherical, with several conic apical horns, very small, constricted near thorax, pores smaller, quasi-circular or elliptical, different sizes, irregularly distributed, bars wider, plate-shape, smooth surface; thorax undeveloped or very short, narrowed, with certain structure changes for different specimens; no lateral wings and feet.

Measurements: Cephalis length 70–125μm, width 90–125μm, thorax length 0–30μm, mouth width 40–75μm.

Holotype: DSDP274-31-2 2 deposited in the South China Sea Institute, CAS, from sample of DSDP274-31-2 in the Ross Sea, pictured in pl. 113, figs. 20–21.

Distribution: Early Eocene–late Pleistocene in Ross Sea, Antarctic.

This new species is distinguished from *Lophophaena capito* Ehrenberg by the latter with obvious thorax, which is slightly extended, with solid lateral wings. Characters of this new species are different with genus *Lophophaena* and *Lithomelissa* for its undeveloped thorax, lattice wall, smooth surface, not thorny.

Genus *Anthocyrtium* Haeckel, 1887

409. *Anthocyrtium quinquepedium* sp. nov.

(Pl. 123, Figs. 16–17)

Shell campanulate, very small; cephalis nearly spherical, partly encompassed by thoracic wall, thin walled, with a apical horn of triangular prism, shorter, shallower constriction between cephalis and thorax; thorax campanulate, slightly thick walled, gradually enlarged from top to bottom, opening mouth, 5–6 terminal feet from peristoma grown downward, all as coniform, relatively short; all the shell pores very small, quasi-circular, sub-regularly distributed, with wider bars, smooth surface.

Measurements: Shell length 85μm, diameter of cephalis 25μm, thorax length 70μm, width 65μm, length of apical horn 15μm, length of terminal feet 15–25μm.

Holotype: DSDP274-32-1 2 deposited in the South China Sea Institute, CAS, from sample of DSDP274-32-1 in the Ross Sea, pictured in pl. 123, figs. 16–17.

Distribution: Late Eocene in Ross Sea, Antarctic.

This new species is similar to *Anthocyrtium reticulatum* Haeckel (1887, p. 1274), but the latter with a longer apical horn and lateral spines, much more terminal feet (more than 12 feet), pores of irregular polygon, thorny surface.

Genus *Corythomelissa* Campbell, 1951

411. *Corythomelissa corysta* sp. nov.

(Pl. 124, Figs. 1–2)

Corythomelissa sp., Petrushevskaya, 1975, p. 590, pl. 11, fig. 11.

Shell helmet shaped, slightly thick walled; larger cephalis, nearly hemispherical, apex horn very small or undeveloped, conical, pores quasi-circular, a little different sizes, generally smaller, irregularly distributed, with coarser bars, or some bars overlapped, approximate spongy structure; main central bar of inner cephalis grown upward, forming apex horn, other lateral bars of inner cephalis extended slantly, forming thoracic ribs and lateral feet; thorax relatively short, broad opening, or under developed, pores sub-regularly distributed, 2–3 rows crosswise, three lateral feet of triangular pyramid grown obliquely (partly broken off), feet fenestrated; with spinous process on bar joints, rough surface.

Measurements: Cephalis length 125μm, width 125μm, thorax length 45μm, width 145μm, length of lateral feet >40μm.

Holotype: DSDP274-29-1 1 deposited in the South China Sea Institute, CAS, from sample of DSDP274-29-1 in the Ross Sea, pictured in pl. 124, figs. 1–2.

Distribution: Paleocene–early Oligocene in Ross Sea, Antarctic.

This new species is similar to *Spongomelissa adunca* Sanfilippo and Riedel, but the latter with a smaller cephalis, which is lobulated (2–3), with a stout apex horn, deep constriction between cephalis and thorax, thorax wider, completely developed. Specimen's feature of this new species is basically the same with *Corythomelissa* sp. (Petrushevskaya, 1975, p. 590, pl. 11, fig. 11), which was reported as unnamed species, appearing in Paleocene stratum of Antarctic Seas.

Genus *Ceratocyrtis* Bütschli, 1882

415. *Ceratocyrtis cylindris* sp. nov.

(Pl. 125, Figs. 1–4)

Shell campanulate or cylindrical, slightly thick walled; cephalis spherical, partly fallen in to thorax, with deeper constriction, 3–4 short apical horns of thin columnar form, pores quasi-circular; thorax nearly cylindrical, shoulders flat, or slightly oblique, with similar width from top to bottom, or narrowed for upper part, pores quasi-circular or elliptical, different sizes, sub-regularly or irregularly arranged, thinner bars; many calthrops or spinous slices on surface, with rather rough surface; thoracic mouth basically flattened, irregular form, no peristoma ring or tooth.

Measurements: Shell length 135–145μm, width 100–110μm, diameter of thorax 45–50μm.

Holotype: DSDP274-10-5 1 deposited in the South China Sea Institute, CAS, from sample of DSDP274-10-5 in the Ross Sea, pictured in pl. 125, figs. 1–2.

Distribution: Early Pliocene in Ross Sea, Antarctic.

This new species is similar to *Ceratocyrtis manumi* Goll and Bjørklund (1989, p. 730, pl. 5, figs. 21–23), main differences are the latter with a broader thorax (more fat), shoulders of declivity, slightly constricted for middle part of shell, and obviously enlarged lower part, the widest position of shell is at the opening mouth, while width of this new species is invariant for its middle-lower parts.

416. *Ceratocyrtis irregularis* sp. nov.

(Pl. 125, Figs. 5–39)

Shell irregularly campanulate, relatively small, moderately thick walled; cephalis shape of brain, with 3–4 or more locellus, stacking, or arranged as mound, flattening strip and irregular form, some located at thoracic top and a deeper constriction with thorax, but most of them of broad flat shape, nearly cladding thoracic top, even extending to periphery of thoracic upper part, causing unclear boundary between cephalis and thorax; 1–3 head spines, very small and short, often laterally growing; thorax campanulate or cylindrical, pores quasi-circular or irregular shape, different sizes, irregularly arranged, different width of bars for different specimens, thoracic mouth opening, nearly flattened, or incomplete, some small terminal teeth formed by bars; smooth surface, no spine of thorax.

Measurements: Shell length 105–150μm, width 95–115μm, cephalis length 25–50μm, width 45–95μm.

Holotype: DSDP274-26-5 2 deposited in the South China Sea Institute, CAS, from sample of DSDP274-26-5 in the Ross Sea, pictured in pl. 125, figs. 5–6.

Distribution: Late Eocene–middle Oligocene in Ross Sea, Antarctic.

This new species is distinguished from other species mainly by it with cephalis of more irregular locellus.

419. *Ceratocyrtis perimashae* sp. nov.

(Pl. 127, Figs. 1–9)

Shell campanulate, relatively larger, thick walled; cephalis spheroidal, very small, lower part caughted in thorax, with deep constriction, pores round, tiny; thorax campanulate or coniform, gradually enlarged from top to bottom, mouth flattened, pores quasi-circular or elliptical, different sizes, irregularly arranged, with hexagonal or polygonal frames, diameter of pores about 1–2 times width of bars; longer apophysis of calthrops arisen from various joints on shell surface, with lateral branches at the ends of calthrops, presented as filamentous, and connected each other, forming secondary structure of arachnoid outside shell, which made up a thinner peripheral layer; rough surface.

Measurements: Shell length 160–245μm, width 130–175μm, diameter of cephalis 30–50μm, distence between shell and peripheral layer 10–25μm.

Holotype: DSDP274-32-1 2 deposited in the South China Sea Institute, CAS, from sample of DSDP274-32-1 in the Ross Sea, pictured in pl. 127, figs. 1–2.

Distribution: Late Eocene–late Oligocene in Ross Sea, Antarctic.

This new species is similar to *Ceratocyrtis mashae* Bjørklund, main differences are the former with a secondary construction of peripheral layer, but the latter with bulges as coniform on joints of shell surface, without terminal lateral spine and peripheral layer, and with thinner bars between pores.

Genus *Dictyocephalus* Ehrenberg, 1860

425. *Dictyocephalus turritus* sp. nov.

(Pl. 128, Figs. 1–15)

Shell smaller, coniform or spire shaped, medium or thick walled; cephalis of brain form, with a slightly larger spherical locellus, partly fallen into interior thorax, not obvious constriction, no apex spine; thorax turriform, gradually wider from top to bottom, the widest shell at mouth, with quasi-circular pores, similar sizes, regularly or sub-regularly arranged, diameter of pores nearly equal width of bars, broad mouth flattened, terminal saw toothing, or with a thin peristoma ring; smooth surface.

Measurements: Shell length 105–120μm, width 75–100μm, cephalis length 25–30μm, width 30–50μm.

Holotype: DSDP274-22-3 2 deposited in the South China Sea Institute, CAS, from sample of DSDP274-22-3 in the Ross Sea, pictured in pl. 128, figs. 1–2.

Distribution: Late Eocene–middle Oligocene in Ross Sea, Antarctic.

This new species is similar to *Dictyocephalus variabilus* sp. nov., main differences are the latter with the widest point at middle shell, lower part or mouth constricted, and with an obvious peristoma ring of lucency.

426. *Dictyocephalus variabilus* sp. nov.

(Pl. 128, Figs. 16–51)

Phormostichoartus fistula Nigrini, Hollis, 1997, pl. 4, figs. 5–7.
Theocampe sp., Ling, 1991b, p. 320, pl. 1, fig. 9.

Shell of two segments, nearly lageniform, middle inflated, lower part slightly constricted, with certainly variable features of shell and pore forms for different specimens; cephalis subglobose or hemispherical, partly caught in intrathoracic, the maximum width of shell at upside near 1/3, gradually constricted downward, mouth flattened; medium thick walled, pores small and many, quasi-circular, similar sizes, regularly or sub-regularly distributed, about 7–9 pores crosswise at the maximum width of shell; smooth surface, with generally hyaline peristoma.

Measurements: Shell length 125–170μm, width 70–105μm, diameter of cephalis 25–45μm.

Holotype: DSDP274-25-1 1 deposited in the South China Sea Institute, CAS, from sample of DSDP274-25-1 in the Ross Sea, pictured in pl. 128, figs. 16–17.

Distribution: Middle Eocene–late Oligocene in Ross Sea, Antarctic.

This new species is similar to specimens of Hollis (1997, pl. 4, figs. 5–7), both shells with only two segments, and no interior lantern ring in secondary segment. Differences of this new species with *Phormostichoartus fistula* Nigrini (1977, p. 253, pl. 1, figs. 11–13) are mainly the latter with shell of 3–4 segments, obviously interior lantern ring, a lateral tube on head, regular shell form, pores larger and less, regularly distributed.

Genus *Carpocanium* Ehrenberg, 1847

429. *Carpocanium spongiforma* sp. nov.

(Pl. 129, Figs. 5–12)

Shell nearly ellipsoidal, no constricting trace on surface, thicker walled, like spongiform; cephalis brain shaped, relatively large, clad in thorax completely; thorax of regular ellipsoid, pores very small, many, quasi-circular, similar sizes of pores and bars, irregularly distributed, even disperse, some small bulges on bars, causing rough surface; shell stenosed toward bottom, mouth flattened, width of mouth about 1/3–1/2 of the shell, some small teeth on peristoma, seemingly rudiment of terminal feet.

Measurements: Shell length 140–175μm, width 90–135μm, cephalis length 38–60μm, width 60–85μm, width of mouth 40–50μm.

Holotype: DSDP274-21-1 2 deposited in the South China Sea Institute, CAS, from sample of DSDP274-21-1 in the Ross Sea, pictured in pl. 129, figs. 5–6.

Distribution: Late Eocene–middle Oligocene in Ross Sea, Antarctic.

This new species is distinguished from other species by it with likeness of spongy wall and larger clad cephalis.

Genus *Carpoglobatus* gen. nov.

Shell is near a global shape, with three segments (cephalis, thorax and abdomen) divided by two interior lantern rings, no constriction on surface, a tube of mouth at the end of abdomen.

Type species: *Carpoglobatus tubulus* sp. nov.

431. *Carpoglobatus tubulus* sp. nov.

(Pl. 129, Figs. 21–22)

Individuals rather larger, ellipsoidal, shell of three segments, no shrink of surface between segments, with two interior lantern rings that separate cephalis, thorax and abdomen, usually the lantern rings curved or inclined; cephalis of simple structure, nearly hemispherical, thin walled, pores quasi-circular or elliptic, different sizes, wider bars; shell of thorax and abdom combined together completely, interior lantern ring of them tilted, thicker walled, pores quasi-circular, with similar sizes, sub-regularly distributed, diameter of pores 2–3 times as broad as bars; mouth obviously constricted, about 1/3 width of shell, the peristoma extended outside as a tube, which length is about 3/4 of width, tube wall thinner, transparent, with some smaller pores, tube opening uneven, no tooth of peristoma; smooth surface.

Measurements: Shell length 230μm, width 175μm, cephalis length 55μm, width 115μm, tube length 45μm, width 55μm.

Holotype: DSDP274-12-5 1 deposited in the South China Sea Institute, CAS, from sample of DSDP274-12-5 in the Ross Sea, pictured in pl. 129, figs. 21–22.

Distribution: Middle Miocene in Ross Sea, Antarctic.

This new species is distinguished from other species of *Carpocanium* by the latter with mouth peristoma, not developed as a tube, with terminal feet, cephalis of complex structure inside thorax.

Family Podocyrtidae Haeckel, 1887

Genus *Pseudodictyophimus* Petrushevskaya, 1971

438. *Pseudodictyophimus gigantospinus* sp. nov.

(Pl. 132, Figs. 1–2)

Larger individuals, two segments, nearly cylindrical, tops of each segment horizontal, vertically joined with side face, two segments enlarged stepwise; cephalis nearly hemispherical, lower part slightly wider, pores quasi-circular, larger difference of pore sizes, smaller pores in upper part, larger pores in lower part, with thinner bars; thorax suddenly magnified, with horizonal shoulders, slightly expanded downward, pores elliptical or irregularly, diameter of pores 1–5 times as broad as bars, pore sizes differ greatly, thinner bars, with broader mouth, no peristoma; shell with a apical horn, a lateral spine and three terminal feet, they are all triangular pyramid, very strong and long; some spines of triangular slice on bars, rough surface.

Measurements: Shell length 180μm, width 175μm, cephalis length 85μm, width 140μm, length of apical horn 165μm, lateral spine 145μm, terminal feet 65–120μm.

Holotype: DSDP274-10-1 1 deposited in the South China Sea Institute, CAS, from sample of DSDP274-10-1 in the Ross Sea, pictured in pl. 132, figs. 1–2.

Distribution: Early Pliocene in Ross Sea, Antarctic.

This new species is similar to *Pseudodictyophimus gracilipes* (Bailey), but the latter with a smaller cephalis, shorter apical horn, no lateral spine, more pores and smooth surface.

441. *Pseudodictyophimus tenellus* sp. nov.

(Pl. 133, Figs. 5–6)

Individuals relatively smaller, cephalis hemispherical, thorax campanulate, with deep constriction, apical horn short and small, thin columnar; whole shell thin walled, with larger pores, circular, elliptical or quadrate, sizes of larger difference, irregularly distributed, thinner bars, diameter of pores 2–6 times as broad as bars, less pores of thorax; three terminal feet relatively small and short, thin columnar, terminal sharp; thoracic mouth incomplete, some spines on surface.

Measurements: Shell length 110μm, width 90μm, cephalis length 30μm, width 45μm, length of apical horn 10μm, terminal feet 30–50μm.

Holotype: DSDP274-7-1 1 deposited in the South China Sea Institute, CAS, from sample of DSDP274-7-1 in the Ross Sea, pictured in pl. 133, figs. 5–6.

Distribution: Late Pliocene in Ross Sea, Antarctic.

Features of this new species is closed to *Pseudodictyophimus gracilipes* (Bailey), but they have evident differences. The latter is thicker walled, with more pores of similar size, thorax is often inflated between lateral ribs, terminal feet and apical horn are relatively stouter, triangular pyramid.

Family Phormocyrtidae Haeckel, 1887

Genus *Cycladophora* Ehrenberg, 1847, emend. Lombari and Lazarus, 1988

448. *Cycladophora tinocampanula* sp. nov.

(Pl. 135, Figs. 1–5)

Shell of two segments; cephalis spherical, smaller, with a apical horn and a lateral spine, short and small, thin columnar, lower part of cephalis slightly fallen into thorax; top of thorax narrowed, gradually increasing until middle-upper part, the lower part with invariant width, presented as cylindrical; pores of cephalis small and sparse, pores of thorax small and dense at top, slightly largen in upper part, pores of middle-lower part quasi-circular, similar sizes, regularly arranged crosswise, with hexagonal frames of bars, thinner bars, diameter of pores about 3–4 times as broad as bars; thoracic mouth opening, flattened, no peristoma; some very small spines on bar joints, slightly rough surface.

Measurements: Shell length 155–175μm, width 100–120μm, diameter of cephalis 25–30μm, length of apical horn 12–20μm.

Holotype: DSDP274-6-3 2 deposited in the South China Sea Institute, CAS, from sample of DSDP274-6-3 in the Ross Sea, pictured in pl. 135, figs. 1–2.

Distribution: Early–late Pliocene in Ross Sea, Antarctic.

Features of this new species is some similar to *Lophocyrtis golli* Chen, main differences are the latter shell of three segments, with a dilated abdominal segment, pores of abdomen is smaller than that of thorax, irregularly arranged.

Family Theocyrtidae Haeckel, 1887, emend. Nigrini, 1967

Genus *Gondwanaria* Petrushevskaya, 1975

459. *Gondwanaria pteroforma* sp. nov.

(Pl. 140, Figs. 1–12)

Gondwanaria japonica (Nakaseko) group, Petrushevskaya, 1975, pl. 21, figs. 4–5.

Shell of two segments, cephalis relatively larger, hemispherical, width of cephalis about 1/2 of thorax, with a apical horn of triangular pyramidal, terminal sharp, length of horn slightly less than length of cephalis; thorax nearly cylindrical, top arc form, with a deep constriction between thorax and cephalis, three lateral wings of triangular pyramid, very strong, base of wings relatively wider, tail end sharp; thoracic mouth opening, generally incomplete, no peristoma; shell pores quasi-circular or elliptical, irregularly distributed, some rises on bars, rough surface.

Measurements: Shell length 125–180μm, width 120–150μm, cephalis length 50–75μm, width 45–85μm, length of apical horn 25–50μm, lateral wing length 45–75μm, base width 25–40μm.

Holotype: DSDP274-29-1 2 deposited in the South China Sea Institute, CAS, from sample of DSDP274-29-1 in the Ross Sea, pictured in pl. 140, figs. 1–2.

Distribution: Late Eocene in Ross Sea, Antarctic.

Features of this new species is different from *Sethocyrtis japonica* Nakaseko by the latter with smaller individuals, thorax nearly spheroidal, mouth coarctated, apical horn undeveloped and generally no lateral wings. This new species is very similar to specimen of Petrushevskaya

(1975, pl. 21, figs. 4–5), which was classified as an atypical type of *Gondwanaria japonica* (Nakaseko) by Petrushevskaya (1975), obviously they have great characteristic difference.

464. *Gondwanaria tenuoria* sp. nov.

(Pl. 142, Figs. 19–28)

Lophophaena (?) sp., Petrushevskaya, 1975, pl. 9, fig. 13.
Gen. and sp. indet., Chen, 1975, pl. 19, fig. 5.

Shell thin and small, nearly calabash-shaped, three segments, thin walled; cephalis spherical, central bar of interior cephalis developed as an apical horn of thin column, pores very small; thorax conoidal, lower part slightly constricted, deep constriction between thorax and abdomen, pores small, quasi-circular, regularly arranged; abdomen cylindrical or variable, slightly broadened toward mouth, pores quasi-circular, similar sizes, larger than pores of thorax, generally regularly arranged, with hexagonal frames, diameter of pores 2–3 times as broad as bars, thinner bars; some small spines on surface, broad mouth incomplete.

Measurements: Shell length 145–160μm, width 75–90μm, diameter of cephalis 25–35μm, thorax length 50–60μm, width 65–70μm, length of apical horn 5–15μm.

Holotype: DSDP274-23-1 1 deposited in the South China Sea Institute, CAS, from sample of DSDP274-23-1 in the Ross Sea, pictured in pl. 142, figs. 19–20.

Distribution: Early-middle Oligocene in Ross Sea, Antarctic.

This new species is similar to *Gondwanaria* cf. *semipolita* (Clark and Campbell), but the latter with larger individuals, thicker walled, no apical horn, smooth surface and no spine; Specimens from Ross Sea are very similar to Gen. and sp. indet. (Chen, 1975, pl. 19, fig. 5), only difference is the latter with several terminal feet on peristoma, probably growing more completely.

Family Podocampidae Haeckel, 1887

Genus *Cyrtopera* Haeckel, 1881

470. *Cyrtopera magnifica* sp. nov.

(Pl. 144, Figs. 1–2)

Shell long conoidal or steepled, relatively larger, with at least 6 segments, similar height for each segment, gradually increased width of shell; cephalis very small, spherical, encompassed by the base of apical horn completely; apical horn very long, wider at its base, cylinder-shaped or long conoidal, terminal sharp; cephalis and abdomen with mutual more than 5 segments, deeper constrictions between every two segments, thin interior lantern ring, all pores quasi-circular, similar sizes, sub-regularly arranged, slightly increasing pore sizes from top to bottom; opening broad, perhaps lacked the last segment; no spine on surface, smoothly, lacking lateral wing.

Measurements: Shell length >300μm, width >170μm, diameter of cephalis 20μm, thorax length 50–60μm, width 65–70μm, length of apical horn 115μm.

Holotype: DSDP274-33-3 2 deposited in the South China Sea Institute, CAS, from sample of DSDP274-33-3 in the Ross Sea, pictured in pl. 144, figs. 1–2.

Distribution: Late Eocene in Ross Sea, Antarctic.

This new species is similar to *Cyrtopera laguncula* Haeckel, but the latter with smaller individuals, cephalis exposed, apical horn shorter and smaller, both pores and frames hexagonal, regularly arranged, thinner bars.

Family Lithocampidae Haeckel, 1887

Genus *Artostrobus* Haeckel, 1887

474. *Artostrobus multiartus* sp. nov.

(Pl. 145, Figs. 31–48)

Lithomitra sp., Bjørklund, 1976, p. 1124, pl. 15, figs. 26, 28(only), pl. 23, figs. 15–21.

Individuals very small, cylindrical, or slightly inflated at middle part, thicker walled; cephalis brain shaped, nearly hemispherical, transparent, no pore; thorax shorter, form of waist drum, clear lantern ring between cephalis and thorax, with deep constriction; abdomen longer, nearly cylindrical, respectively narrowed lower part, more than 3 segments of abdomen, with interior lantern rings, but no constriction on surface; all pores quasi-circular, similar sizes, regularly arranged crosswise, smooth surface; opening mouth, flattened, no peristoma.

Measurements: Shell length 70–95μm, width 30–40μm, diameter of cephalis 12–20μm.

Holotype: DSDP274-42-2 1 deposited in the South China Sea Institute, CAS, from sample of DSDP274-42-2 in the Ross Sea, pictured in pl. 145, figs. 31–32.

Distribution: Early Paleocene–late Eocene in Ross Sea, Antarctic.

This new species is similar to *Artostrobus pachyderma* (Ehrenberg), but the latter with very thick shell wall, the thickest at middle shell, the wall coated both thorax and abdomen continuously, no constriction, smooth surface. Differences of this new species and *Artostrobus missilis* O'Connor are the latter with similar thorax and abdomen, they are difficult to be divided, between them no obvious constriction and lantern ring.

Genus *Eucyrtidium* Ehrenberg, 1847, emend. Nigrini, 1967

479. *Eucyrtidium anomurum* sp. nov.

(Pl. 146, Figs. 37–42)

Eucyrtidium sp. A, Petrushevskaya, 1975, p. 581, pl. 14, figs. 21–22.

Individuals larger, shell of 5–8 segments or more, abnormally grown, some as distorted shape, interior lantern rings curved irregularly or nearly spiral; cephalis spherical, very small, a interior bar may extended as small apical spine; thorax conoidal, with less width than abdomen, deeper constriction; abdomen of multisection, each segment with different form, variable width, deep constrictions between every two segments; all pores quasi-circular, similar sizes, sub-regularly arranged lengthways or irregularly distributed; opening mouth incomplete, slightly rough surface.

Measurements: Shell length 180–300μm, width 105–120μm, diameter of cephalis 20–30μm, thorax length 35–50μm, width 75–95μm, apical spine 5–20μm.

Holotype: DSDP274-33-3 1 deposited in the South China Sea Institute, CAS, from sample of DSDP274-33-3 1 in the Ross Sea, pictured in pl. 146, figs. 39–40.

Distribution: Late Eocene–early Oligocene in Ross Sea, Antarctic.

This new species is similar to *Eucyrtidium antiquum* Caulet, but the latter with regularly form, no shell distorted, similar shape and size for various segments of abdomen, all interior lantern rings horizonal and mutual parallel.

482. *Eucyrtidium gelidium* sp. nov.

(Pl. 147, Figs. 21–24)

Shell fat bullet shaped, enlarged from top to bottom, thicker walled; cephalis spherical, very small, with a small apical horn or rise; thorax conoidal, abdomen nearly cylindrical, 2–3 segments, gradually increasing width, unconspicuous constriction between segments, with thinner lantern rings; pores quasi-circular or elliptical, different sizes, irregularly distributed, with wider bars; seemly some rises of conic or triangular slice on surface, and sticking several dendritic sundries; mouth opening, not shrinked, incomplete margin, slightly rough surface.

Measurements: Shell length 140–150μm, width 110–125μm, diameter of cephalis 15–20μm, thorax length 30–55μm, width 50–75μm, apical horn 5–15μm.

Holotype: DSDP274-25-1 2 deposited in the South China Sea Institute, CAS, from sample of DSDP274-25-1 in the Ross Sea, pictured in pl. 147, figs. 21–22.

Distribution: Middle Oligocene in Ross Sea, Antarctic.

Features of this new species is similar to *Eucyrtidium annulatum* (Popofsky), but the latter with thinner wall, smaller pores, similar sizes, regularly arranged, and smooth surface.

Genus *Siphocampe* Haeckel, 1881

490. *Siphocampe mesinflatis* sp. nov.

(Pl. 151, Figs. 5–13)

Shell cylindrical, four segments; cephalis relatively larger, brain shaped, nearly hemispherical, apical horn very small, with a lateral tube; shallower constriction between cephalis and thorax, thorax cylindrical, slightly amplified downward; the first segment of abdomen inflated, shape of waist drum, arc side, with the widest segment of shell; the second segment of abdomen cylindrical, relatively narrowed, similar width with thorax, straight side; similar height for various segments of thorax and abdomen, clear interior lantern rings, deeper outside constrictions; all pores very small, hexagonal or quasi-circular, regularly arranged crosswise, 7–10 pores in a horizontal row on each segment; not shrinked mouth obviously, flattened, no peristoma ring.

Measurements: Shell length 105–130μm, width 55–75μm, cephalis length 20–25μm, width 40–45μm.

Holotype: DSDP274-1-3 1 deposited in the South China Sea Institute, CAS, from sample of DSDP274-1-3 in the Ross Sea, pictured in pl. 151, figs. 5–6.

Distribution: Middle Miocene–Holocene in Ross Sea, Antarctic.

Features of this new species is some similar to *Siphocampe corbula* (Harting), but the latter with spindle-shaped shell, different height for various segments, shallower constrictions, shrinked mouth, seems with a peristoma ring.

Genus *Spirocyrtis* Haeckel, 1881, emend.

493. *Spirocyrtis bellulis* sp. nov.

(Pl. 152, Figs. 1–4)

Spirocyrtis scalaris Haeckel, Petrushevskaya, 1971d, p. 211, figs. 126i, vii.

Individuals very small, shell of 3 segments, medium thick walled; cephalis brain shaped, nearly spherical, both apical horn and lateral tube very small; thorax conoidal, gradually enlarged downward, with deeper constriction between thorax and abdomen, horizontal; abdomen cylindrical, top of abdomen distinctly broaden comparing to thorax, length of abdomen about 1.3 times of thorax, mouth opening, no peristoma ring; all pores quasi-circular or hexagonal, similar sizes, relatively regularly arranged, with hexagonal frames, 5–7 rows of pores on each segment of thorax and abdomen respectively; smooth surface.

Measurements: Shell length 85–95μm, diameter of cephalis 15μm, thorax length 25–30μm, width 30–38μm, abdomen length 40–50μm, width 50–55μm.

Holotype: DSDP274-23-3 2 deposited in the South China Sea Institute, CAS, from sample of DSDP274-23-3 in the Ross Sea, pictured in pl. 152, figs. 1–2.

Distribution: Early-middle Oligocene in Ross Sea, Antarctic.

This new species is some similar to *Spirocyrtis scalaris* Haeckel, both with similar structure of cephalis, apical horn and lateral tube, main difference is the latter with spiral and oblique constrictions, not horizontal.

495. *Spirocyrtis rectangulis* sp. nov.

(Pl. 152, Figs. 9–11)

Individuals broader, shell of three segments, thin walled; cephalis larger, multi-lobulate of brain shaped, irregular, lower part fallen into thorax, with a small lateral horn; thorax conoidal at upper part, inflated at middle part, slightly narrowed and straight lower part, joined vertically with larger abdomen, with clear interior lantern ring; abdomen promptly enlarged width, cylindrical or rectangular, broad, top of abdomen horizontal and straight, orthogonal with its side, mouth flattened, not constricted; all pores hexagonal, similar sizes, regularly arranged, more than 8 rows of pores on thorax and abdomen respectively, with hexagonal frames, thinner bars, diameter of pores 2–3 times as broad as bars; smooth surface, no spine.

Measurements: Shell length 155μm, cephalis length 30μm (part falling in thorax), width 38μm, thorax length 75μm, width 100μm (middle point), abdomen length 65μm, width 130μm.

Holotype: DSDP274-30-6 2 deposited in the South China Sea Institute, CAS, from sample of DSDP274-30-6 in the Ross Sea, pictured in pl. 152, figs. 9–11.

Distribution: Late Eocene in Ross Sea, Antarctic.

This new species is some similar to *Spirocyrtis scalaris* Haeckel, but the latter with shell of turriform, gradually enlarged from top to bottom, with more spiral segments. Specimens of this new species from the Ross Sea is similar to that of Petrushevskaya (1971d, p. 211, figs. 126i–iv).

图版及说明

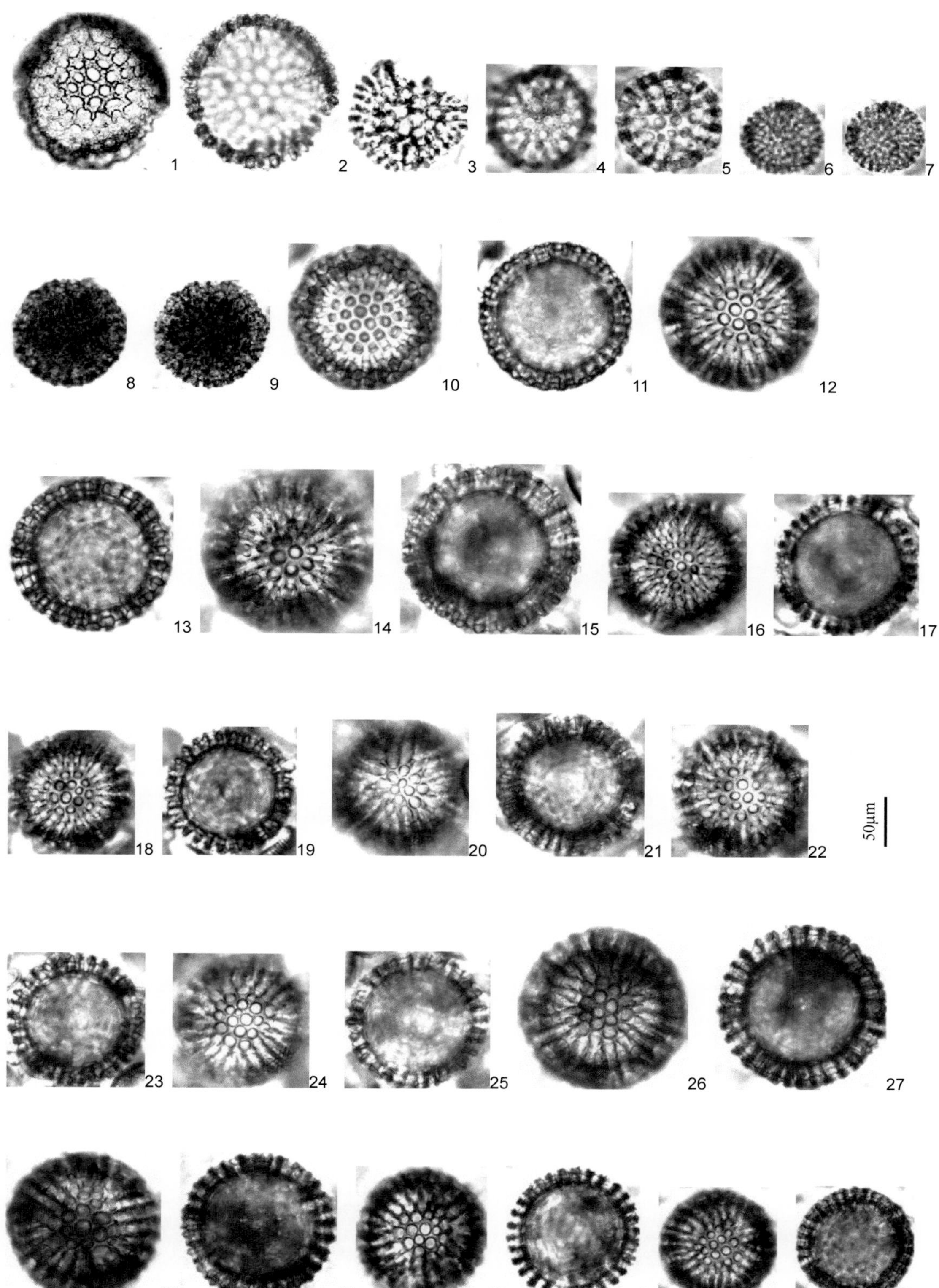

1–9. *Cenosphaera archantarctica* sp. nov.；10–33. *Cenosphaera compacta* Haeckel

图版 2

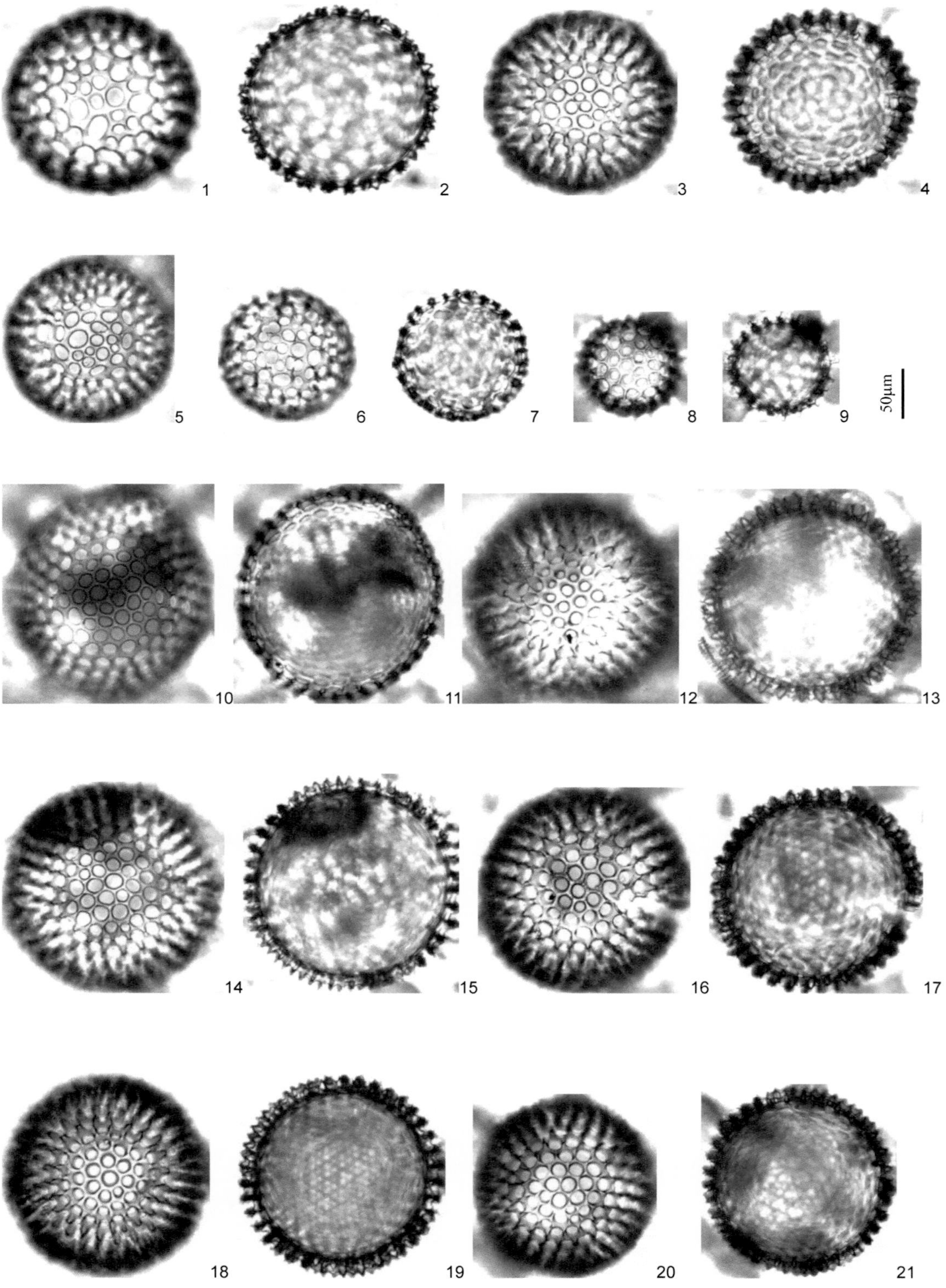

1–9. *Cenosphaera cristata* Haeckel；10–11. *Cenosphaera elysia* Haeckel；12–21. *Cenosphaera favosa* Haeckel

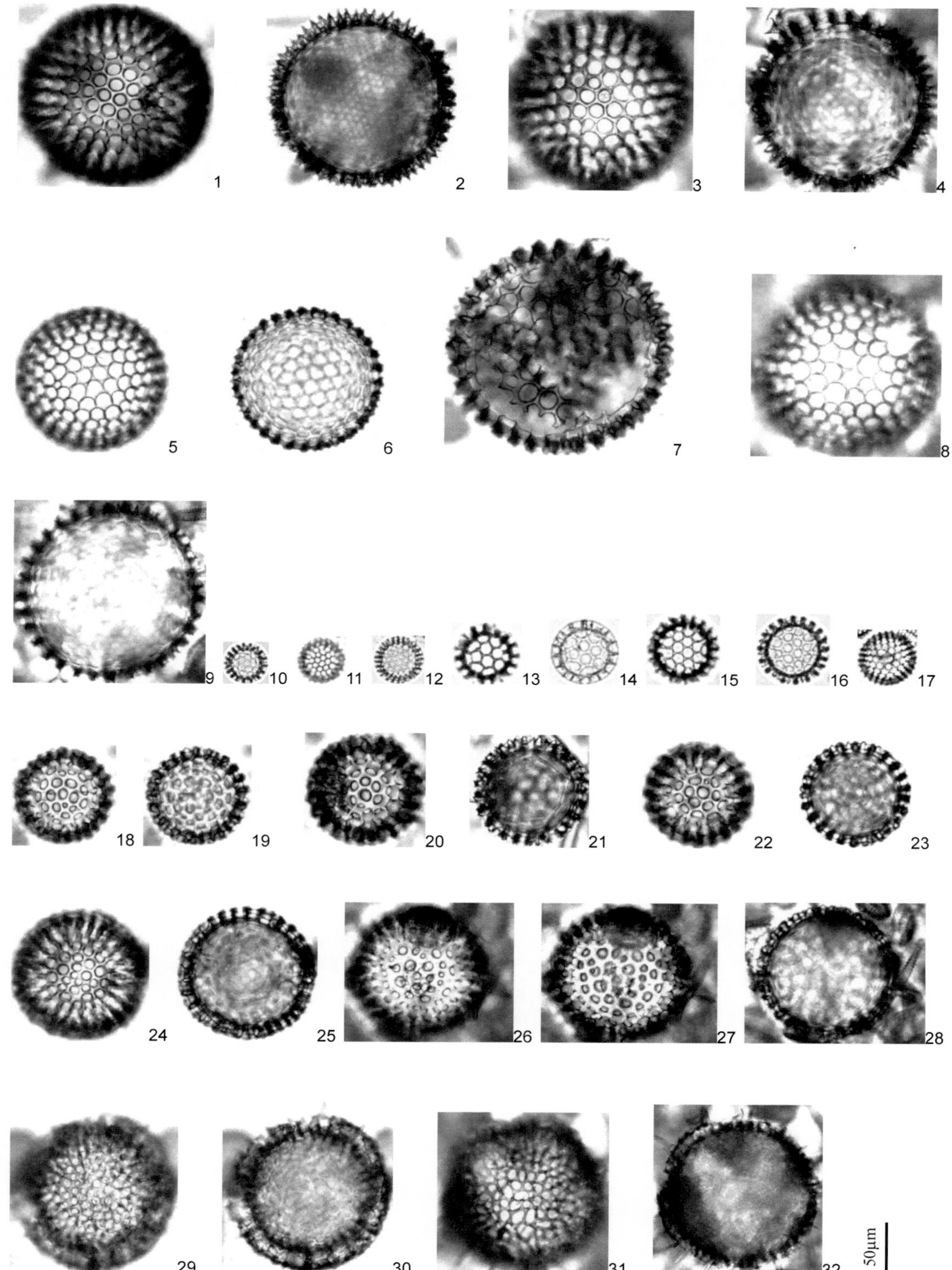

1–4. *Cenosphaera favosa* Haeckel; 5–9. *Cenosphaera mellifica* Haeckel; 10–17. *Cenosphaera miniantarctica* sp. nov.; 18–28. *Cenosphaera nagatai* Nakaseko; 29–32. *Cenosphaera pseudocoela* sp. nov.

图版 4

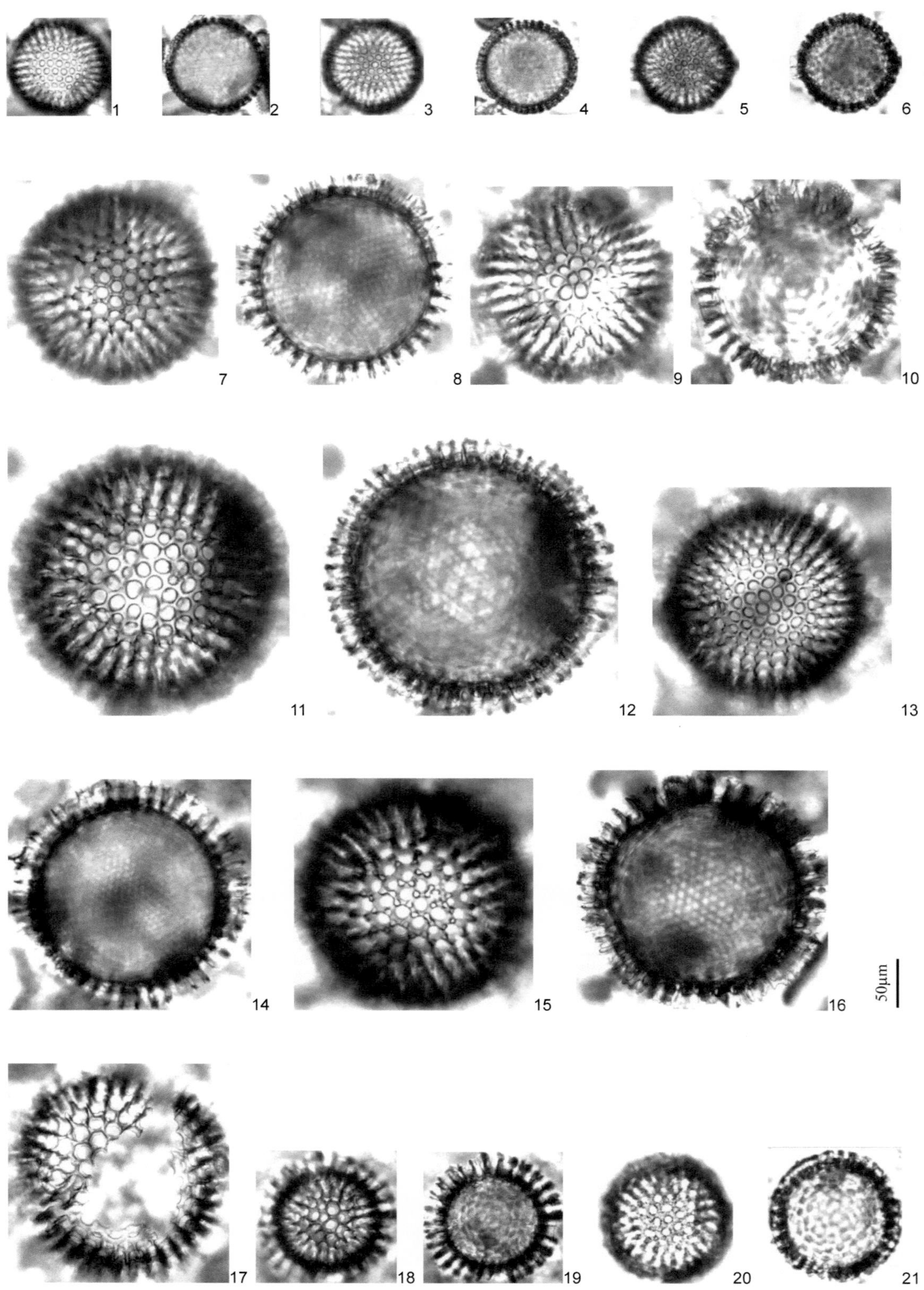

1–6. *Cenosphaera riedeli* Blueford；7–21. *Cenosphaera solantarctica* sp. nov.

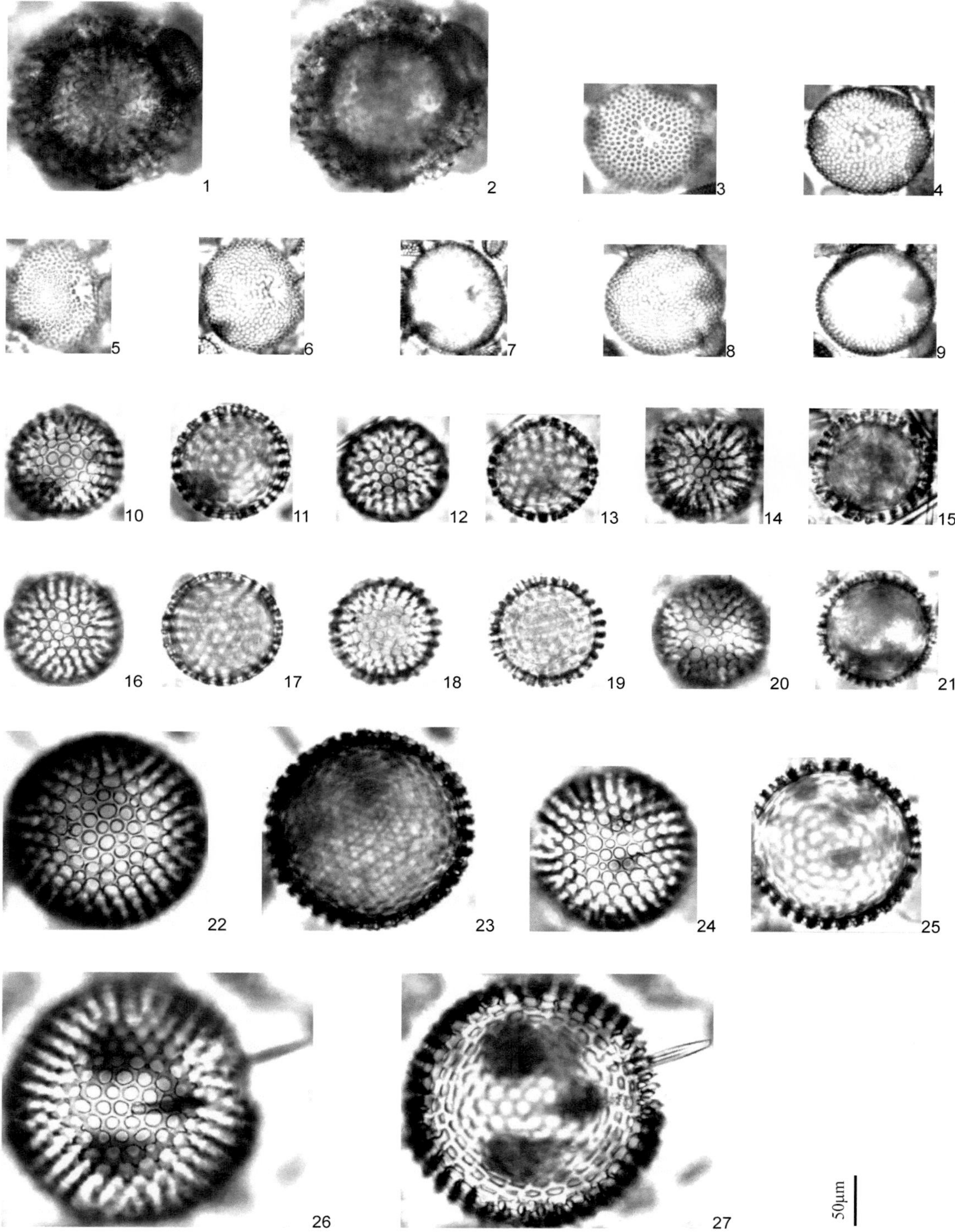

1–2. *Cenosphaera spongiformis* sp. nov.; 3–9. *Cenosphaera veneris* Clark and Campbell; 10–25. *Cenosphaera vesparia* Haeckel; 26–27. *Cenosphaera xiphacantha* sp. nov.

图版 6

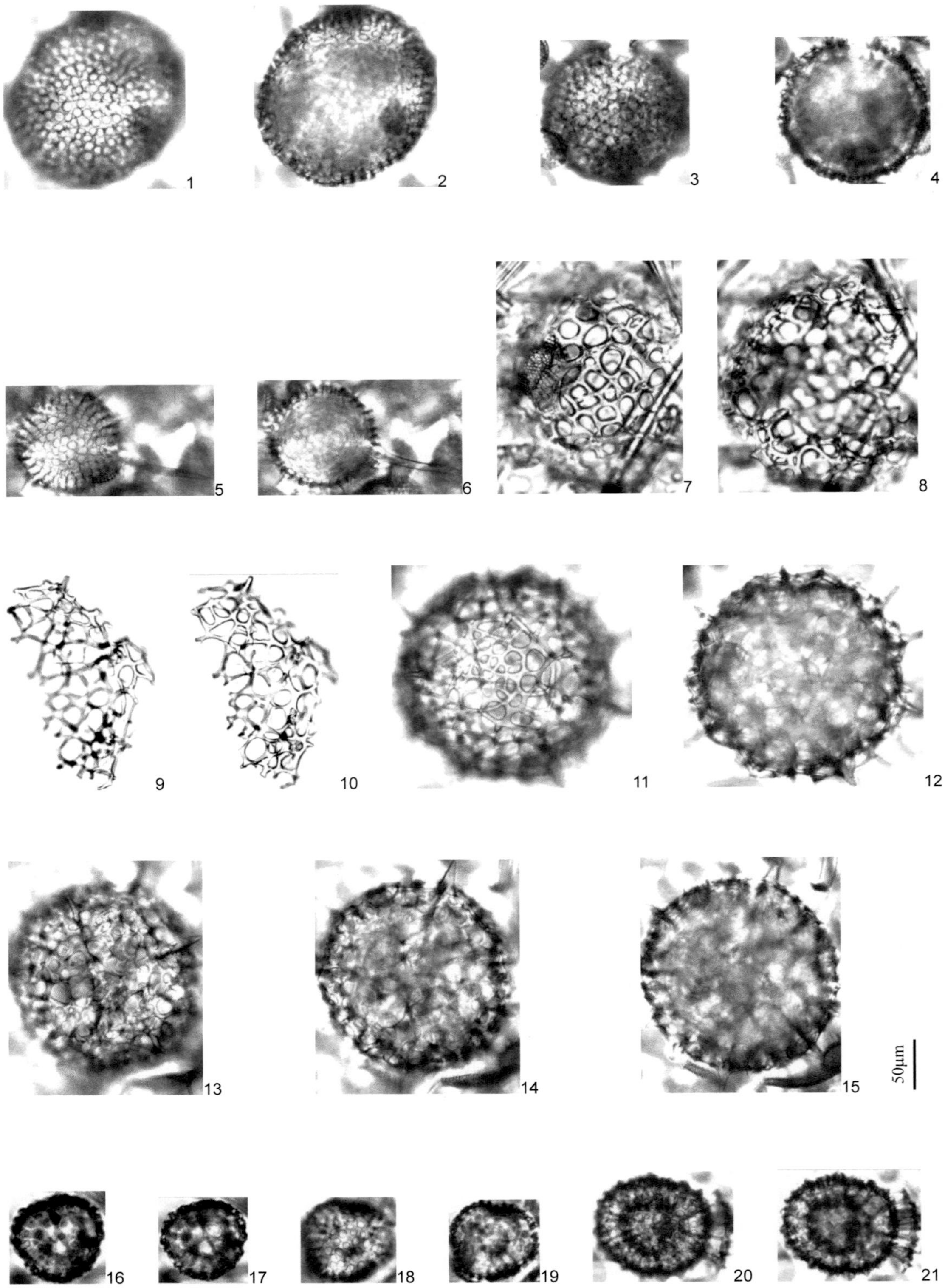

1–2. *Cenosphaera* sp. 1； 3–4. *Cenosphaera* sp. 2； 5–6. *Cenosphaera* sp. 3； 7–10. *Cenosphaera* sp. 4；
11–15. *Carposphaera angulata* (Haeckel)； 16–21. *Carposphaera anomala* sp. nov.

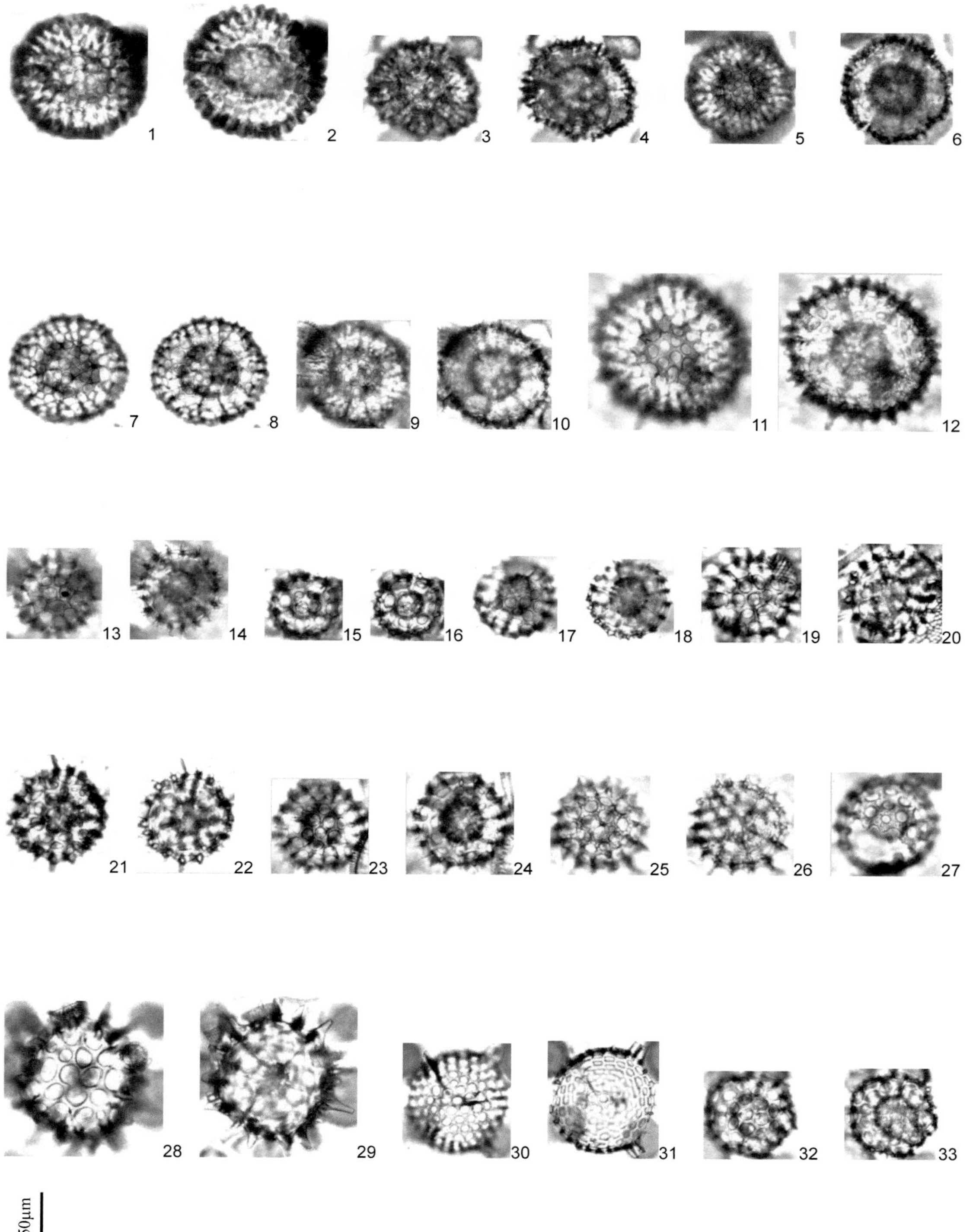

1–12. *Carposphaera globosa* Clark and Campbell；13–27. *Carposphaera magnaporulosa* Clark and Campbell；28–29. *Carposphaera sterrmona* sp. nov.；30–31. *Carposphaera* sp. 1；32–33. *Carposphaera* sp. 2

图版 8

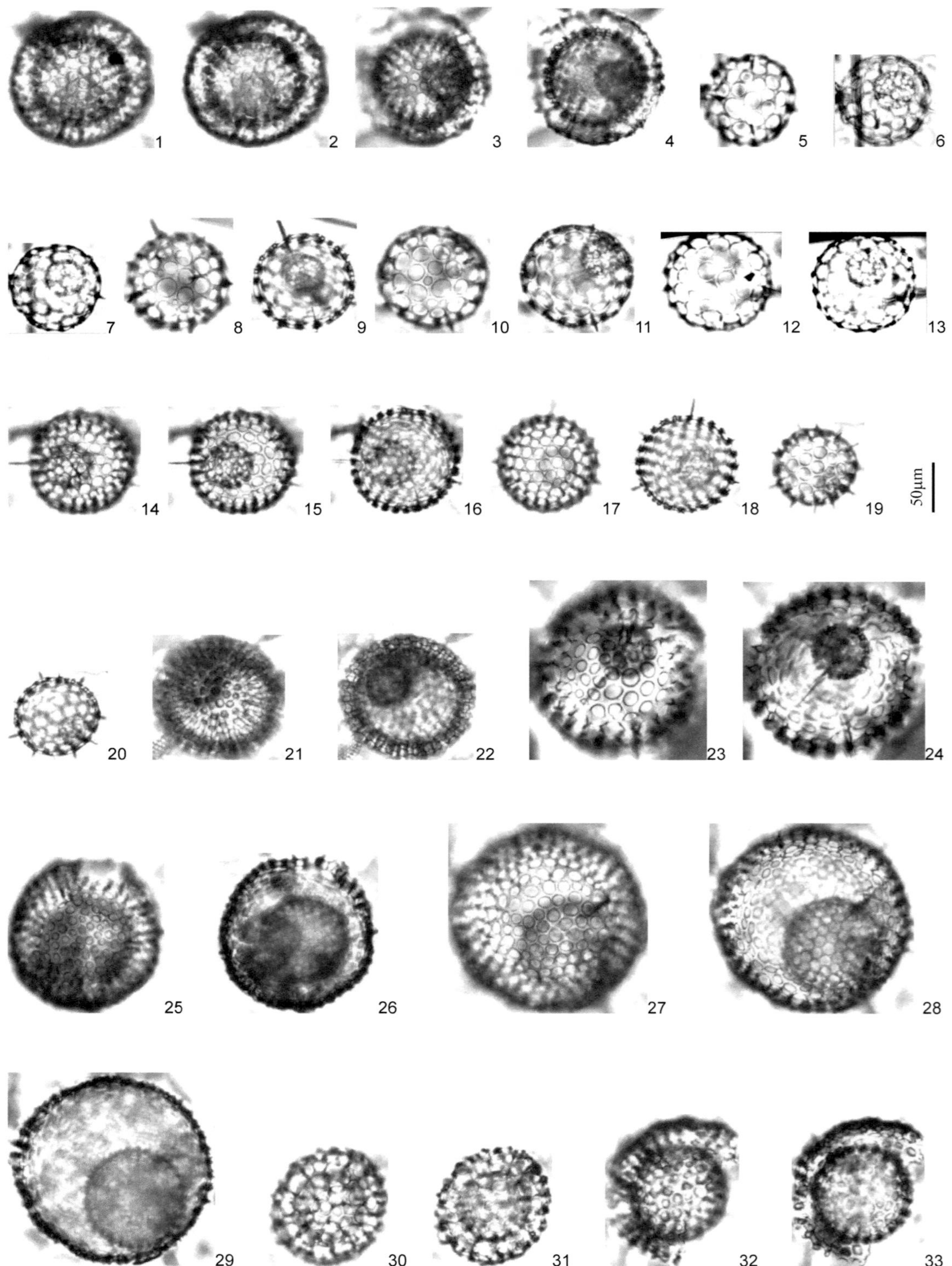

1–4. *Eccentrisphaera biderma* sp. nov.；5–20. *Eccentrisphaera bimedullaris* sp. nov.；
21–24. *Eccentrisphaera minima* (Clark and Campbell)；25–26. *Eccentrisphaera porolaris* sp. nov.；
27–29. *Eccentrisphaera trimedullaris* sp. nov.；30–31. *Liosphaera carpolaria* sp. nov.；32–33. *Liosphaera* sp. 1

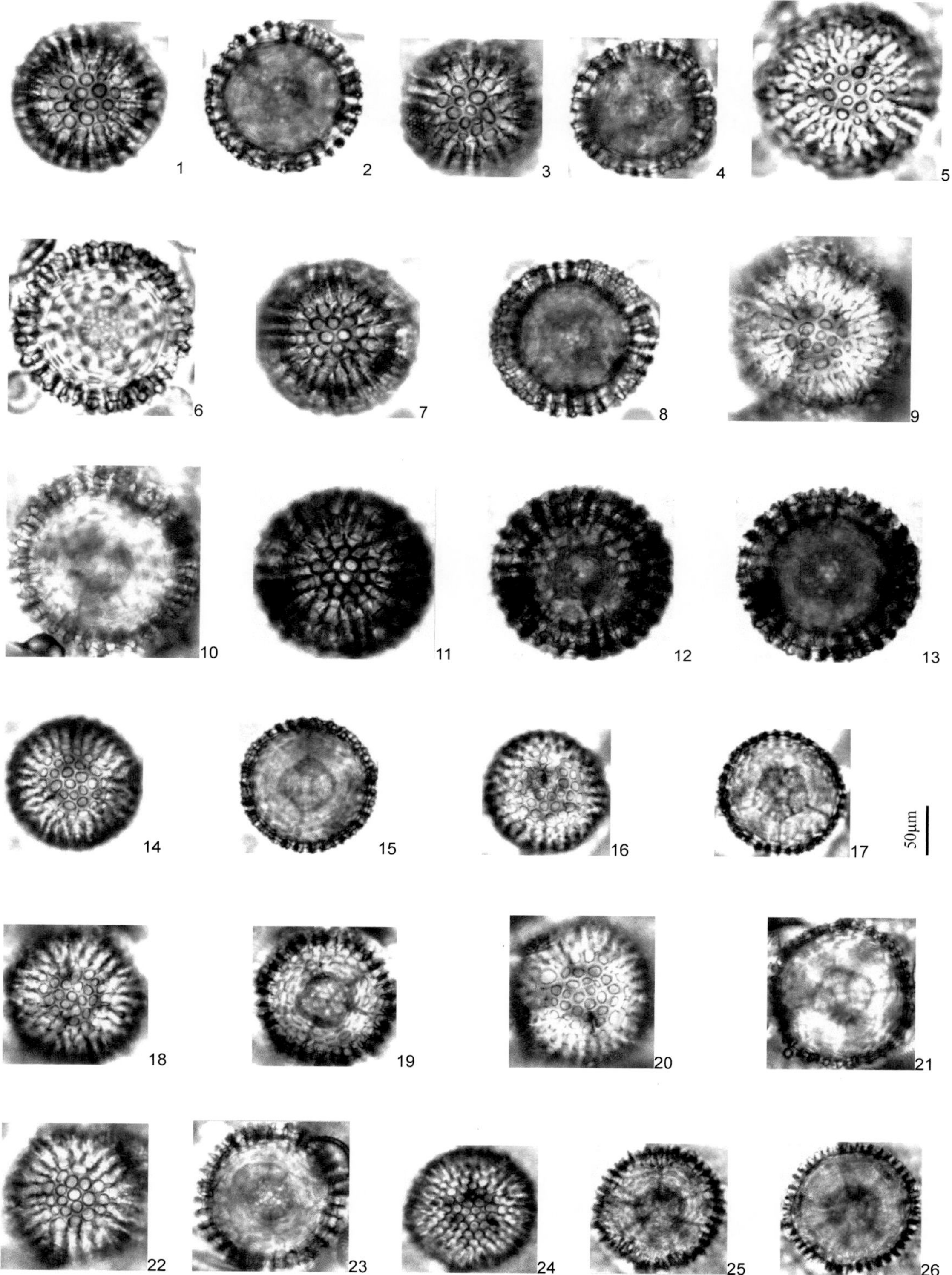

1–13. *Thecosphaera akitaensis* Nakaseko；14–21. *Thecosphaera entocuba* Chen et al.；
22–26. *Thecosphaera glebulenta* (Sanfilippo and Riedel)

图版 10

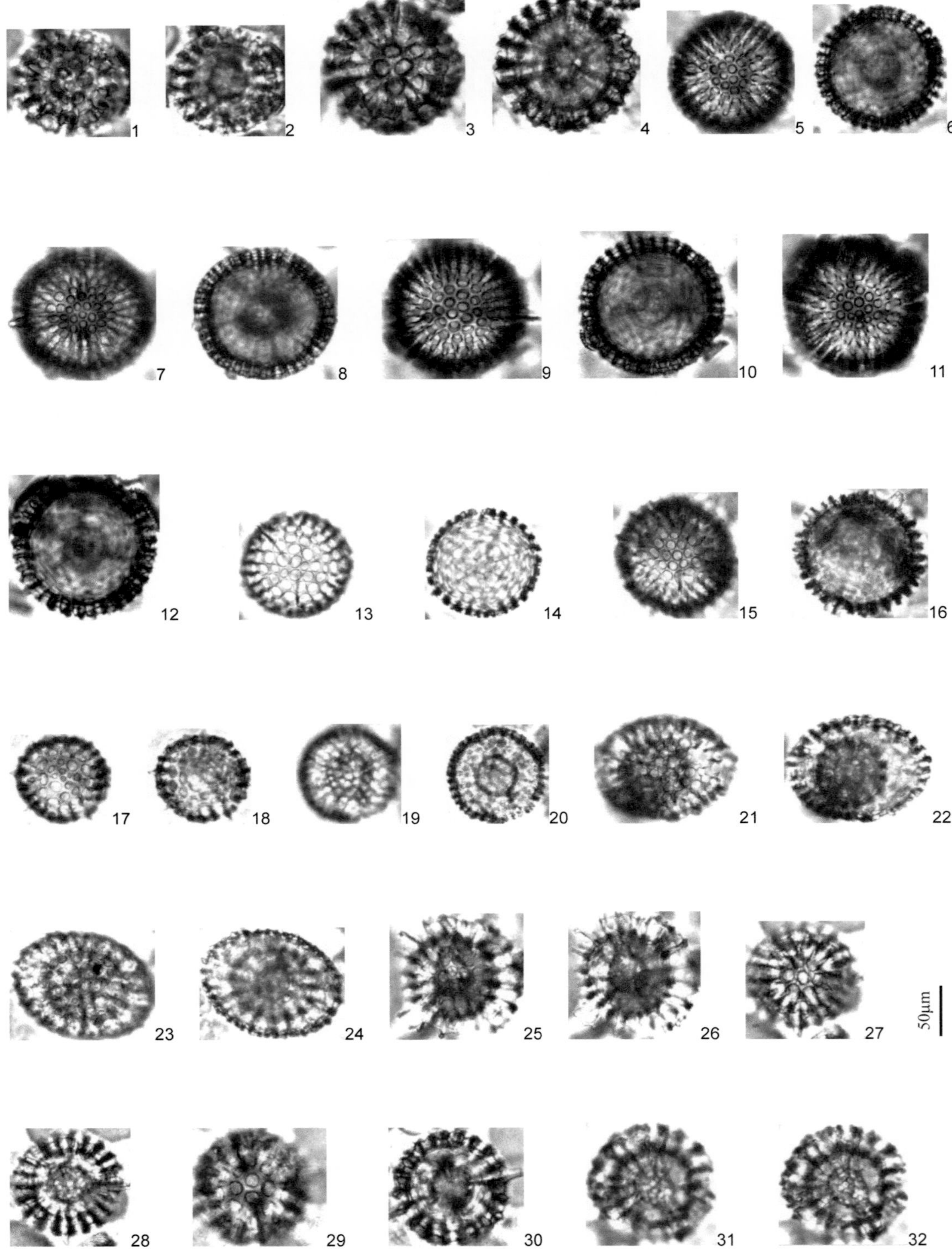

1–4. *Thecosphaera japonica* Nakaseko；5–12. *Thecosphaera minutapora* sp. nov.；13–20. *Thecosphaera nobile* (Ehrenberg)；21–24. *Thecosphaera ovata* sp. nov.；25–32. *Thecosphaera pachycortica* sp. nov.

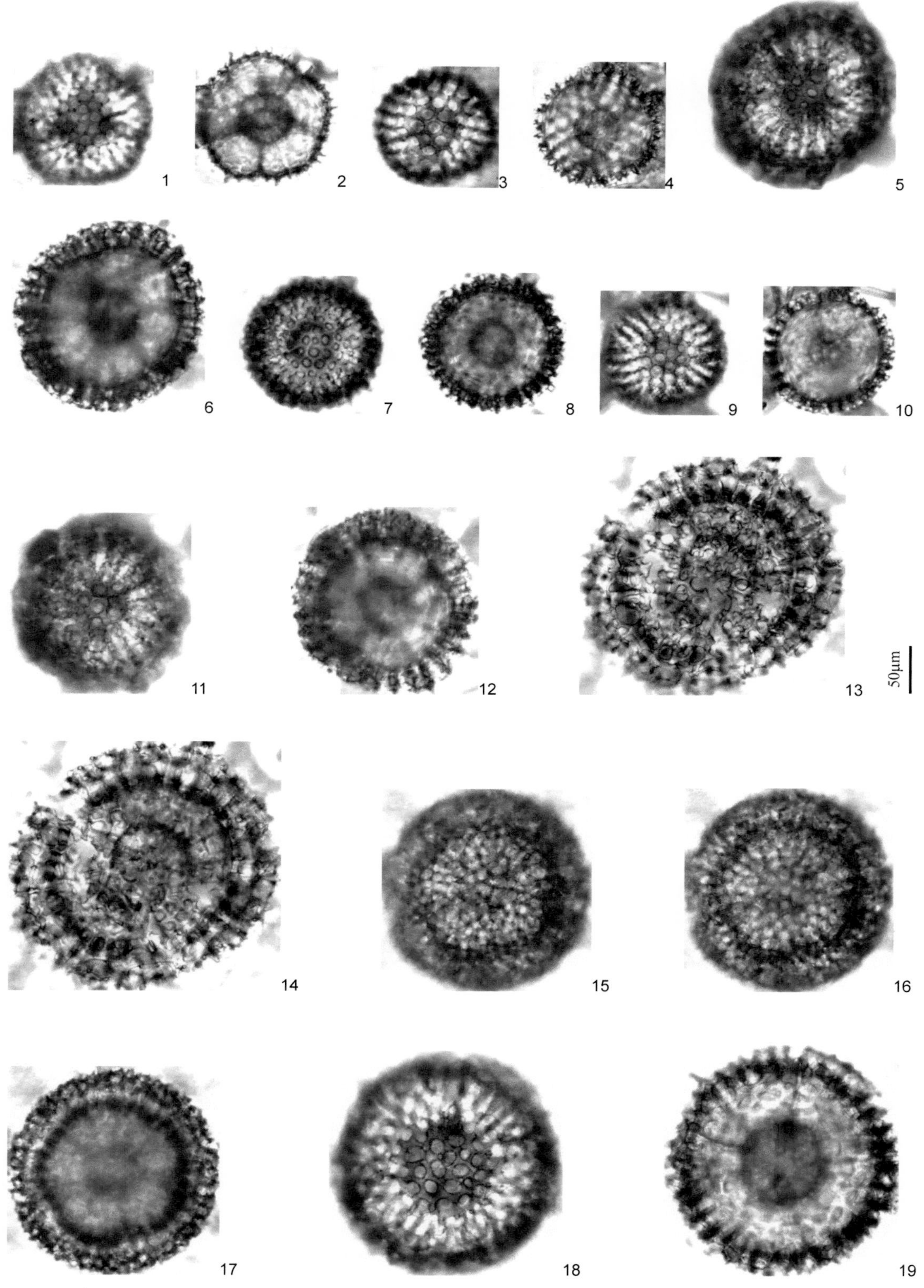

1–4. *Thecosphaera parviakitaensis* (Kamikuri)；5–19. *Thecosphaera reticularis* sp. nov.

图版 12

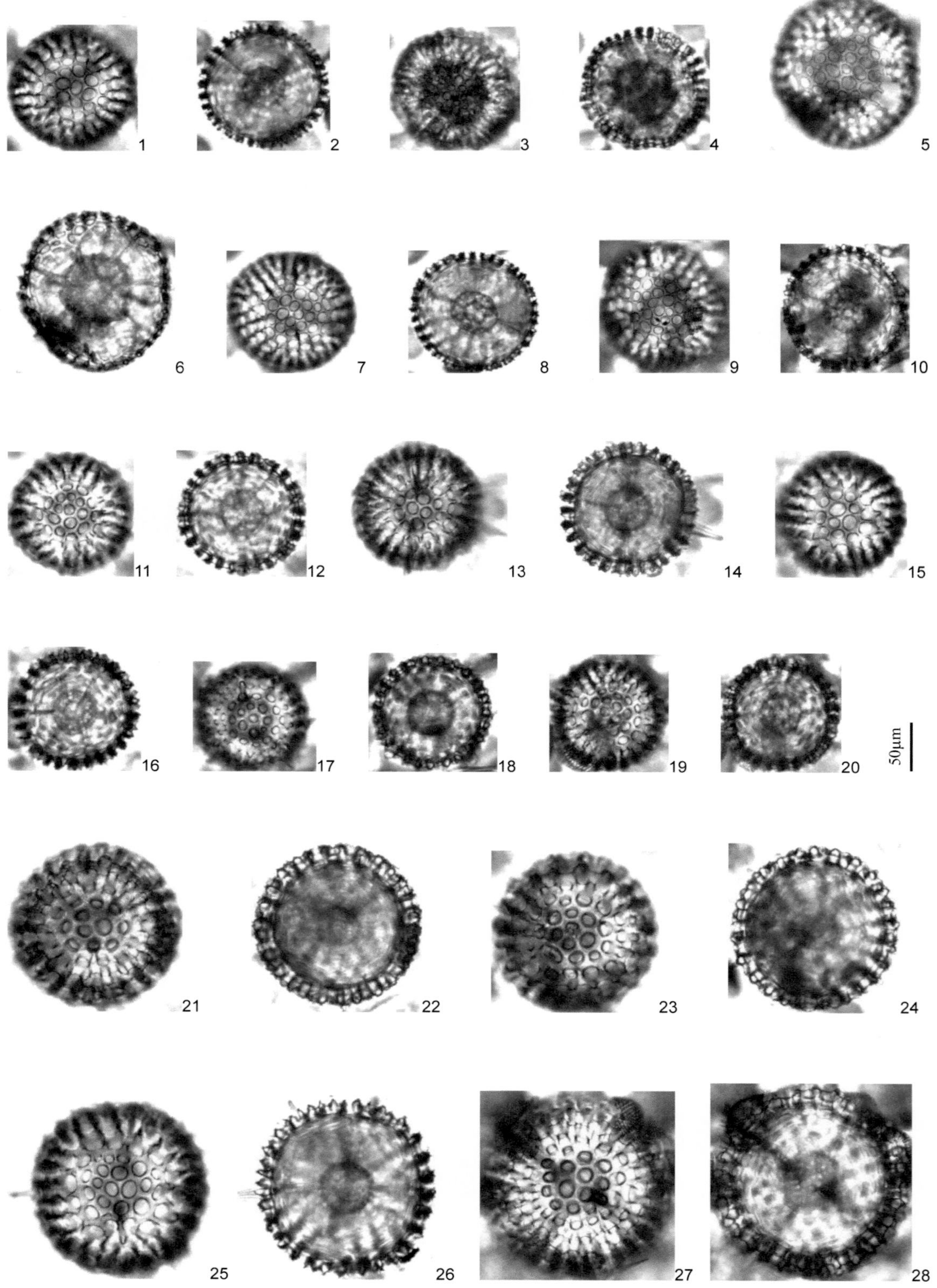

1–10. *Thecosphaera sanfilippoae* Blueford；11–20. *Thecosphaera tochigiensis* Nakaseko；21–28. *Thecosphaera zittelii* Dreyer

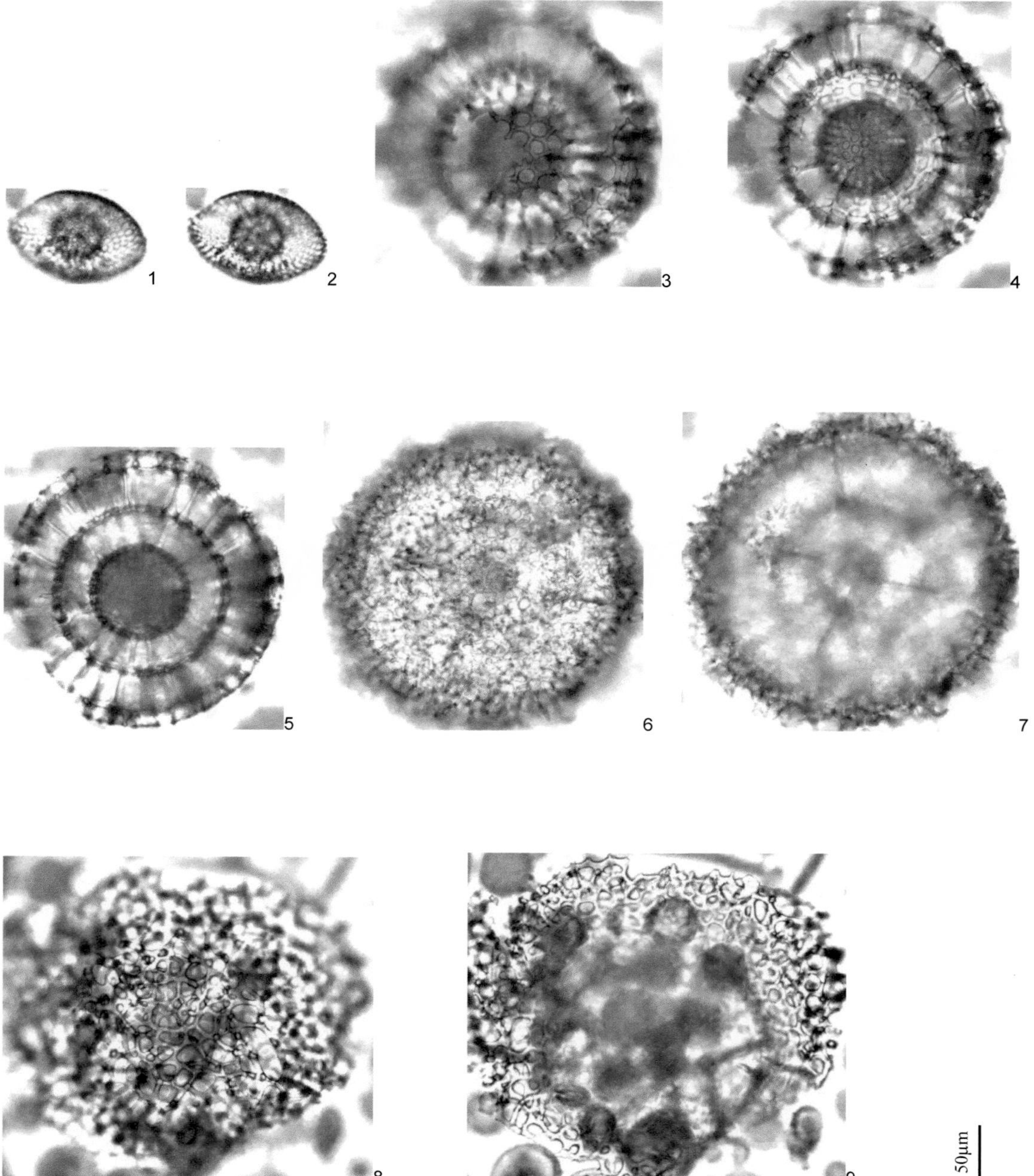

1–2. *Thecosphaera* sp. 1；3–5. *Cromyosphaera quadruplex* Haeckel；6–7. *Cromyosphaera asperata* sp. nov.；8–9. *Plegmosphaera asperula* sp. nov.

图版 14

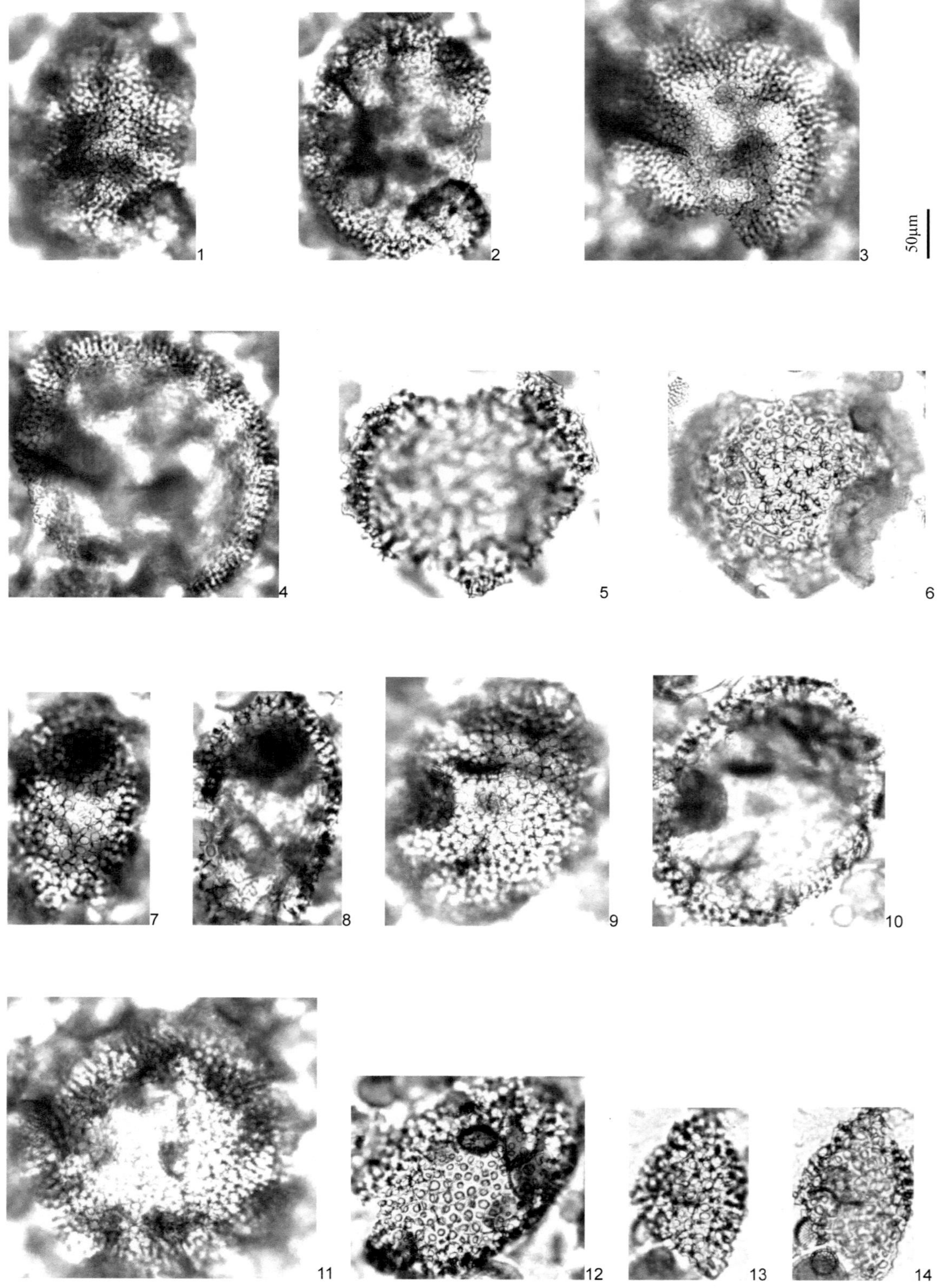

1–14. *Plegmosphaera asperula* sp. nov.

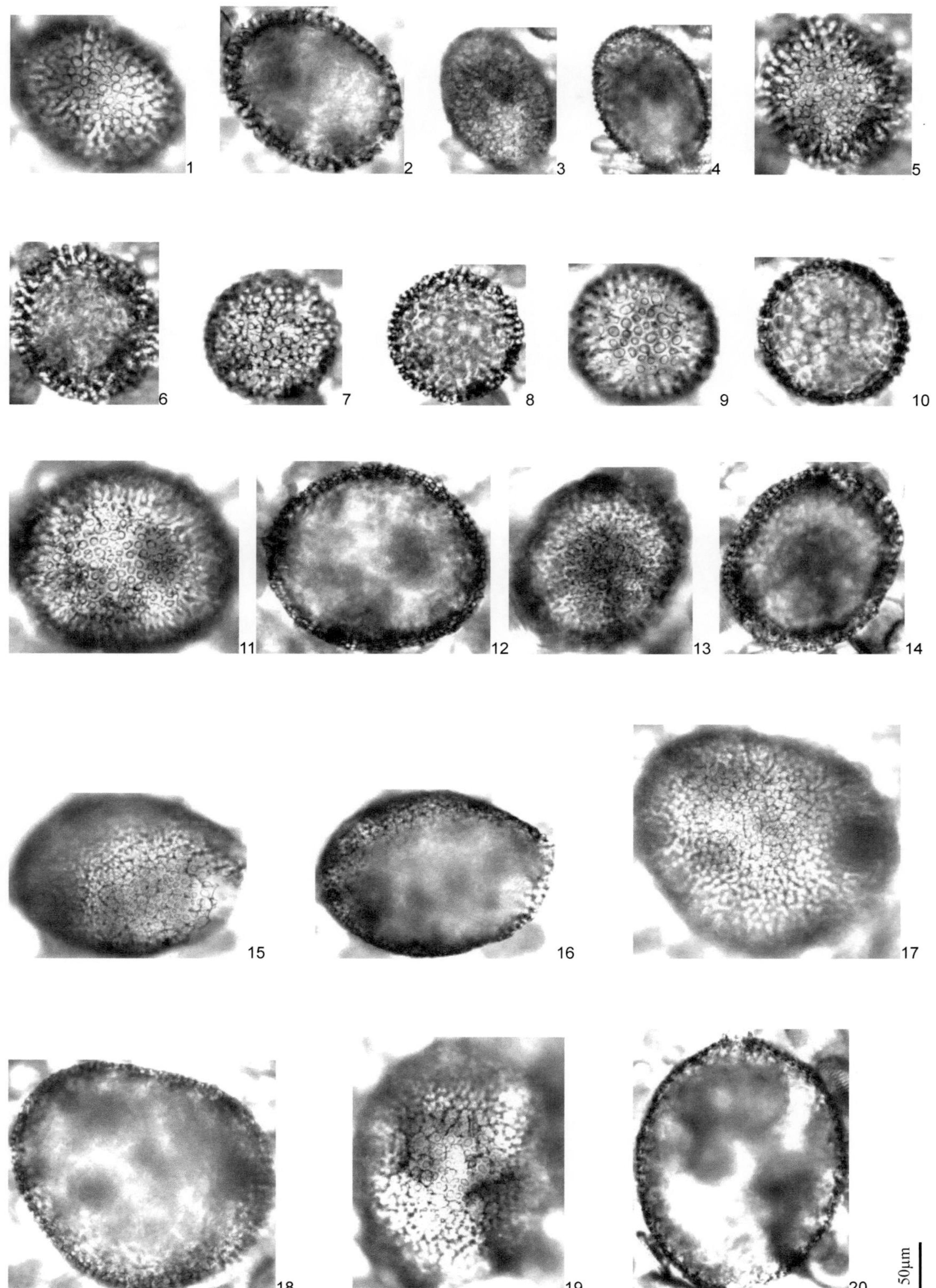

1–14. *Plegmosphaera globula* sp. nov.；15–20. *Plegmosphaera petrushevia* sp. nov.

图版 16

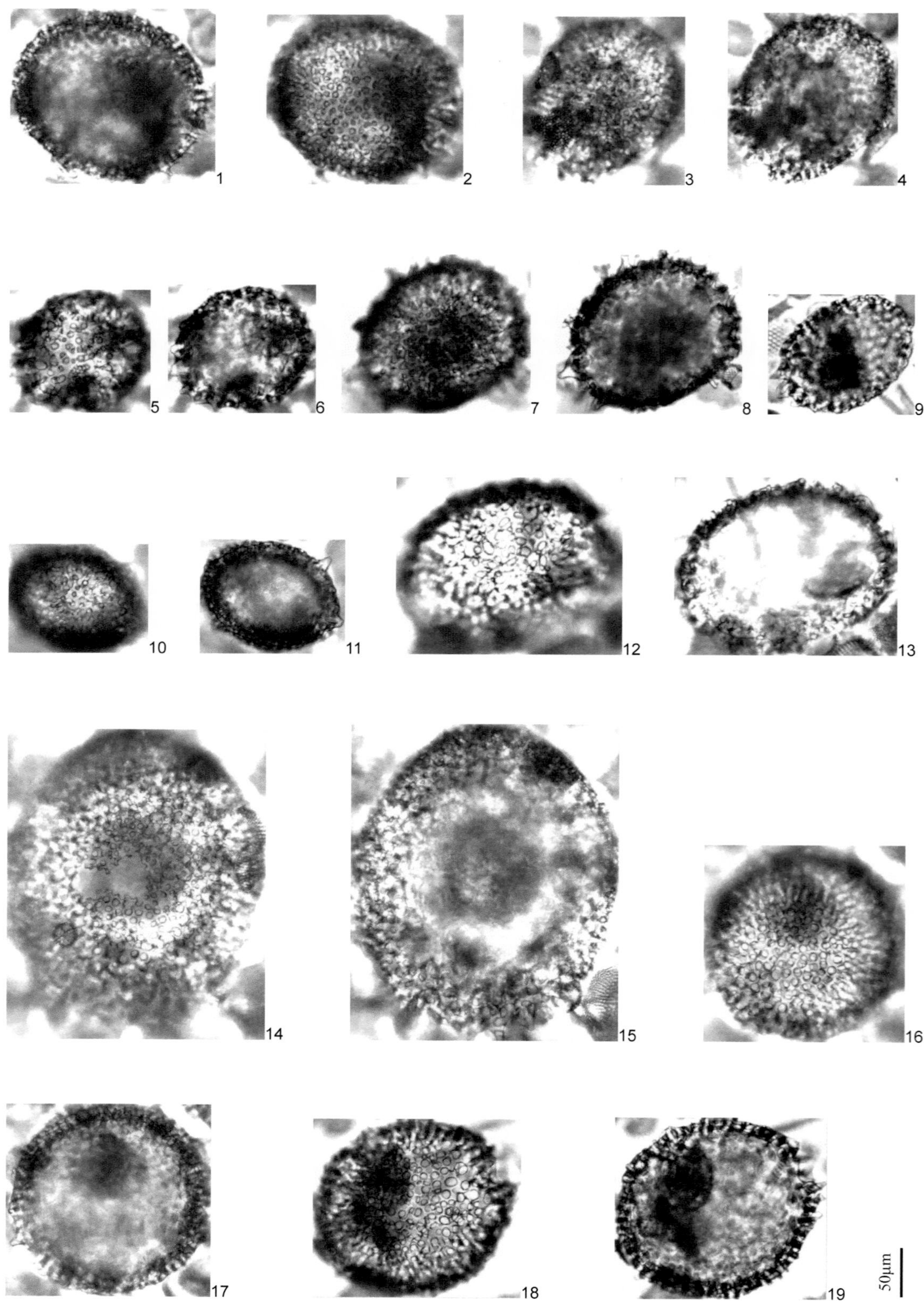

1–19. *Plegmosphaera spiculata* sp. nov.

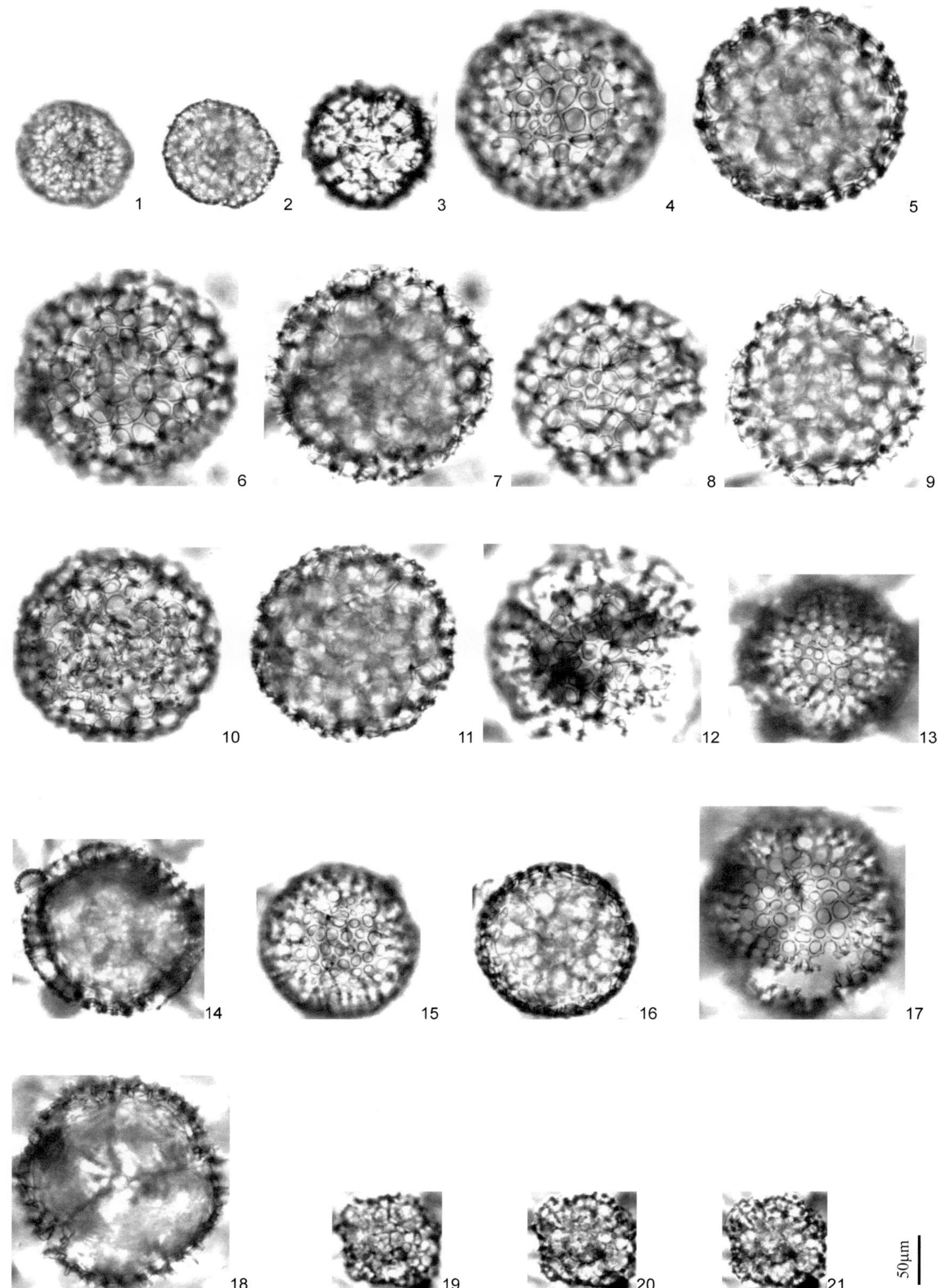

1–16. *Spongoplegma antarcticum* Haeckel；17–18. *Spongoplegma* aff. *antarcticum* Haeckel；19–21. *Spongoplegma quadratum* sp. nov.

图版 18

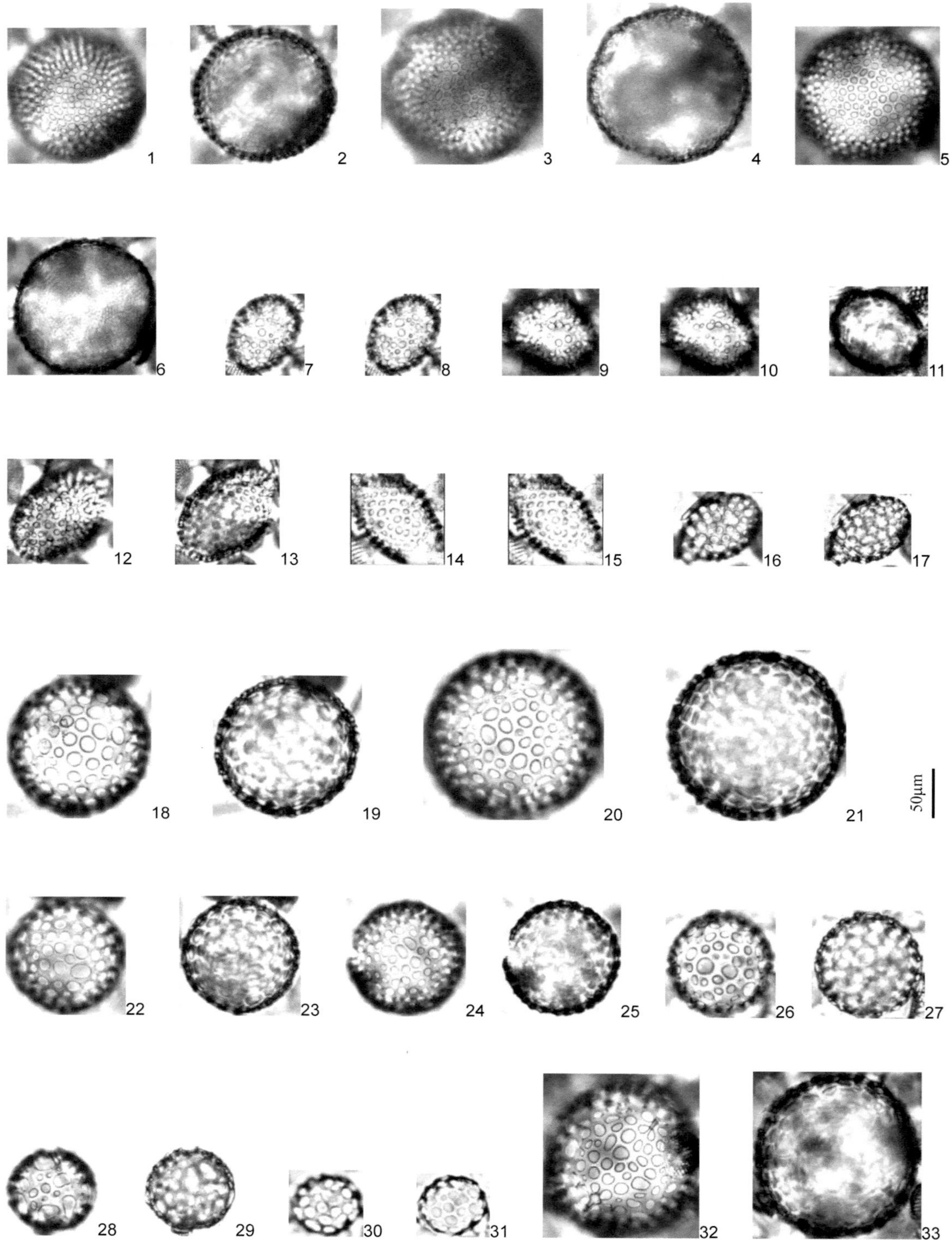

1–6. *Collosphaera confossa* Takahashi；7–17. *Collosphaera elliptica* Chen and Tan；18–27. *Collosphaera huxleyi* Müller；28–29. *Collosphaera macropora* Popofsky；30–31. *Collosphaera* sp. 1；32–33. *Collosphaera* sp. 2

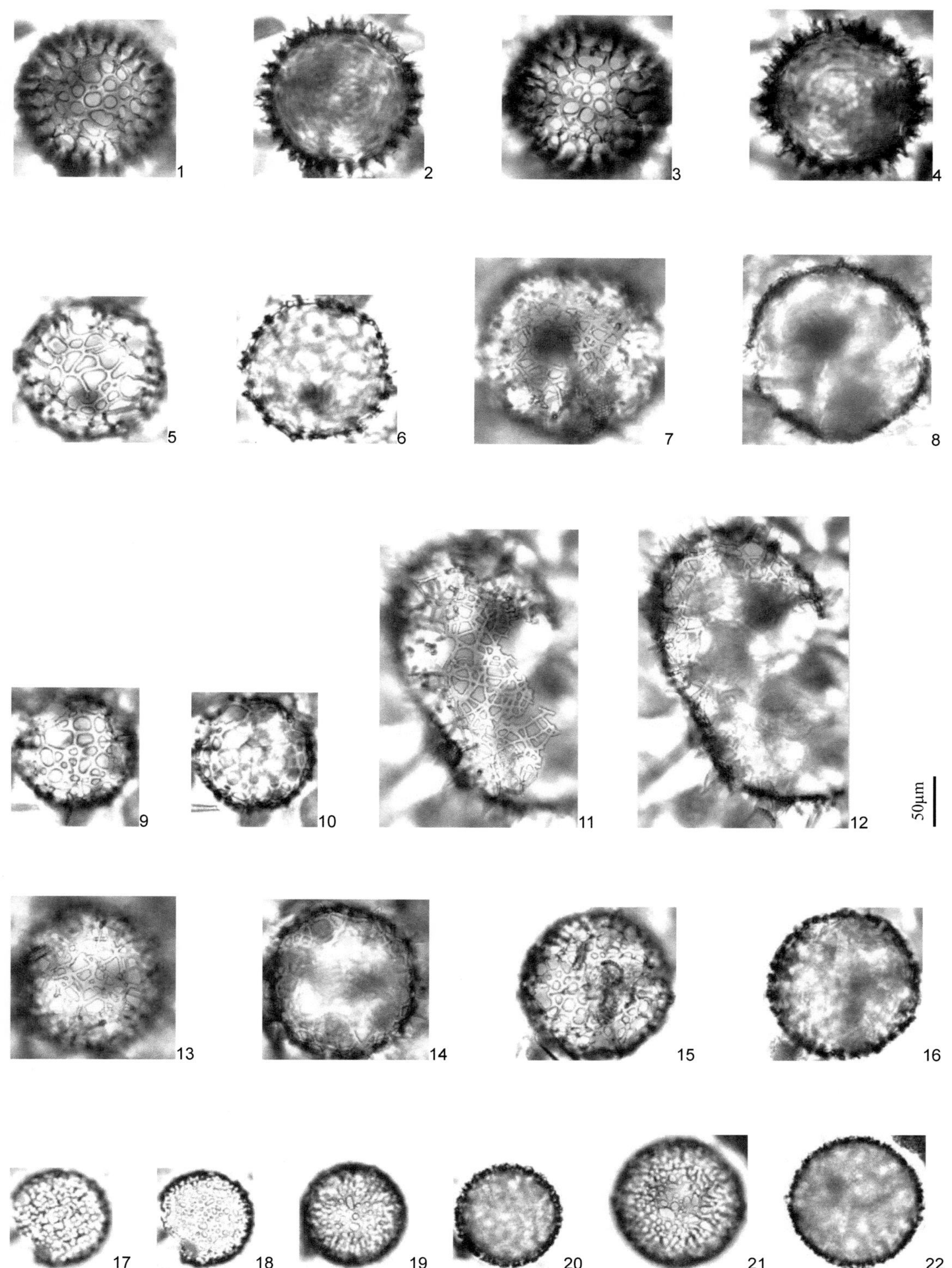

1–4. *Acrosphaera acuconica* sp. nov.；5–14. *Acrosphaera arachnodictyna* Chen et al.；15–16. *Acrosphaera hirsuta* Perner；17–22. *Acrosphaera*? *mercurius* Lazarus

图版 20

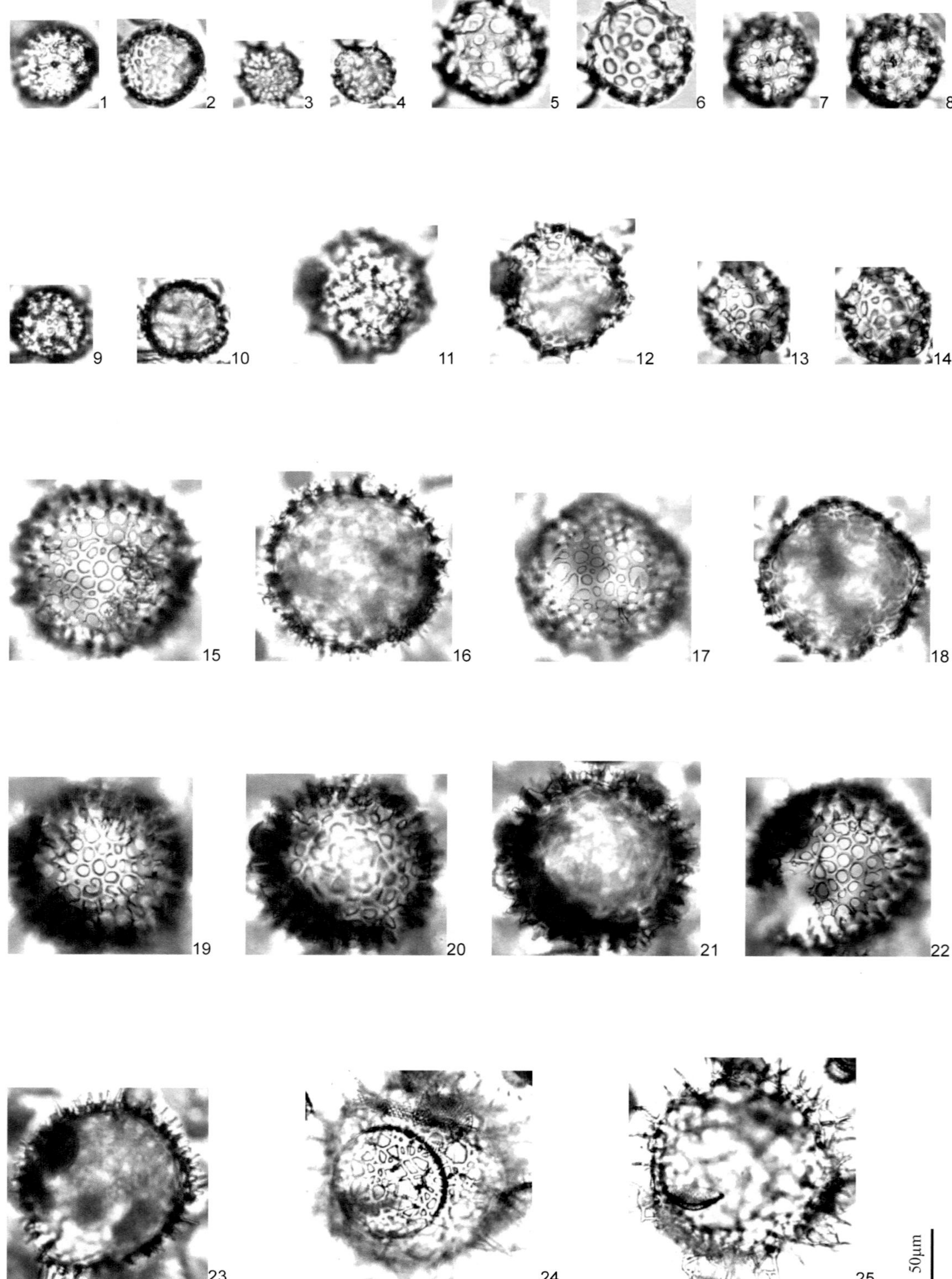

1–2. *Acrosphaera*? *mercurius* Lazarus；3–14. *Acrosphaera spinosa* (Haeckel)；15–23. *Acrosphaera spinosa echinoides* Haeckel；24–25. *Acrosphaera* sp. 1

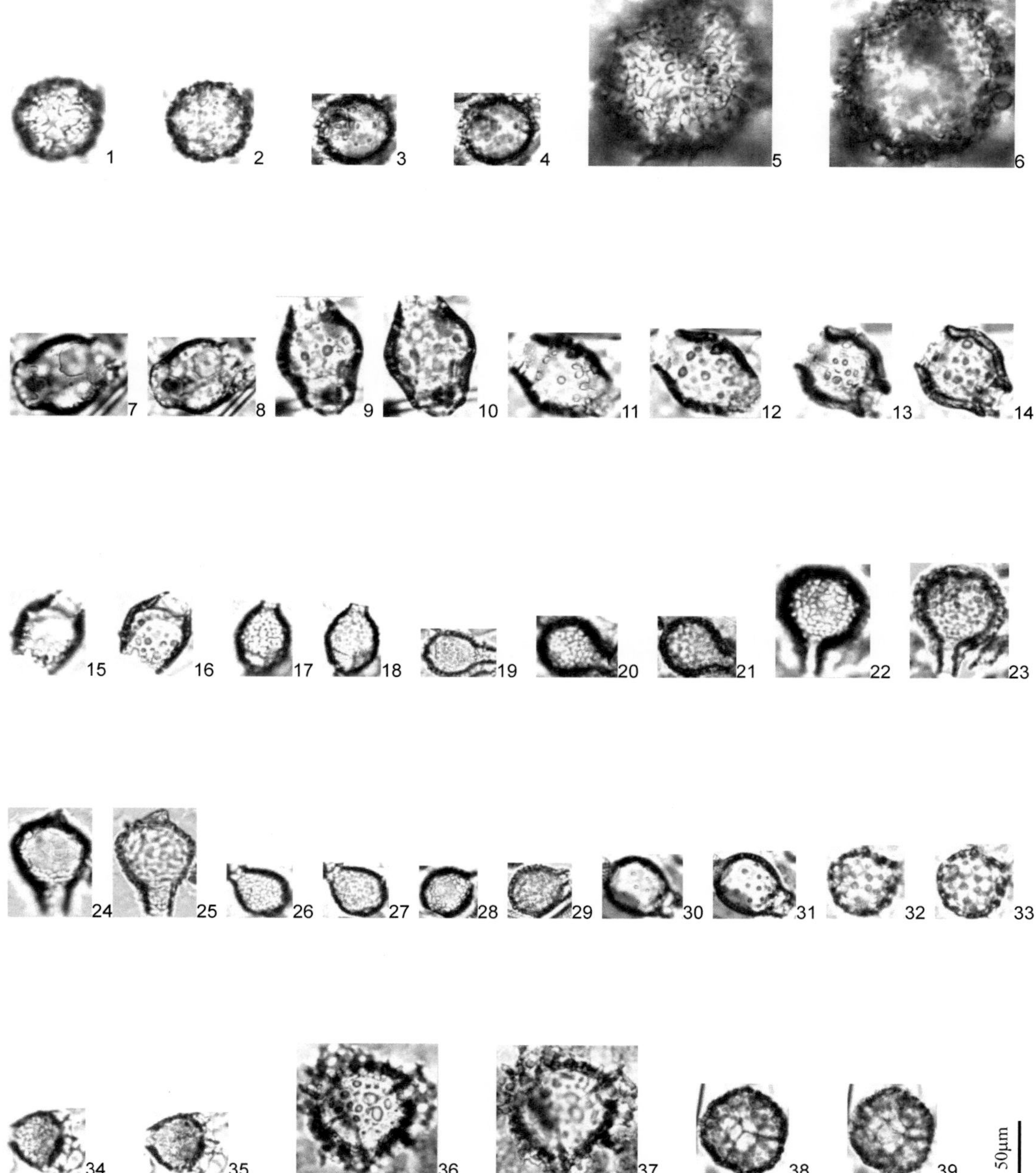

1–4. *Siphonosphaera circumtexta* (Haeckel)；5–6. *Siphonosphaera mixituba* sp. nov.；7–16. *Solenosphaera bitubula* sp. nov.；17–33. *Solenosphaera monotubulosa* (Hilmers) group；34–35. *Trisolenia megalactis* Ehrenberg；36–37. *Trisolenia* sp. 1；38–39. *Lonchosphaera cauleti* Dumitrică

图版 22

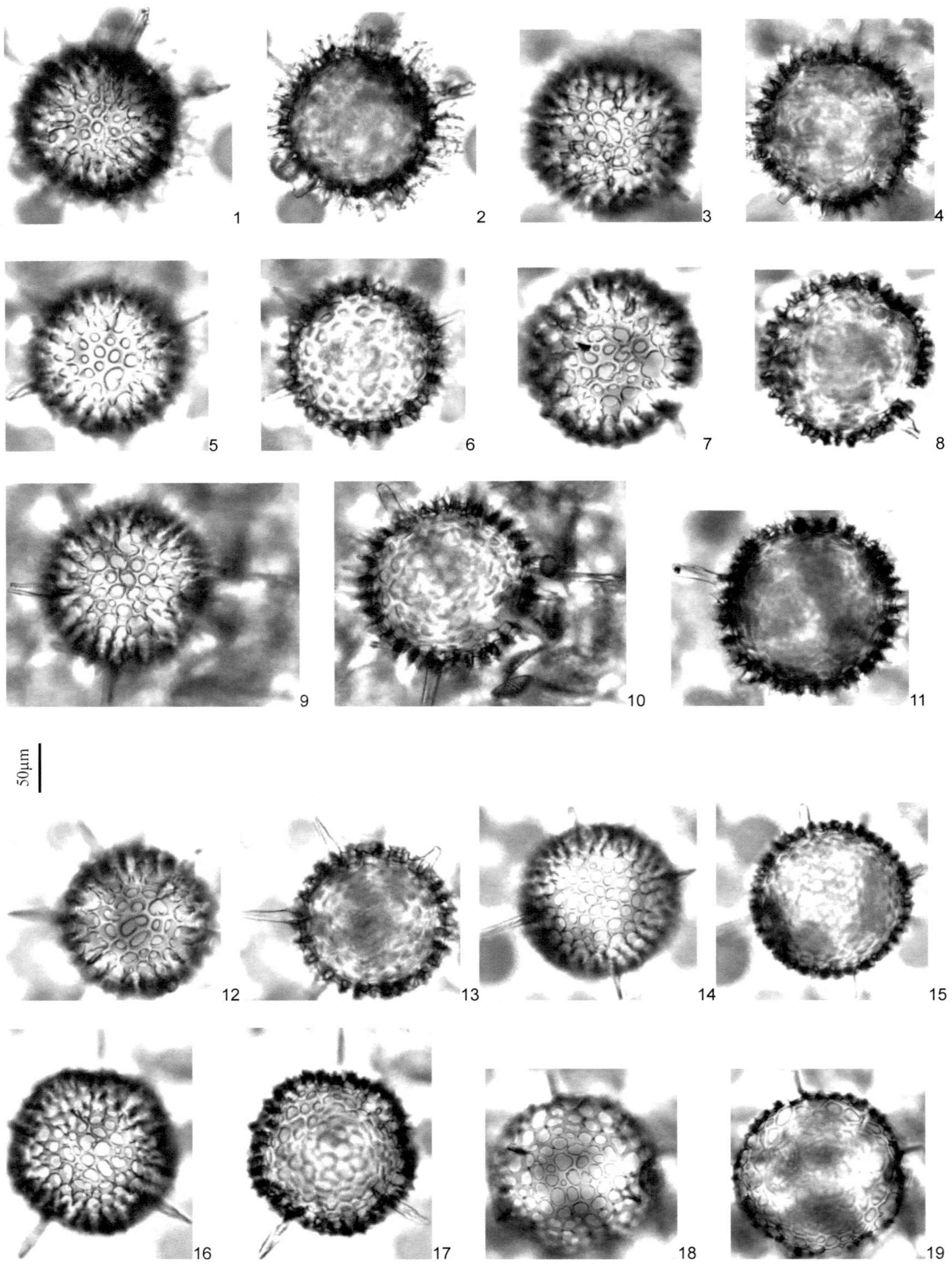

1–11. *Lonchosphaera scabrata* sp. nov.; 12–19. *Lonchosphaera spicata* Popofsky group

图版 23

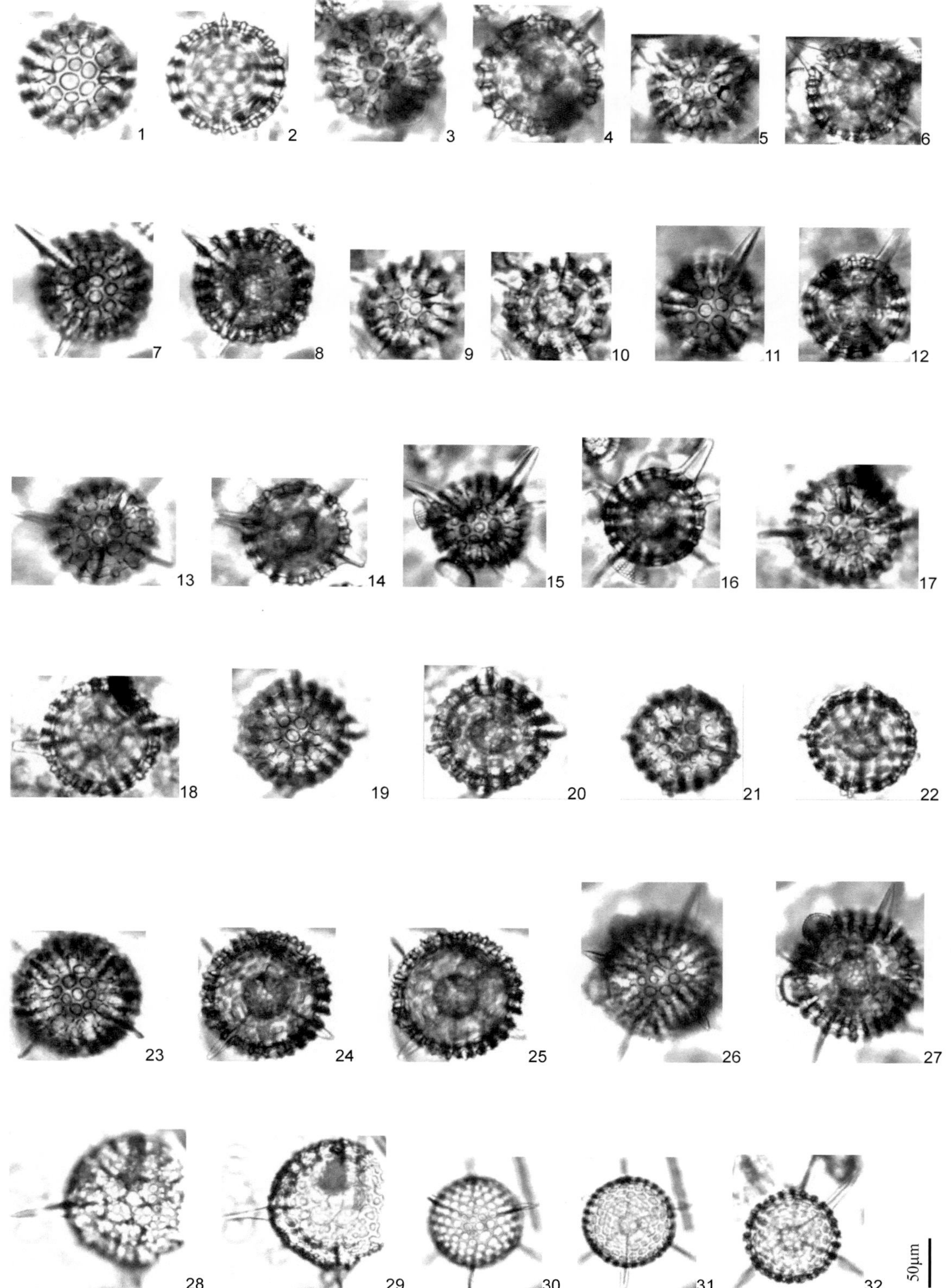

1–2. *Lonchosphaera* sp. 1；3–22. *Hexalonche anaximensis* Haeckel；23–27. *Hexalonche esmarki* Goll and Bjørklund；28–29. *Hexalonche gelidis* sp. nov.；30–32. *Hexalonche philosophica* Haeckel

图版 24

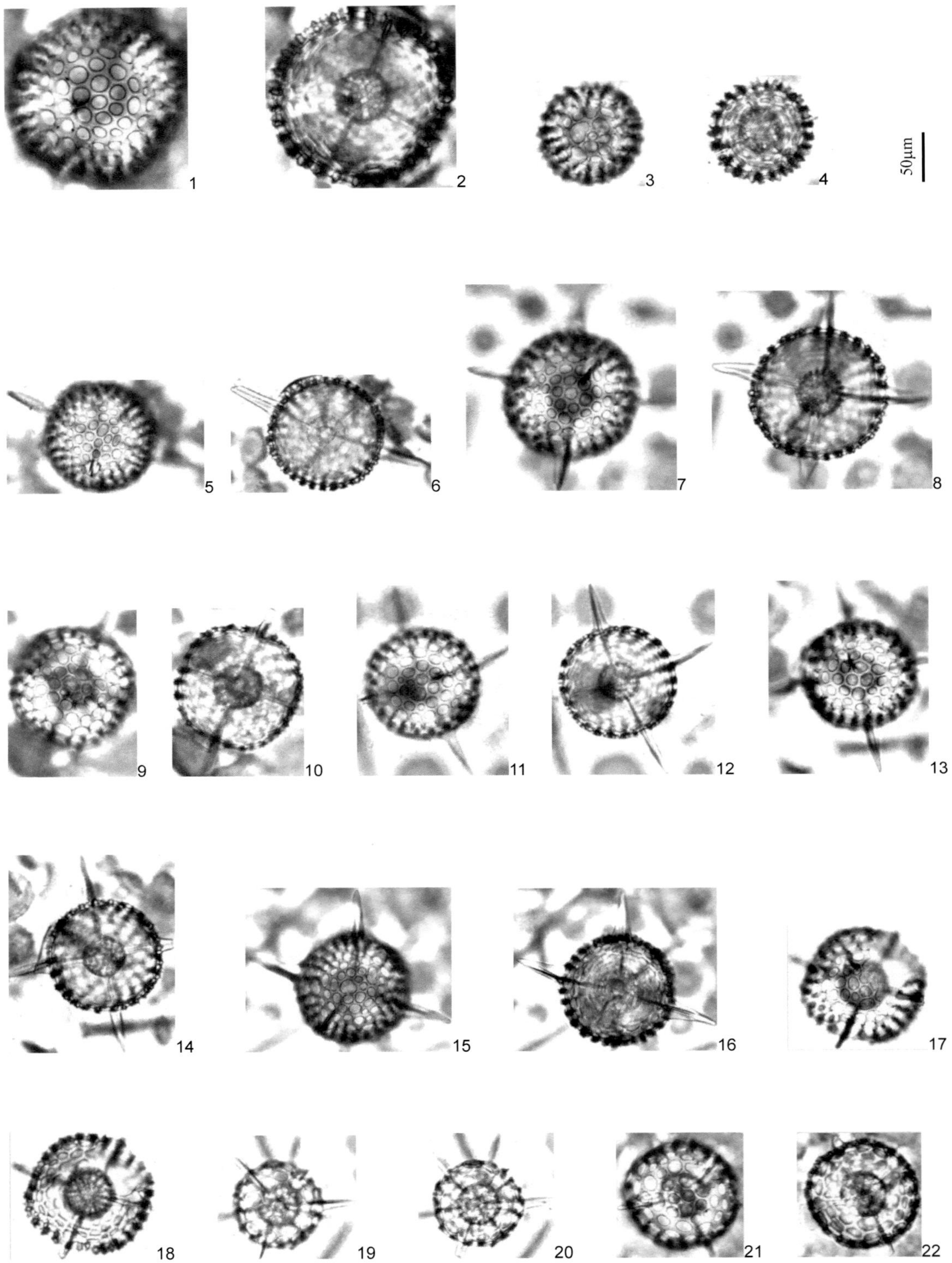

1–2. *Hexalonche philosophica* Haeckel；3–4. *Hexacontium*? *akitaensis* (Nakaseko)；5–18. *Hexacontium axotrias* Haeckel；19–22. *Hexacontium enthacanthum* Jørgensen

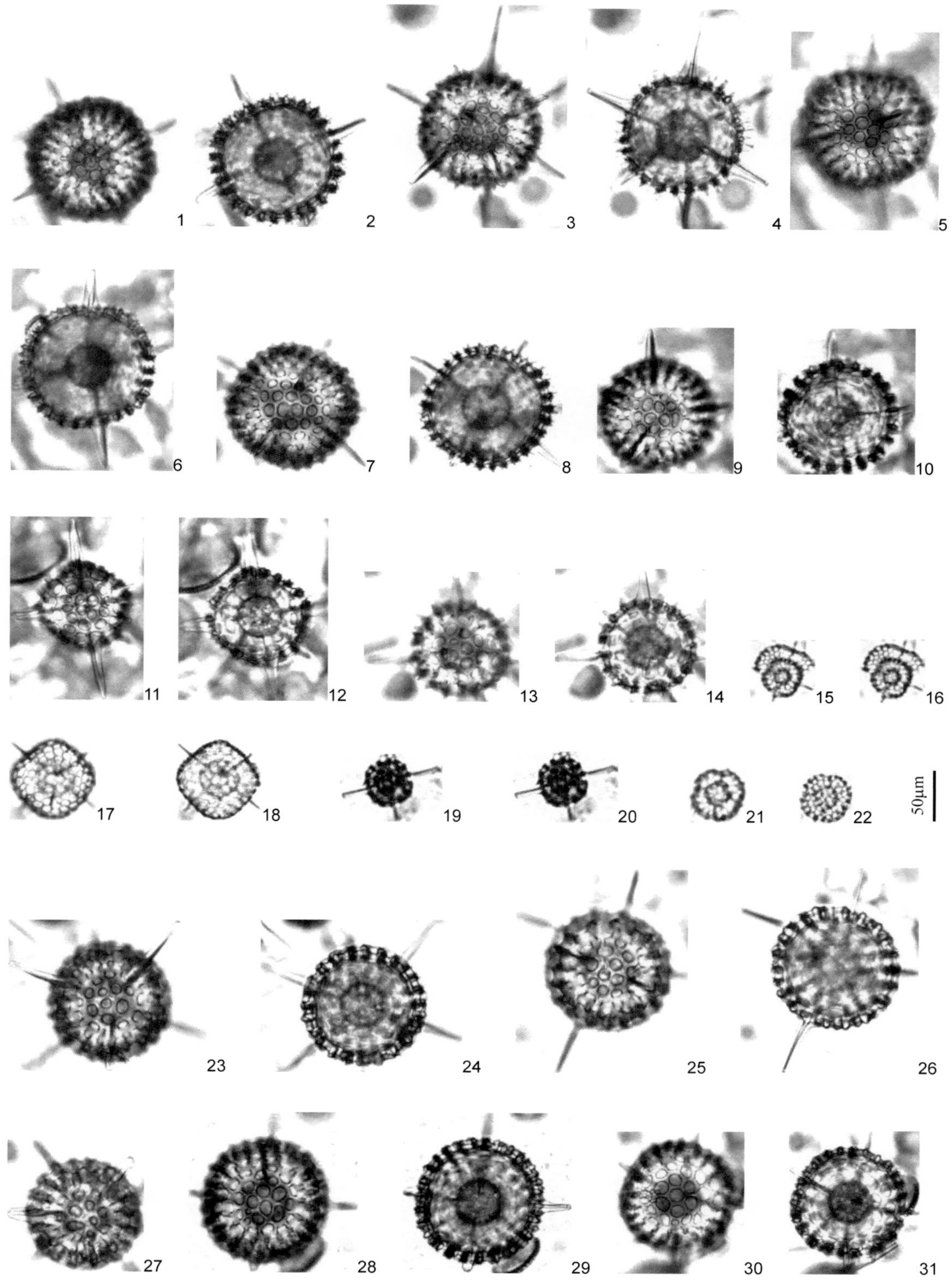

1–10. *Hexacontium favosum* Haeckel；11–14. *Hexacontium hostile* Cleve；15–22. *Hexacontium melpomene* (Haeckel)；23–31. *Hexacontium pachydermum* Jørgensen

图版 26

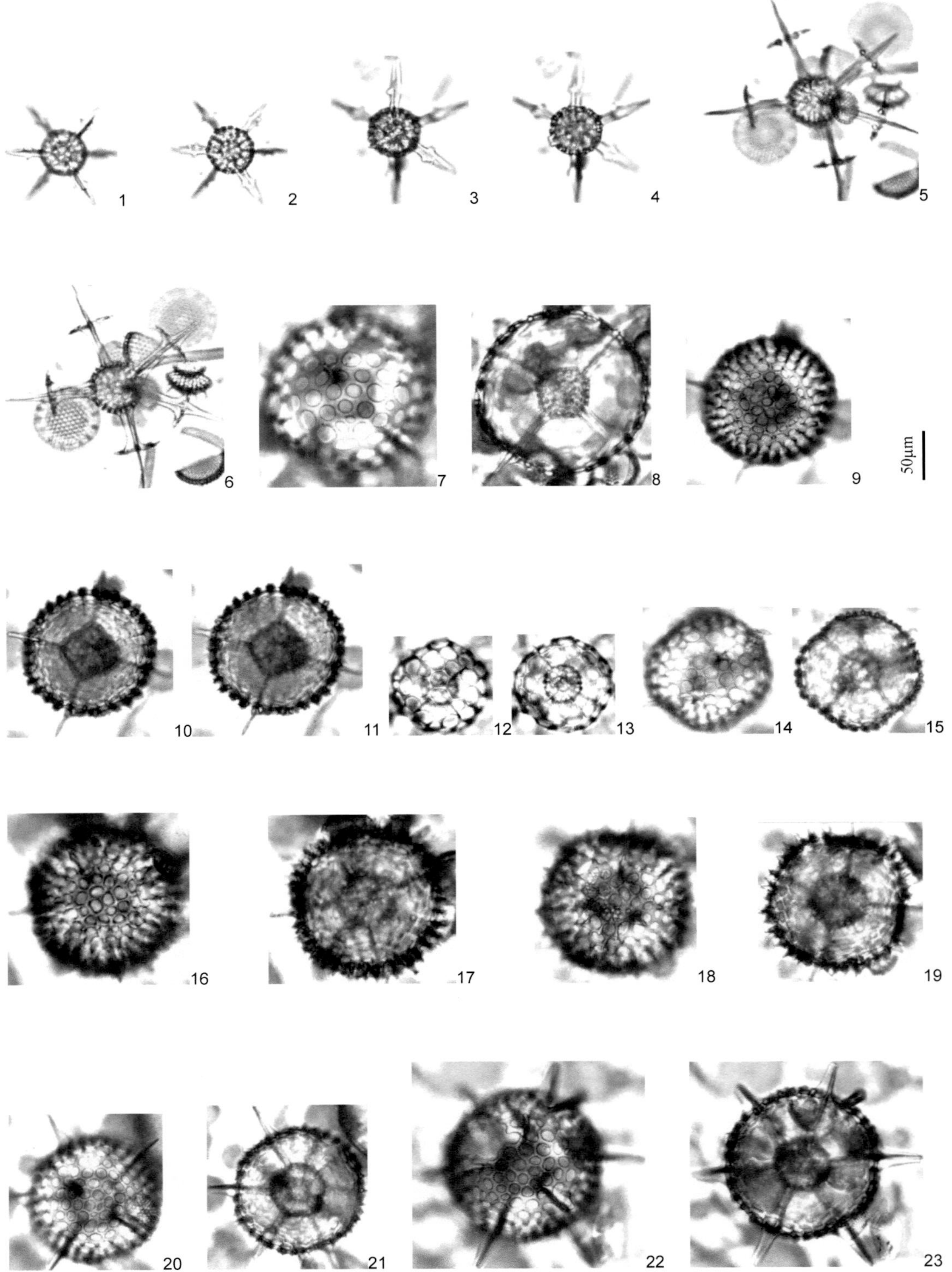

1–6. *Hexacontium parallelum* sp. nov.; 7–11. *Hexacontium quadrangulum* sp. nov.; 12–15. *Hexacontium quadratum* Tan; 16–19. *Hexacontium sceptrum* Haeckel; 20–23. *Hexacontium senticetum* Tan and Su

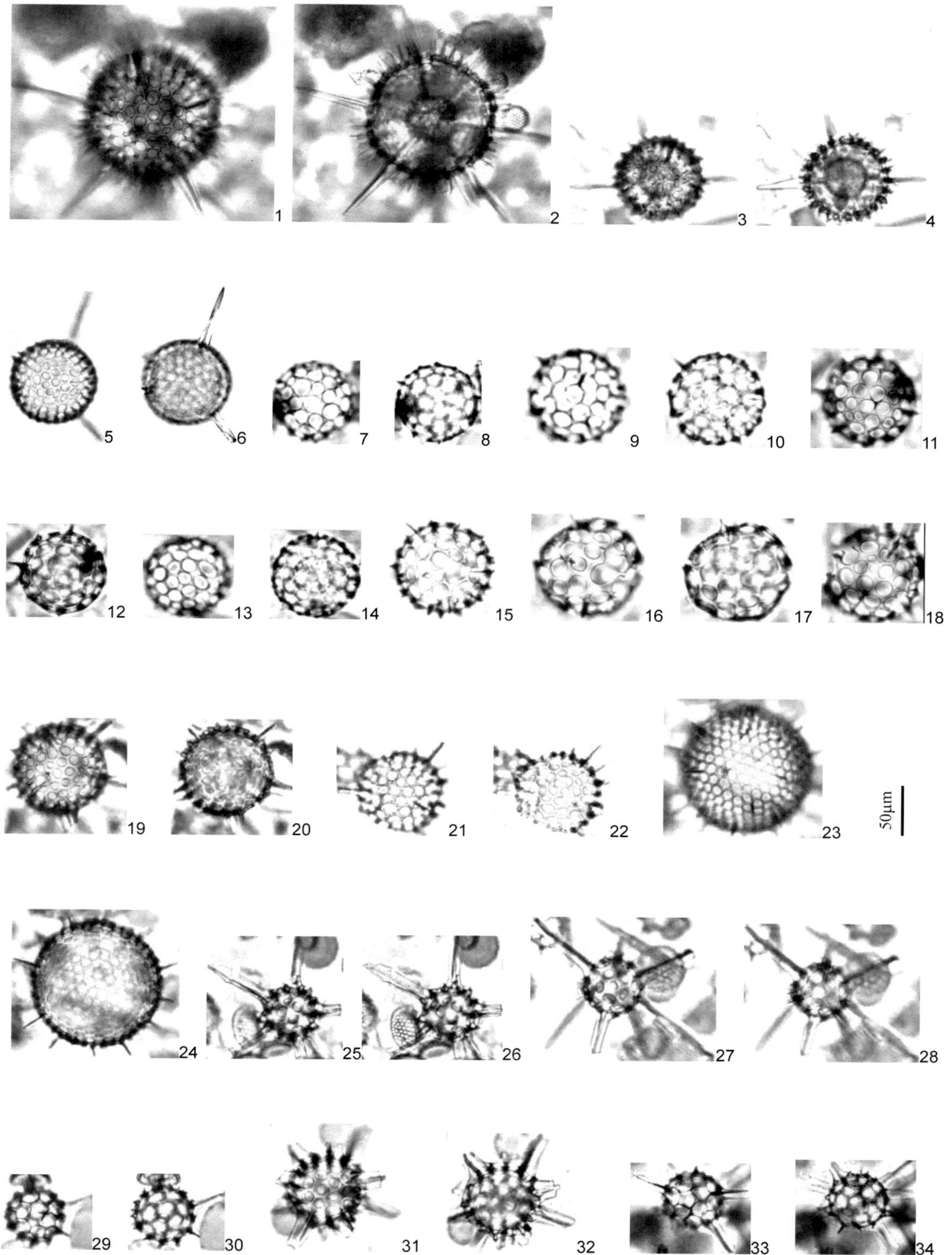

1–2. *Hexacromyum elegans* Haeckel；3–4. *Hexacromyum rara* (Carnevale) group；
5–6. *Acanthosphaera enigmaticus* (Hollande and Enjumet)；7–18. *Acanthosphaera marginata* Popofsky；
19–22. *Acanthosphaera polageota* sp. nov.；23–24. *Acanthosphaera* sp. 1；25–34. *Heliosphaera dentata* (Mast)

图版 28

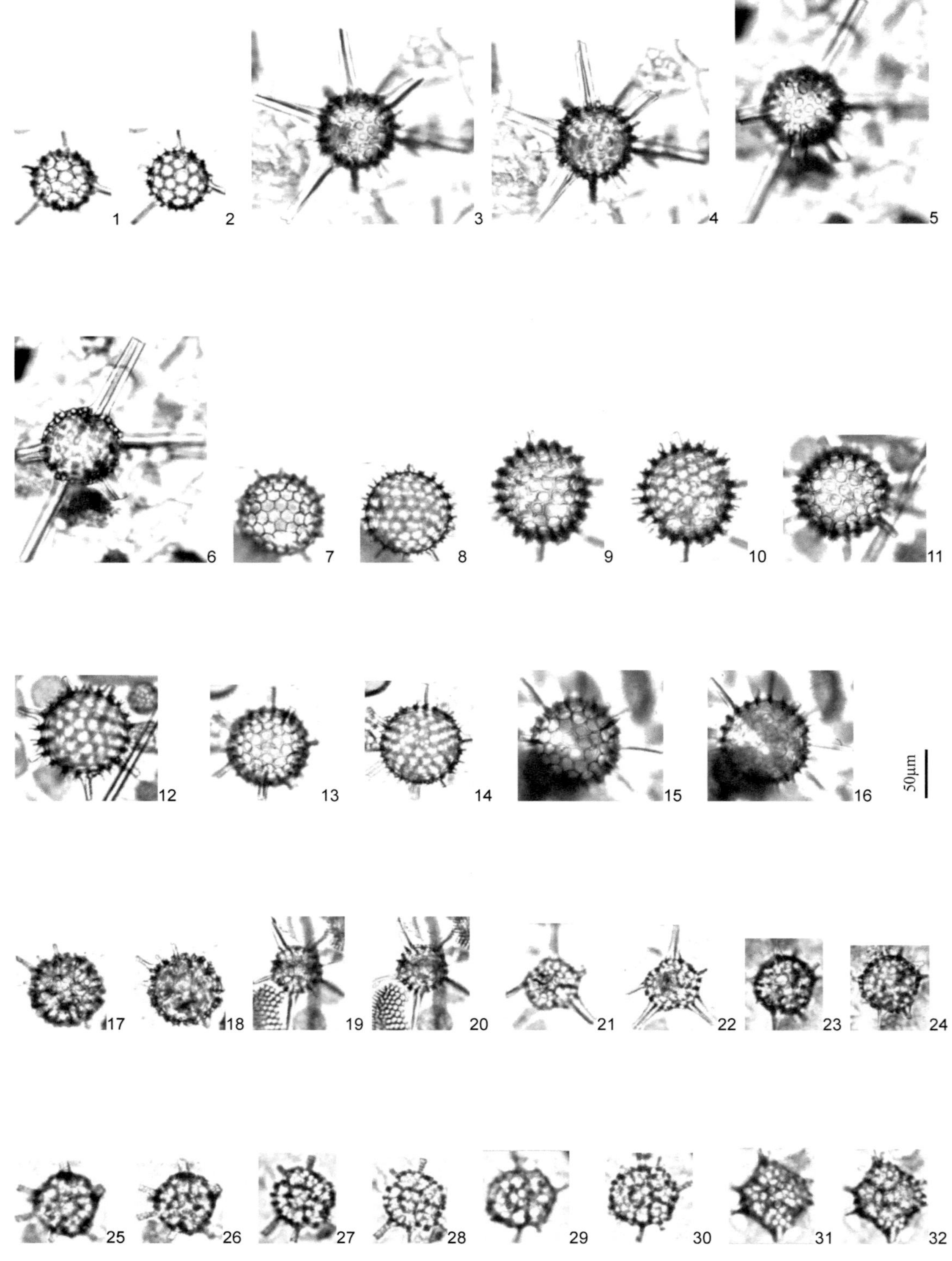

1–2. *Heliosphaera macrohexagonaria* Tan； 3–6. *Heliosphaera miniopora* sp. nov.； 7–16. *Cladococcus viminalis* Haeckel；
17–18. *Haliomma acanthophora* Popofsky； 19–22. *Haliomma asteranota* sp. nov.； 23–32. *Haliomma microlaris* sp. nov.

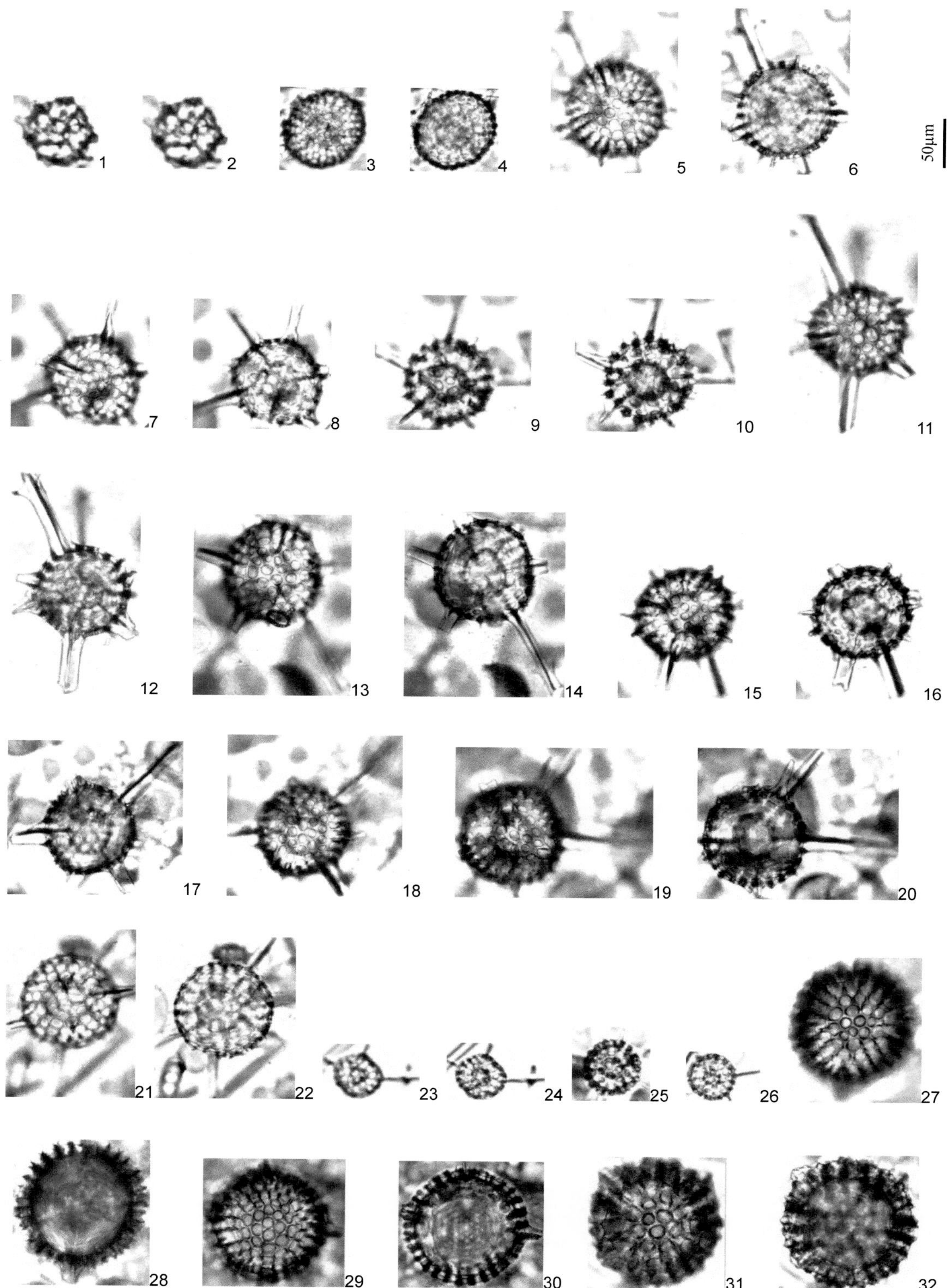

1–2. *Haliomma microlaris* sp. nov.; 3–4. *Haliomma ovatum* Ehrenberg; 5–20. *Haliomma stylota* sp. nov.; 21–22. *Haliomma* sp. 1; 23–26. *Haliomma* sp. 2; 27–32. *Haliommetta hadraspina* sp. nov.

图版 30

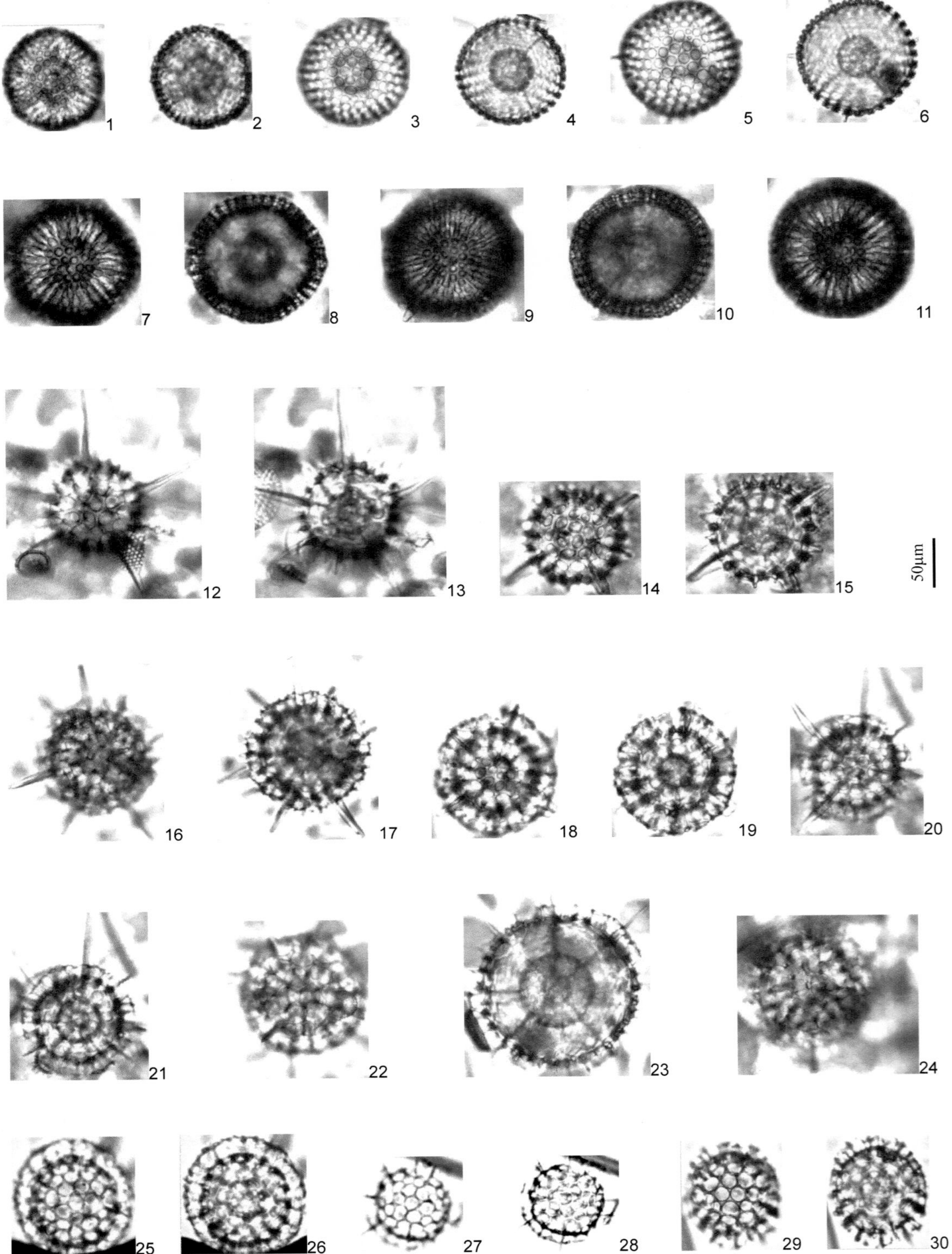

1–11. *Haliommetta miocenica* (Campbell and Clark) group； 12–15. *Haliommetta rossina* sp. nov.； 16–24. *Actinomma boreale* Cleve； 25–30. *Actinomma cocles* Renaudie and Lazarus

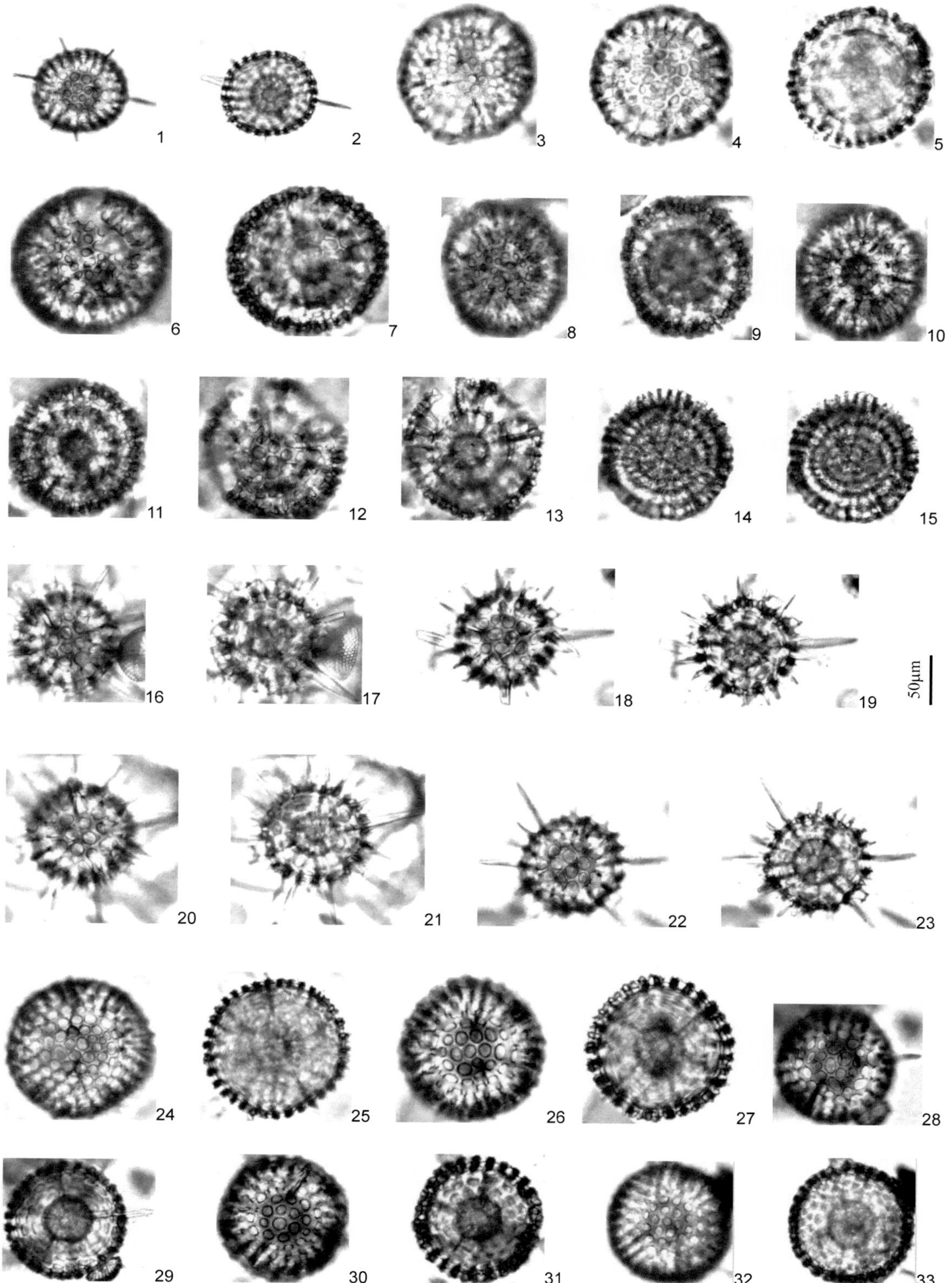

1–2. *Actinomma delicatulum* (Dogiel)；3–15. *Actinomma dumitricanis* sp. nov.；16–17. *Actinomma golownini* Petrushevskaya；18–23. *Actinomma hastatum* (Haeckel)；24–33. *Actinomma henningsmoeni* Goll and Bjørklund

图版 32

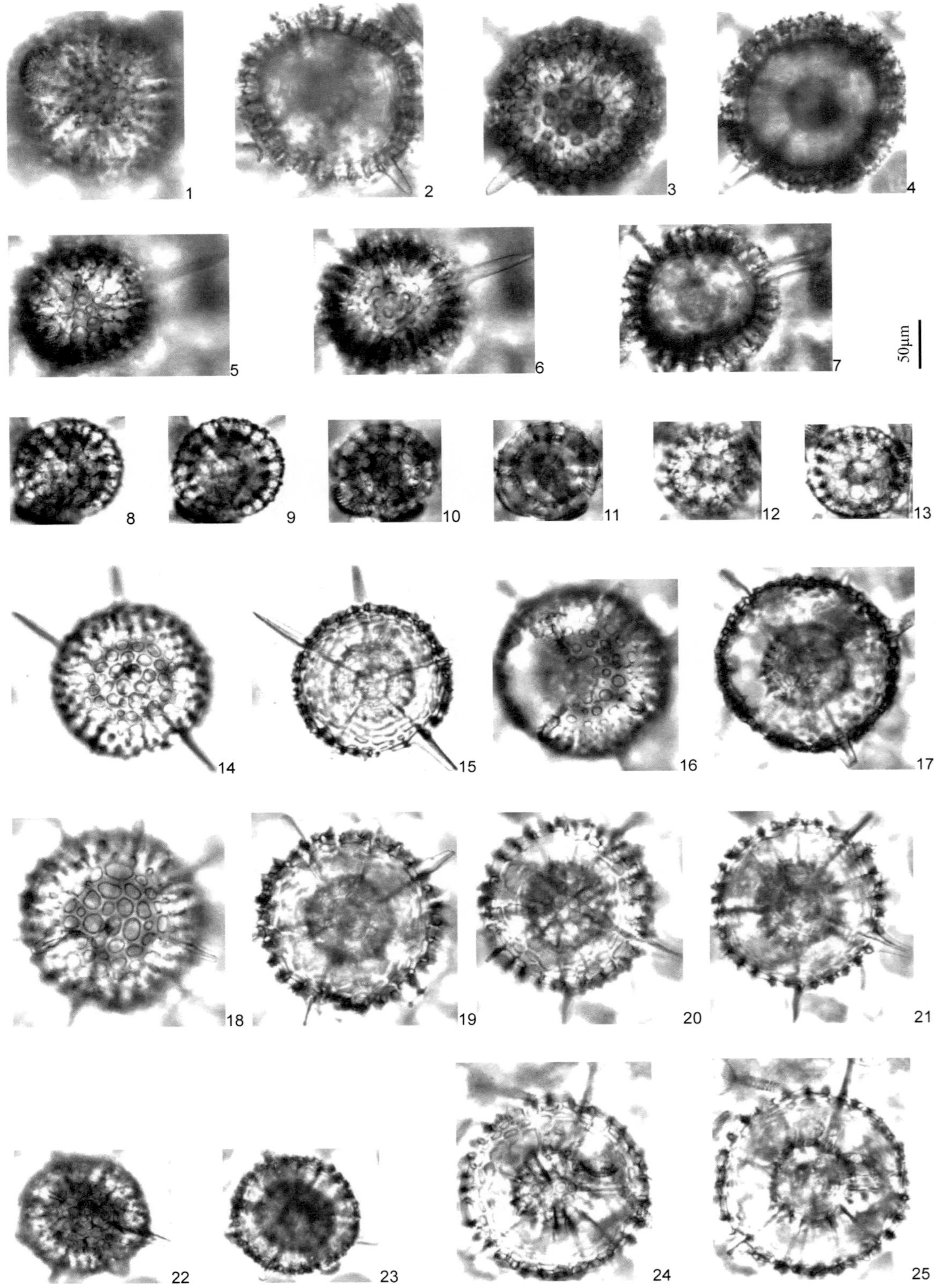

1–7. *Actinomma hexactis* Stöhr；8–13. *Actinomma holtedahli* Bjørklund；14–25. *Actinomma impolita* sp. nov.

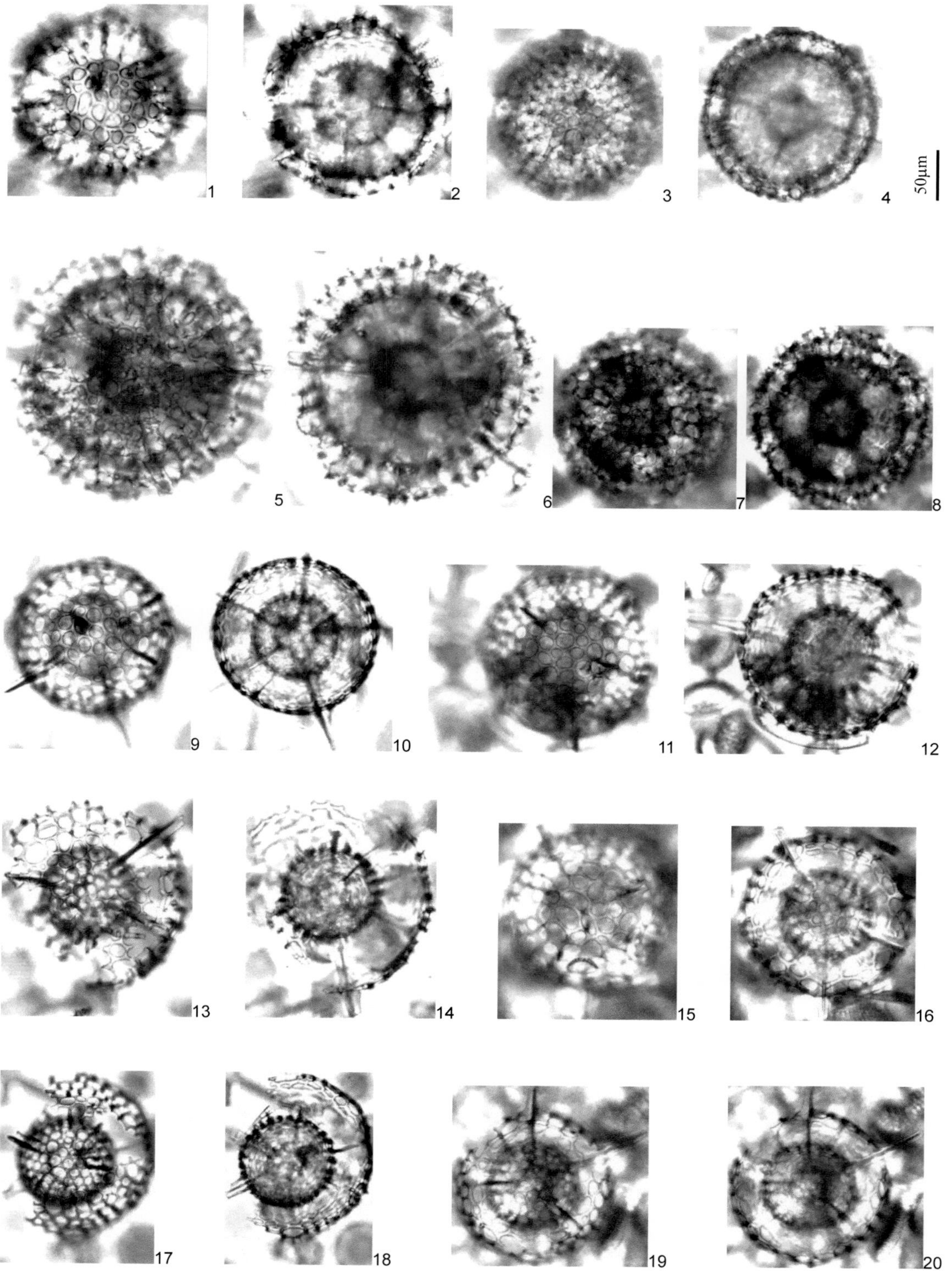

1–8. *Actinomma kerguelenensis* Caulet；9–20. *Actinomma laminata* sp. nov.

图版 34

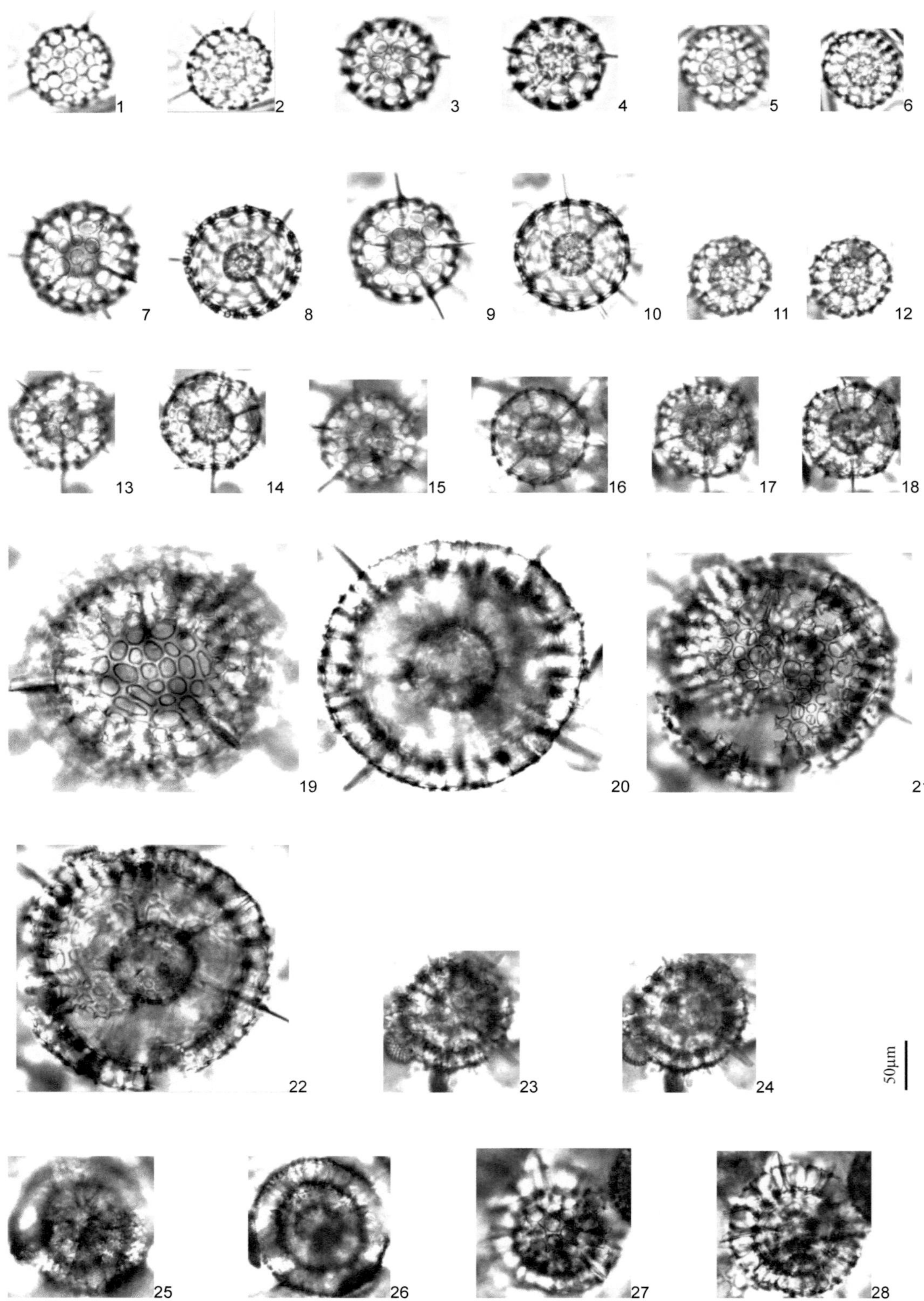

1–18. *Actinomma leptoderma* (Jørgensen)；19–28. *Actinomma livae* Goll and Bjørklund

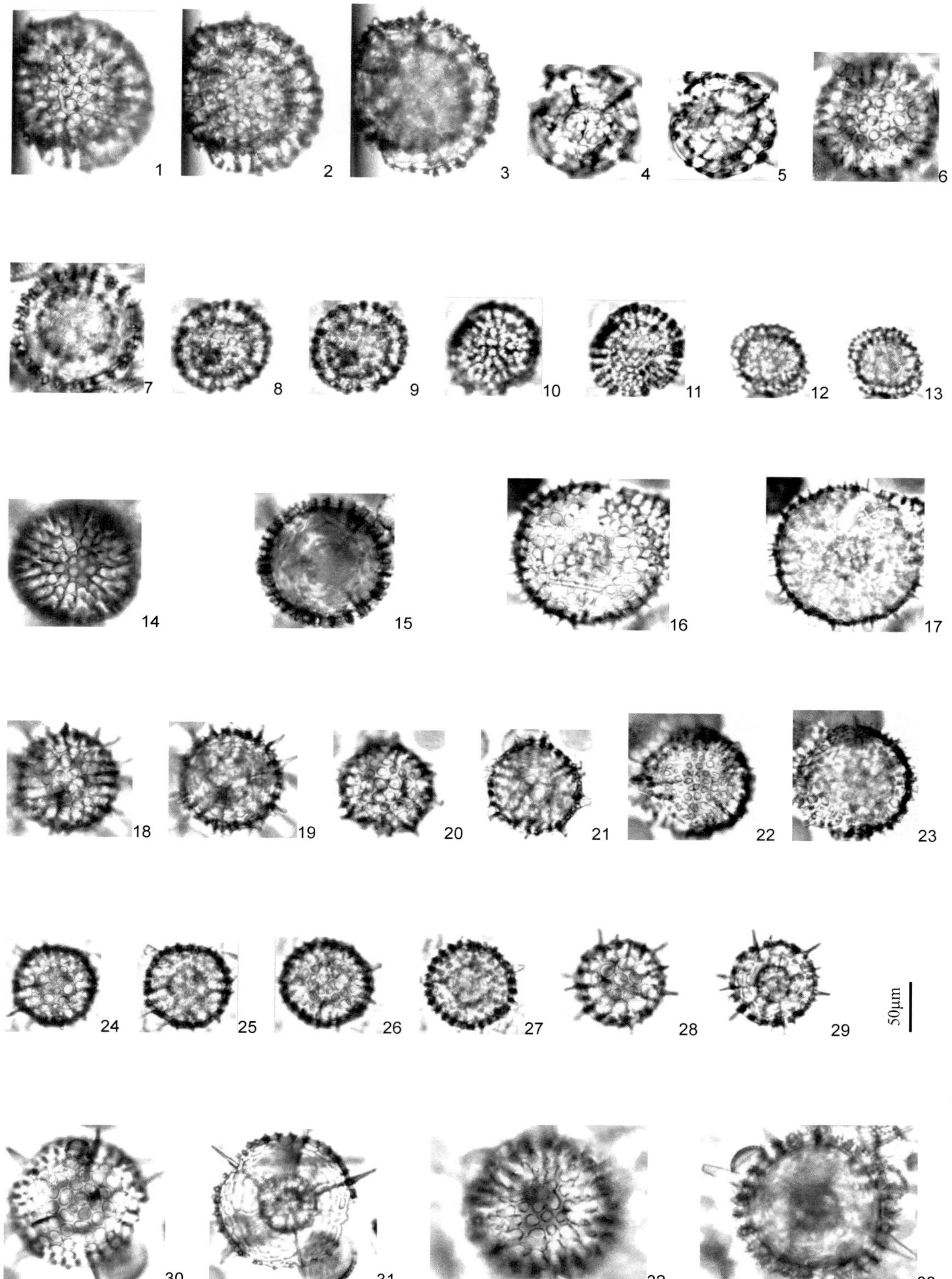

1–13. *Actinomma magicula* sp. nov.；14–15. *Actinomma medusa* (Ehrenberg)；16–27. *Actinomma mirabile* Goll and Bjørklund；28–29. *Actinomma pachyderma* Haeckel；30–33. *Actinomma plasticum* Goll and Bjørklund

图版 36

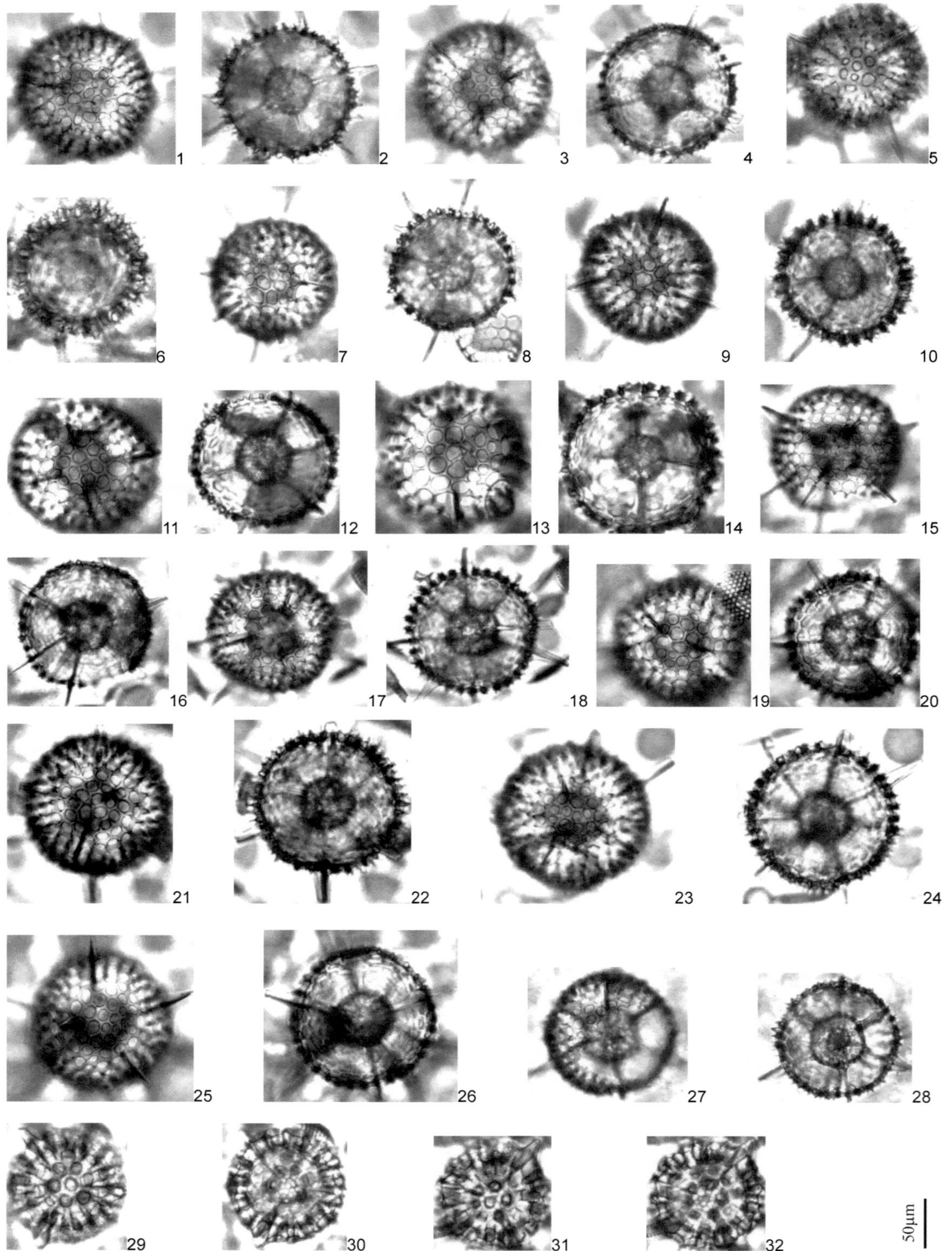

1–28. *Actinomma plasticum* Goll and Bjørklund；29–32. *Actinomma solidula* sp. nov.

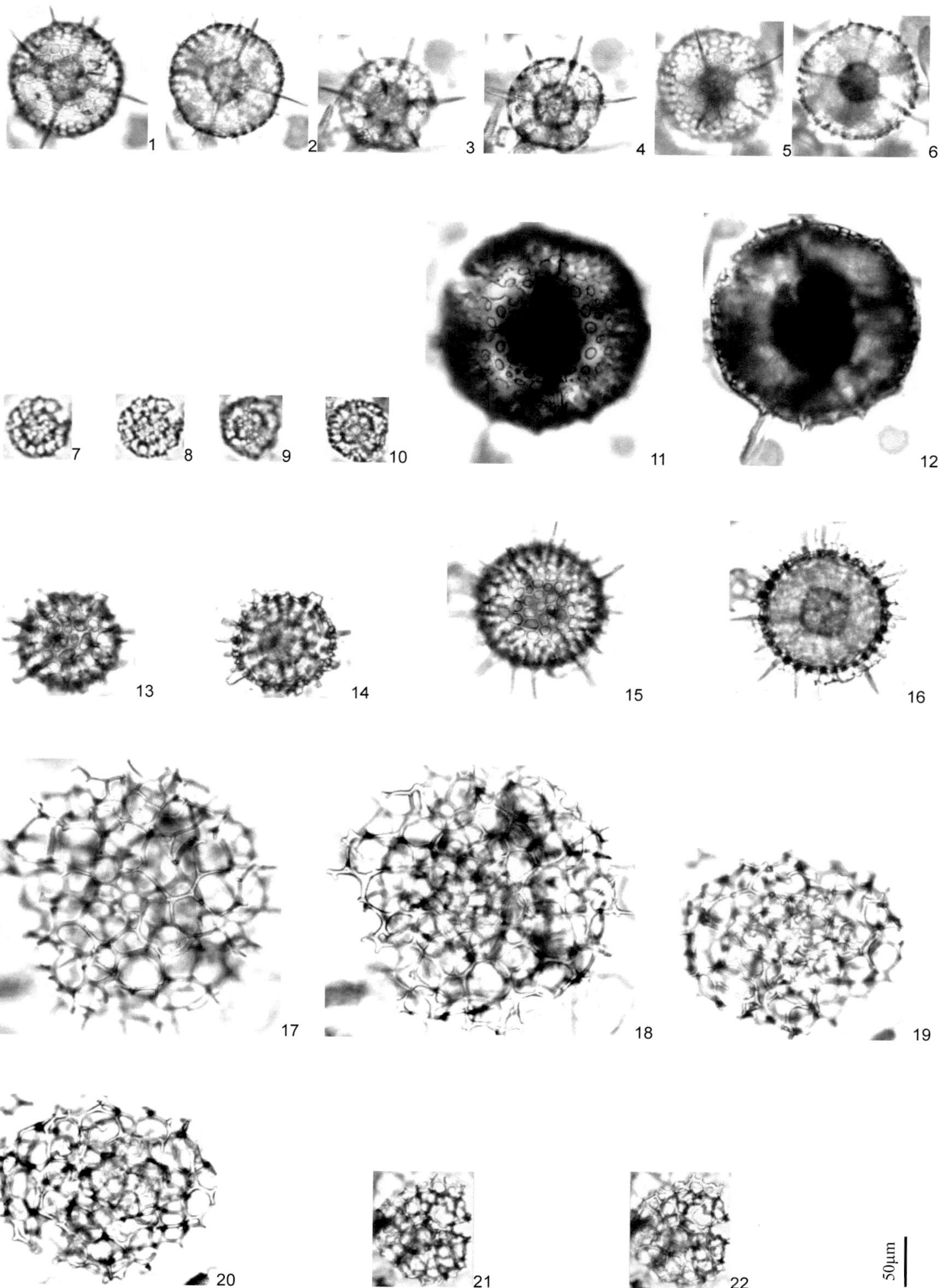

1–6. *Actinomma yosii* Nakaseko；7–10. *Actinomma* sp. 1；11–12. *Actinomma* sp. 2；13–14. *Echinomma popofskii* Petrushevskaya；15–16. *Cromyechinus* sp. 1；17–20. *Spongodrymus elaphococcus* Haeckel；21–22. *Rhizosphaera paradoxa* Popofsky

图版 38

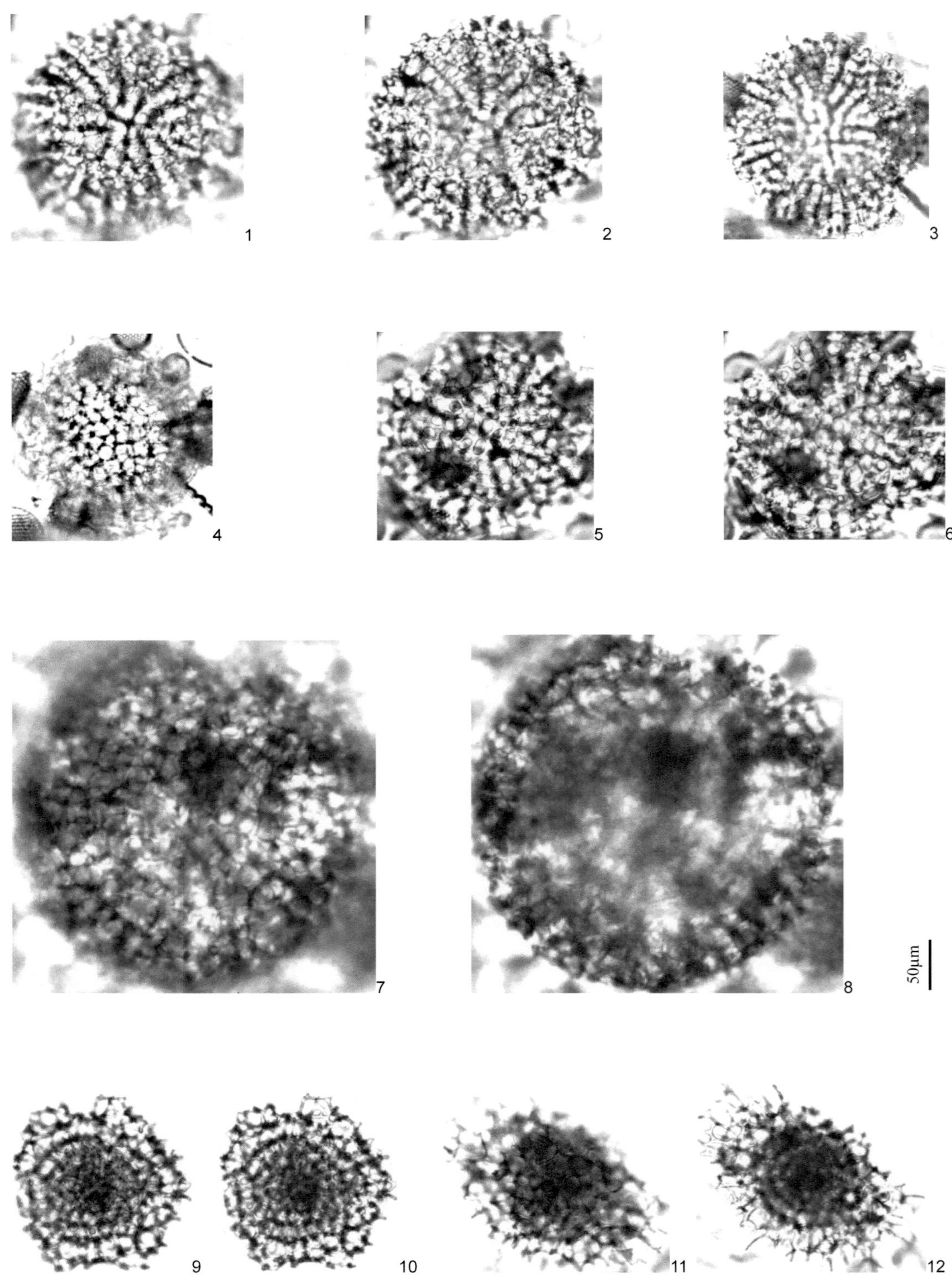

1–8. *Rhizosphaera reticulata* (Haeckel)；9–12. *Spongosphaera spongiosum* (Müller) group

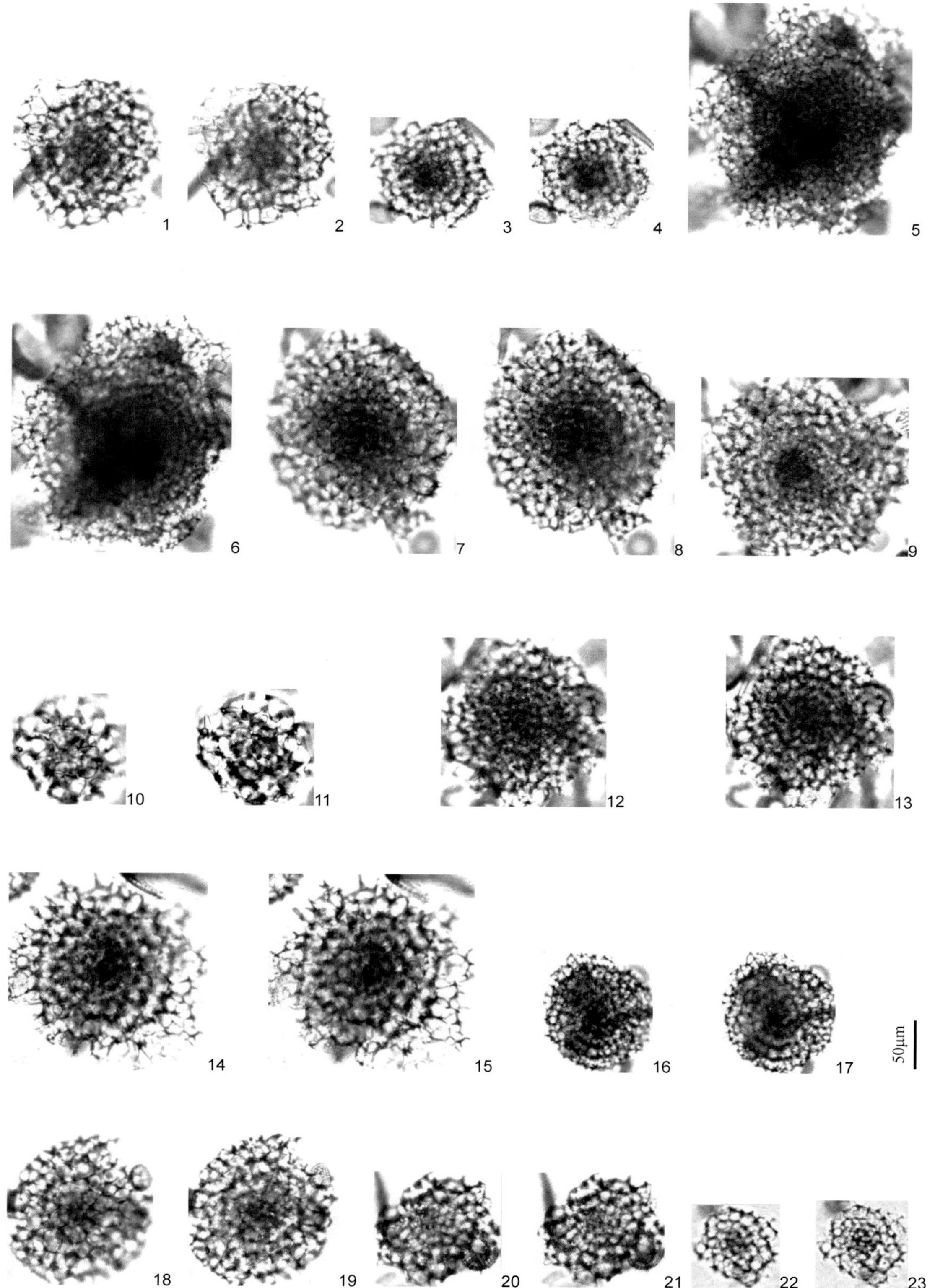

1–23. *Spongosphaera spongiosum* (Müller) group

图版 40

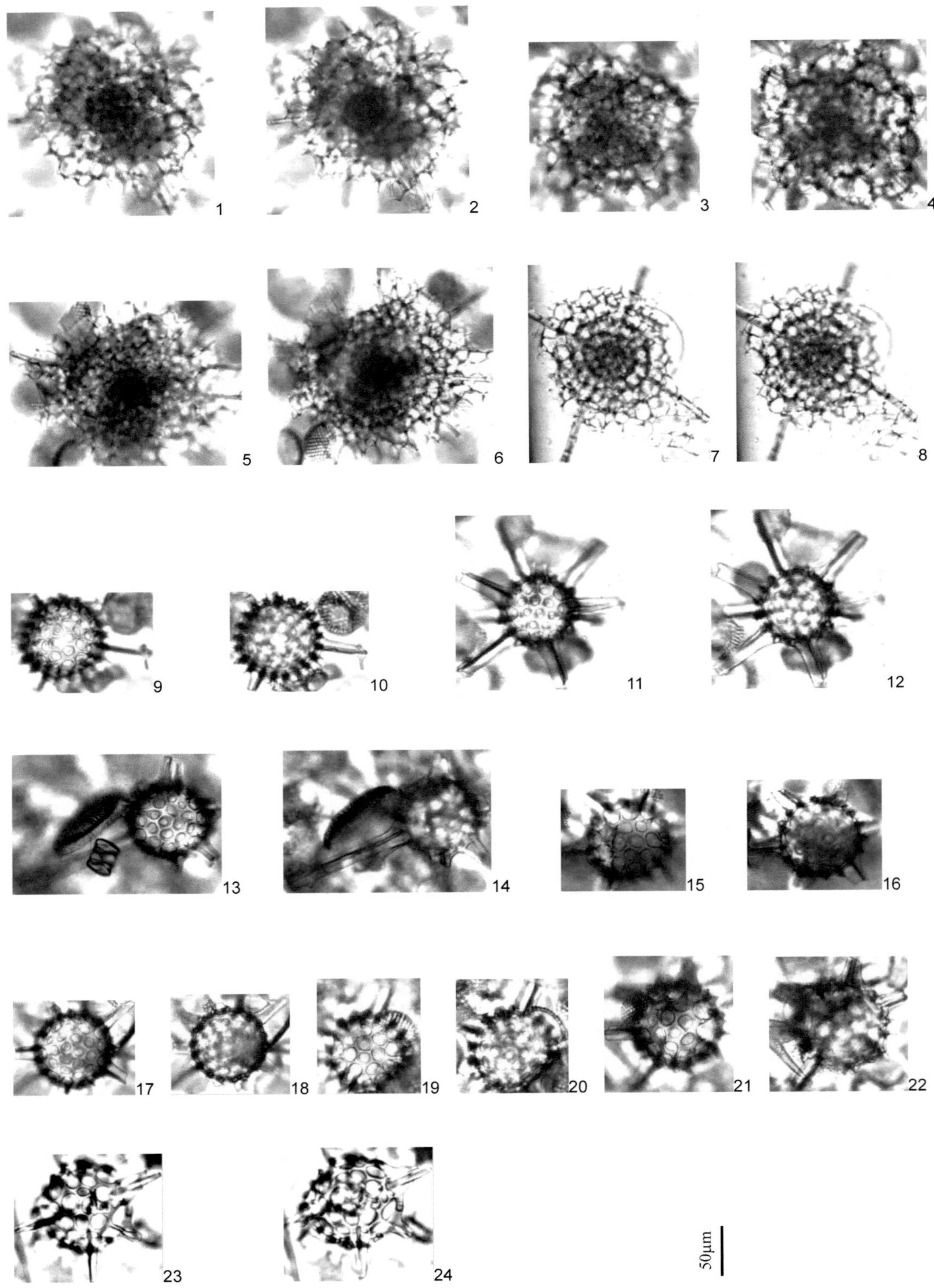

1–8. *Spongosphaera streptacantha* Haeckel；9–22. *Lychnosphaera regina* Haeckel；23–24. *Lychnosphaera* sp. 1

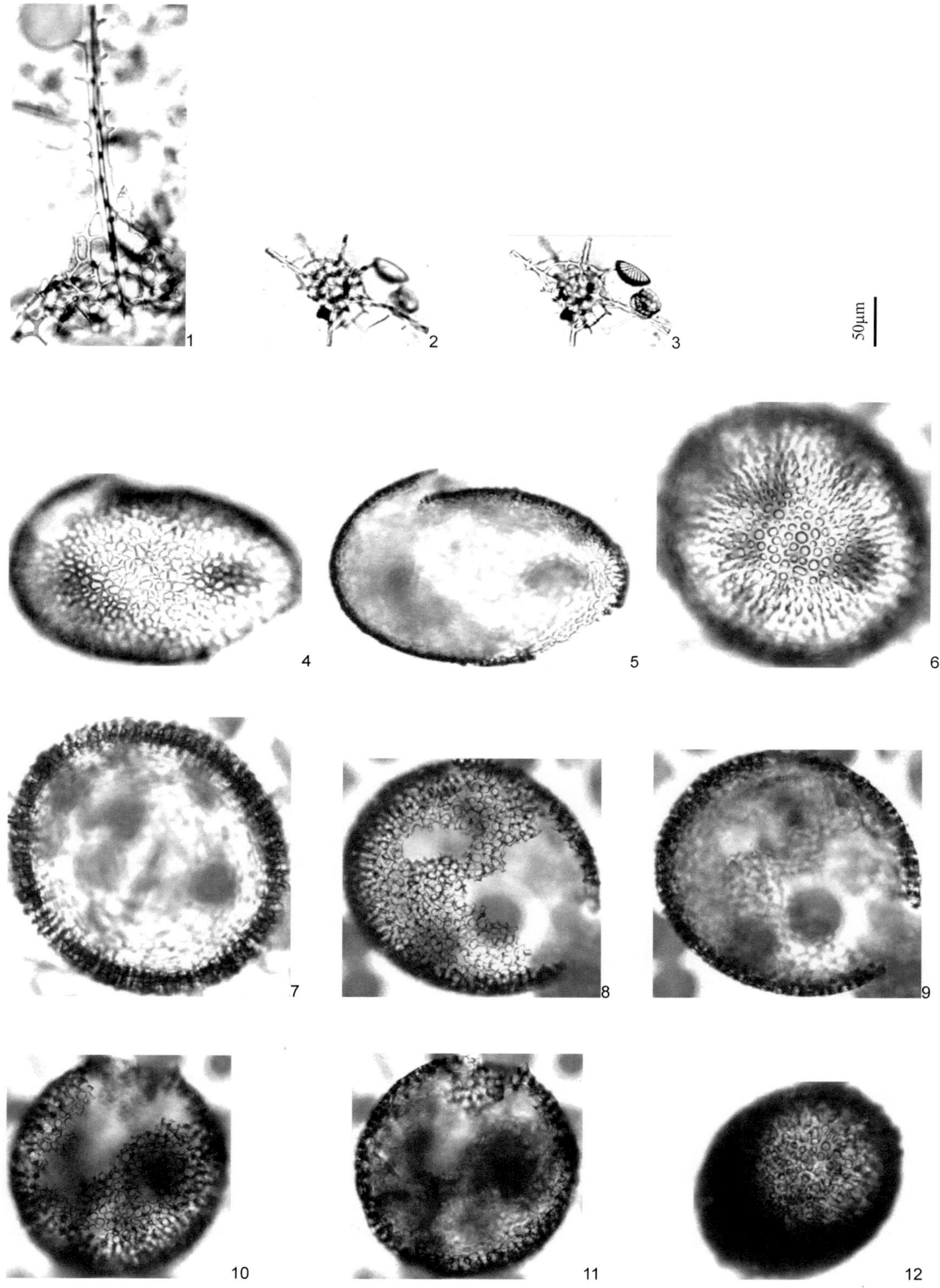

1–3. *Rhizoplegma boreale* (Cleve)；4–5. *Cenellipsis bergontianus* Carnevale group；6–12. *Cenellipsis monikae* (Petrushevskaya)

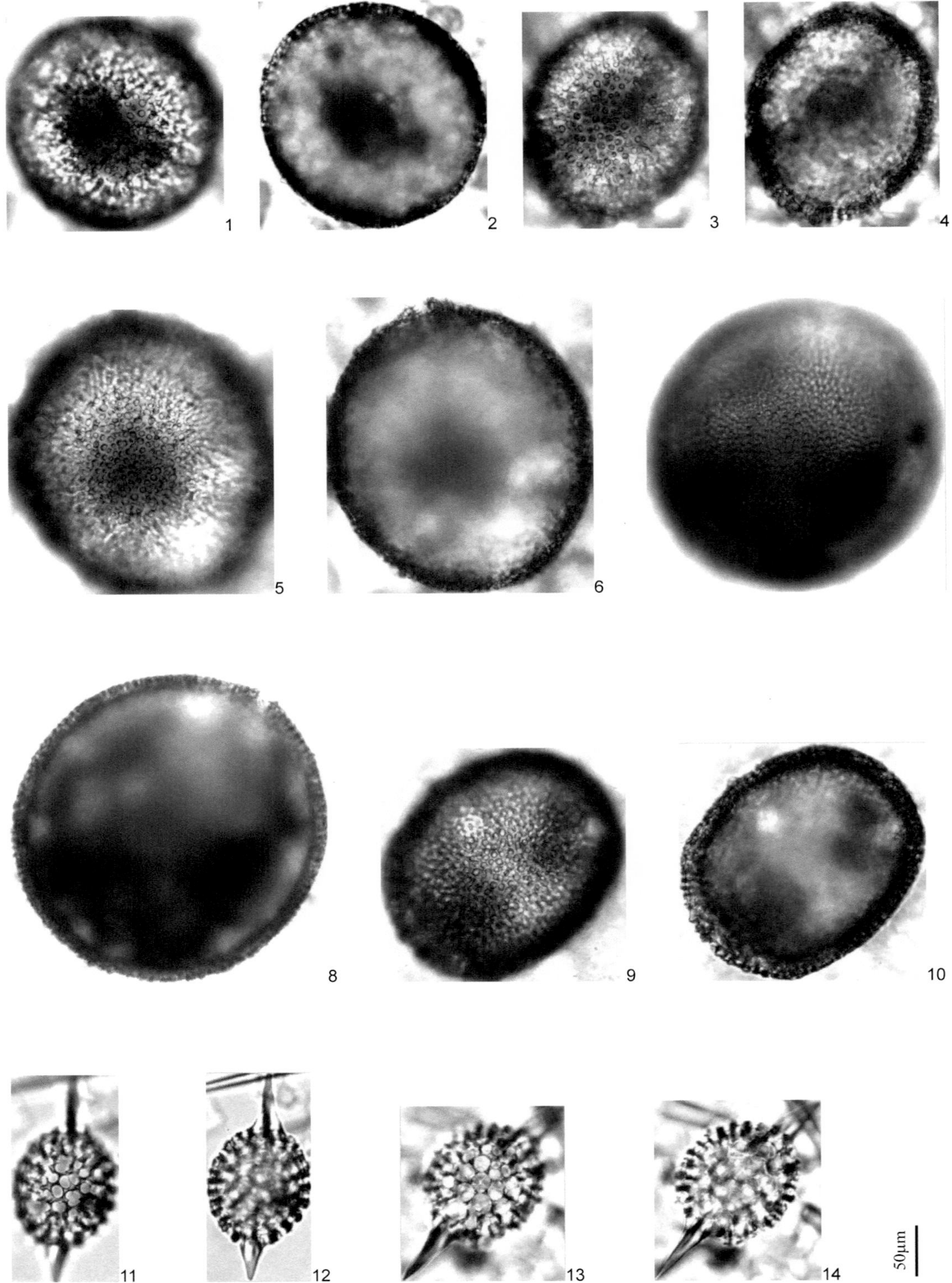

1–10. *Cenellipsis monikae* (Petrushevskaya)；11–14. *Ellipsoxiphus atractus* Haeckel

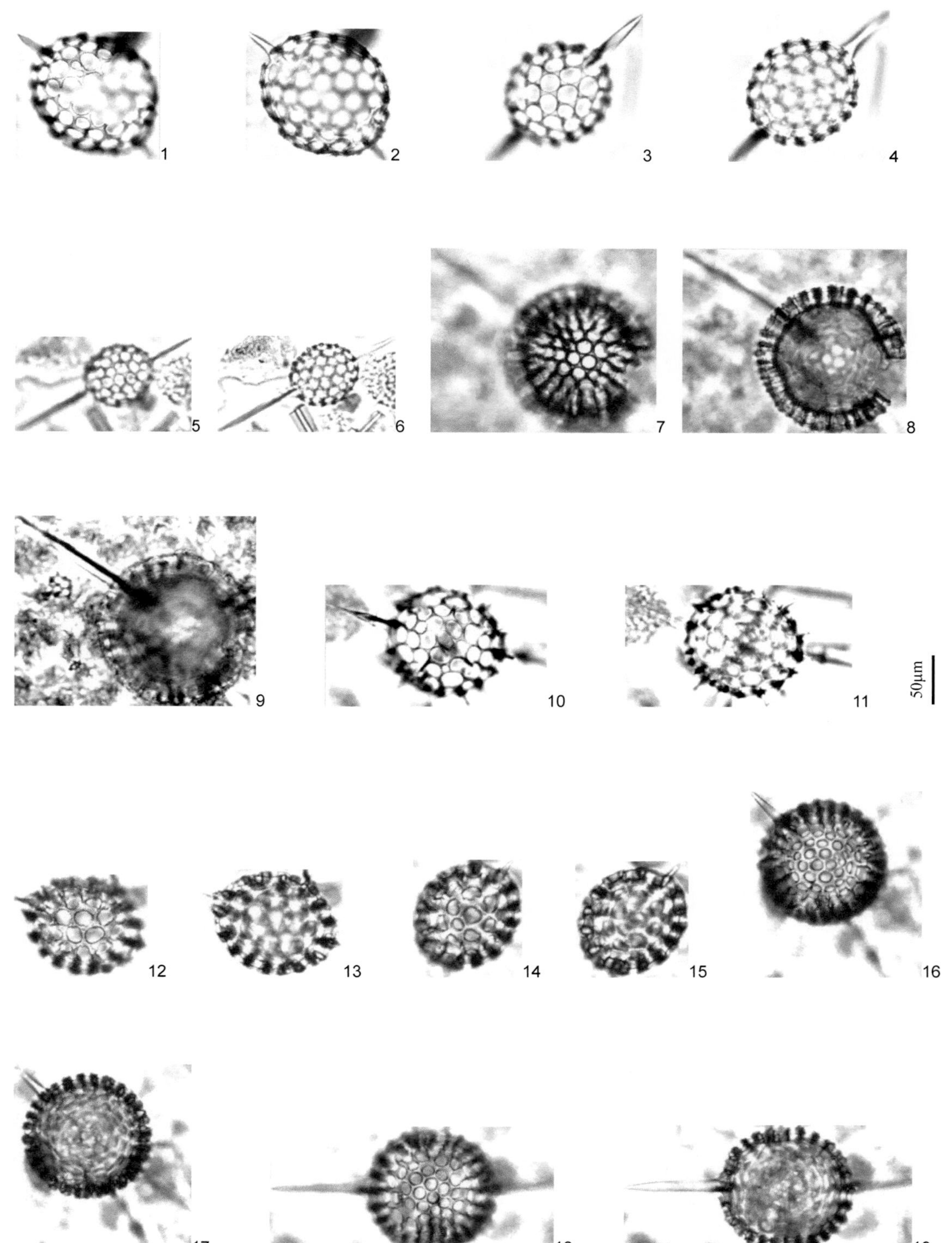

1–6. *Ellipsoxiphus leptodermicus* sp. nov.; 7–9. *Ellipsoxiphus rotundalus* sp. nov.; 10–11. *Ellipsoxiphus echinatus* sp. nov.; 12–13. *Lithapium pyriforme* Haeckel; 14–15. *Lithapium halicapsa* Haeckel; 16–19. *Axoprunum stauraxoniurn* Haeckel

图版 44

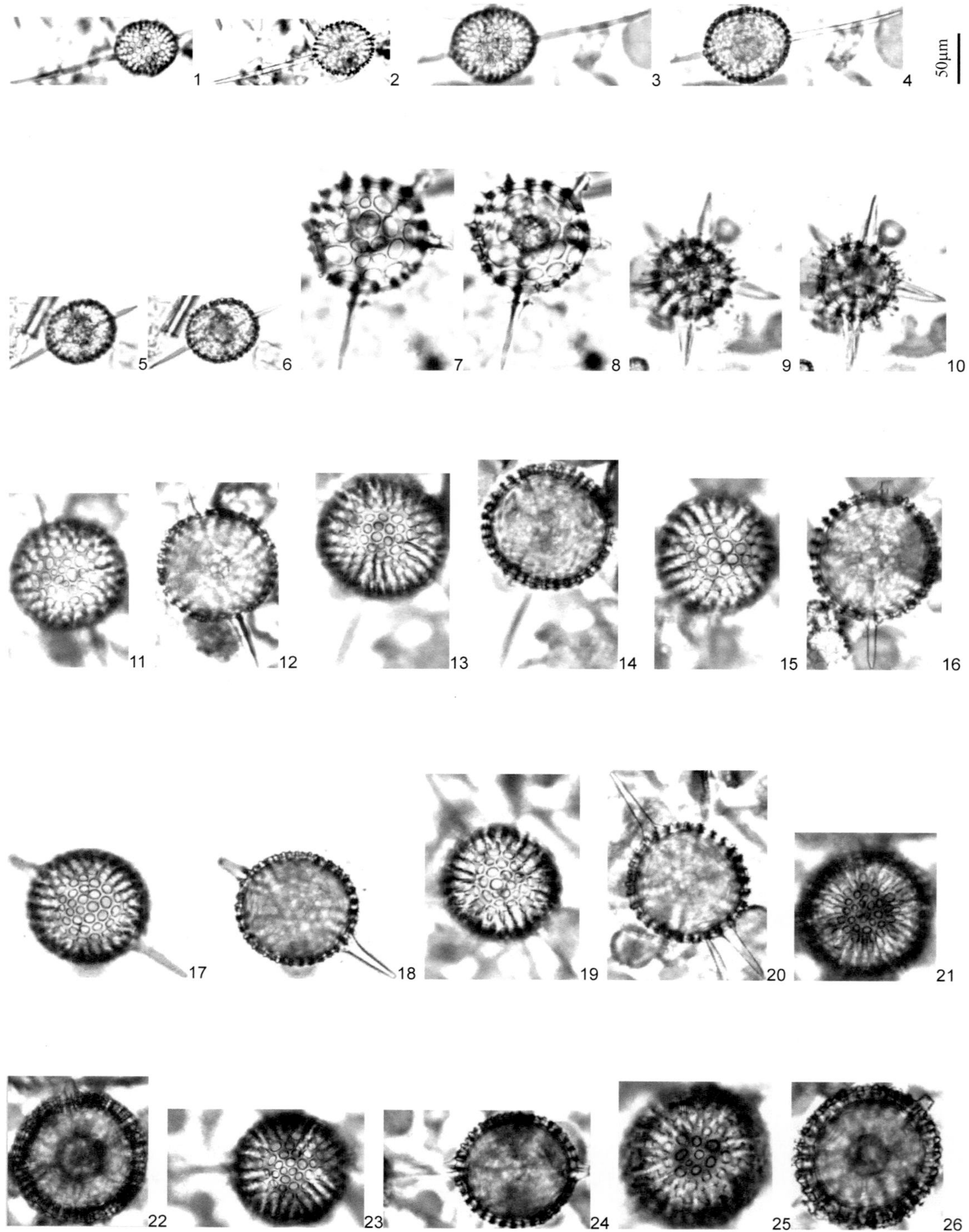

1–6. *Stylosphaera coronata laevis* Ehrenberg；7–10. *Stylosphaera goruna* Sanfilippo and Riedel；
11–26. *Stylosphaera minor* Clark and Campbell

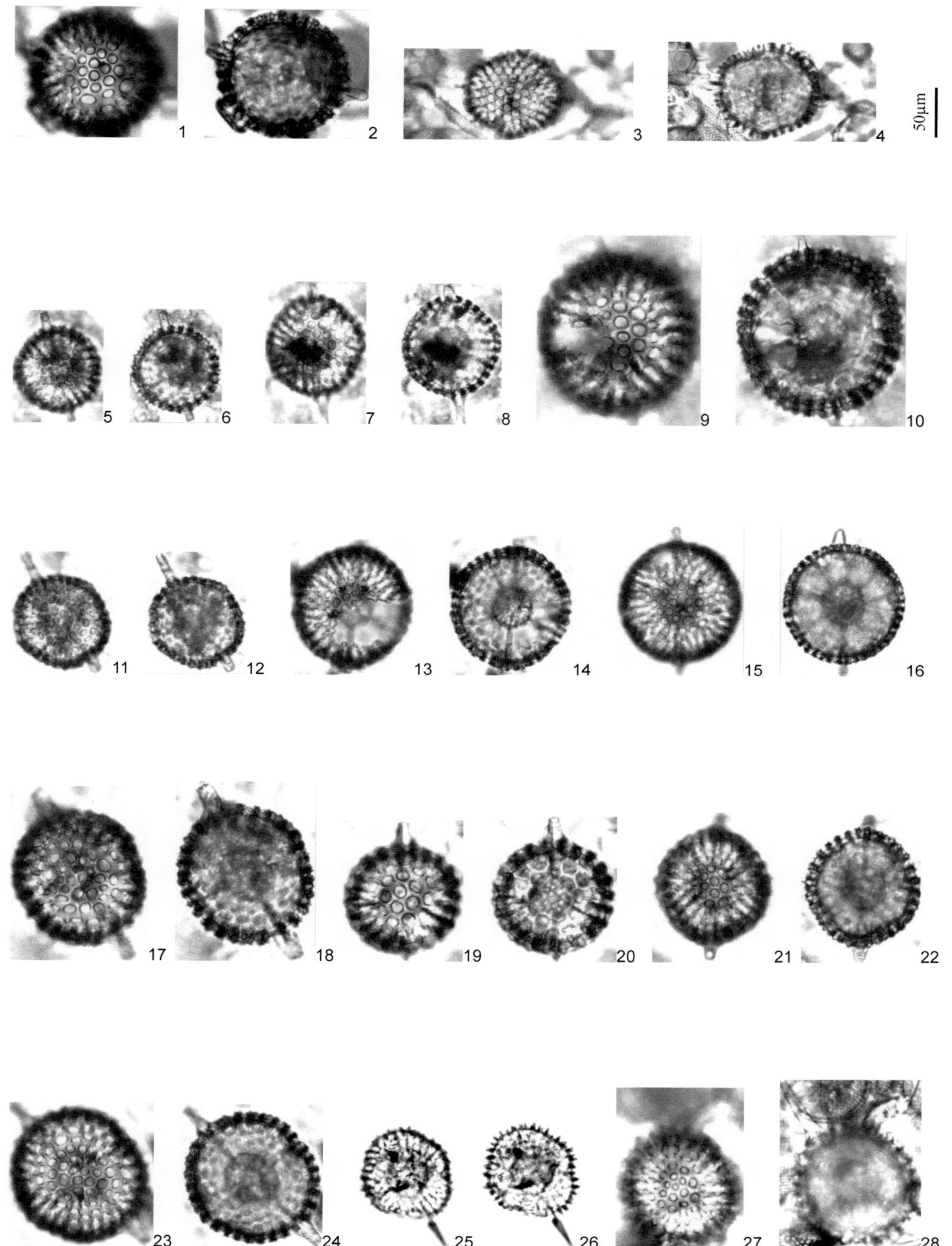

1–24. *Stylosphaera minor* Clark and Campbell；25–26. *Stylosphaera pyriformis* (Bailey)；27–28. *Stylosphaera* sp. 1

图版 46

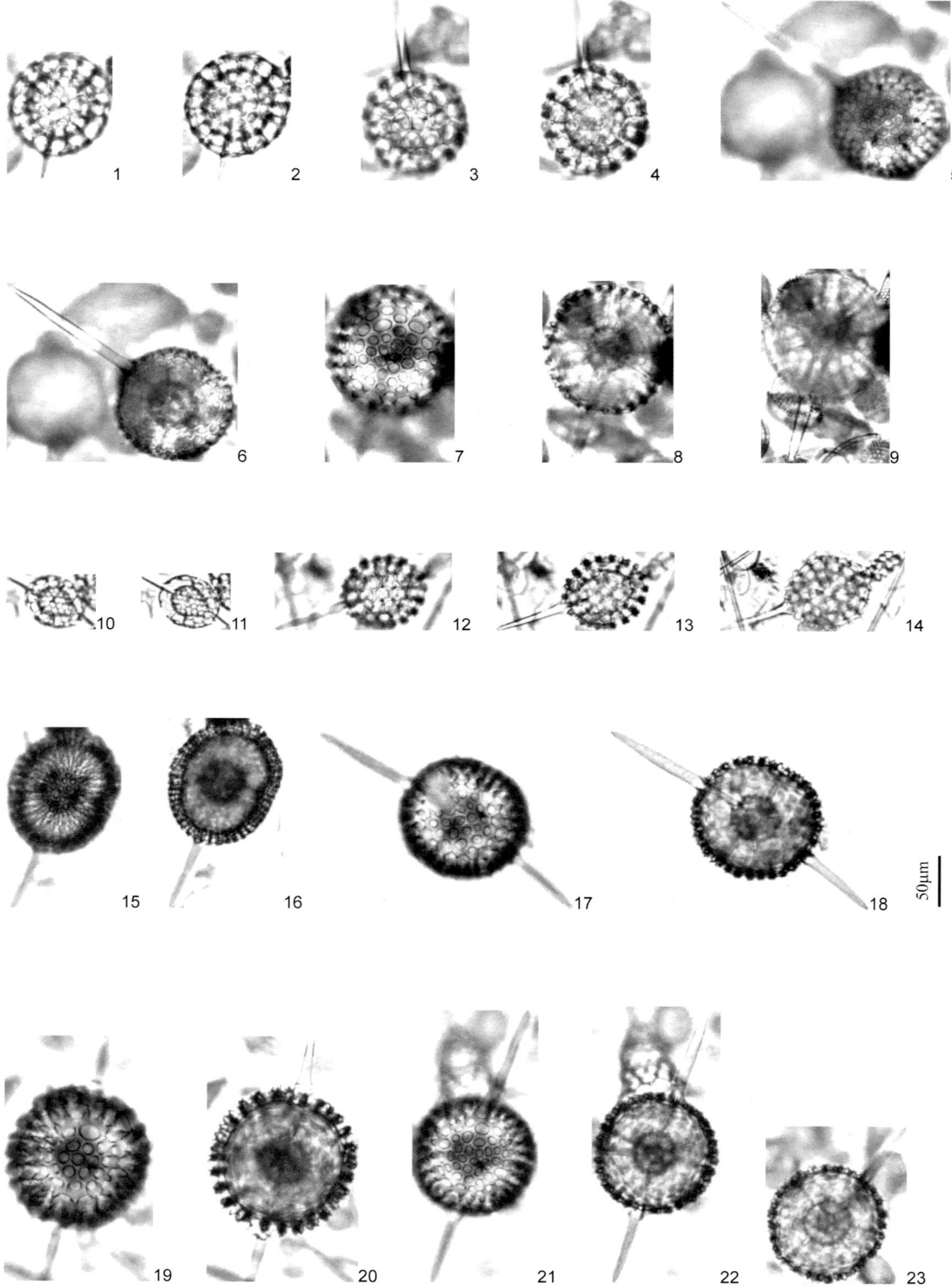

1–4. *Druppatractus carpocus* sp. nov.; 5–9. *Druppatractus hastatus* Blueford group; 10–11. *Druppatractus irregularis* Popofsky; 12–14. *Druppatractus pyriformus* sp. nov.; 15–16. *Stylatractus angelina* (Campbell and Clark); 17–23. *Stylatractus gracilus* sp. nov.

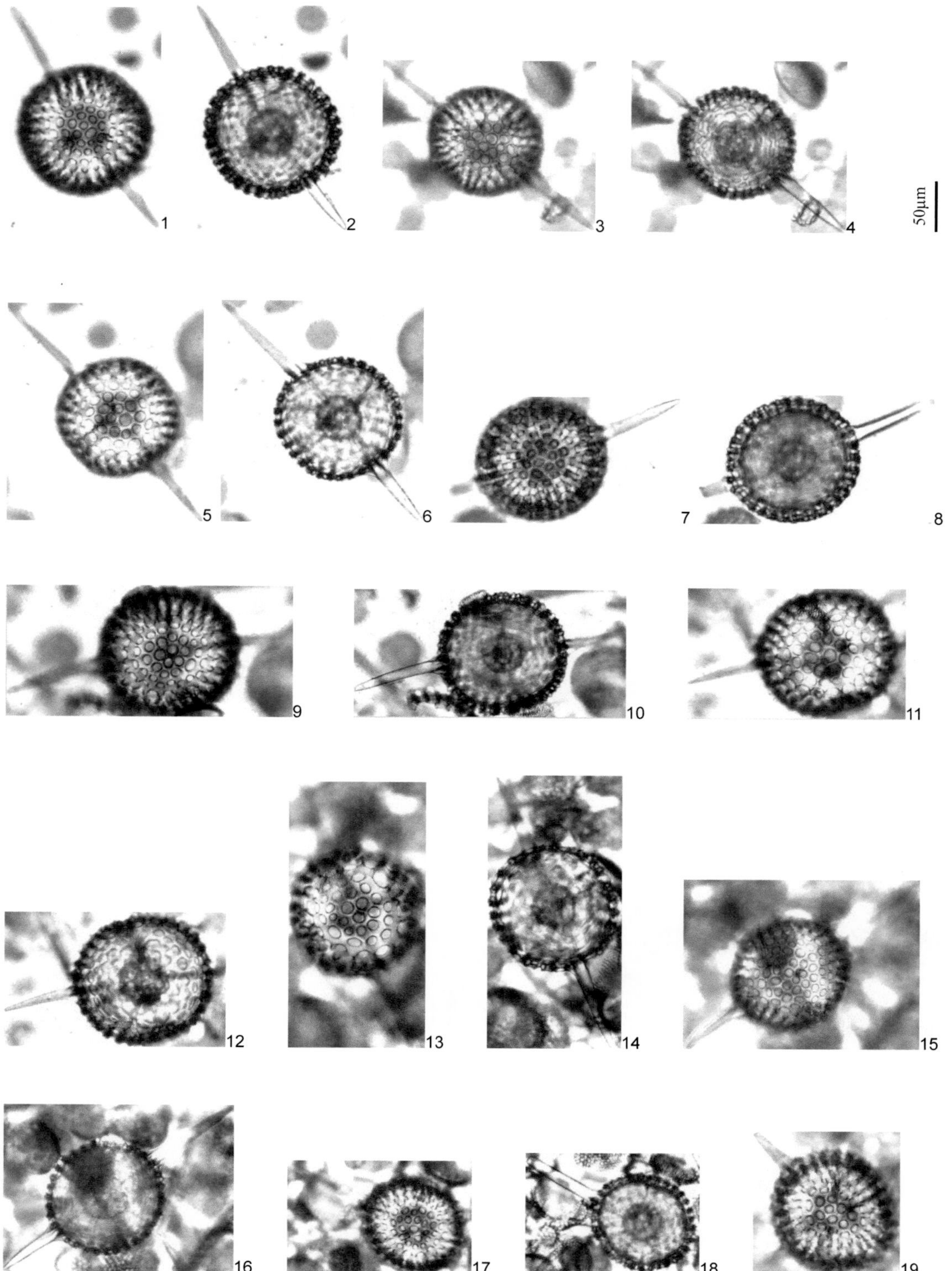

1–19. *Stylatractus gracilus* sp. nov.

图版 48

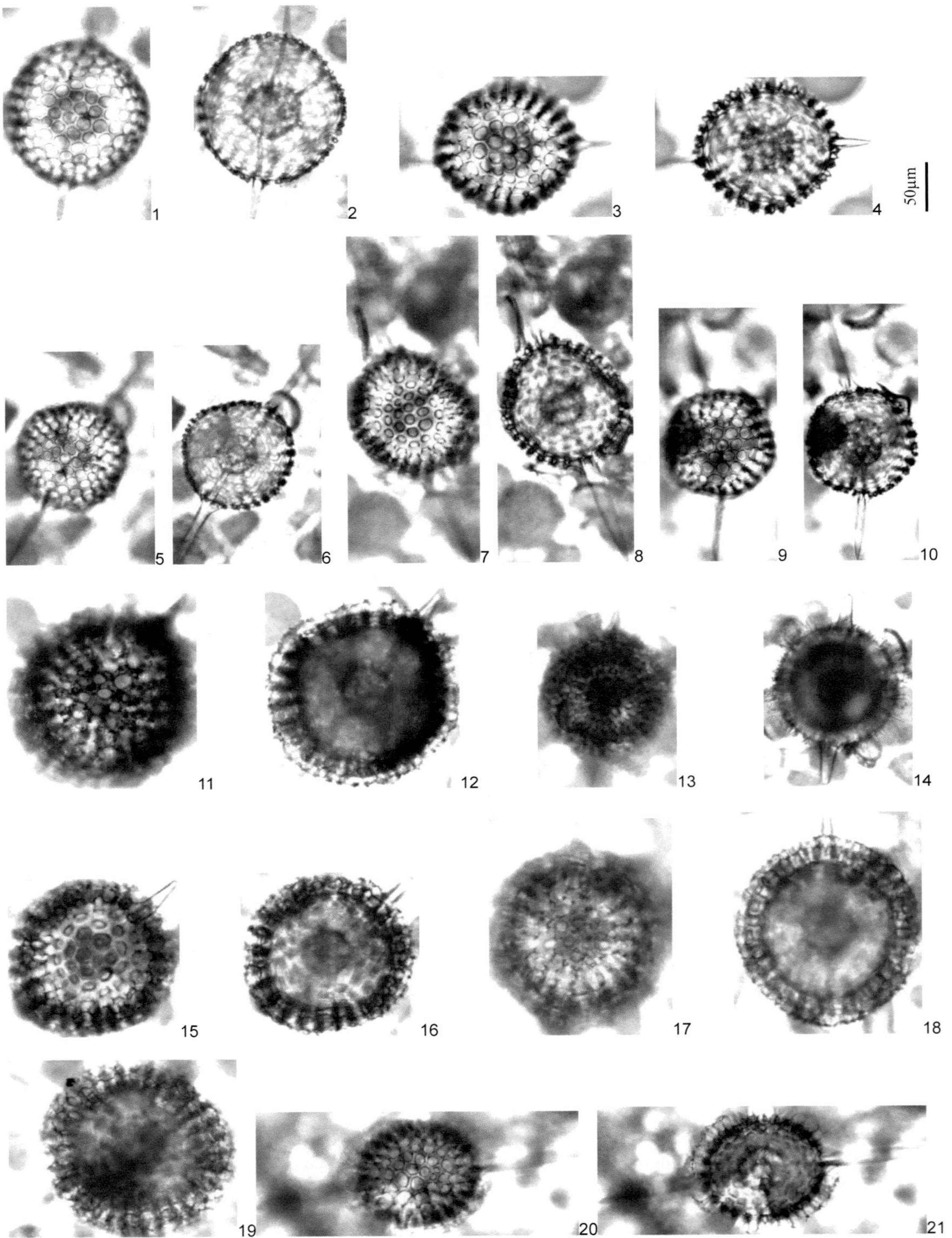

1–10. *Stylatractus irregularis* (Takemura)；11–21. *Stylatractus palliatum* (Haeckel)

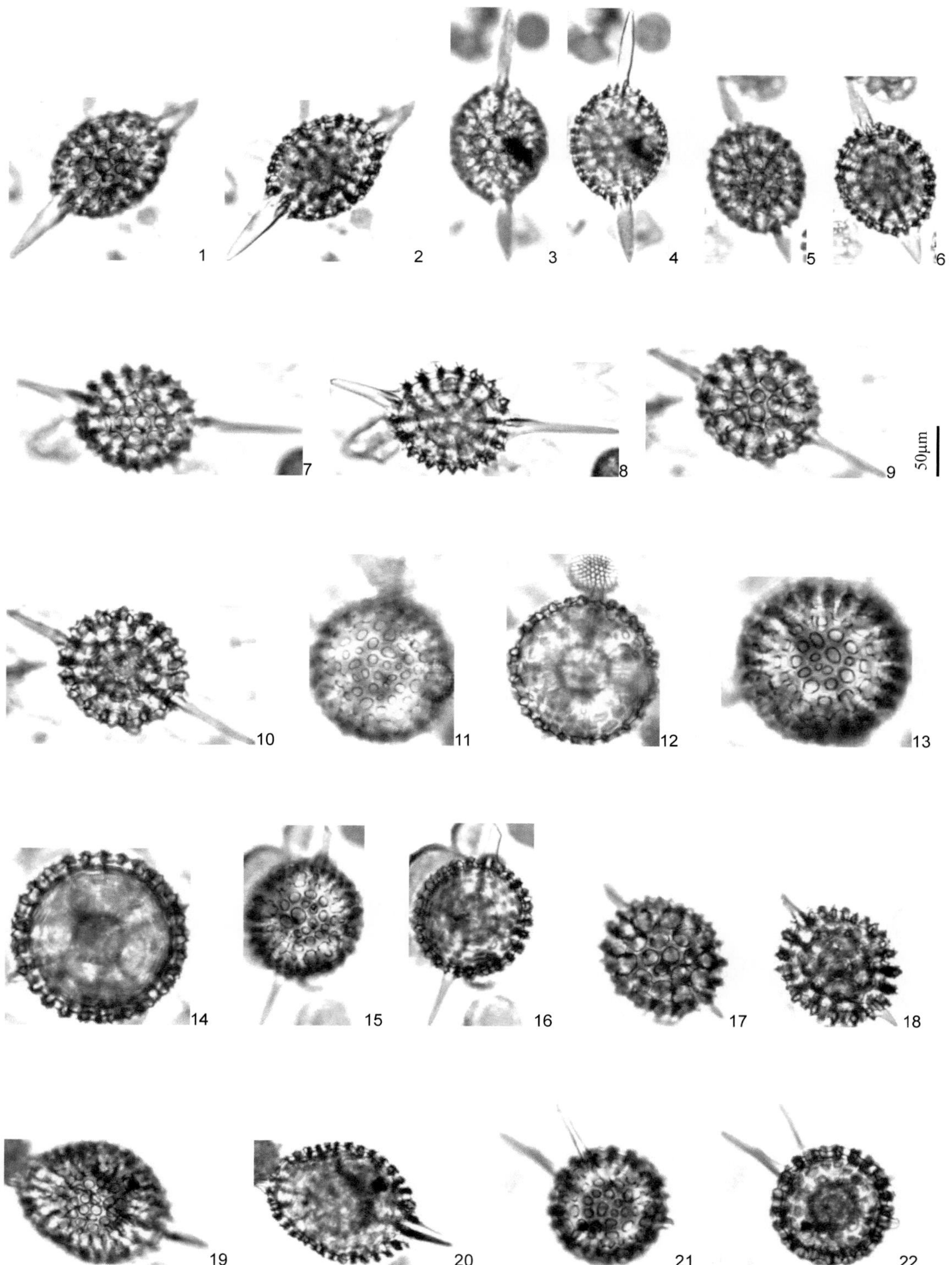

1–6. *Stylatractus pluto* (Haeckel)；7–10. *Xiphatractus longostylus* sp. nov.；
11–16. *Xiphatractus megaxyphos megaxyphos* (Clark and Campbell)；17–18. *Xiphatractus spumeus* Dumitrica；
19–20. *Xiphatractus* sp. 1；21–22. *Xiphatractus* sp. 2

图版 50

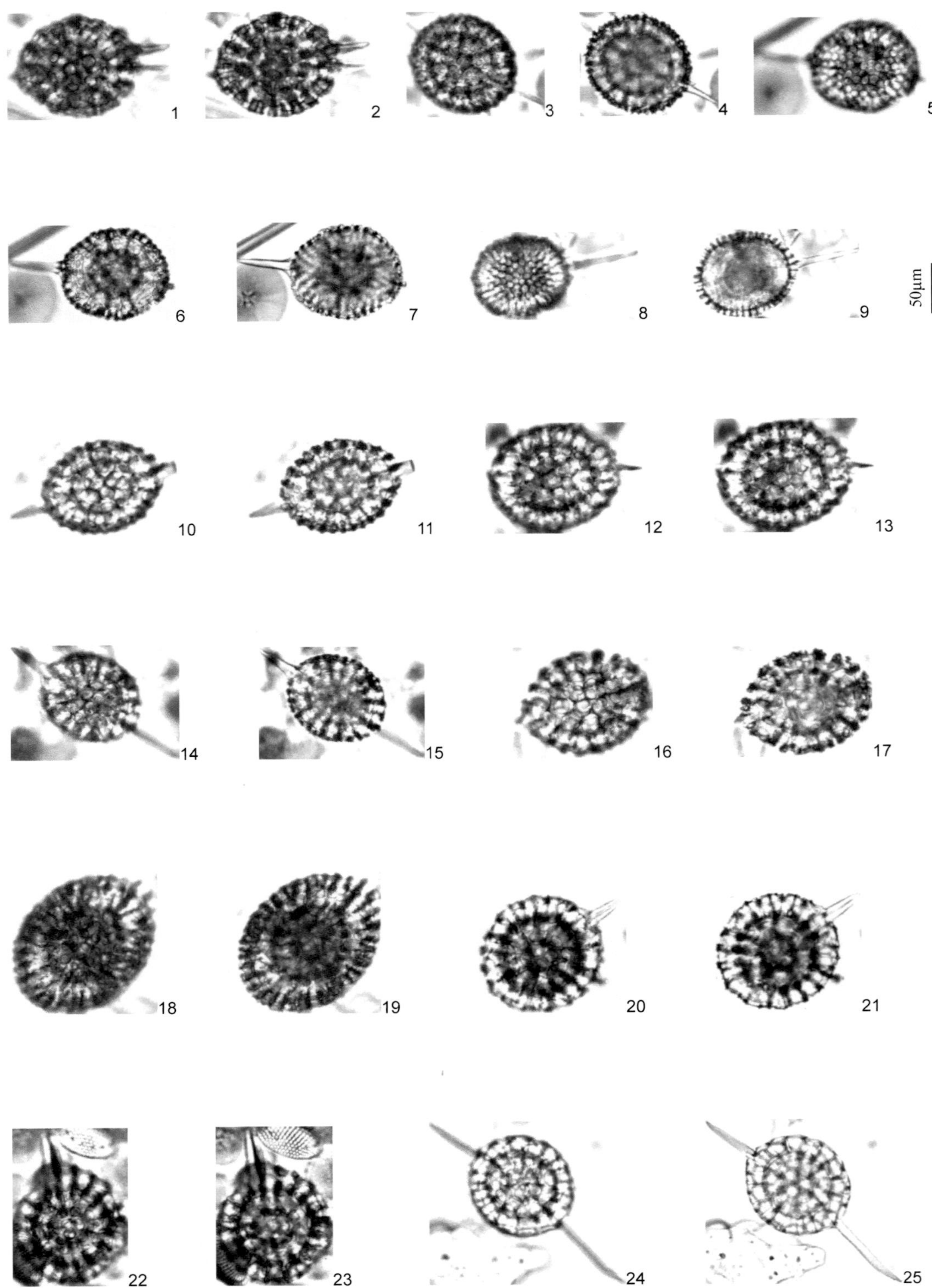

1–2. *Amphisphaera aotea* Hollis；3–4. *Amphisphaera dixyphos* (Ehrenberg)；5–9. *Amphisphaera nigriniae* Kamikuri；10–23. *Amphisphaera radiosa* (Ehrenberg) group；24–25. *Amphisphaera sperlita* sp. nov.

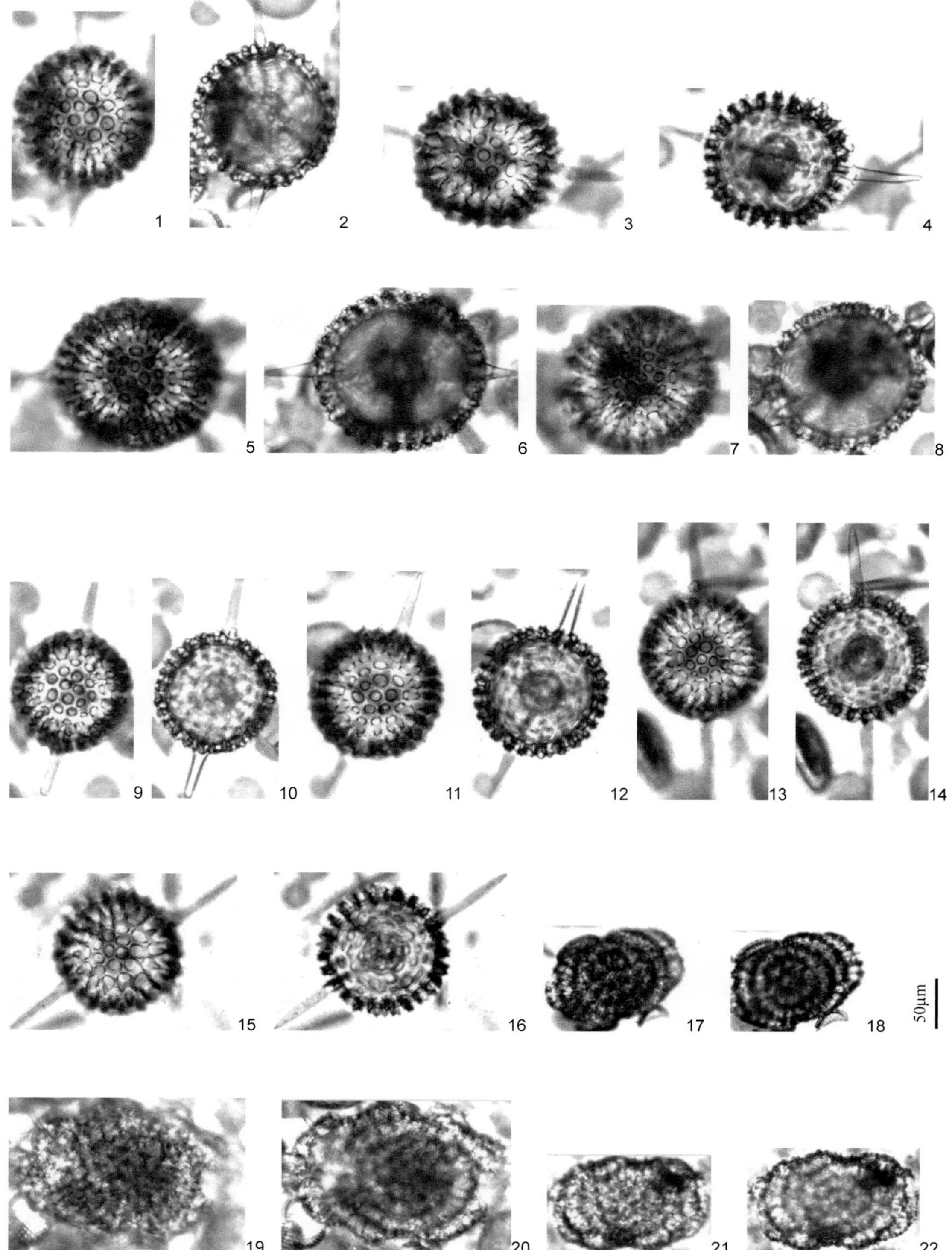

1–16. *Stylacontarium acquilonium* (Hays) group；17–22. *Spongurus cauleti* Goll and Bjørklund

图版 52

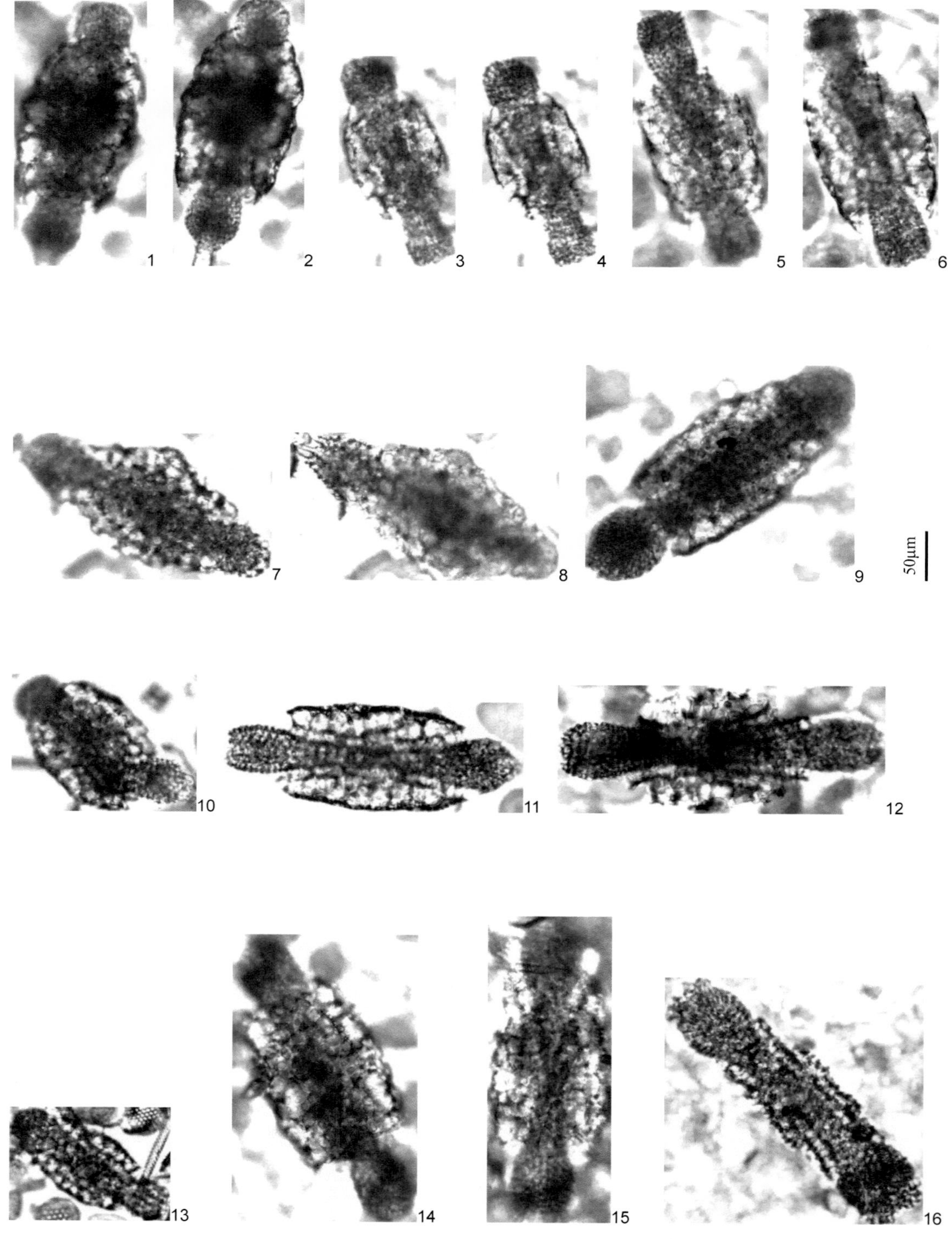

1–16. *Spongocore cylindricus* Haeckel

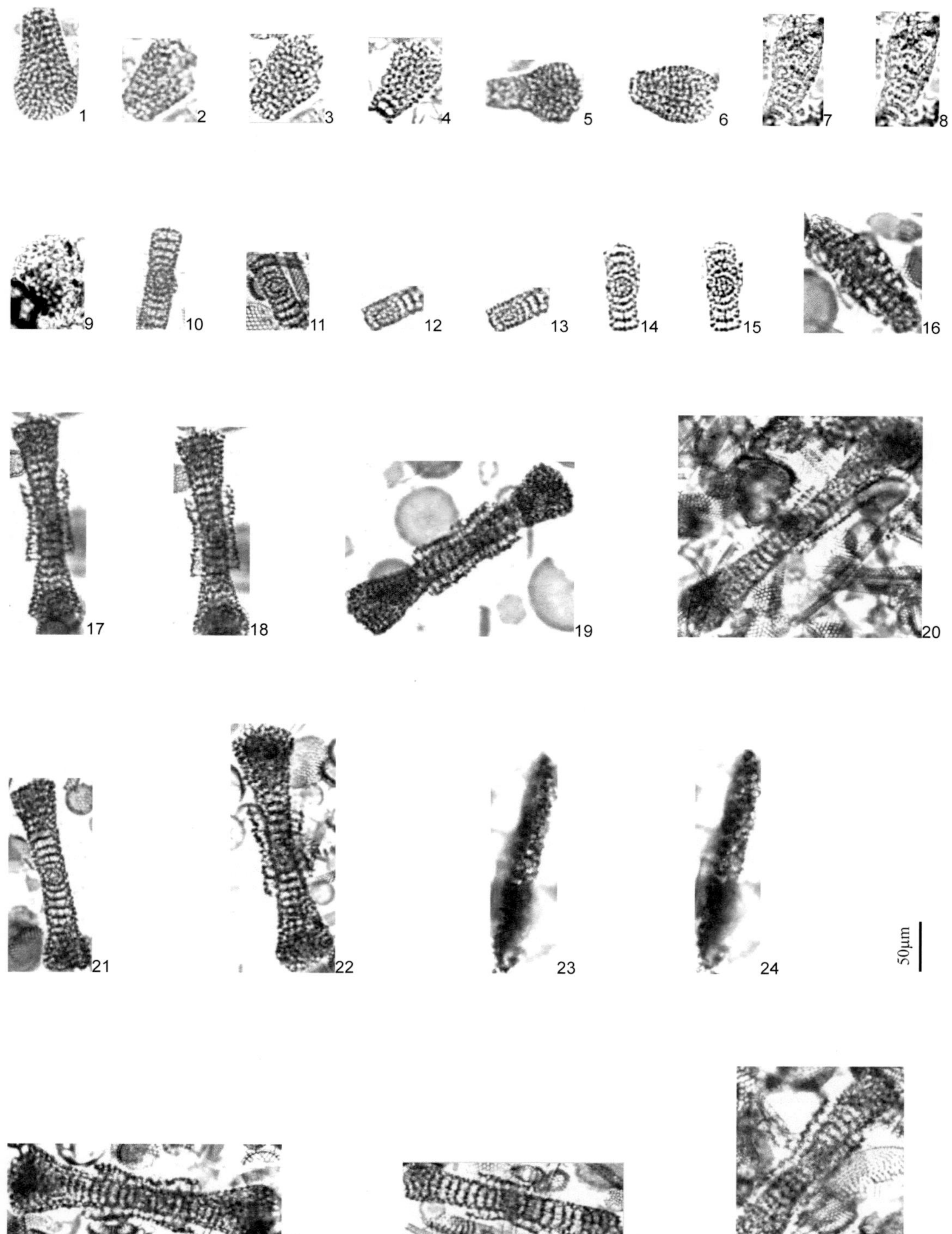

1–9. *Spongocore* sp. 1；10–15. *Amphymenium monstrosum* (Popofsky)；16–27. *Amphymenium* cf. *splendiarmatum* Clark and Campbell

图版 54

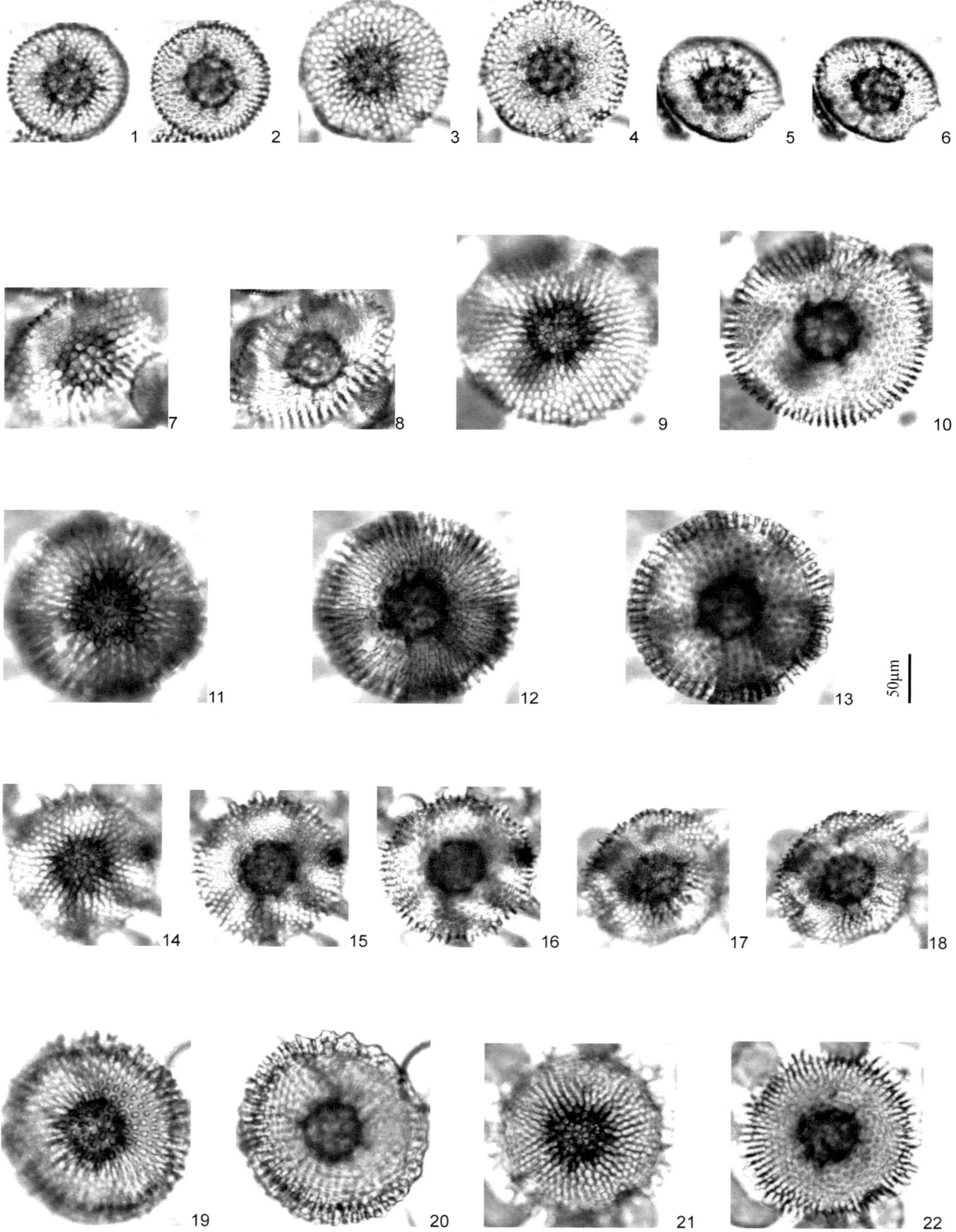

1–6. *Phacodiscus clypeus* Haeckel；7–8. *Phacodiscus rotula* Haeckel；9–13. *Phacodiscus lentiformis* Haeckel；14–22. *Astrophacus perplexus* (Clark and Campbell)

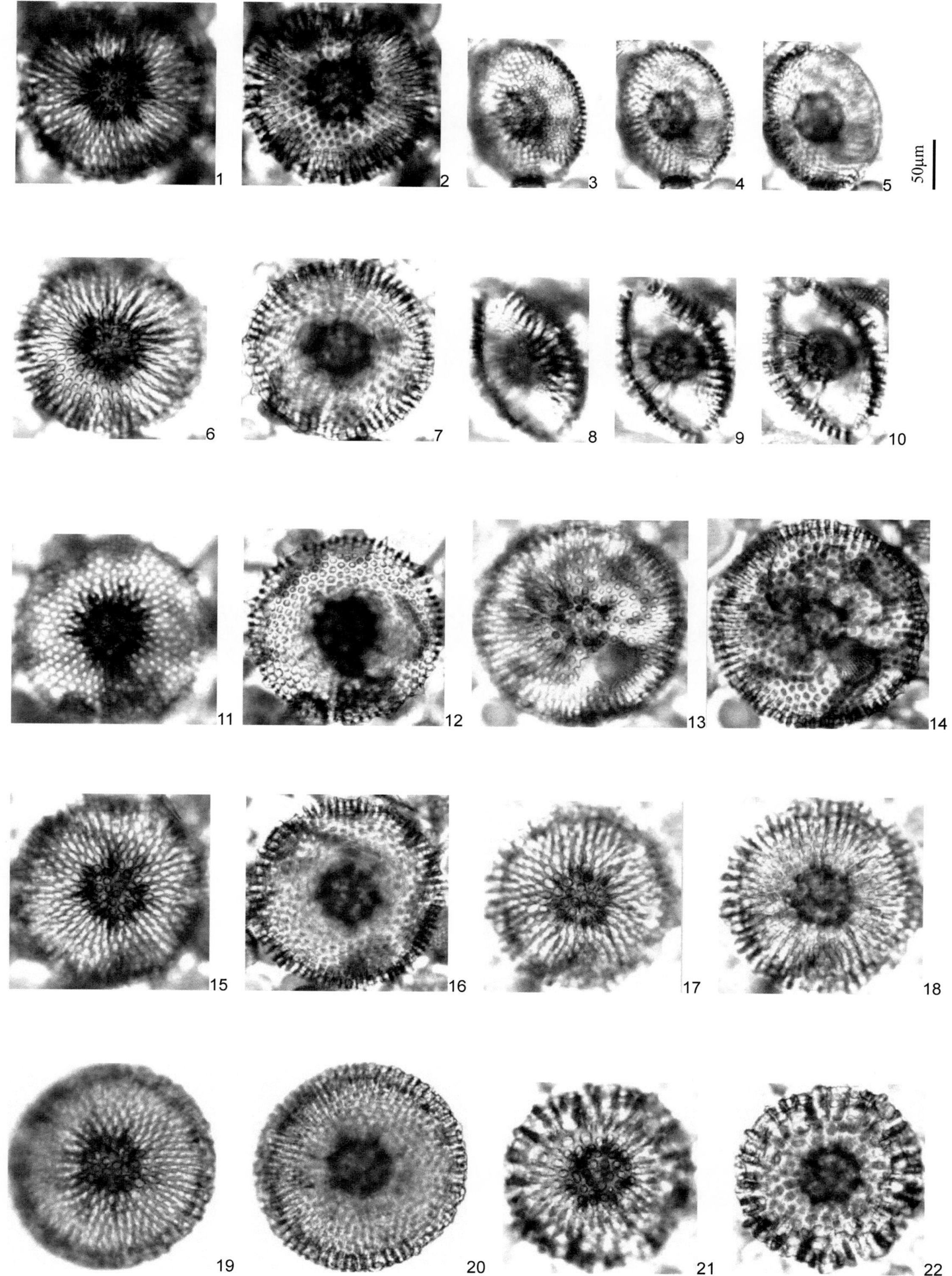

1–22. *Astrophacus perplexus* (Clark and Campbell)

图版 56

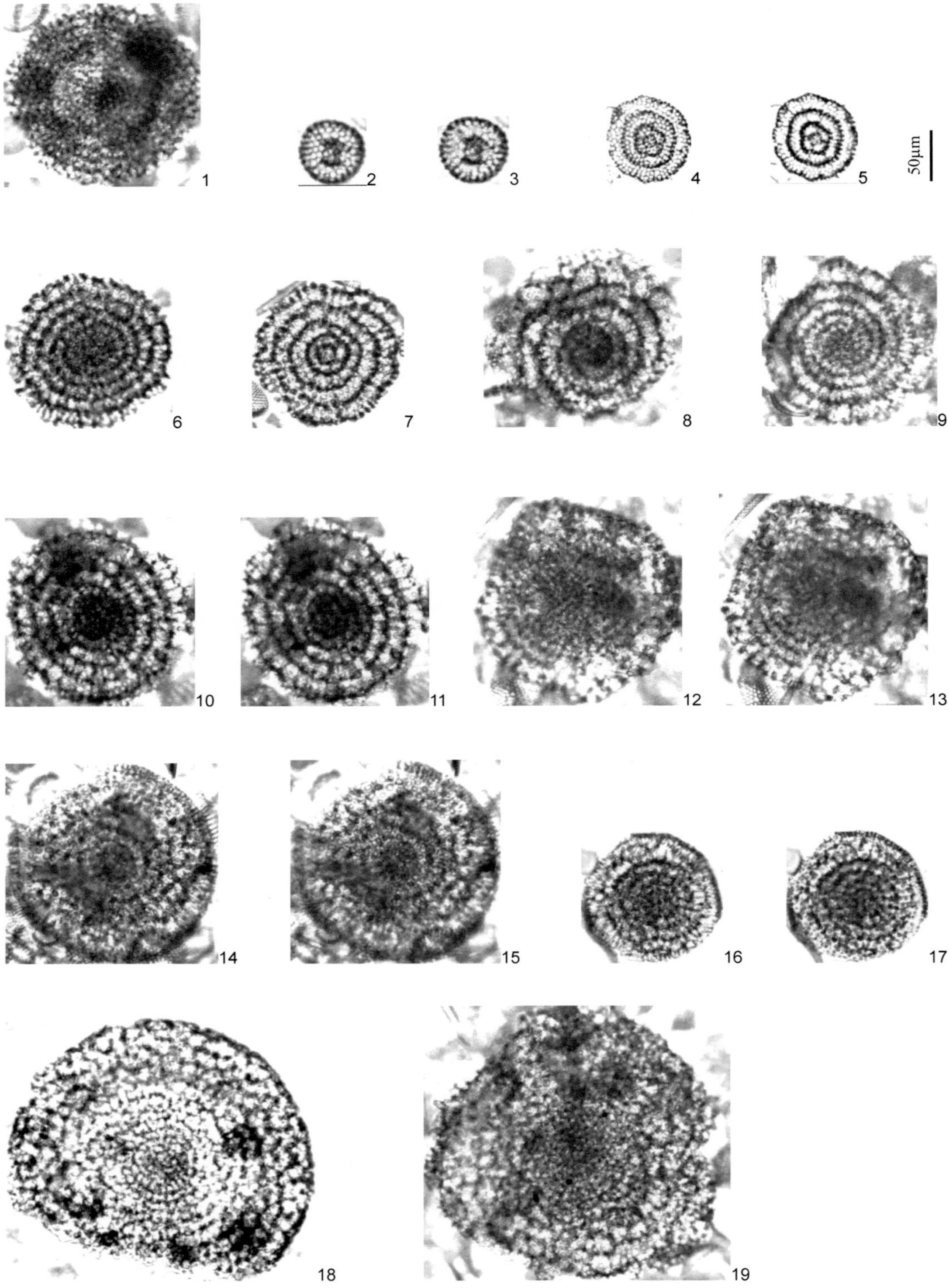

1. *Lithocyclia ocellus* Ehrenberg；2–3. *Archidiscus stauroniscus* Haeckel；4–5. *Porodiscus circularis* Clark and Campbell；6–7. *Porodiscus micromma* (Harting)；8–9. *Porodiscus perispira* Haeckel；10–11. *Porodiscus quadrigatus* Haeckel；12–19. *Perichlamydium irregularmus* sp. nov.

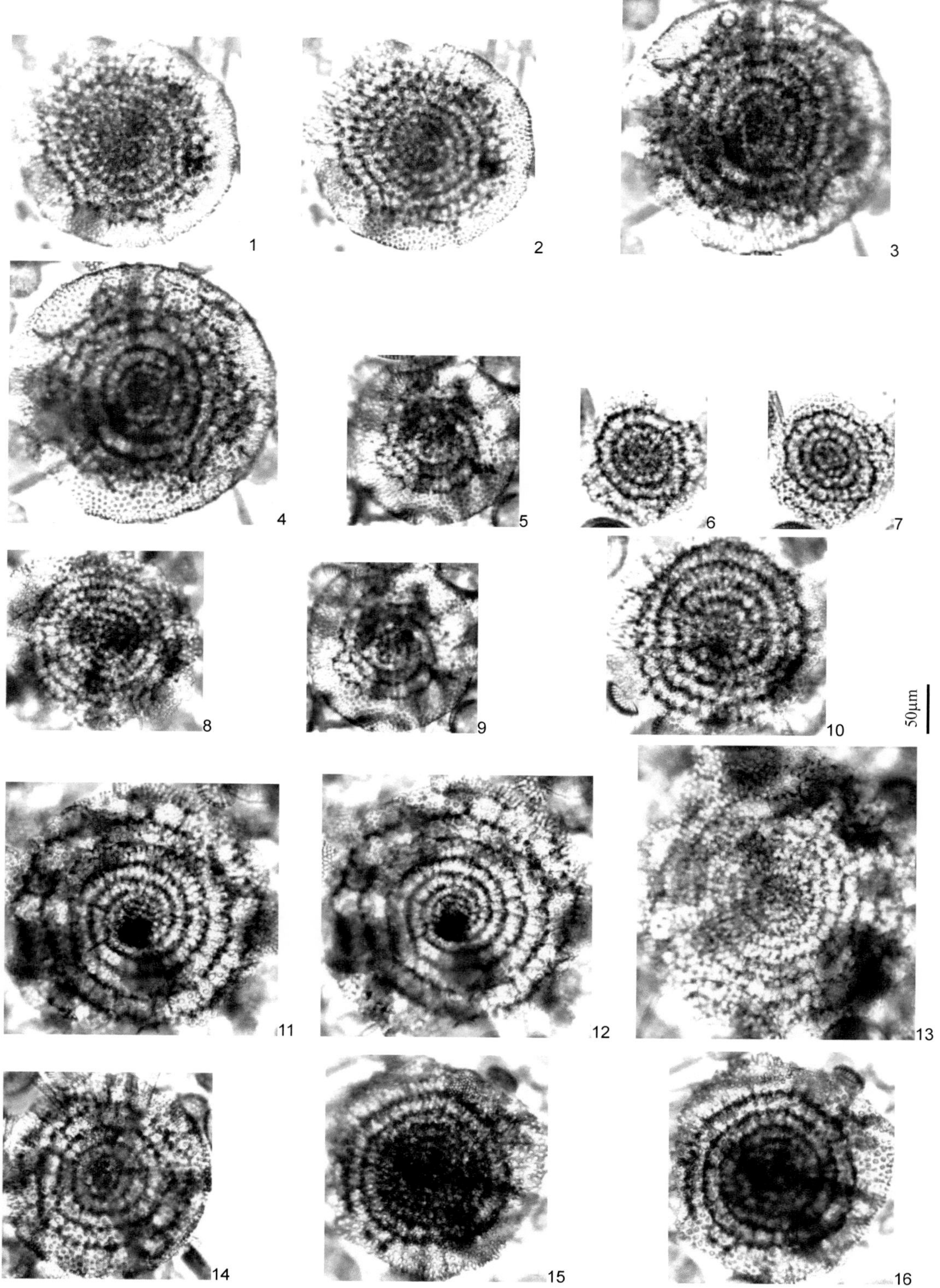

1–16. *Perichlamydium limbatum* Ehrenberg

图版 58

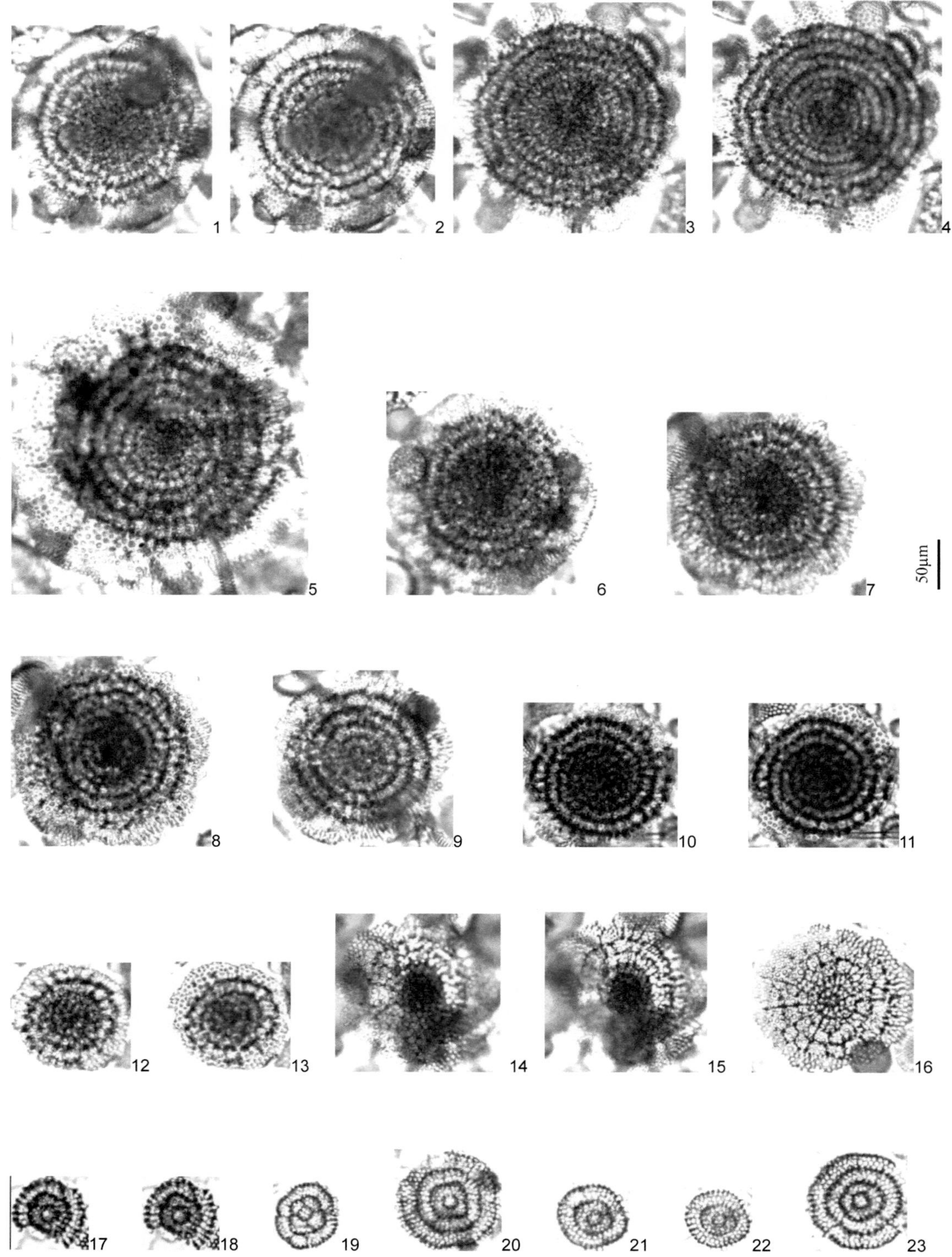

1–11. *Perichlamydium praetextum* (Ehrenberg) group；12–16. *Perichlamydium saturnus* Haeckel；17–23. *Stylodictya aculeata* Jørgensen

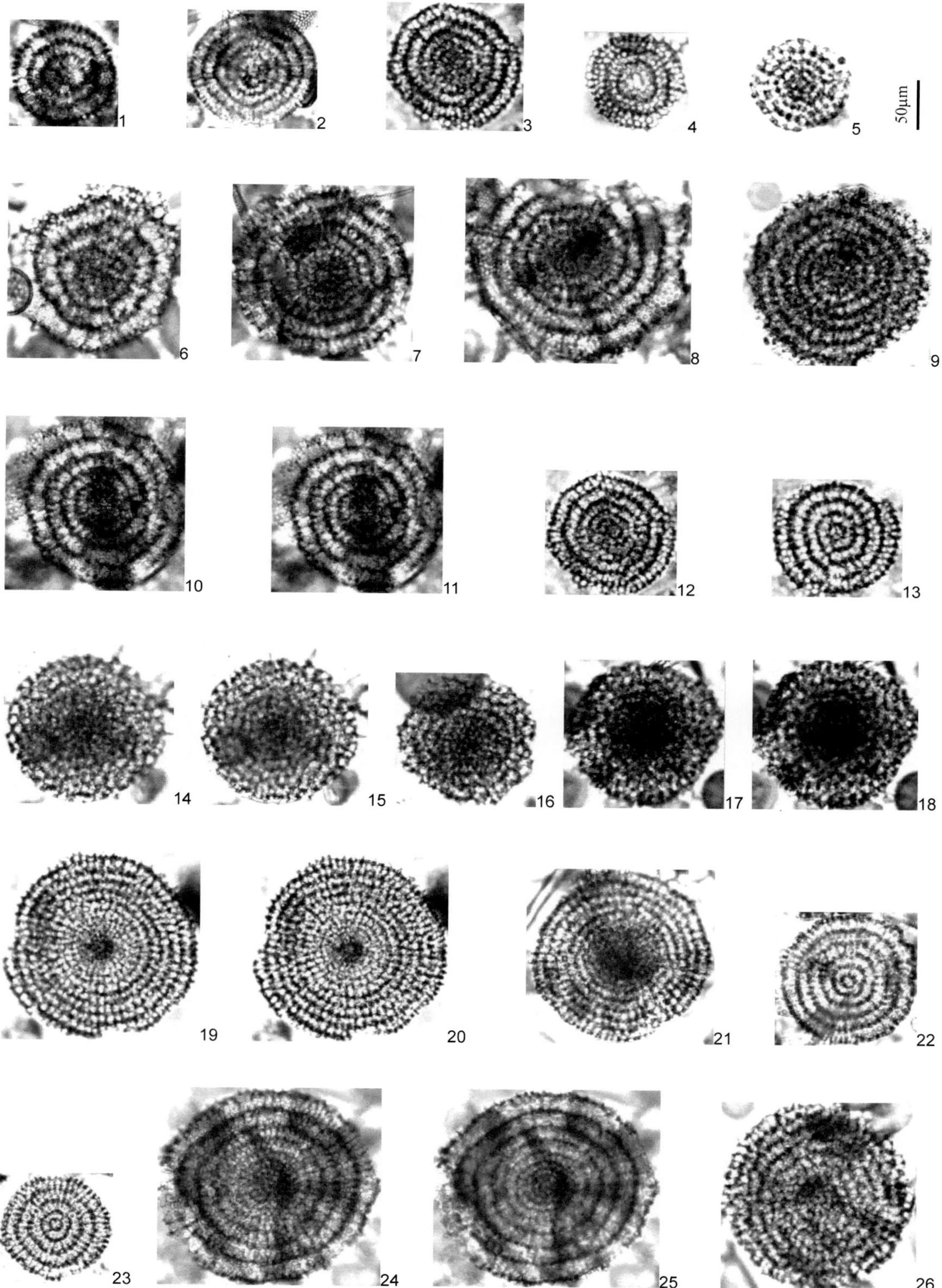

1–5. *Stylodictya circularis* (Clark and Campbell); 6–13. *Stylodictya heliospira* Haeckel; 14–18. *Stylodictya minima* Clark and Cambell; 19–26. *Stylodictya validispina* Jørgensen

图版 60

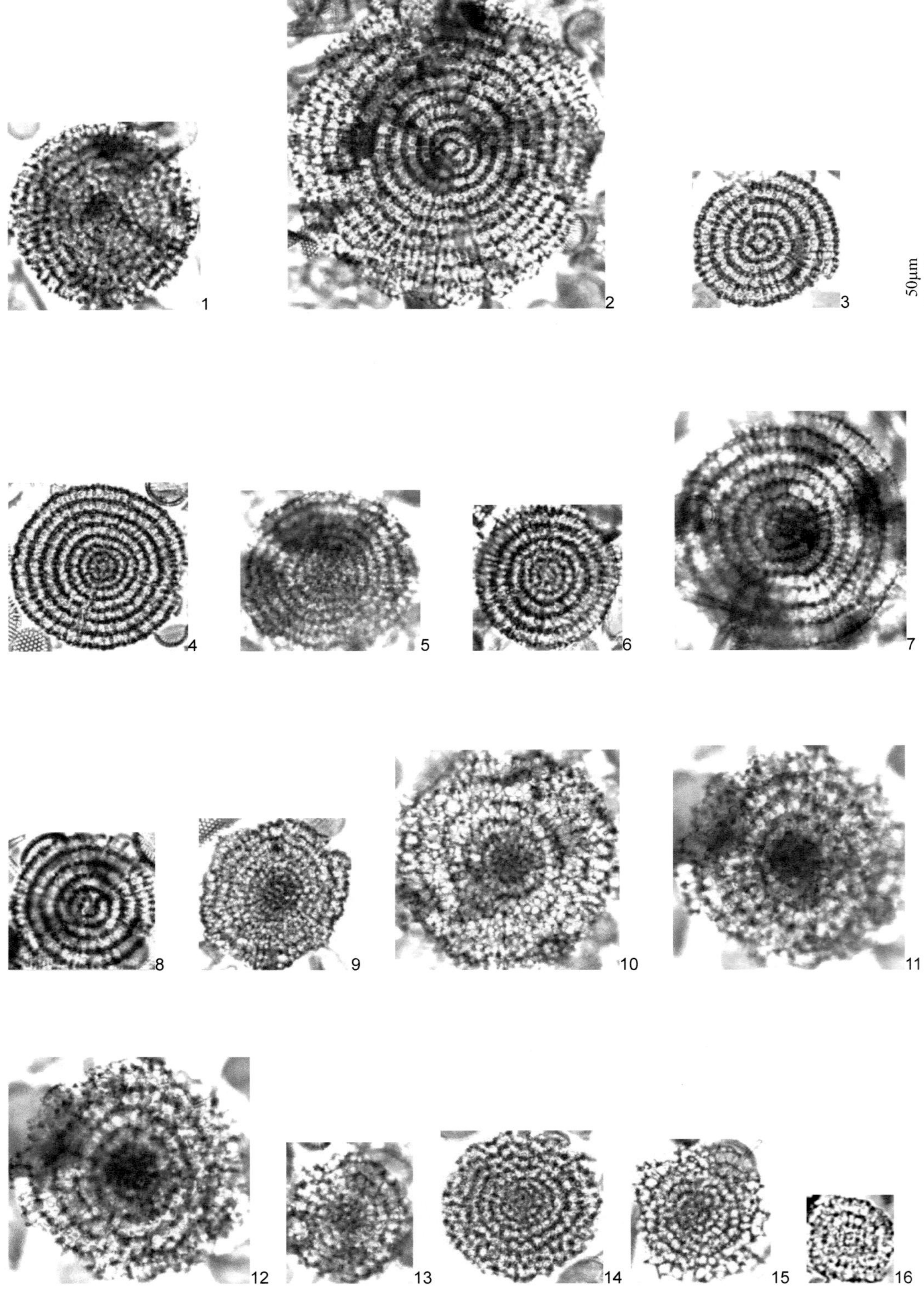

1–8. *Stylodictya validispina* Jørgensen；9–12. *Stylodictya variata* sp. nov.；13–14. *Stylodictya* sp. 1；15. *Stylodictya* sp. 2；16. *Staurodictya medusa* Haeckel

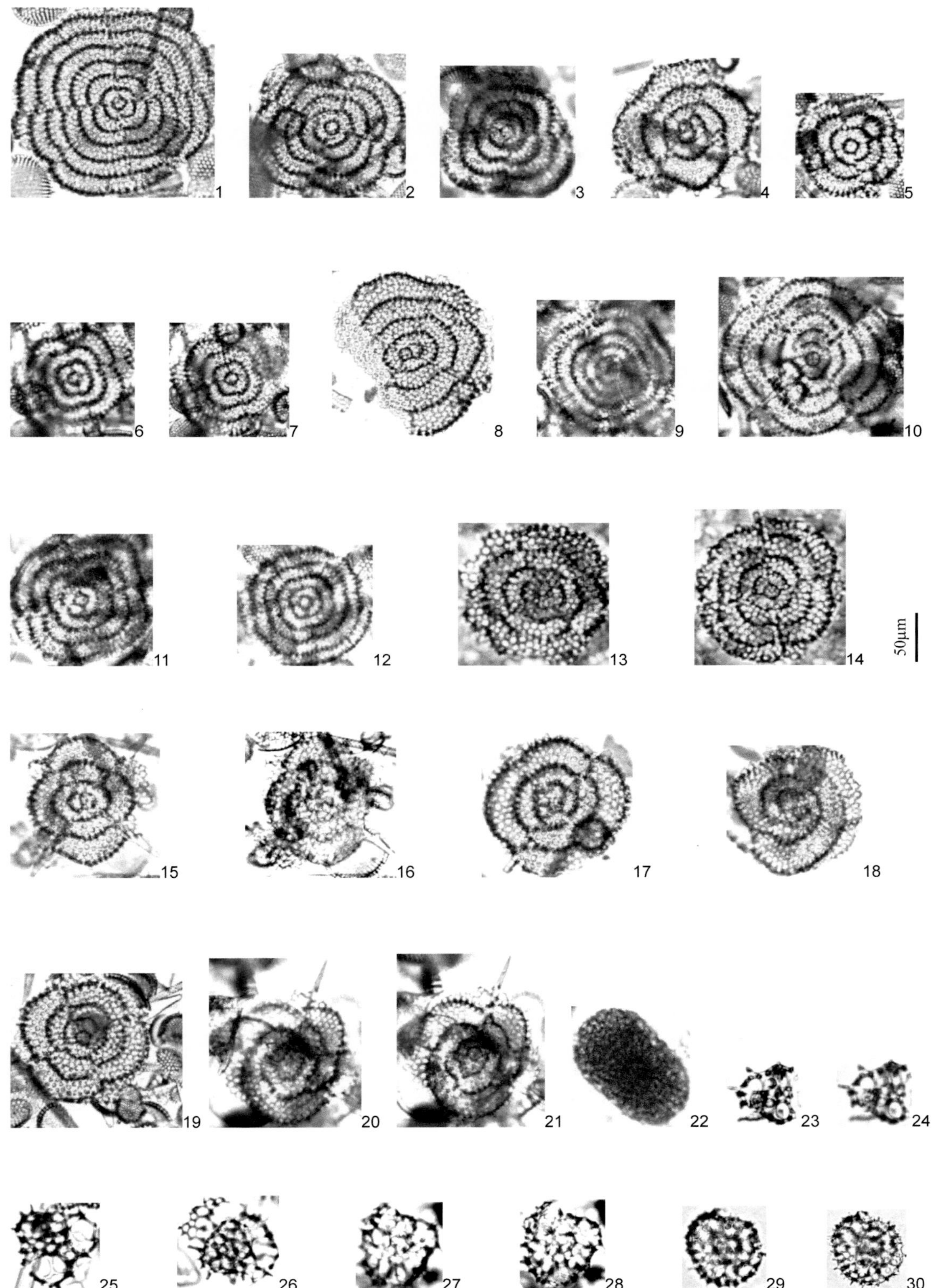

1–14. *Staurodictya medusa* Haeckel；15–21. *Staurodictya ocellata* (Ehrenberg)；22. *Amphibrachium* sp. 1；23–28. *Hexapyle spinulosa* Chen and Tan；29–30. *Discopyle* sp. 1

图版 62

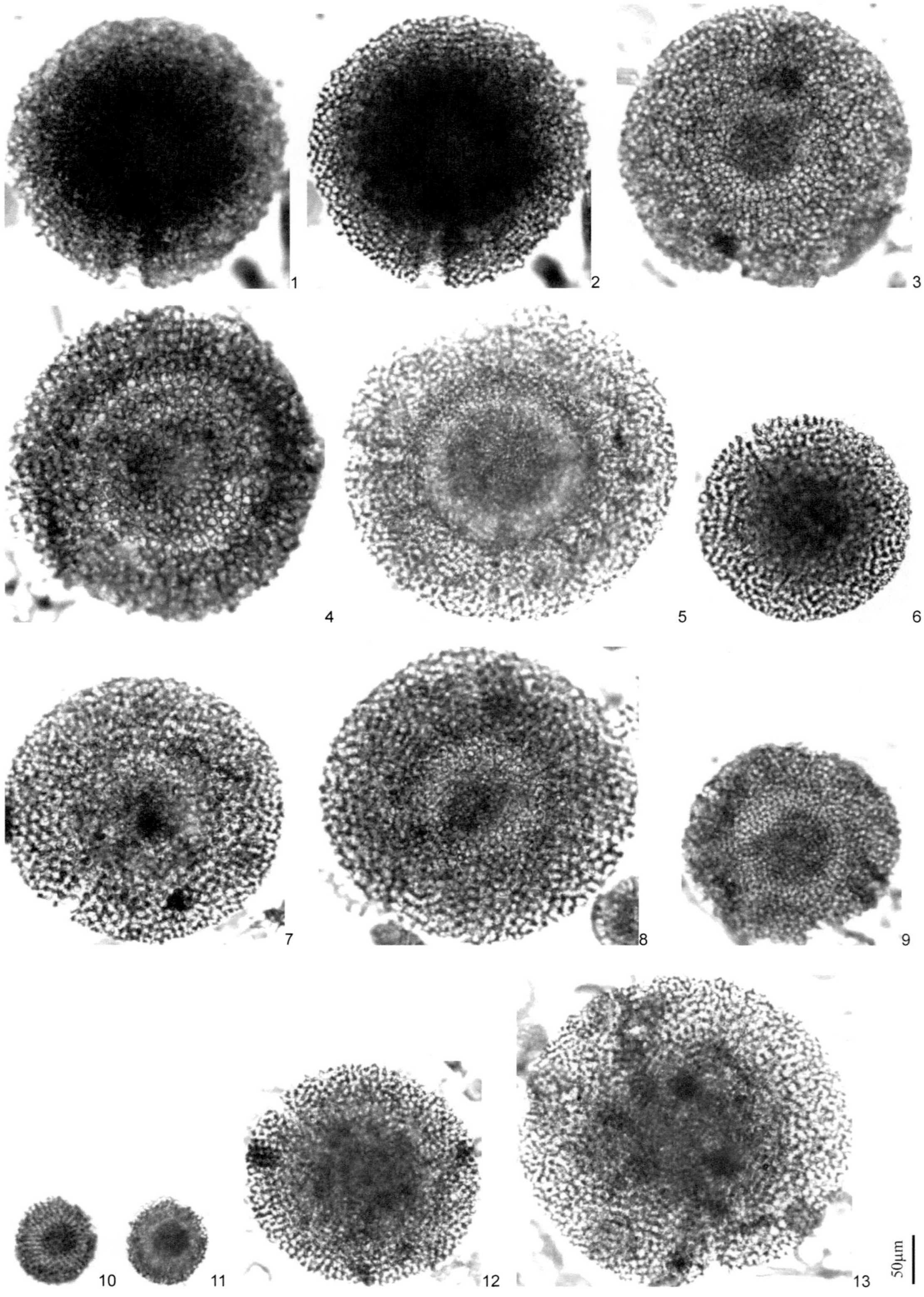

1–13. *Spongodiscus biconcavus* Haeckel, emend. Chen et al.

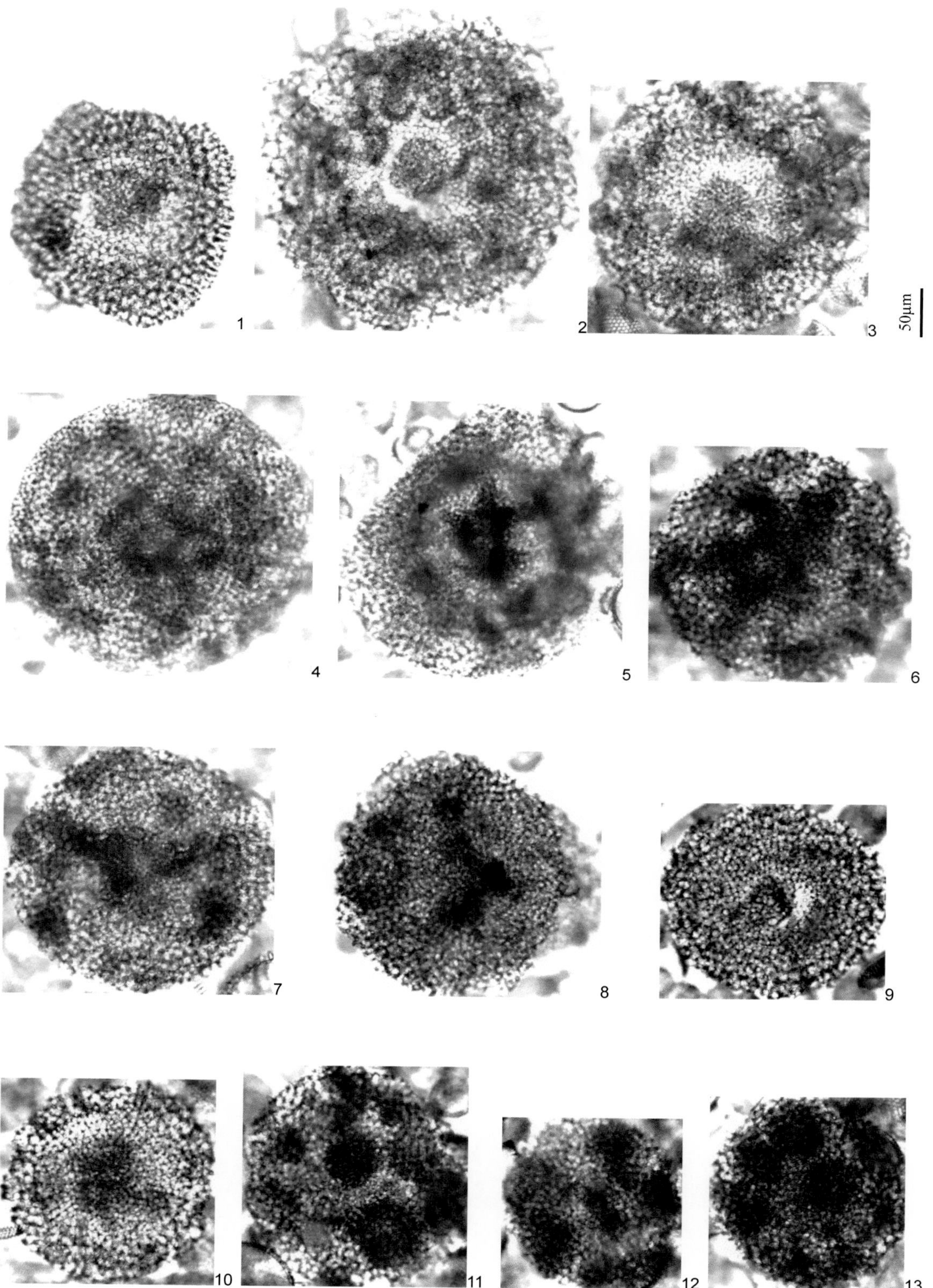

1–13. *Spongodiscus communis* Clark and Campbell group

图版 64

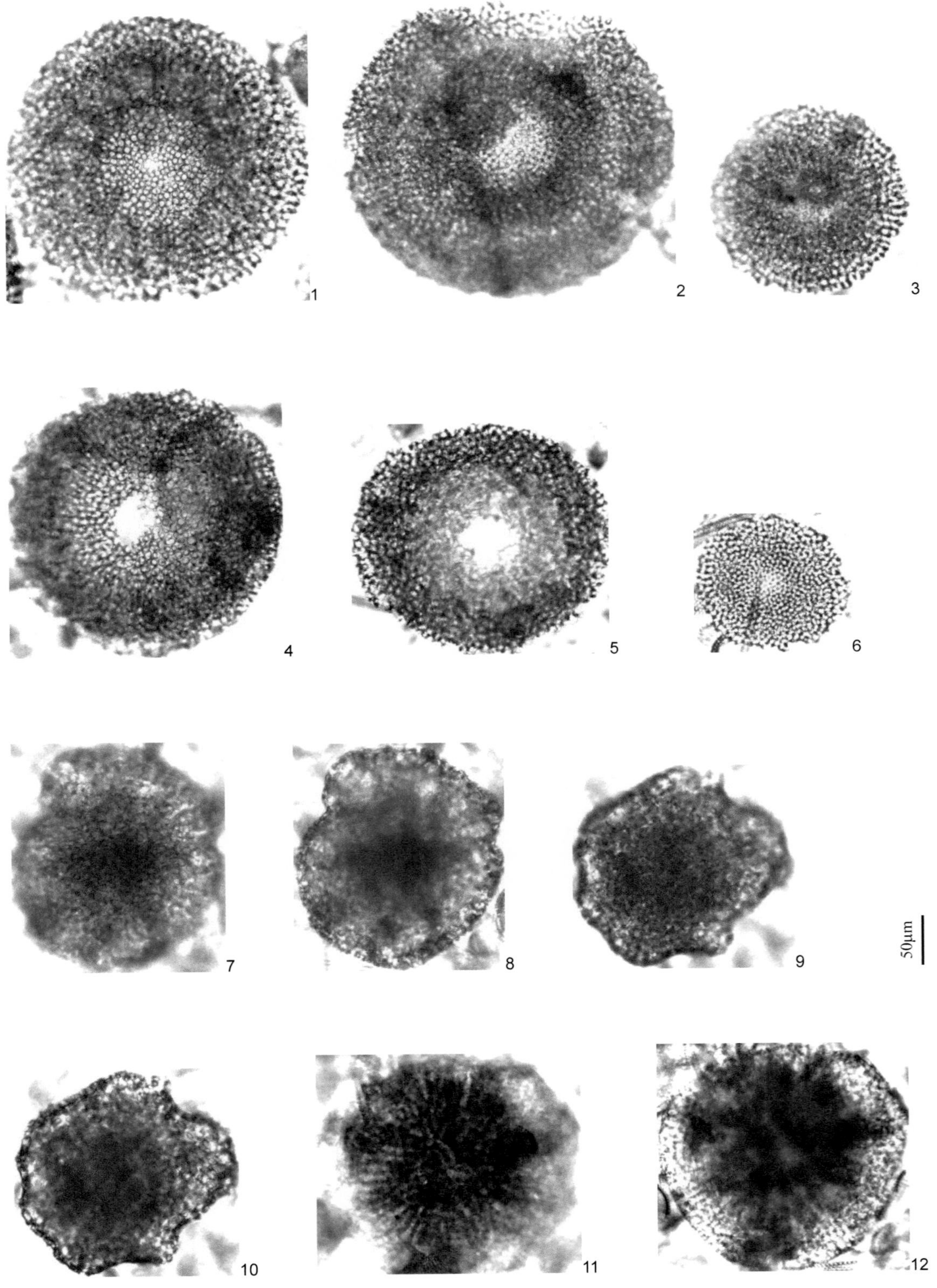

1–6. *Spongodiscus inconcavus* sp. nov.; 7–12. *Spongodiscus perizonatus* sp. nov.

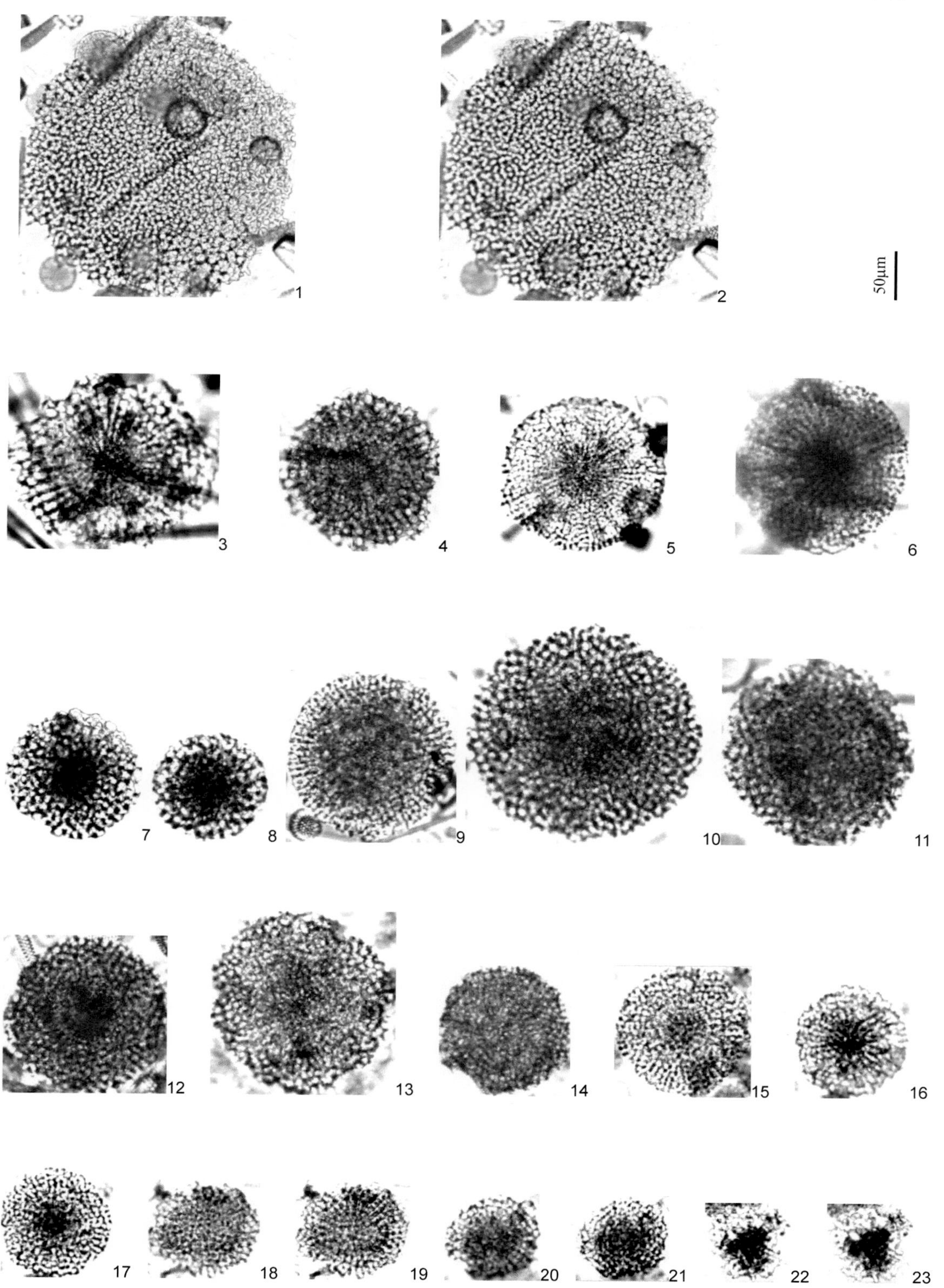

1–2. *Spongodiscus planarius* sp. nov.； 3–6. *Spongodiscus radialinus* sp. nov.； 7–23. *Spongodiscus resurgens* Ehrenberg group

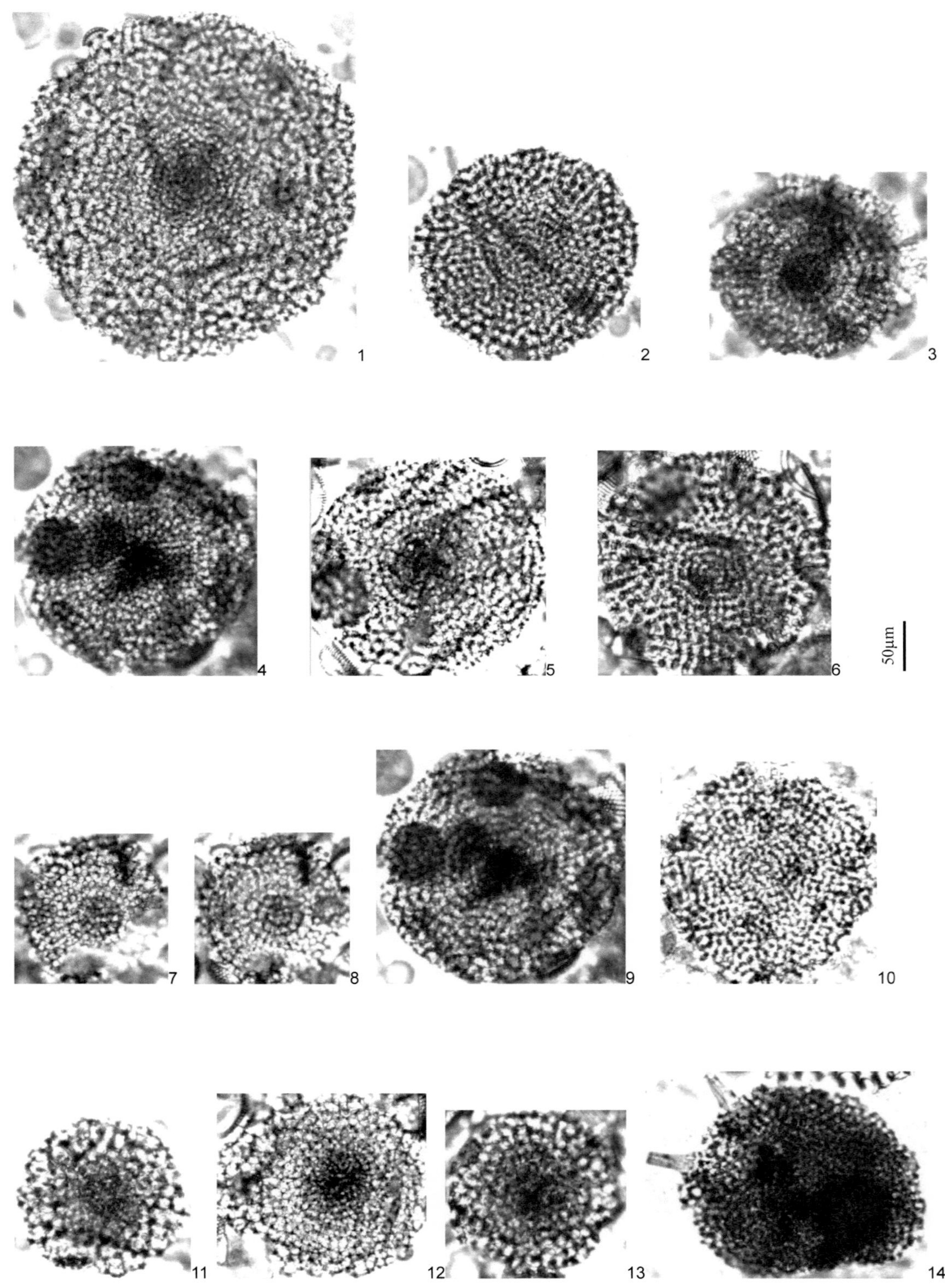

1–10. *Spongodiscus trachodes* (Renz)；11–13. *Spongodiscus* sp. 1；14. *Spongodiscus* sp. 2

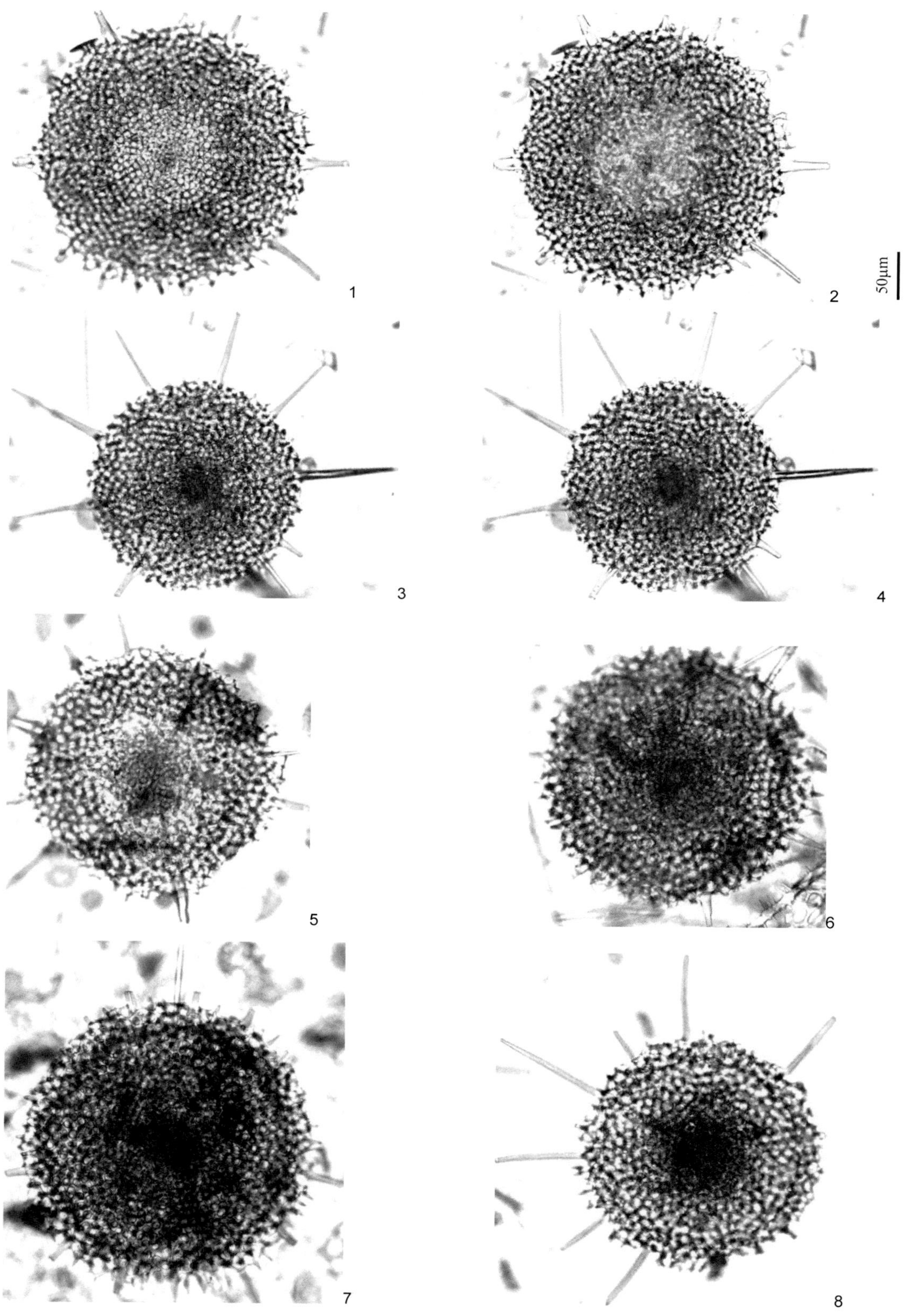

1–8. *Spongotrochus glacialis* Popofsky group

图版 68

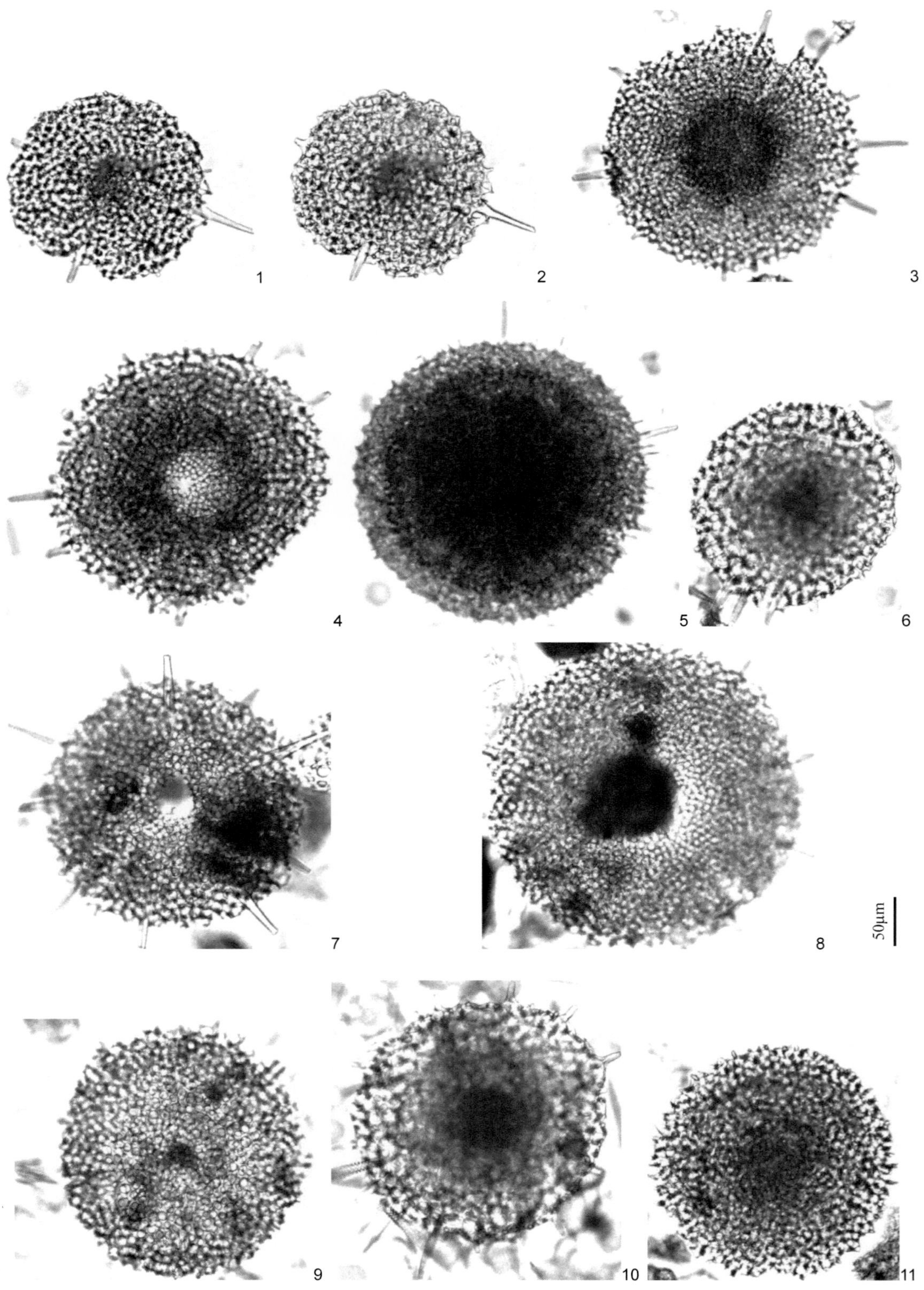

1–11. *Spongotrochus glacialis* Popofsky group

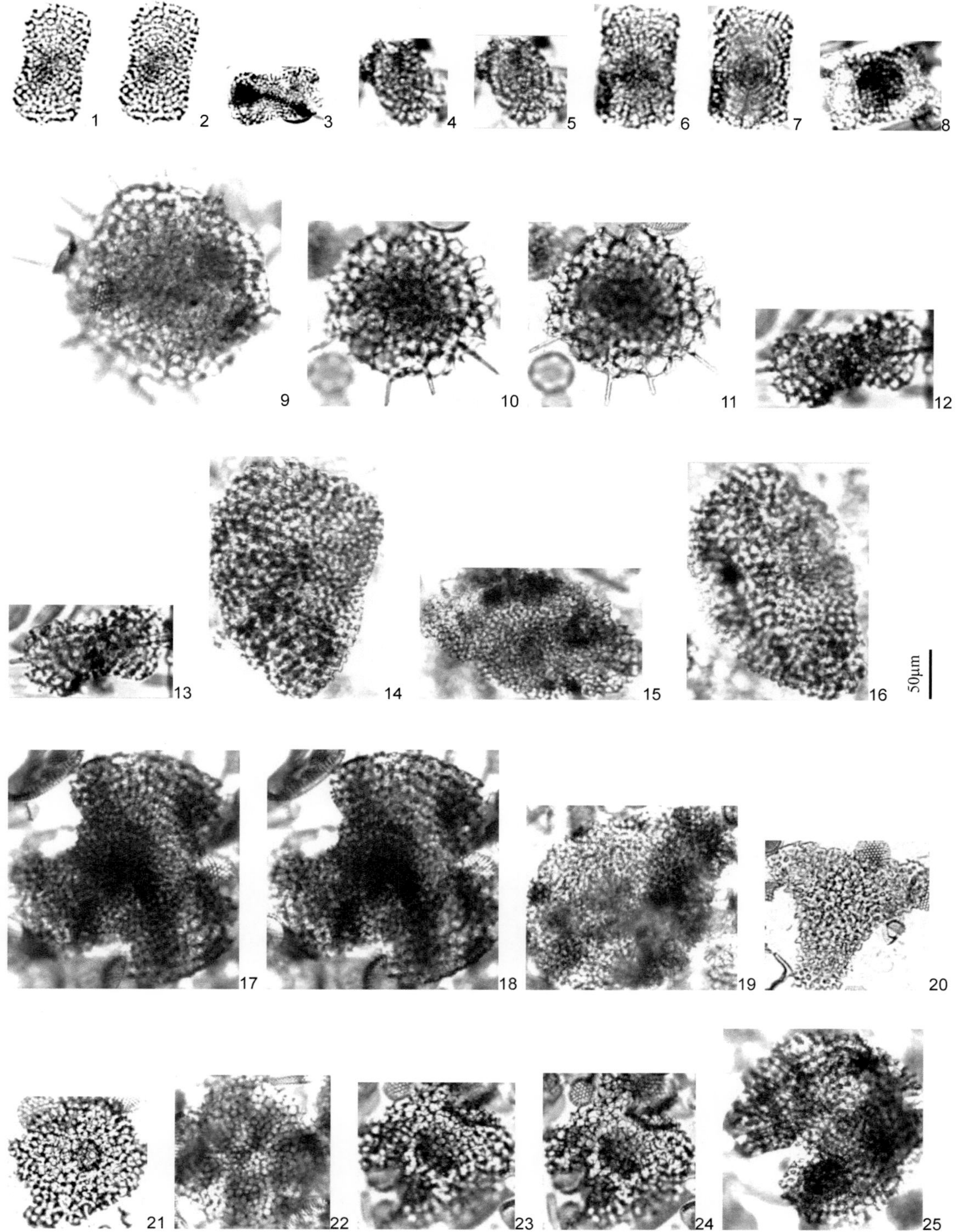

1–8. ?*Spongotrochus rhabdostyla* Bütschli；9–11. *Spongotrochus vitabilis* Goll and Bjørklund；12–13. *Spongotrochus* sp. 1；14–16. *Spongobrachium antoniae* (O'Connor)；17–18. *Dictyocoryne truncatum* (Ehrenberg)；19. *Dictyocoryne* sp. 1；20. *Dictyocoryne* sp. 2；21–22. *Spongaster tetras* Ehrenberg；23–24. *Spongaster* sp. 1；25. *Spongaster* sp. 2

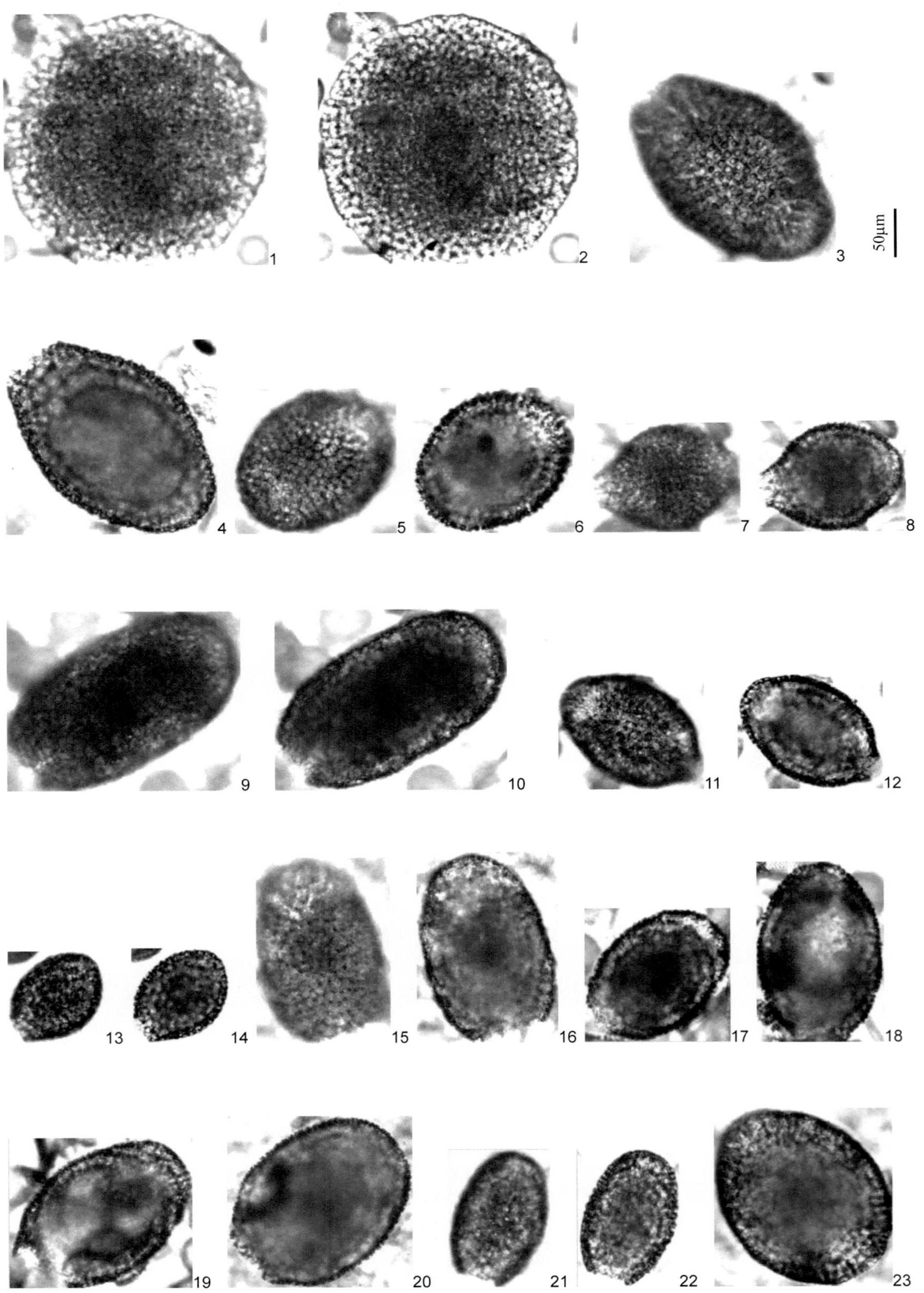

1–2. *Spongopyle osculosa* Dreyer；3–23. *Larcopyle adelstoma* (Kozlova and Gobovets)

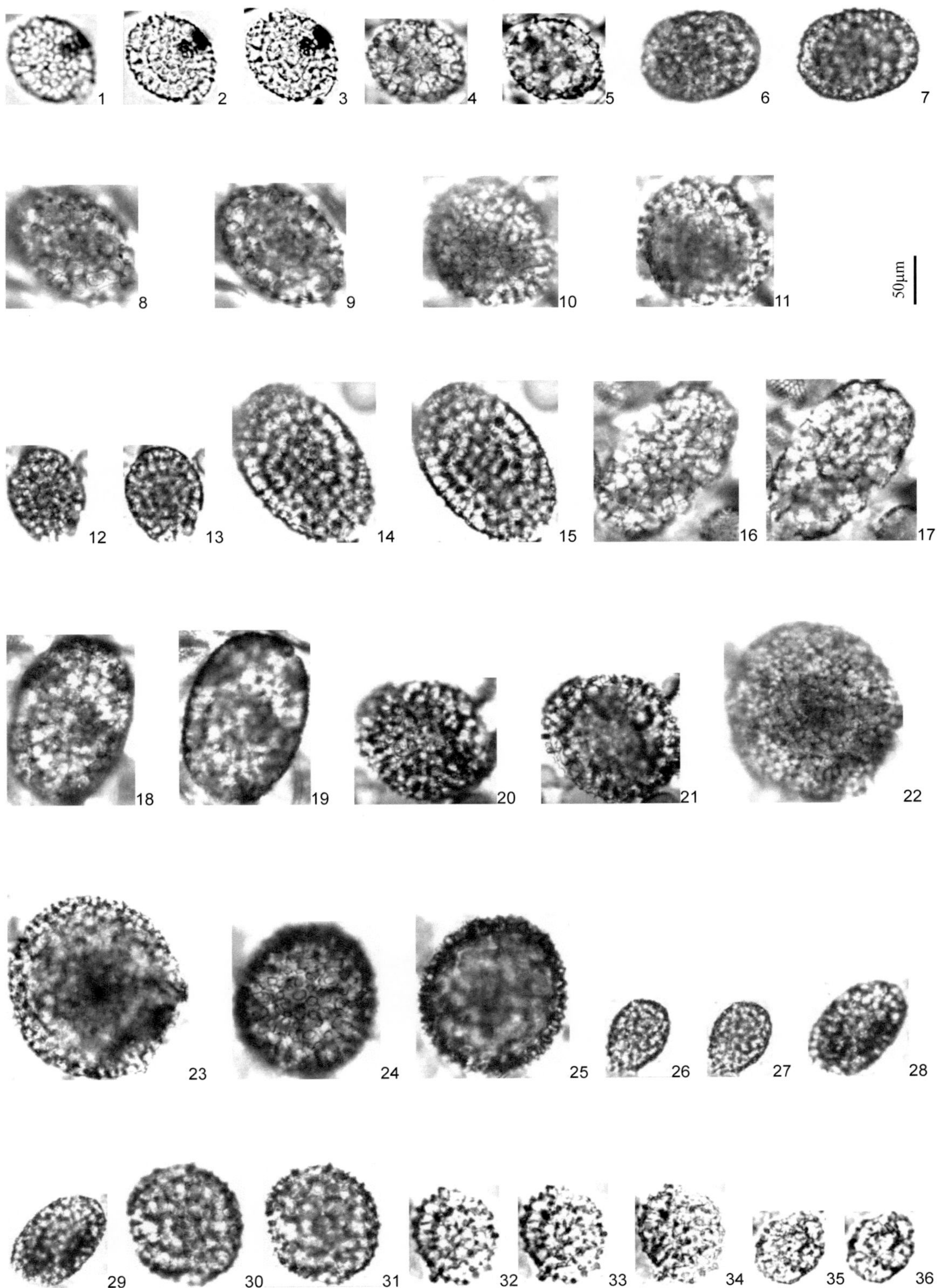

1–36. *Larcopyle butschlii* Dreyer

图版 72

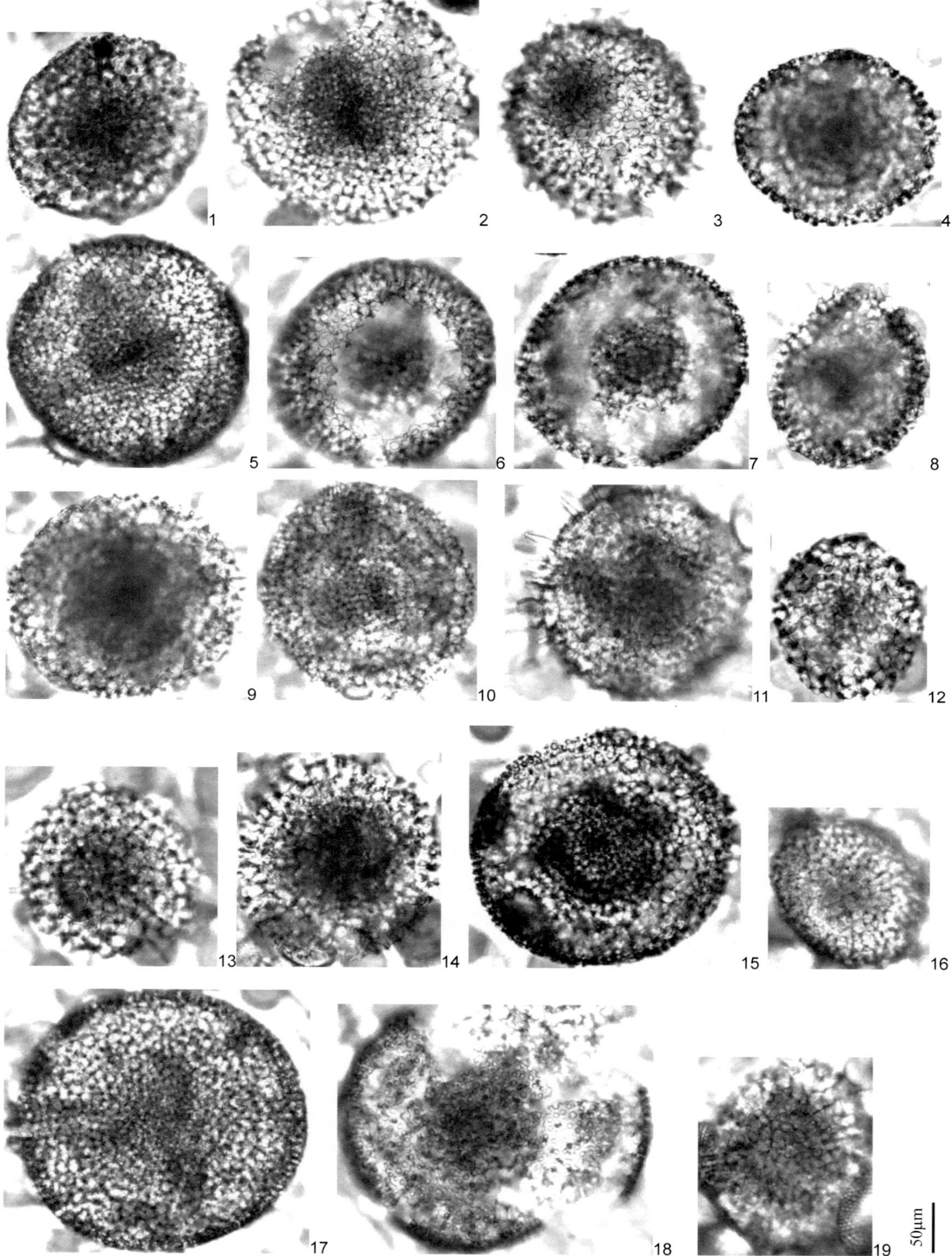

1–19. *Larcopyle frakesi* (Chen) group

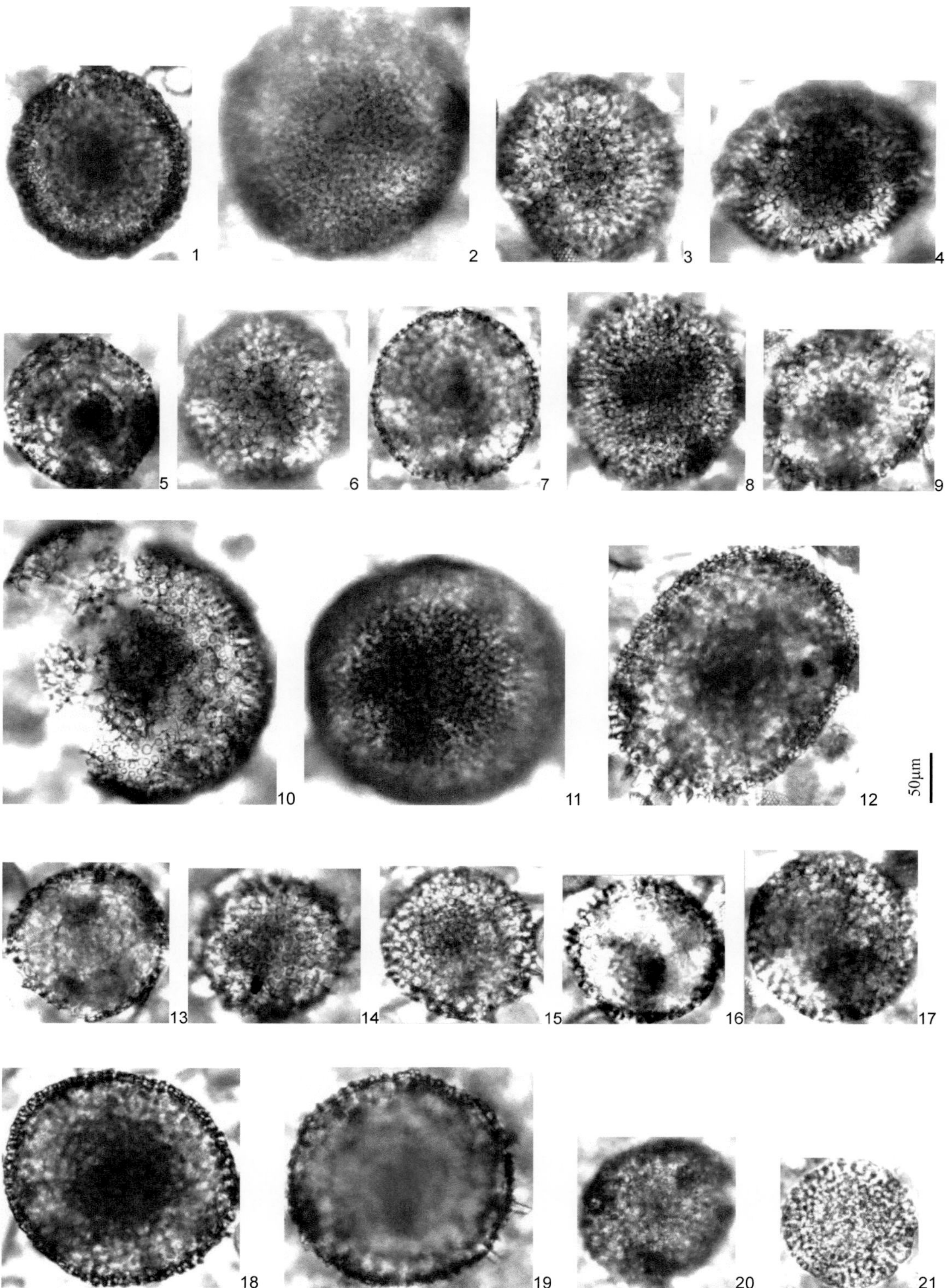

1–21. *Larcopyle frakesi* (Chen) group

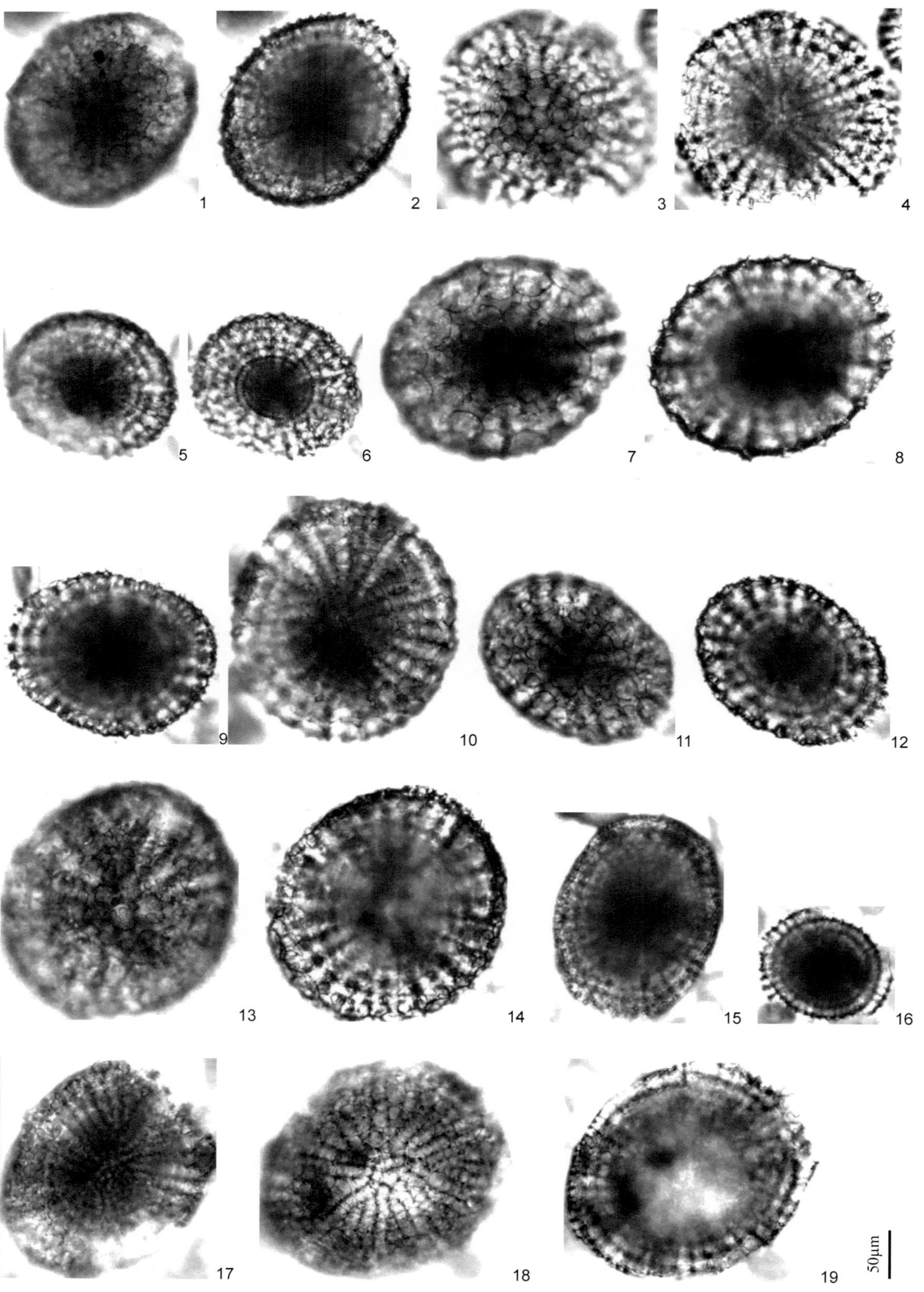

1–19. *Larcopyle hayesi* (Chen) group

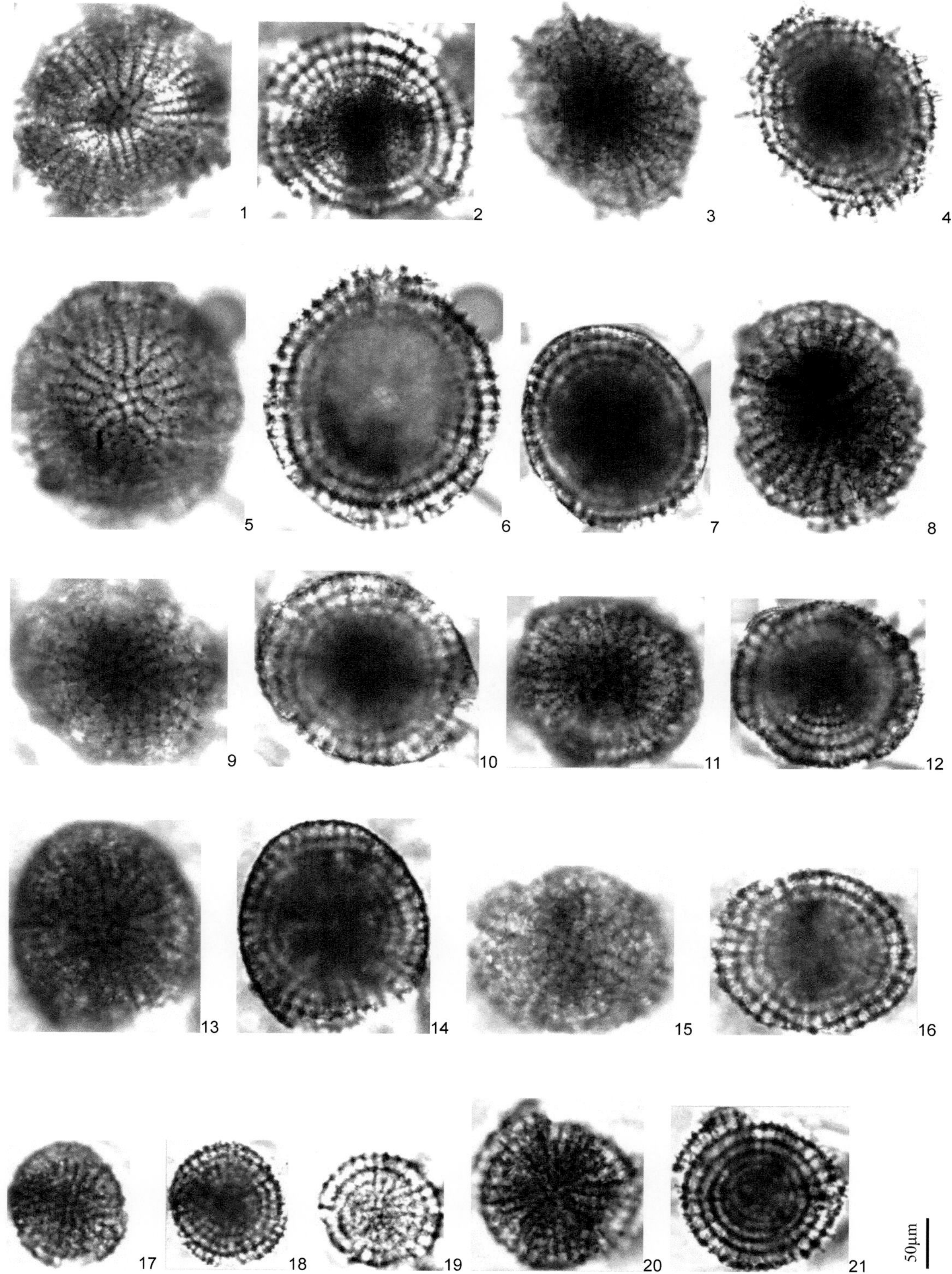

1–21. *Larcopyle hayesi* (Chen) group

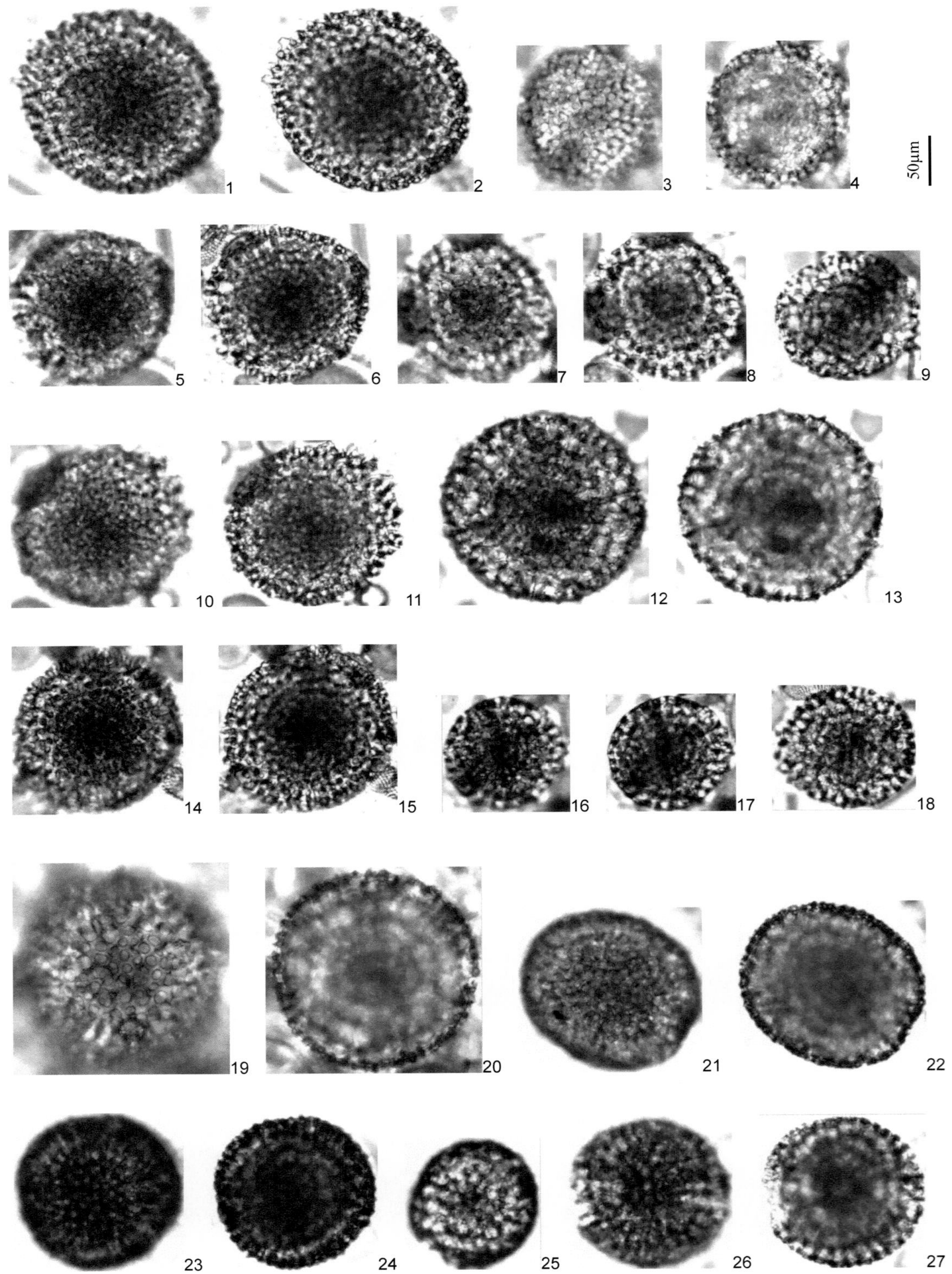

1–27. *Larcopyle ovata* (Kozlova and Gorbovetz)

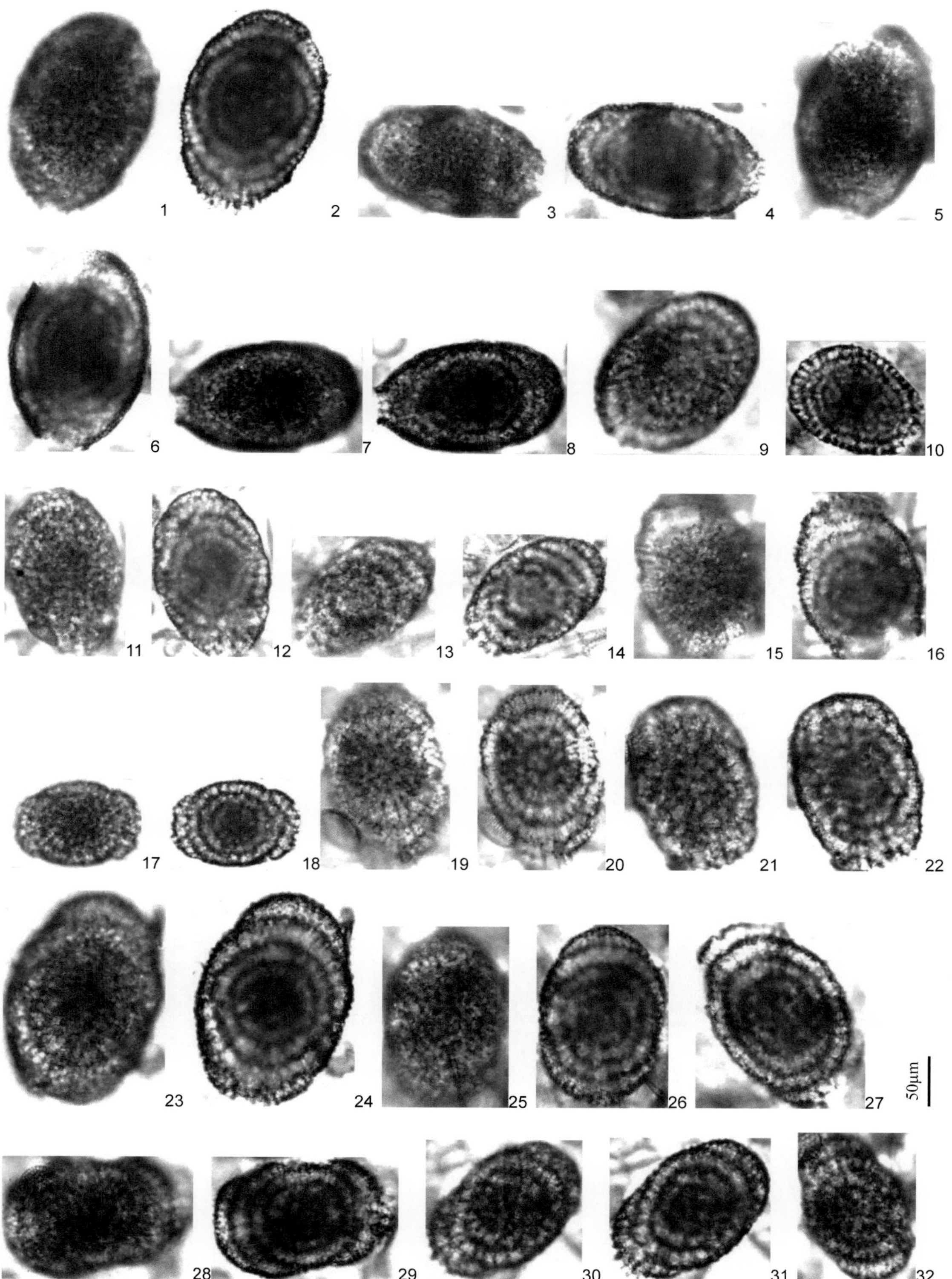

1–10. *Larcopyle polyacantha* (Campbell and Clark)；11–32. *Larcopyle polyacantha titan* Lazarus et al.

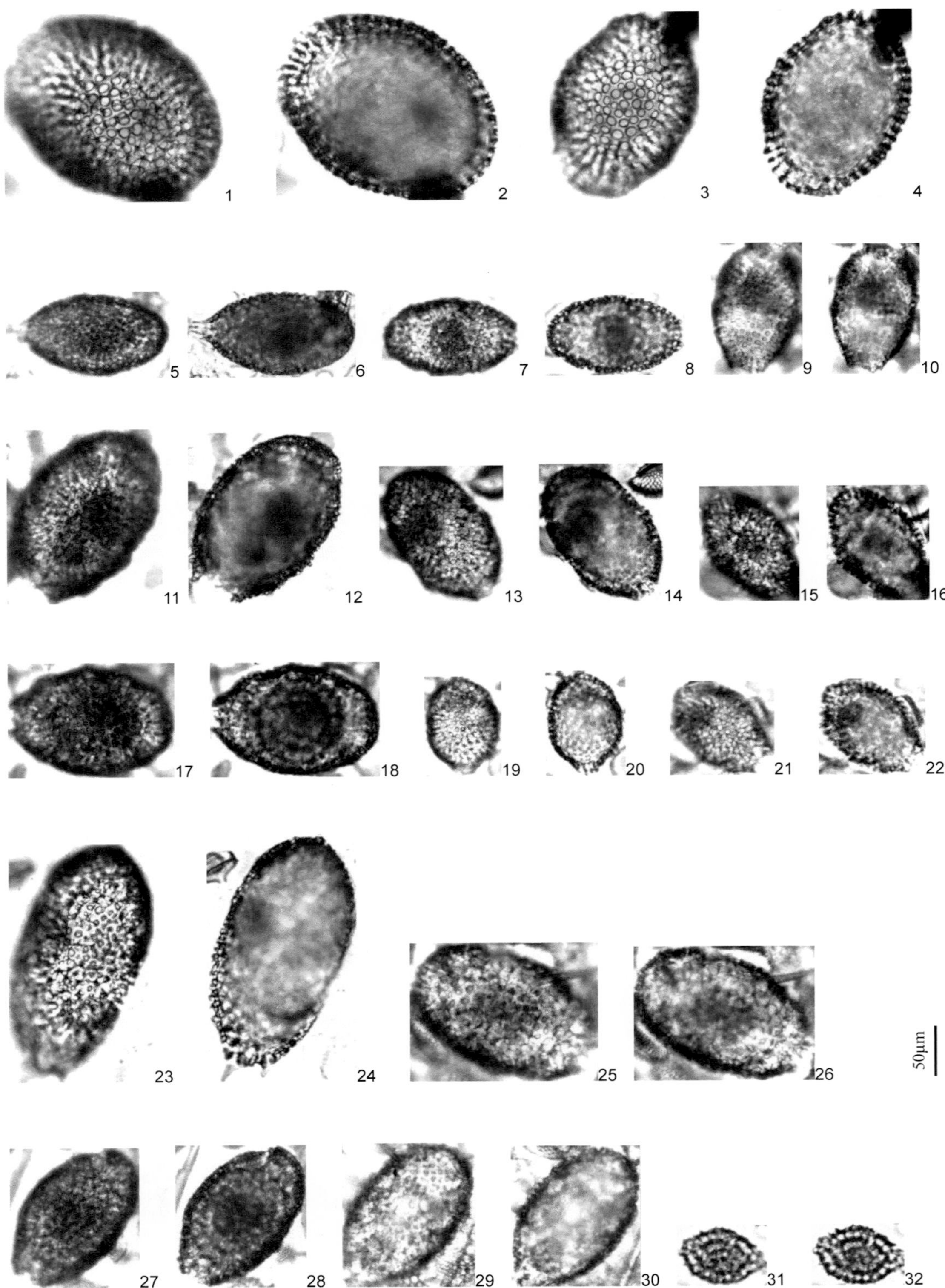

1–30. *Larcopyle titan* (Campbell and Clark) group；31–32. *Larcopyle* sp. 1

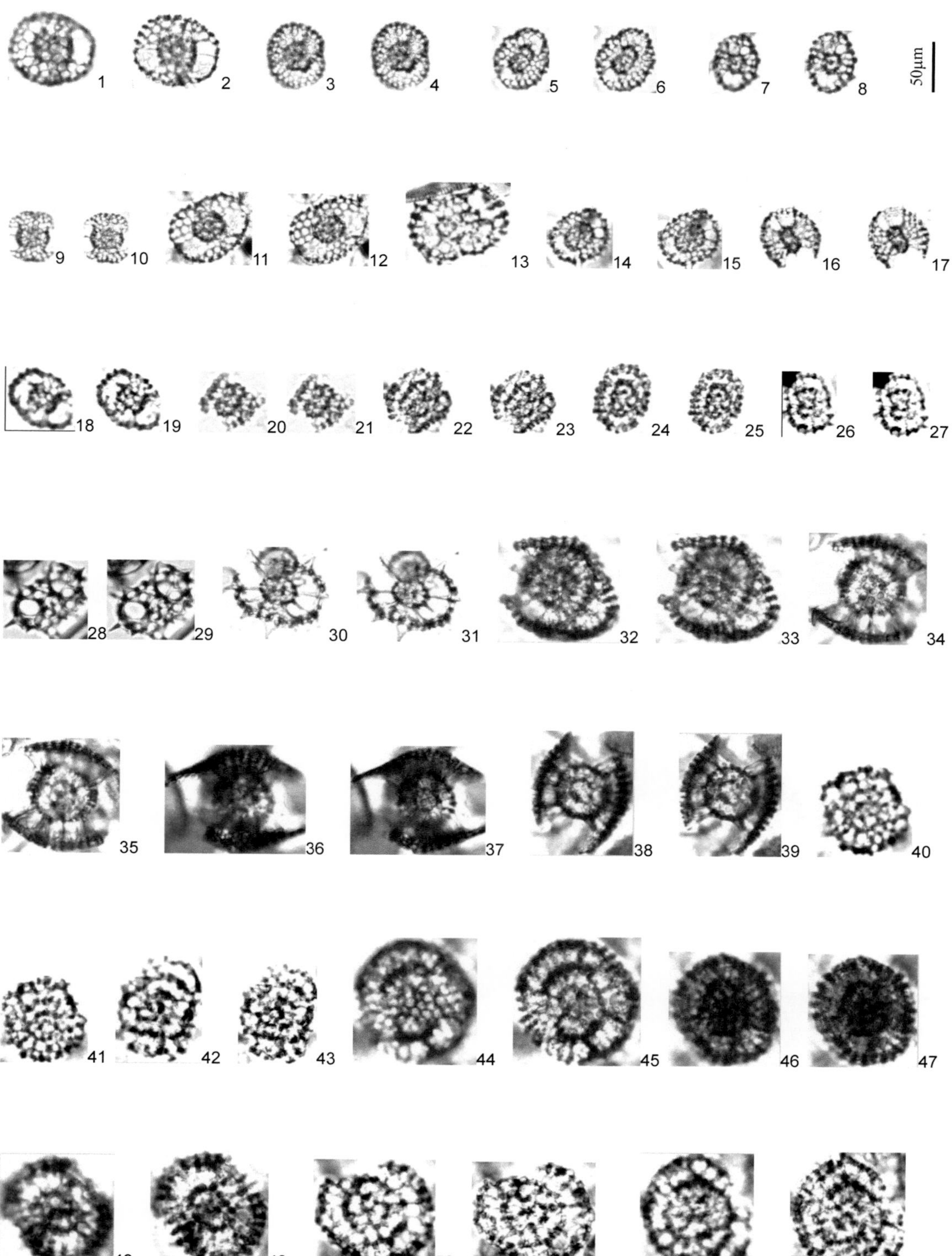

1–4. *Larnacalpis lentellipsis* Haeckel；5–27. *Tetrapyle quadriloba* (Ehrenberg)；28–29. *Tetrapyle turrita* Haeckel；30–31. *Octopyle sexangulata* Haeckel；32–33. *Pylonium palaeonatum* sp. nov.；34–39. *Pylonium neoparatum* sp. nov.；40–53. *Pylozonium saxitalum* sp. nov.

图版 80

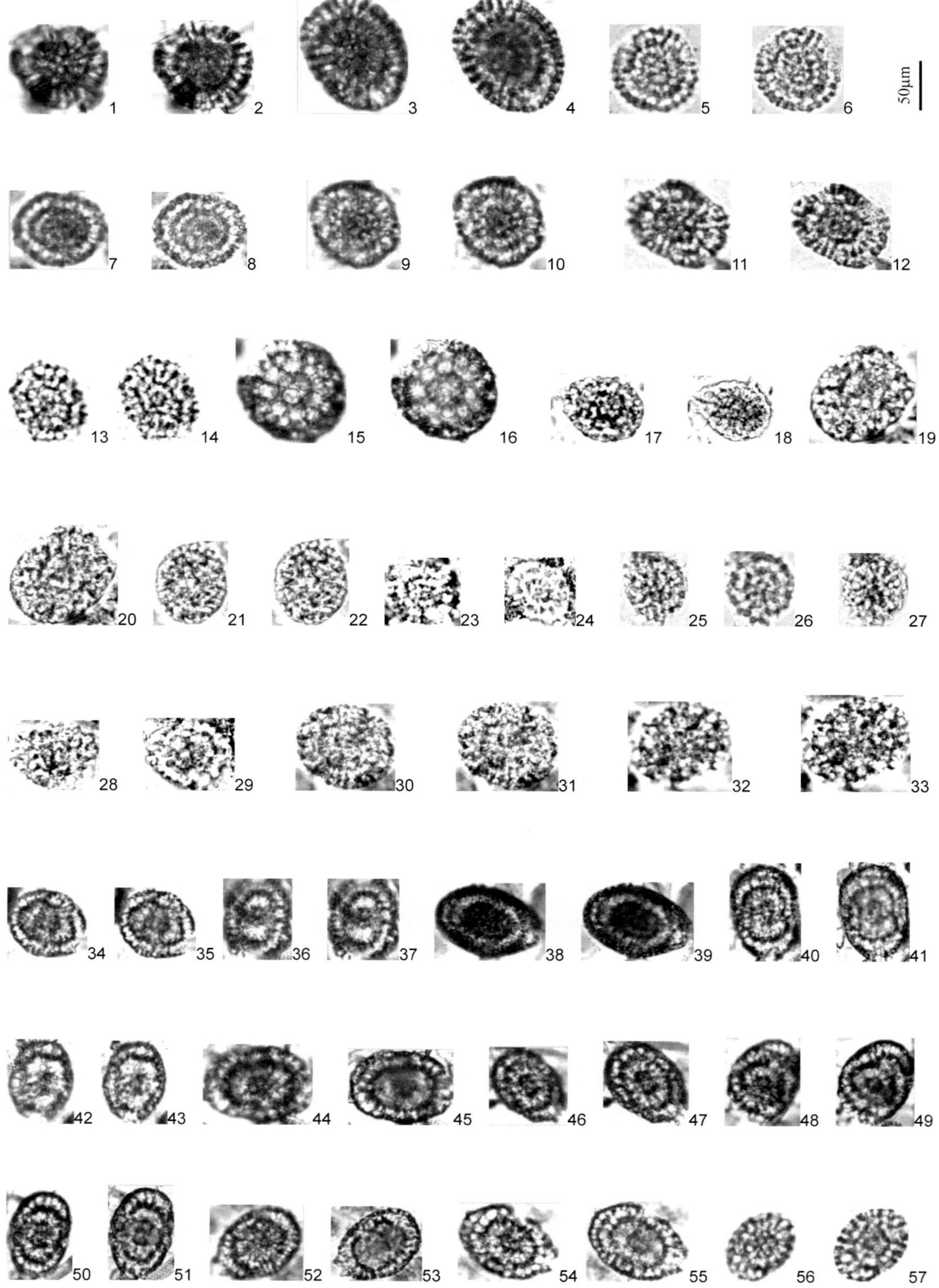

1–14. *Pylozonium* sp. 1；15–16. *Prunopyle antarctica* Dreyer；17–33. *Prunopyle archaeota* sp. nov.；34–57. *Prunopyle minuta* sp. nov.

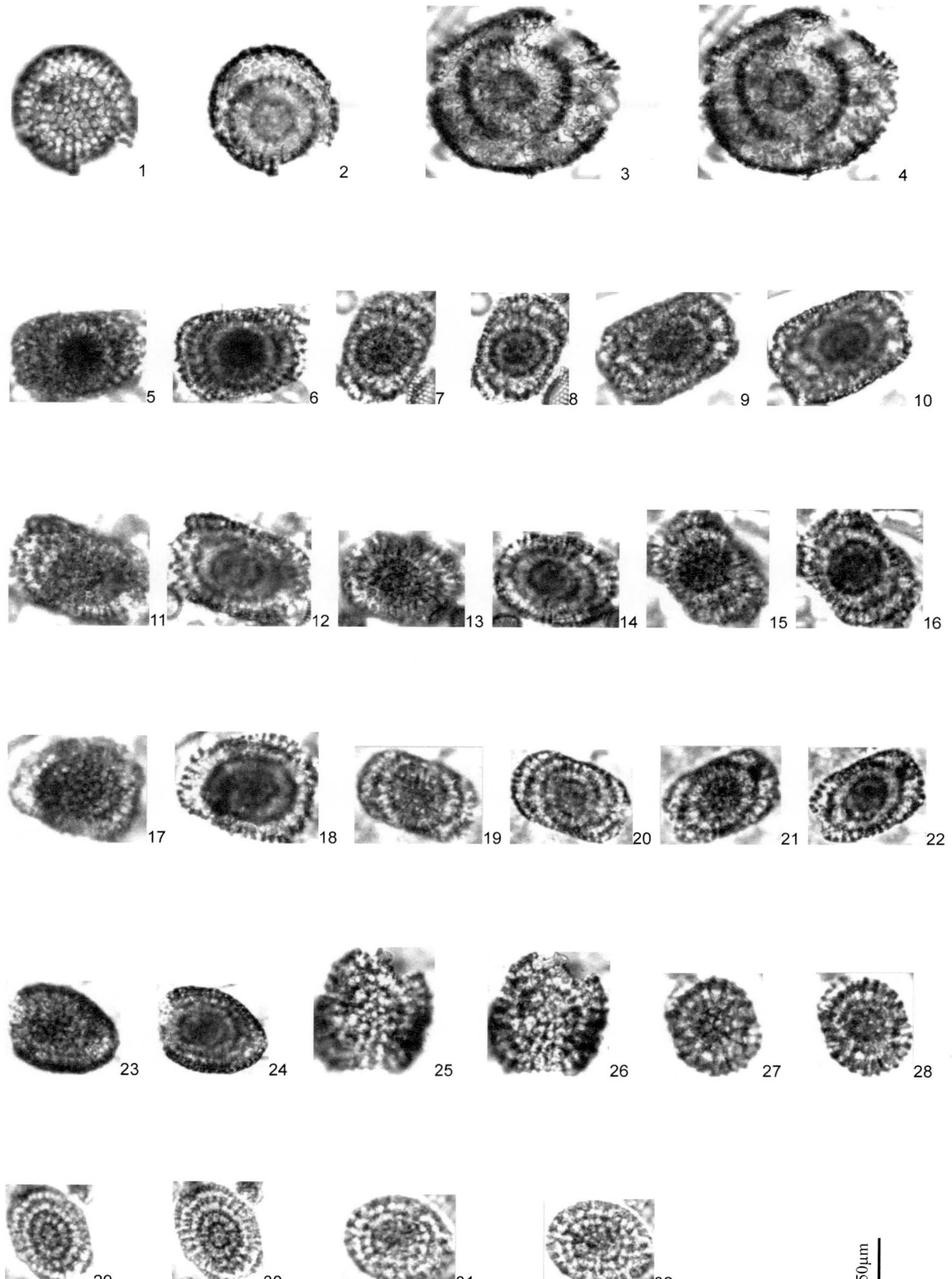

1–4. *Prunopyle opeocila* sp. nov.； 5–32. *Prunopyle quadrata* sp. nov.

图版 82

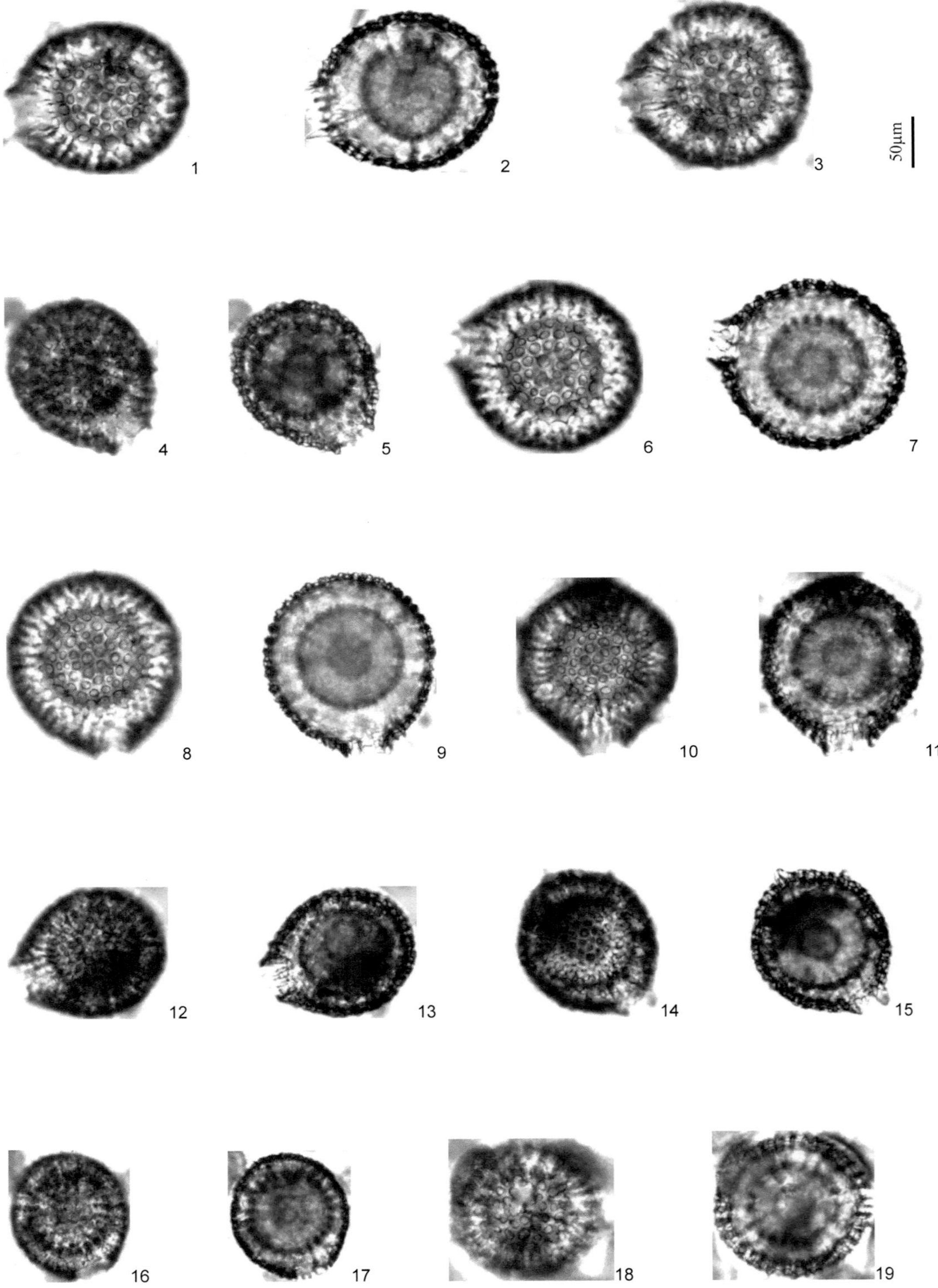

1–19. *Prunopyle tetrapila* Hays group

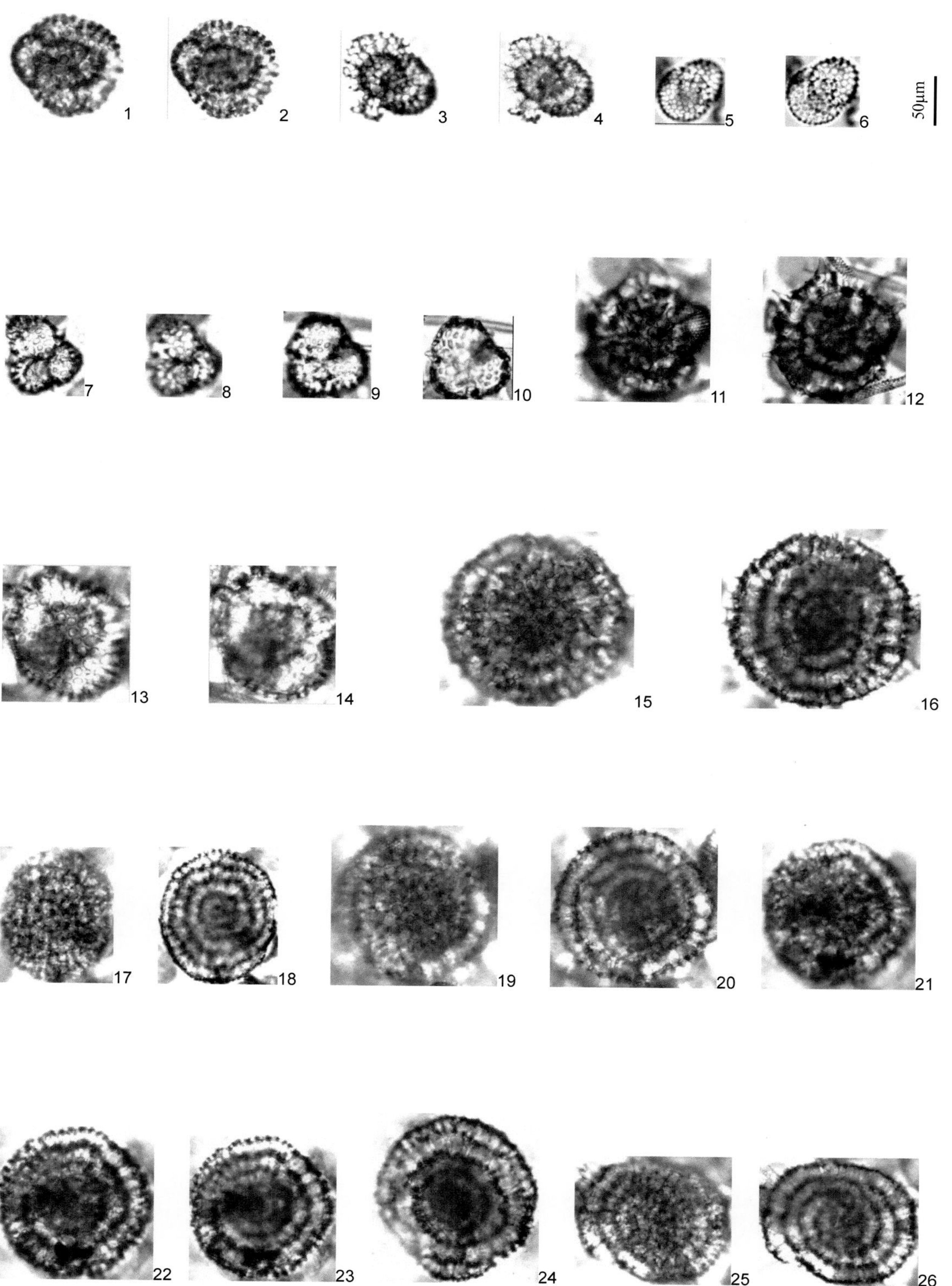

1–2. *Cubotholus regularis* Haeckel；3–6. *Dipylissa bensoni* Dumitrica；7–10. *Triosphaera bulloidis* sp. nov.；
11–14. *Spirema flustrella* Haeckel；15–26. *Spirema lentellipsis* Haeckel

图版 84

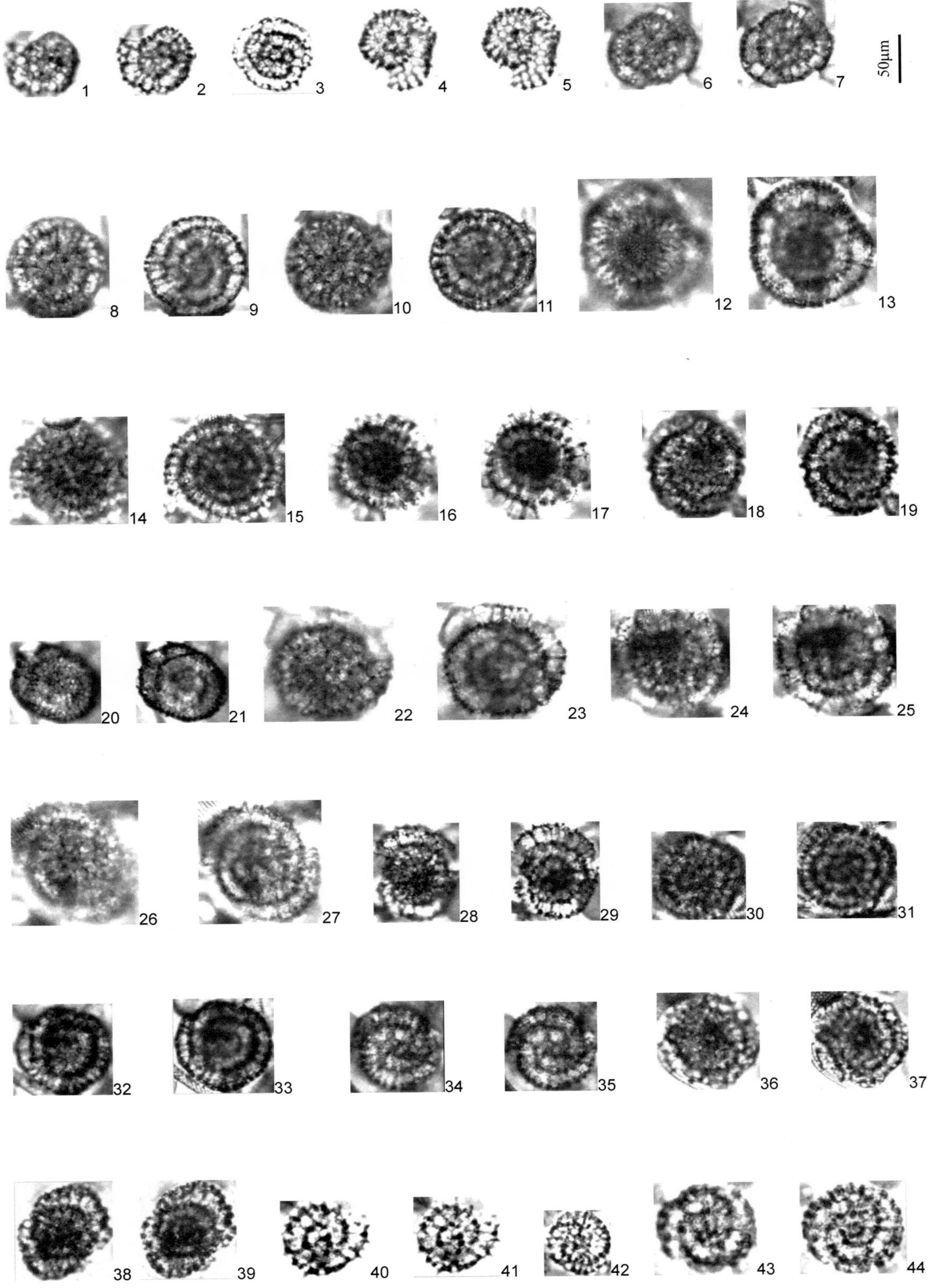

1–44. *Spirema melonia* Haeckel

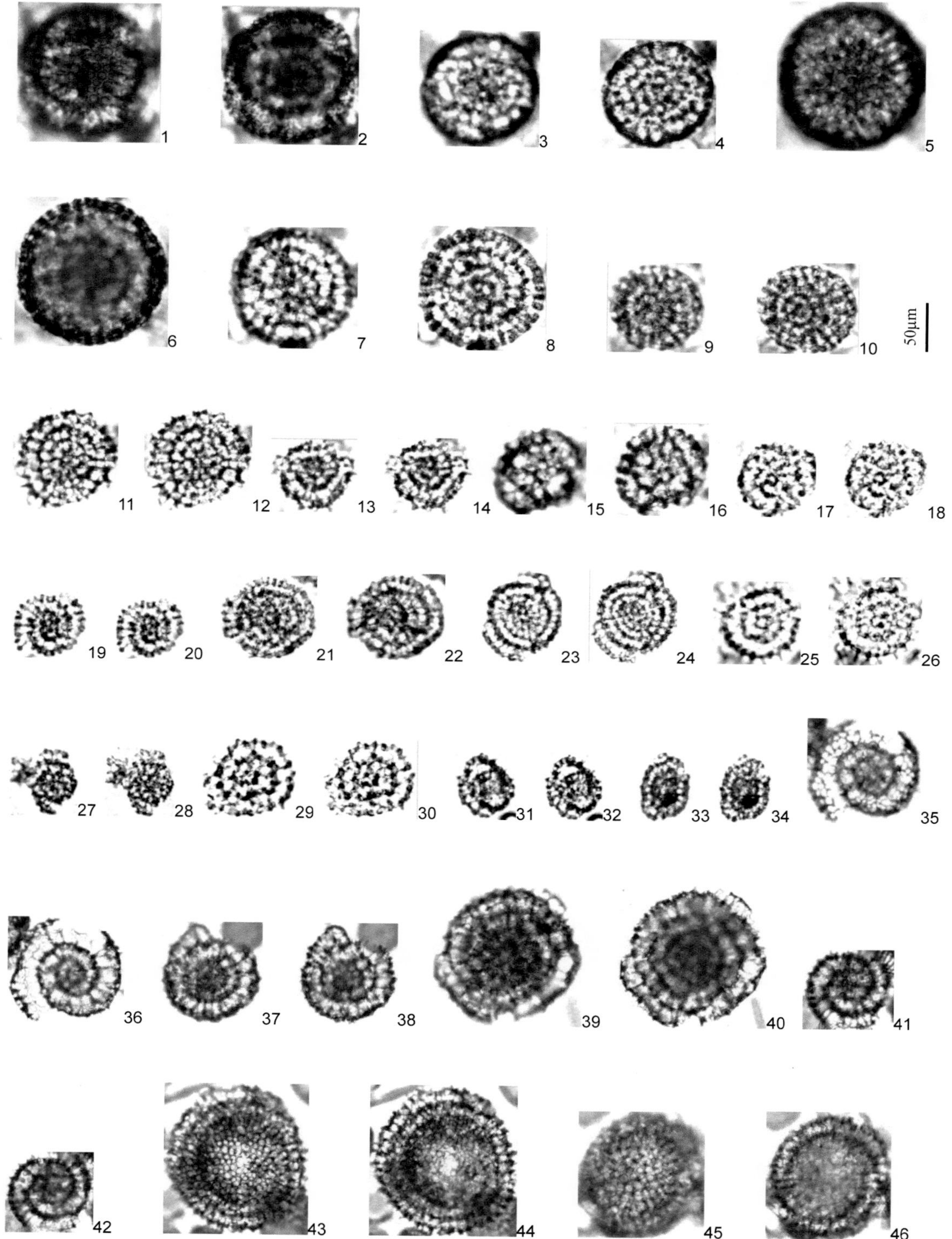

1–10. *Spirema vetula* sp. nov.；11–34. *Spirema* sp. 1；35–42. *Lithelius alveolina* Haeckel；43–46. *Lithelius coelomalis* sp. nov.

图版 86

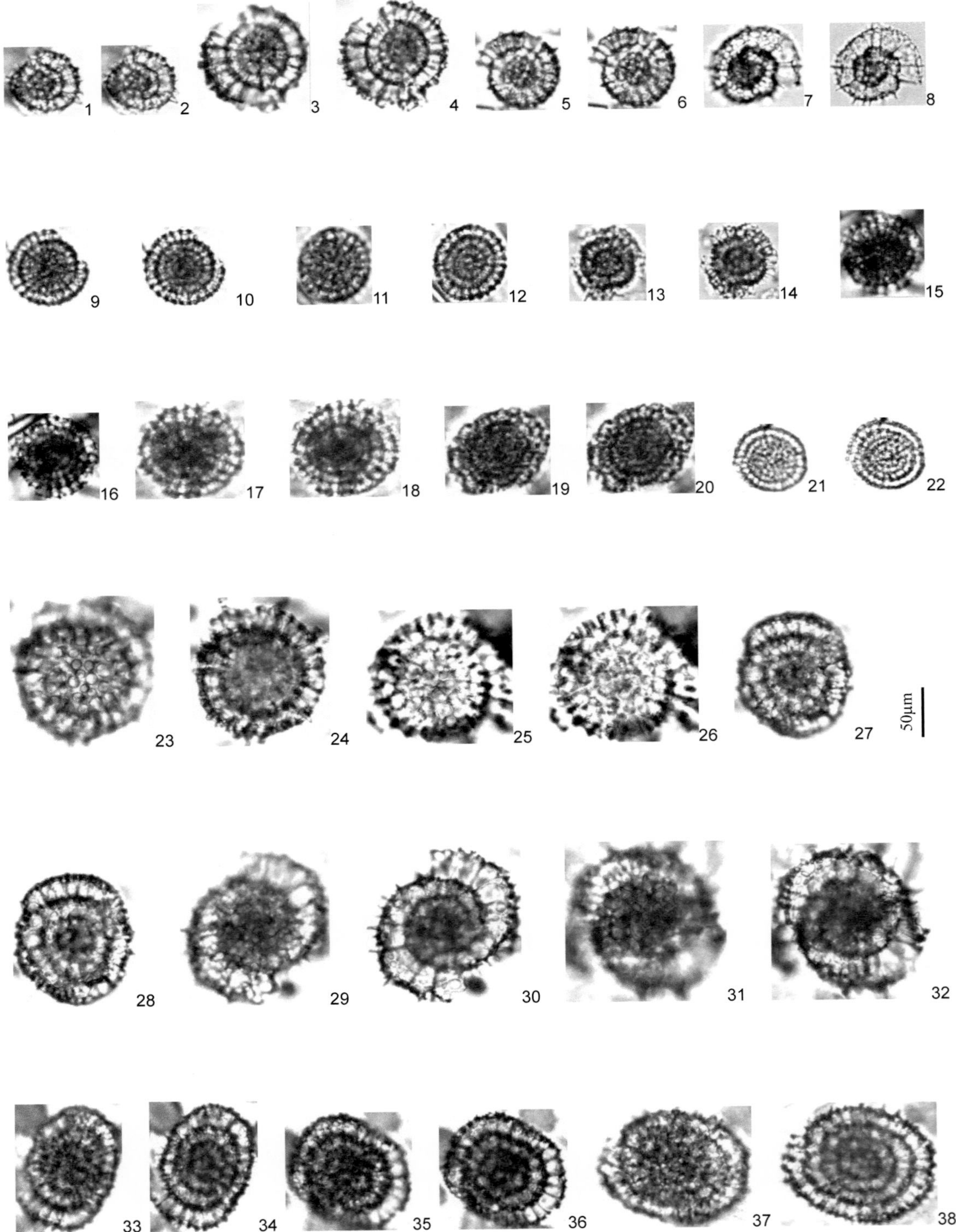

1–8. *Lithelius klingi* Kamikuri；9–22. *Lithelius minor* Jørgensen；23–28. *Lithelius nautiloides* Popofsky；29–32. *Lithelius nerites* Tan and Su；33–38. *Lithelius regularis* sp. nov.

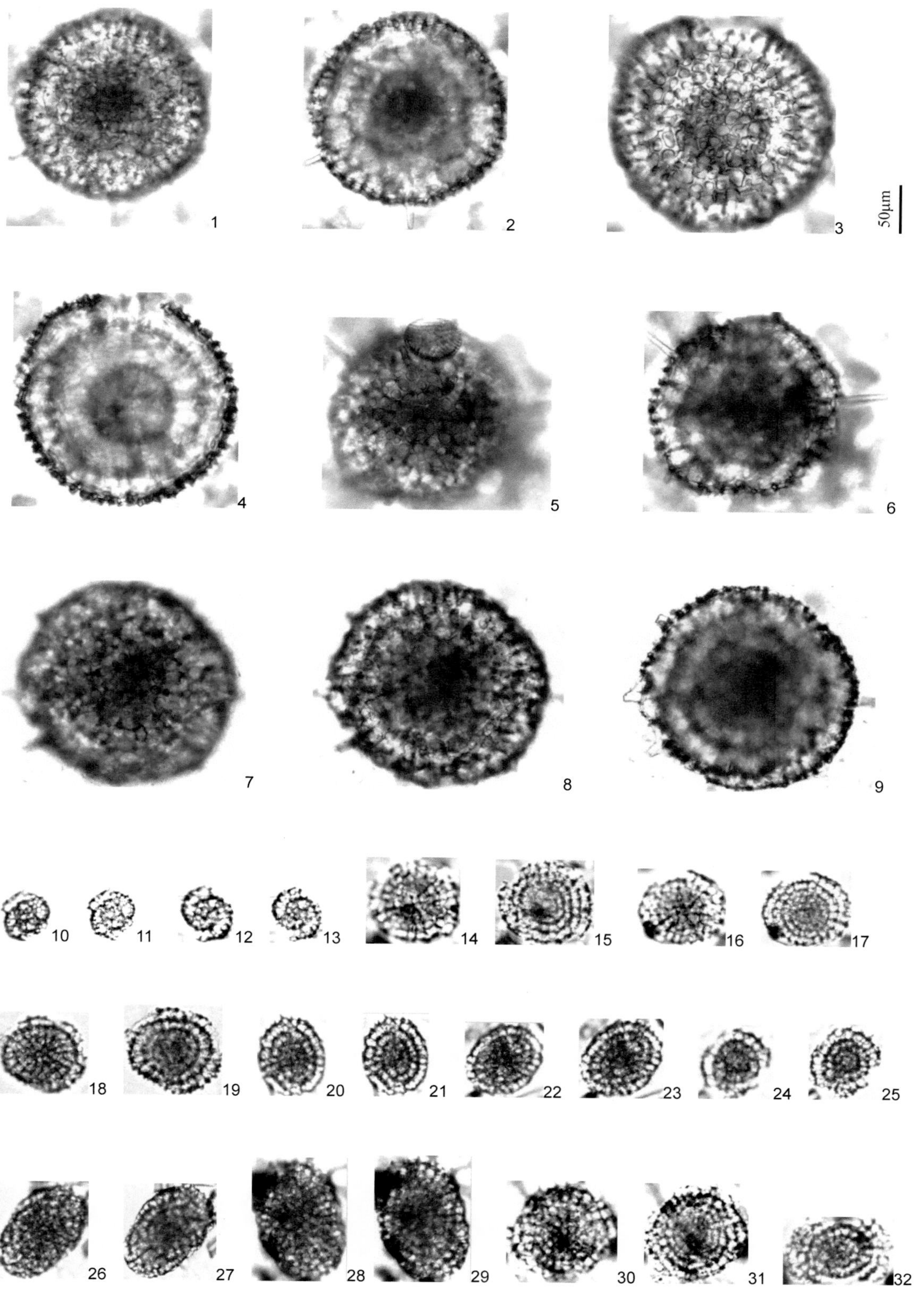

1–9. *Lithelius rotalarius* sp. nov.；10–13. *Lithelius solaris* Haeckel；14–32. *Lithelius spiralis* Haeckel

图版 88

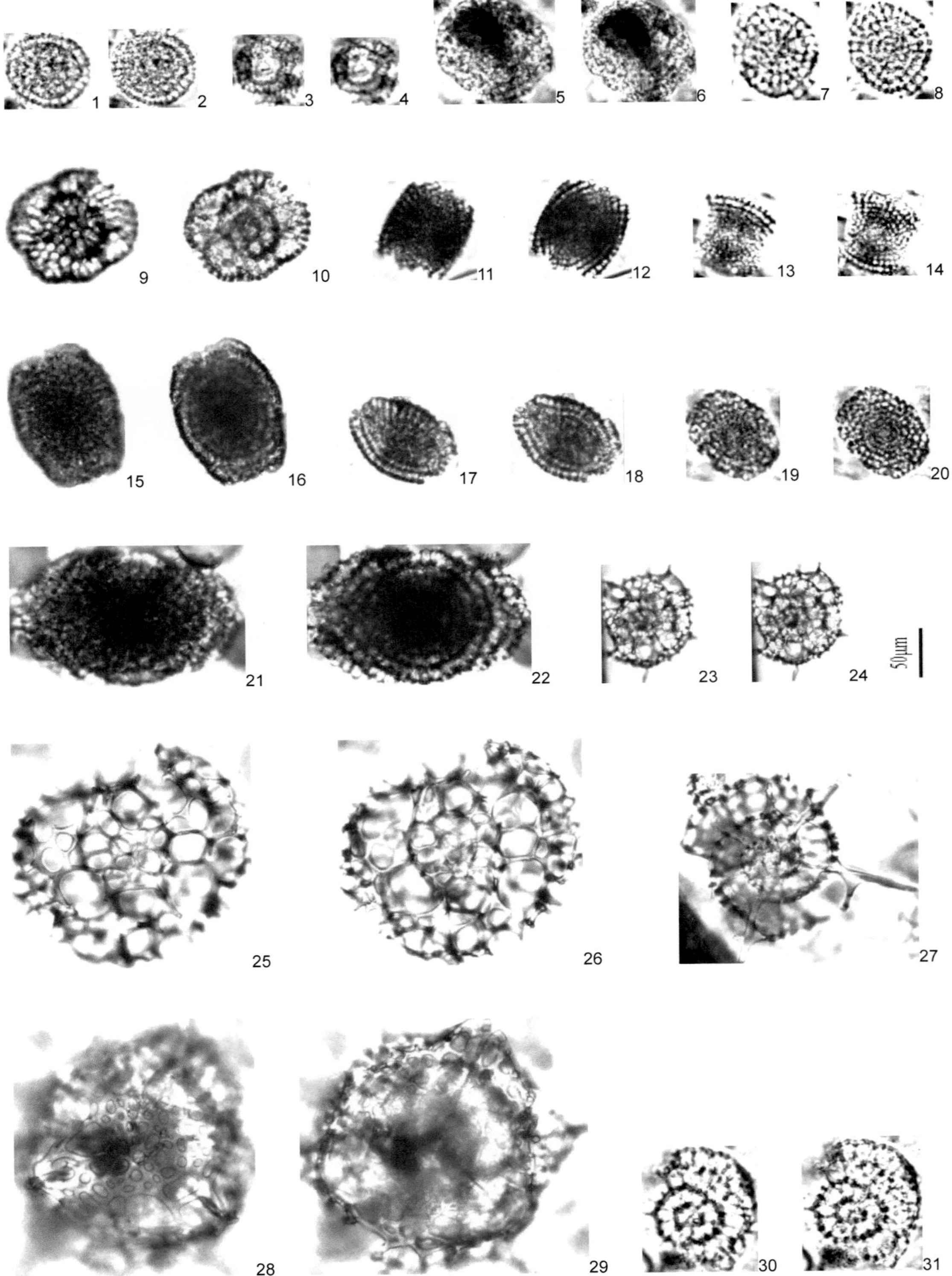

1–8. *Lithelius* sp. 1；9–10. *Larcospira bulbosa* Goll and Bjørklund；11–20. *Lithocarpium polyacantha* (Campbell and Clark) group；21–22. *Cromyodruppa*? *concentrica* Lipman；23–31. *Streblacantha circumyexta* (Jørgensen)

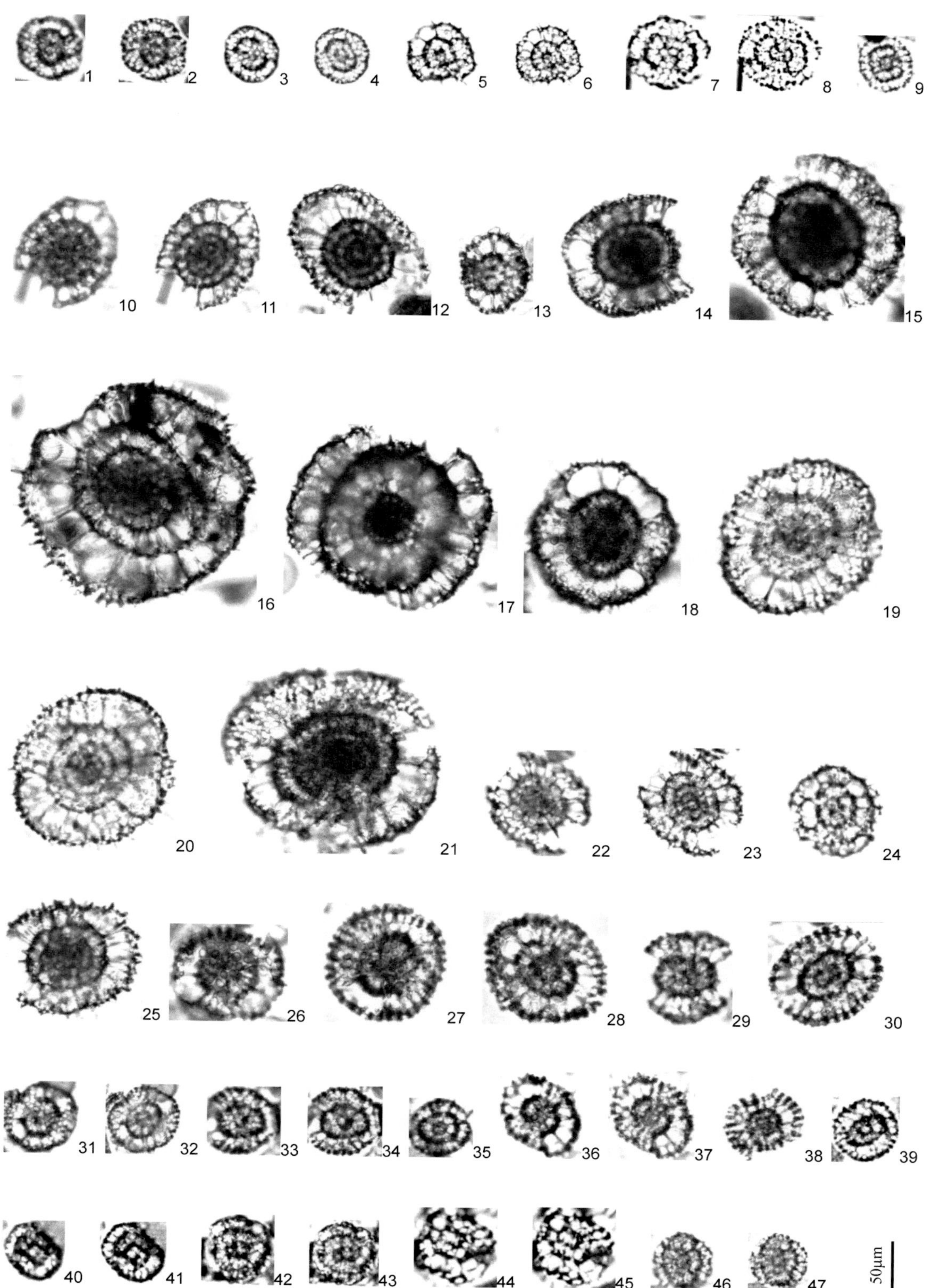

1–39. *Phorticium pylonium* Haeckel group；40–47. *Phorticium* sp. 1

图版 90

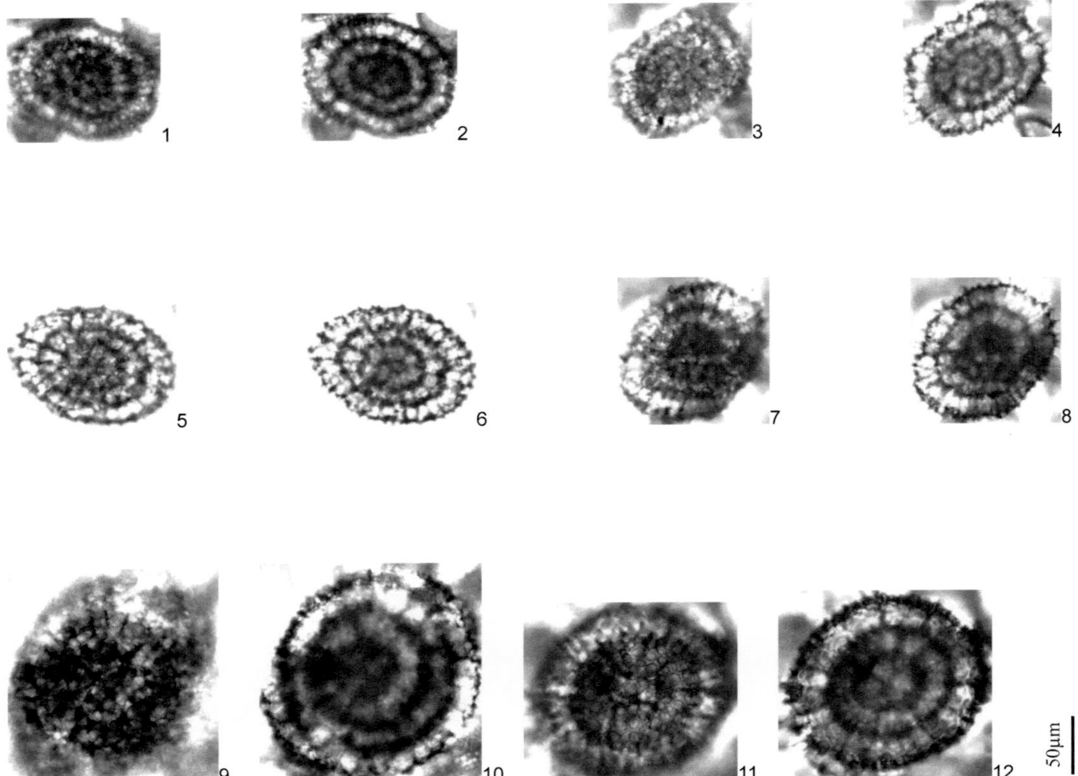

1–8. *Carpodrupa rosseala* sp. nov.; 9–12. *Carpodrupa* sp. 1

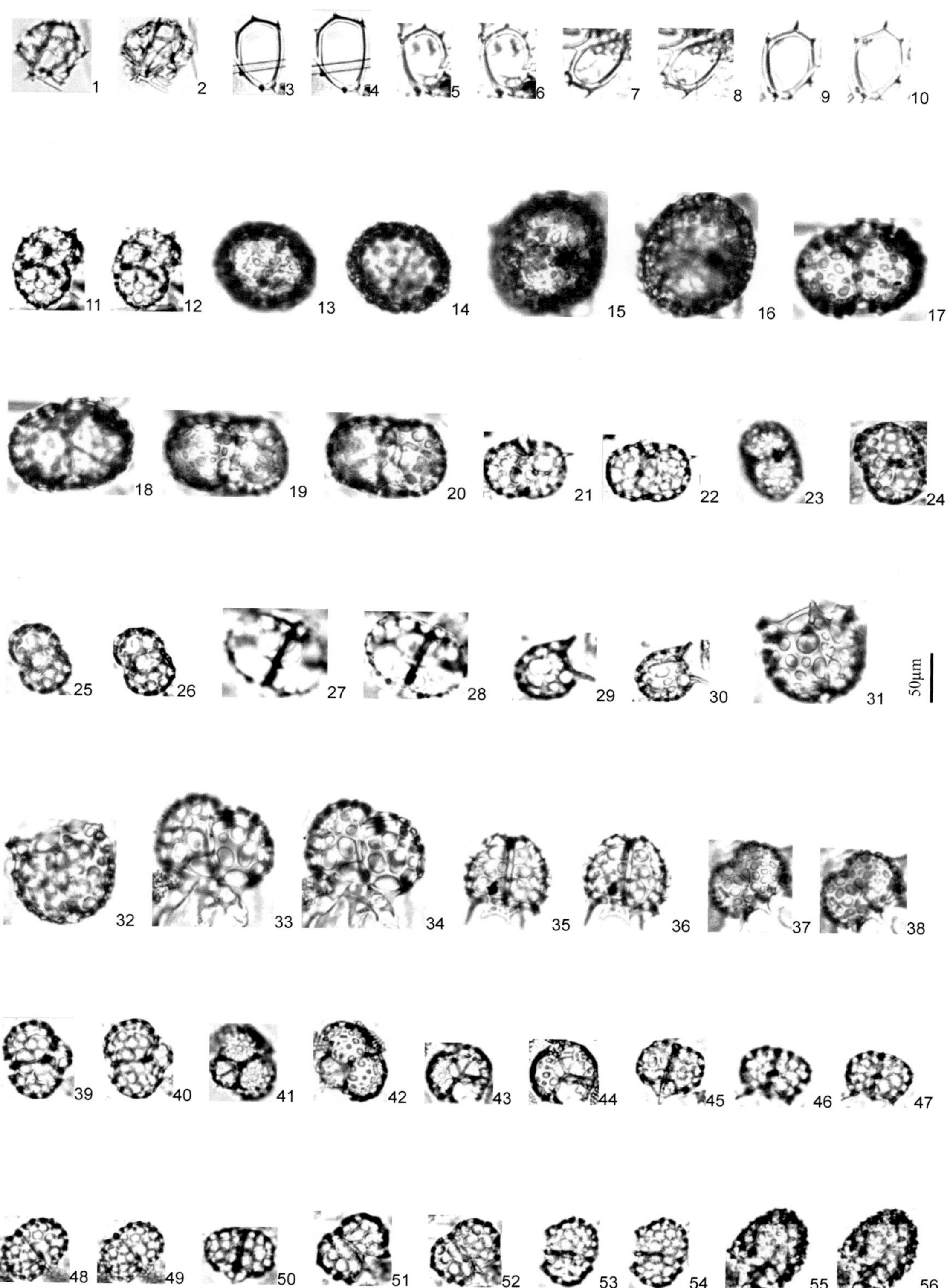

1–2. *Plectacantha oikiskos* Jørgensen；3–10. *Zygocircus productus* (Hertwig)；11–26. *Liriospyris glabra* sp. nov.；27–30. *Liriospyris* sp. 1；31–38. *Tholospyris antarctica* (Haecker) group；39–54. *Tholospyris peltella* sp. nov.；55–56. *Desmospyris coryatis* sp. nov.

图版 92

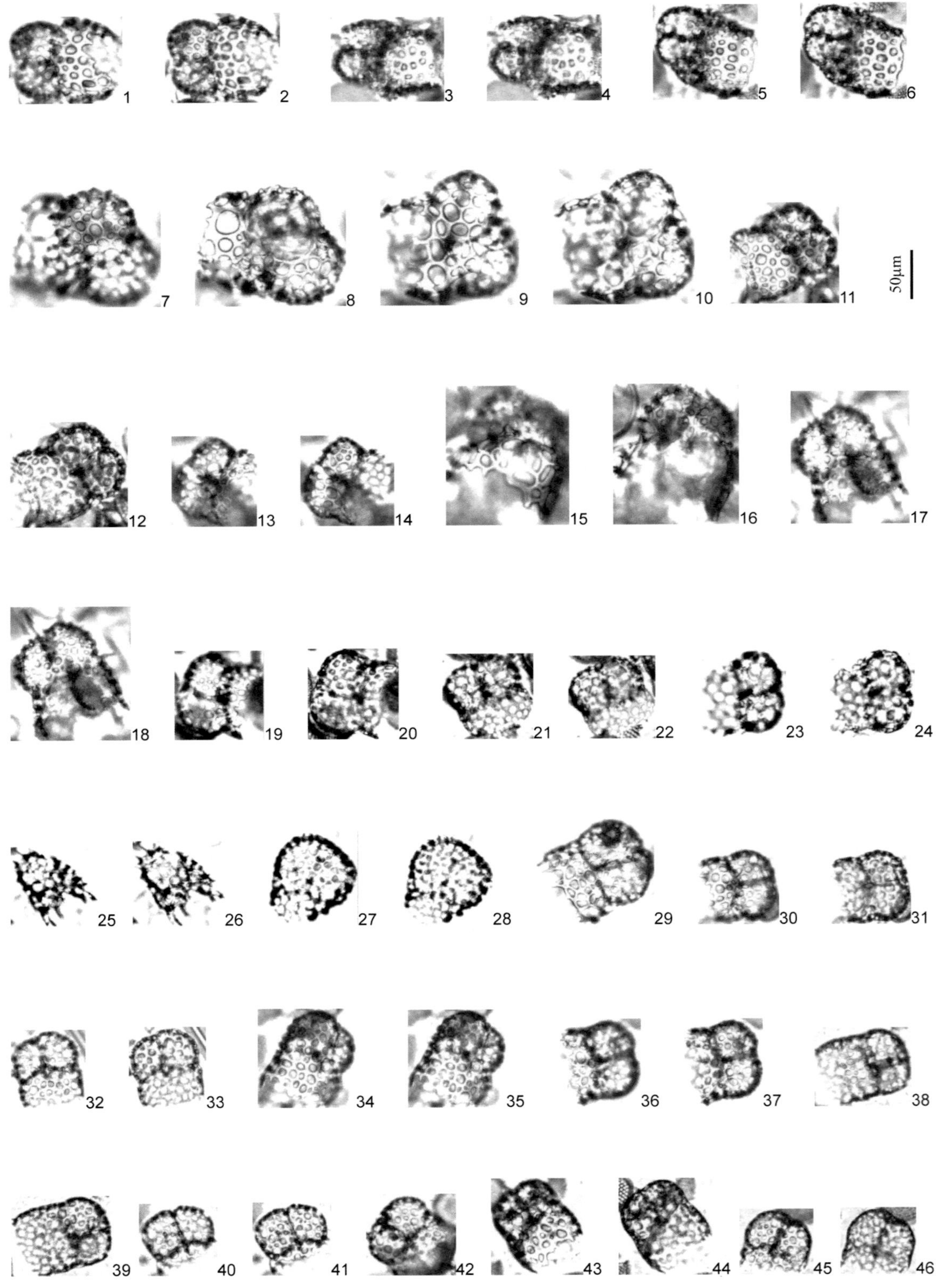

1–6. *Desmospyris futaba* Kamikuri；7–24. *Desmospyris haysi* (Chen)；25–26. *Desmospyris multichyparis* sp. nov.；27–46. *Desmospyris rhodospyroides* Petrushevskaya

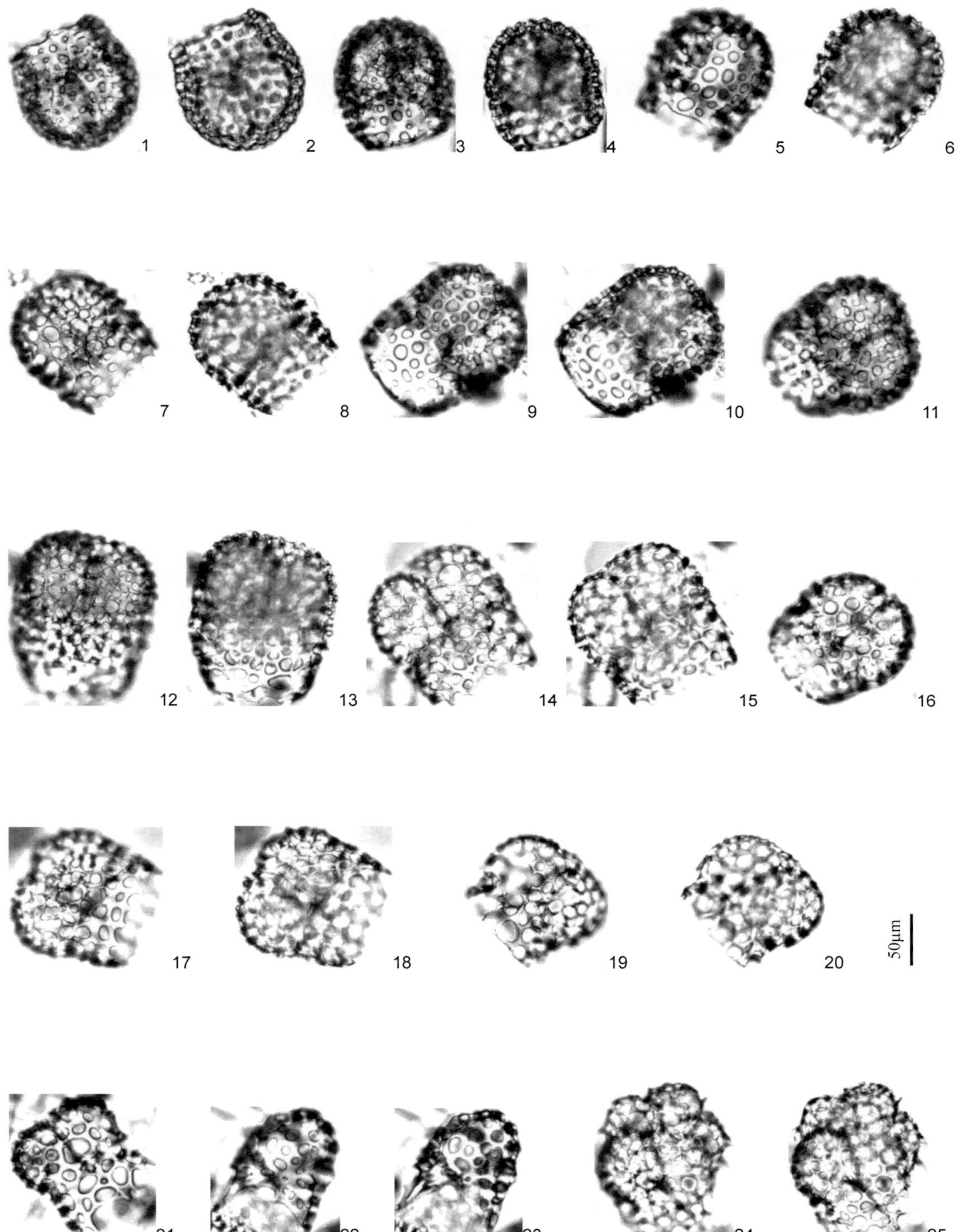

1–20. *Desmospyris spongiosa* Hays；21–23. *Desmospyris trichyparis* sp. nov.；24–25. *Amphimelissa quadrata* sp. nov.

图版 94

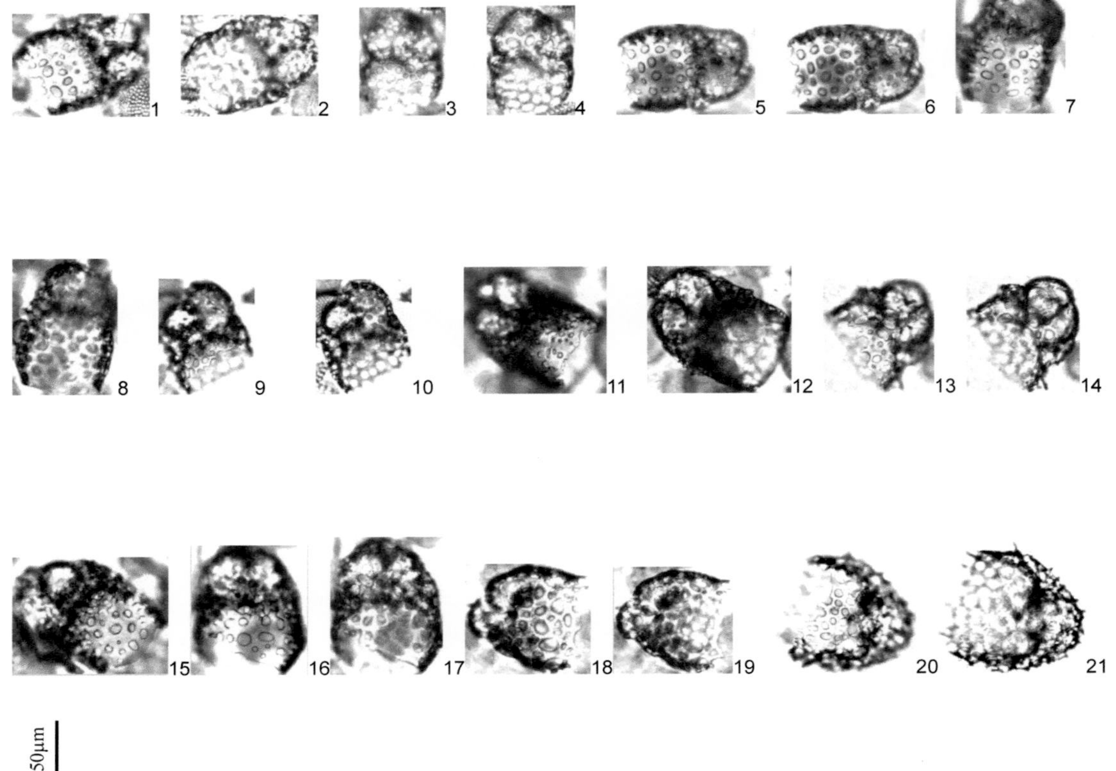

1–19. *Amphimelissa setosa* (Cleve)；20–21. *Amphimelissa* sp. 1

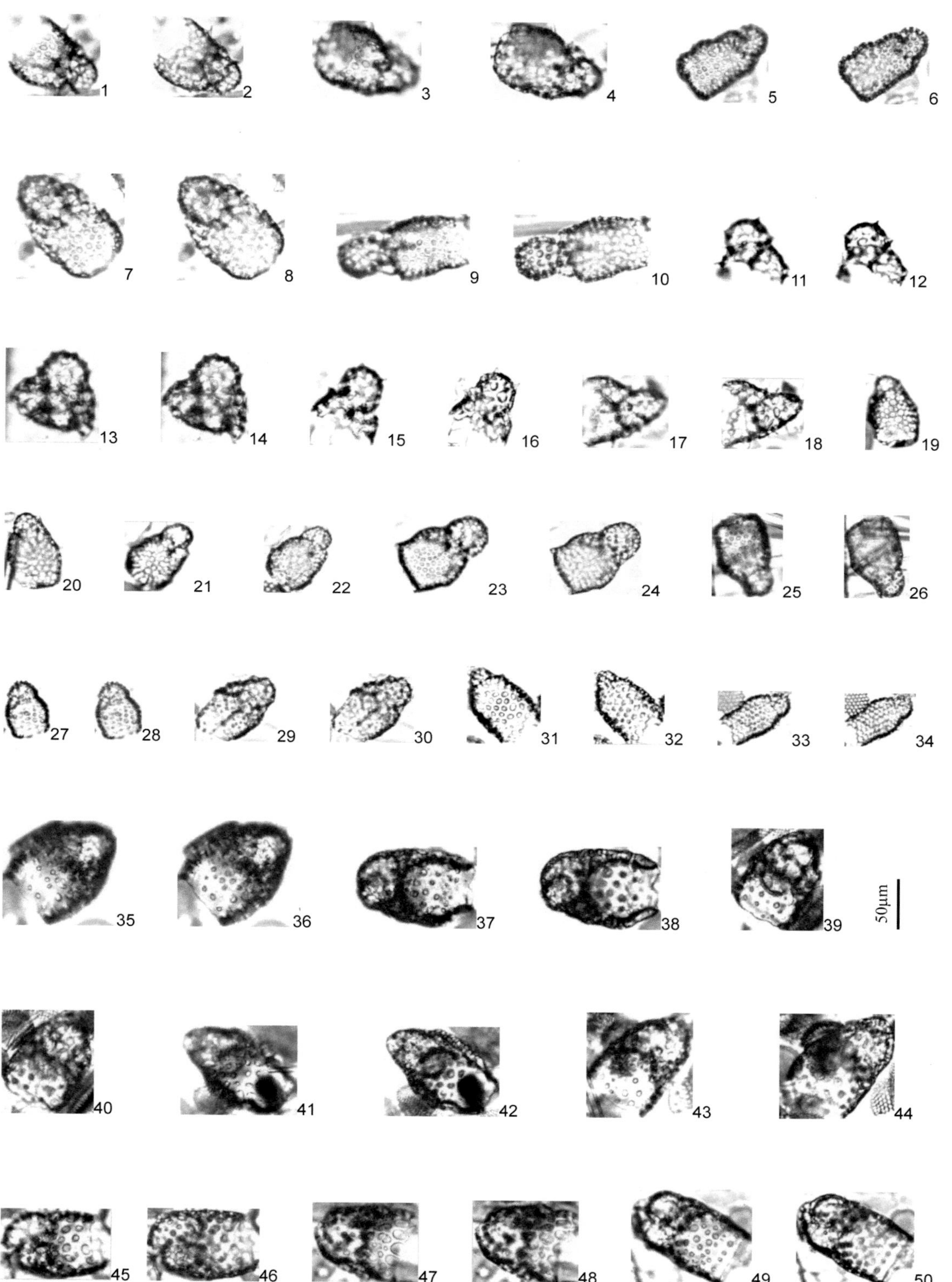

1–4. *Botryopera conradae* (Chen)；5–10. *Botryopera cylindrita* sp. nov.；11–30. *Botryopera triloba* (Ehrenberg) group；31–34. *Botryopera* sp. 1；35–50. *Botryopyle cribrosa* (Ehrenberg)

图版 96

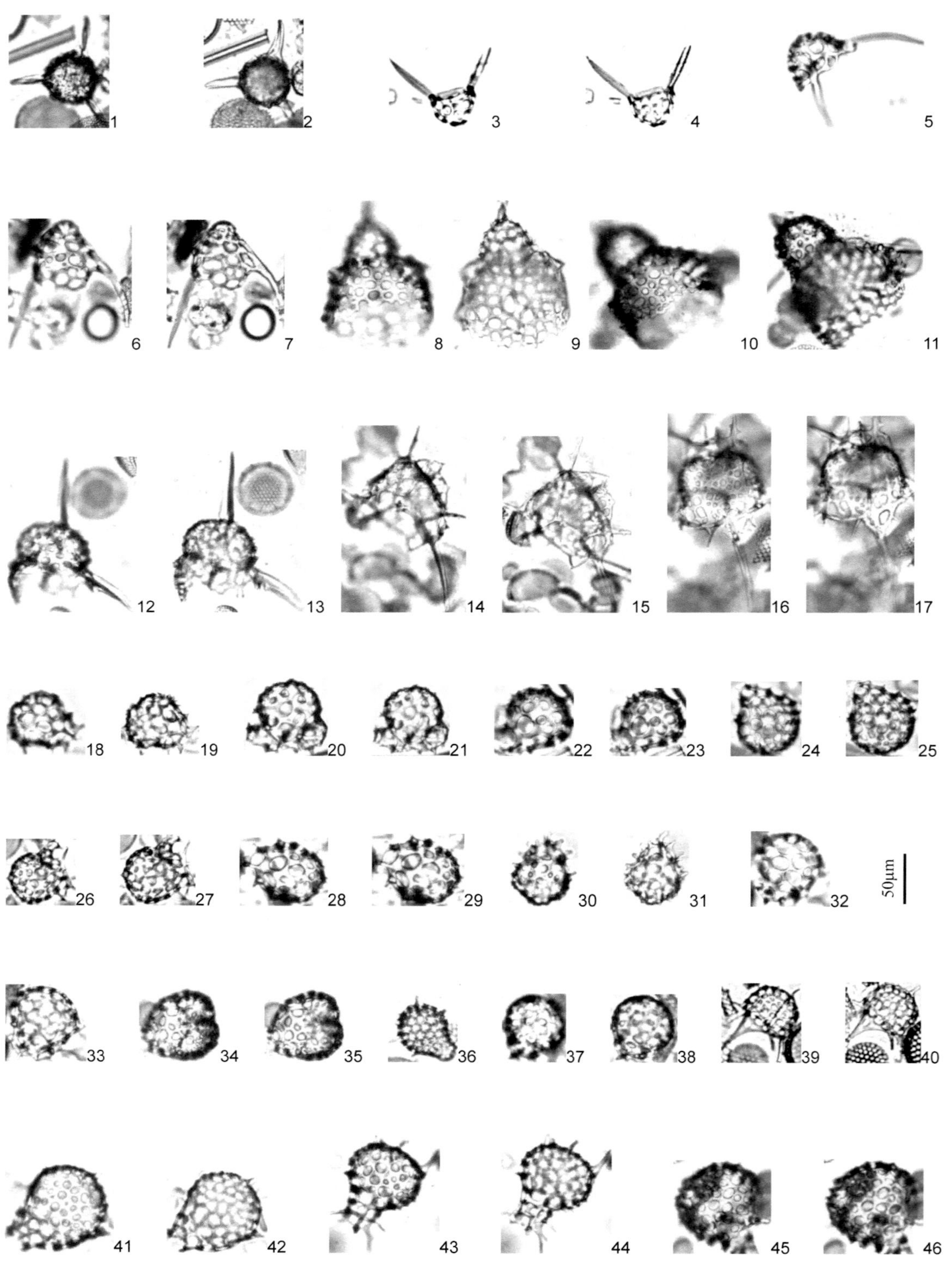

1–2. *Archipilium orthopterum* Haeckel; 3–5. *Archipilium* sp. aff. *A. orthopterum* Haeckel; 6–7. *Archipilium vicinun* sp. nov.; 8–11. *Tripilidium*? *clavipes* Clark and Campbell; 12–13. *Cladoscenium advena* (Clark and Campbell) group; 14–17. *Cladoscenium tricolpium* (Haeckel); 18–38. *Peridium longispinum* Jørgensen; 39–40. *Peridium* sp. 1; 41–46. *Archipera dipleura* Tan and Tchang

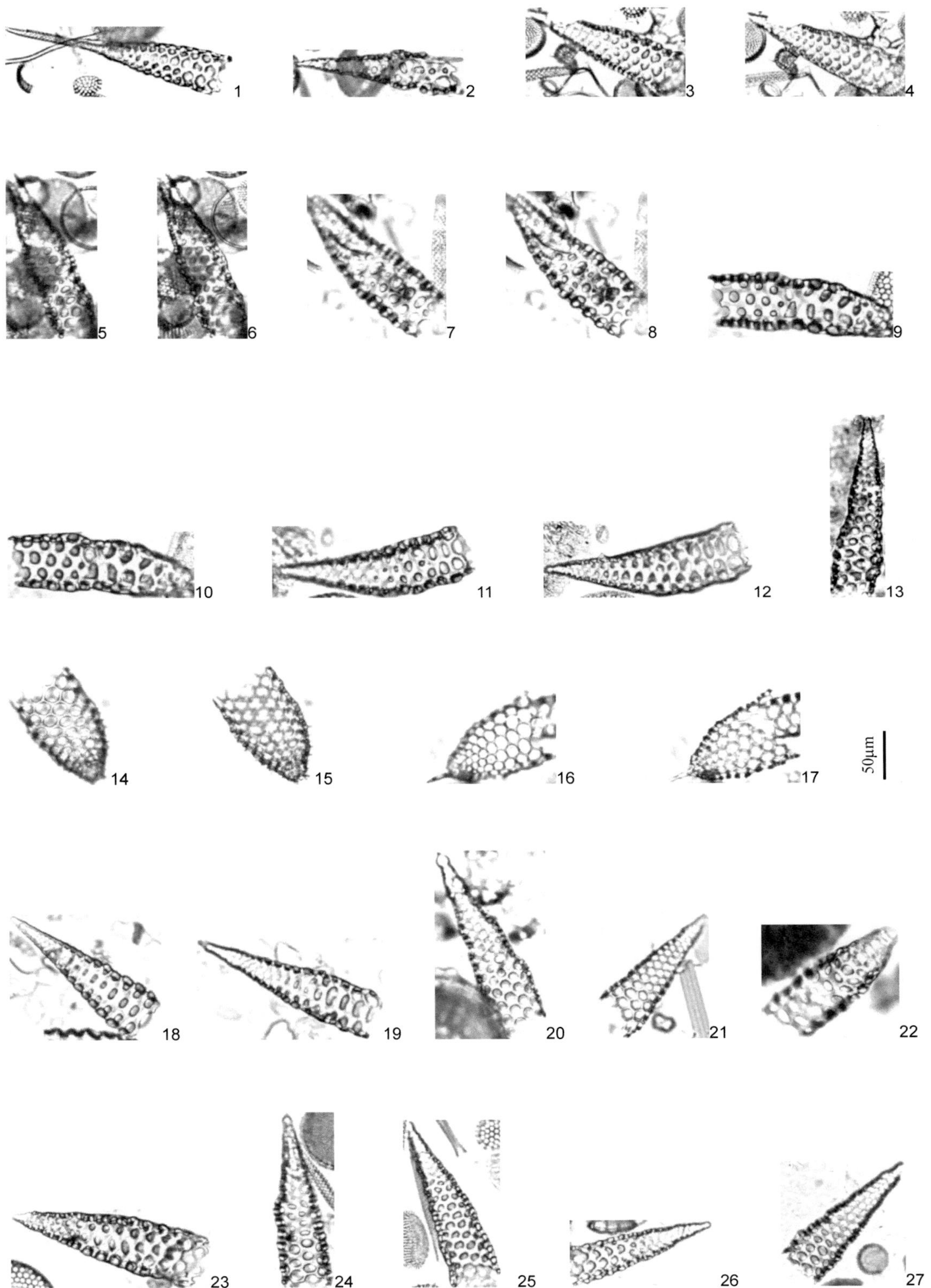

1–6. *Cornutella californica* Campbell and Clark；7–13. *Cornutella clathrata* Ehrenberg；14–17. *Cornutella mitra* Ehrenberg；18–27. *Cornutella profunda* Ehrenberg

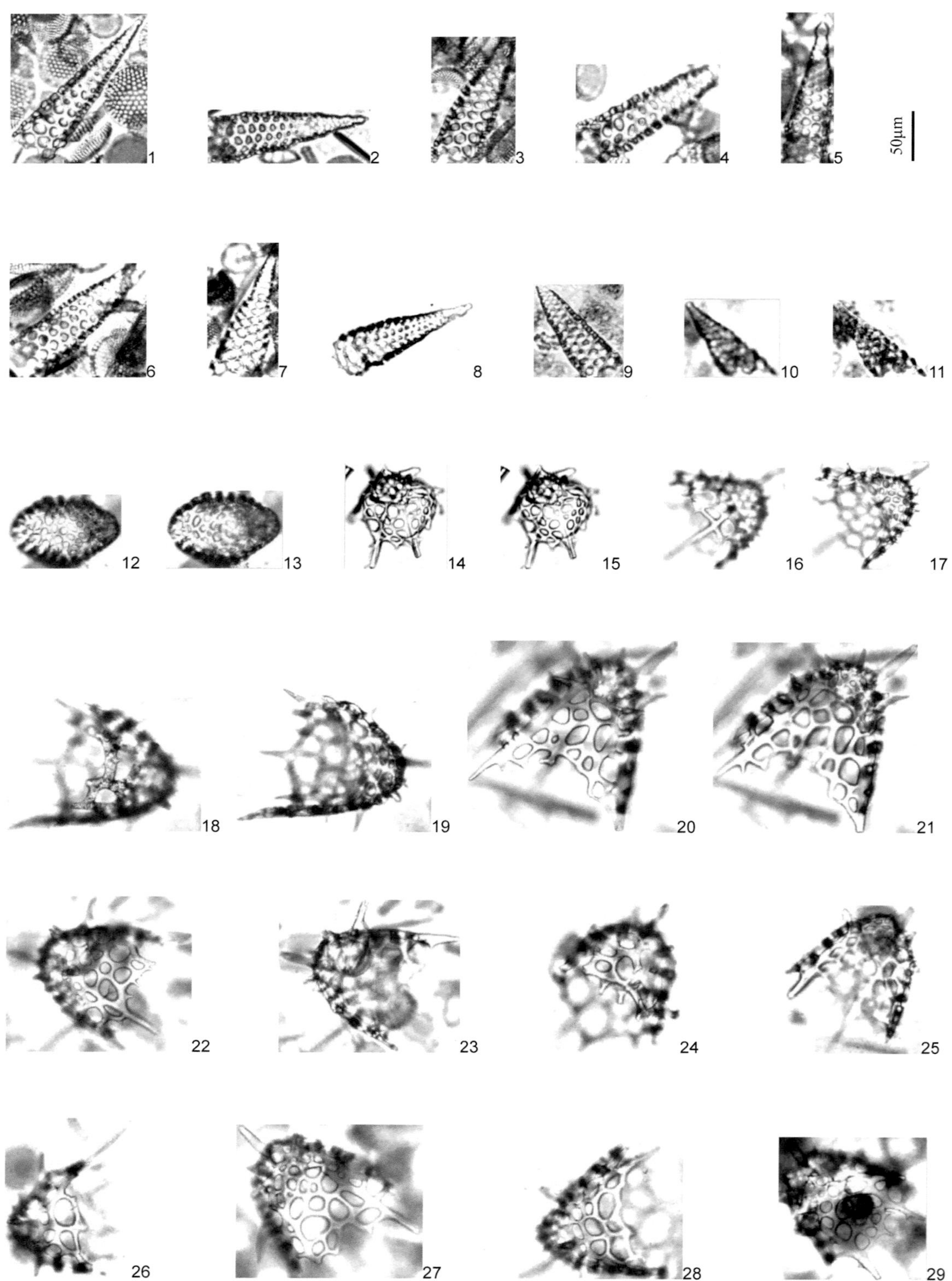

1–11. *Cornutella profunda* Ehrenberg；12–13. *Archicapsa olivaeformis* sp. nov.；14–29. *Dictyophimus echinatus* sp. nov.

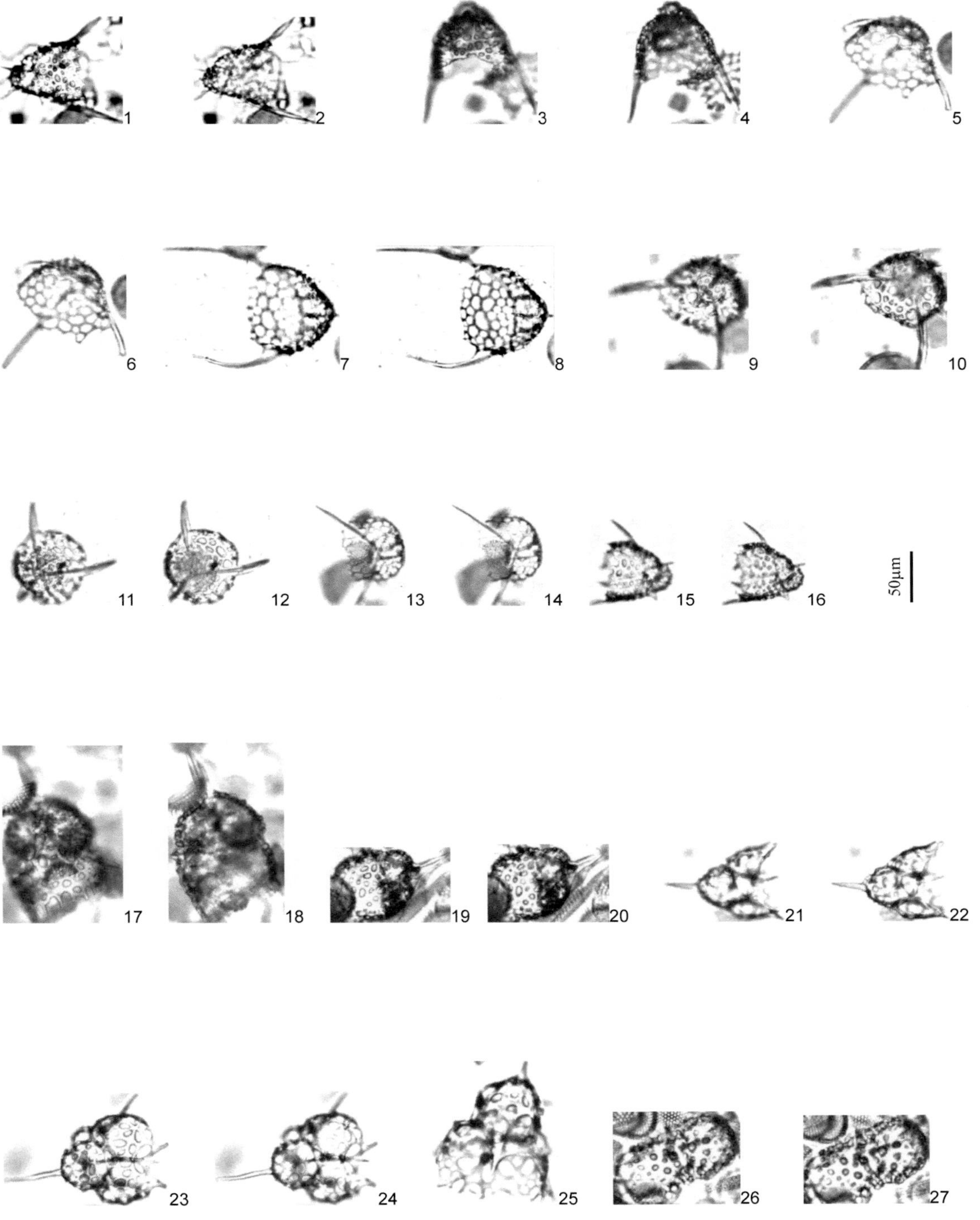

1–2. *Dictyophimus hirundo* (Haeckel) group；3–12. *Dictyophimus histricosus* Jørgensen；13–14. *Dictyophimus lubricephalus* sp. nov.；15–16. *Dictyophimus platycephalus* Haeckel；17–20. *Lithomelissa arrhencorna* sp. nov.；21–25. *Lithomelissa callifera* sp. nov.；26–27. *Lithomelissa capito* Ehrenberg

图版 100

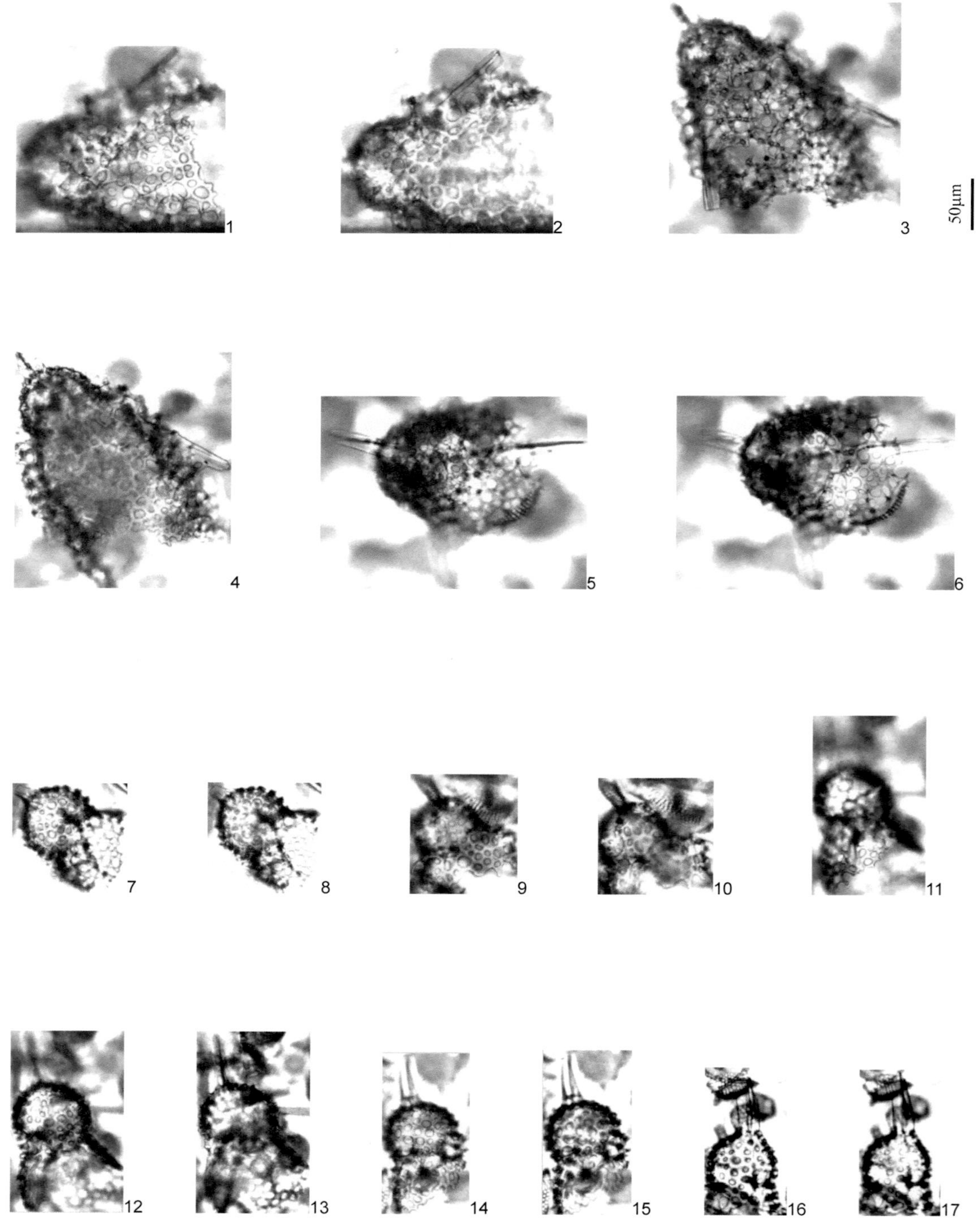

1–6. *Lithomelissa challengerae* Chen；7–17. *Lithomelissa cratospina* sp. nov.

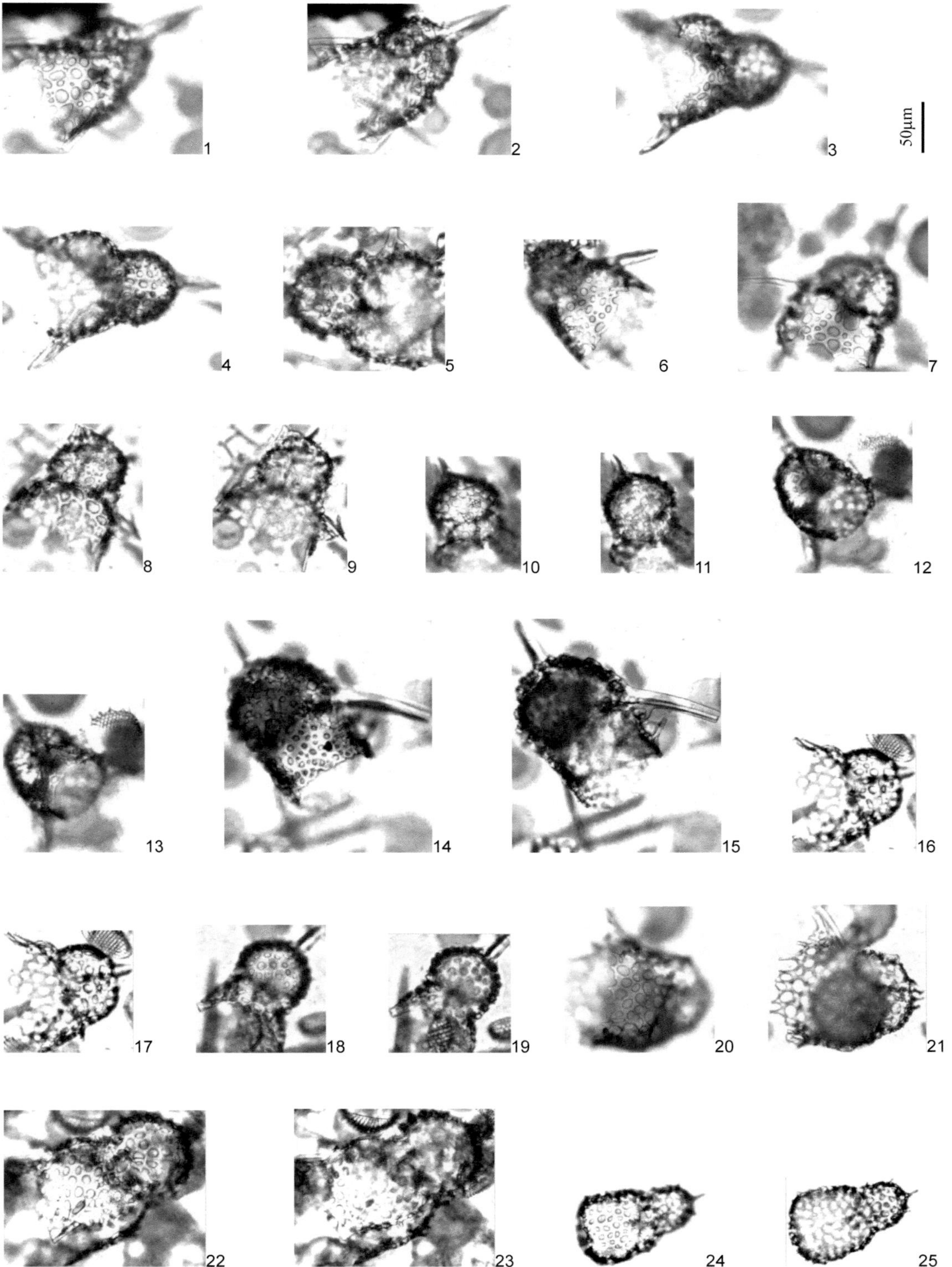

1–23. *Lithomelissa ehrenbergii* Bütschli；24–25. *Lithomelissa* sp. A aff. *L. ehrenbergi* Bütschli

图版 102

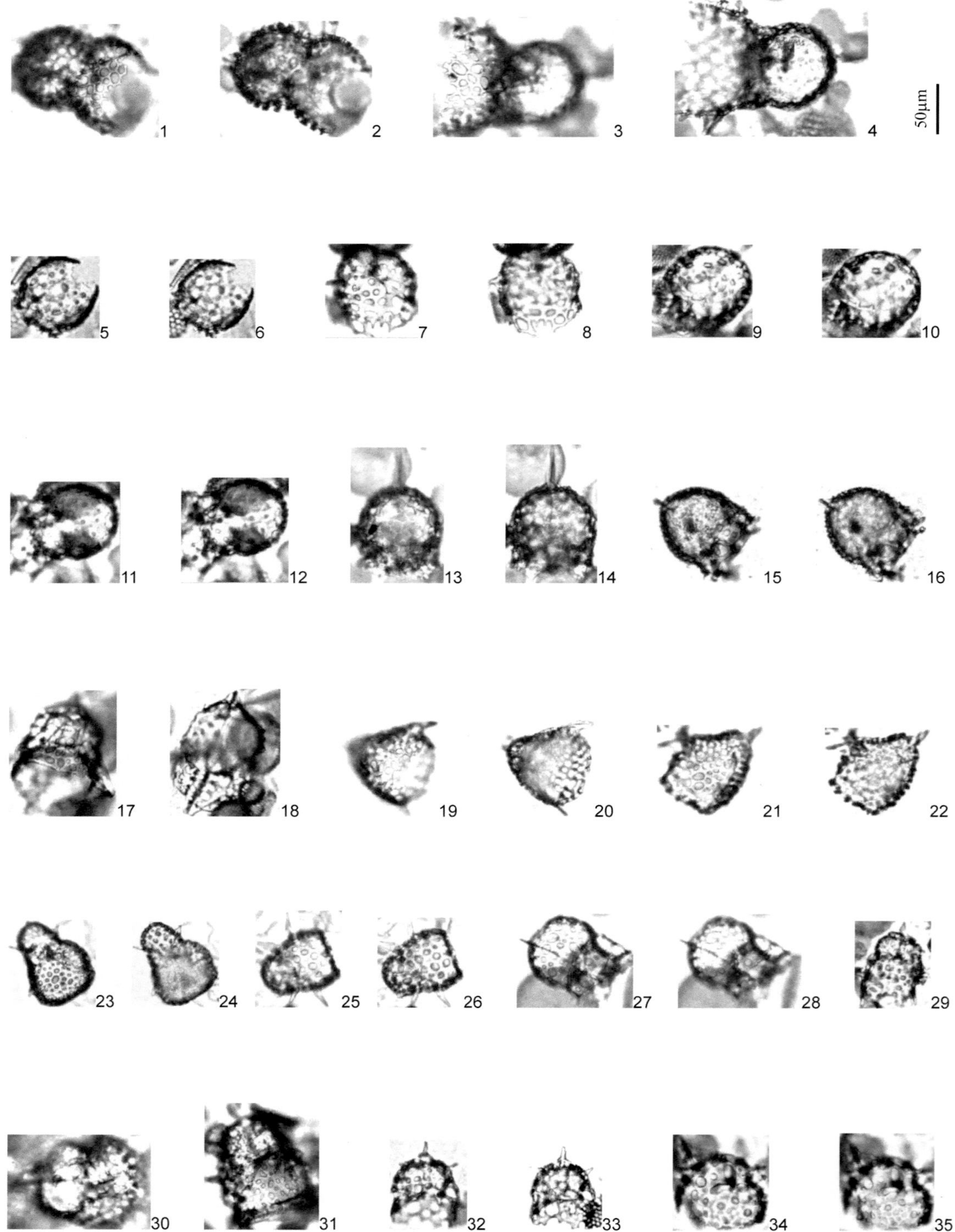

1–4. *Lithomelissa gelasinus* O'Connor；5–12. *Lithomelissa globicapita* sp. nov.；13–18. *Lithomelissa* cf. *haeckeli* Bütschli；19–22. *Lithomelissa hirundiforma* sp. nov.；23–26. *Lithomelissa laticeps* Jørgensen；27–35. *Lithomelissa ministyla* sp. nov.

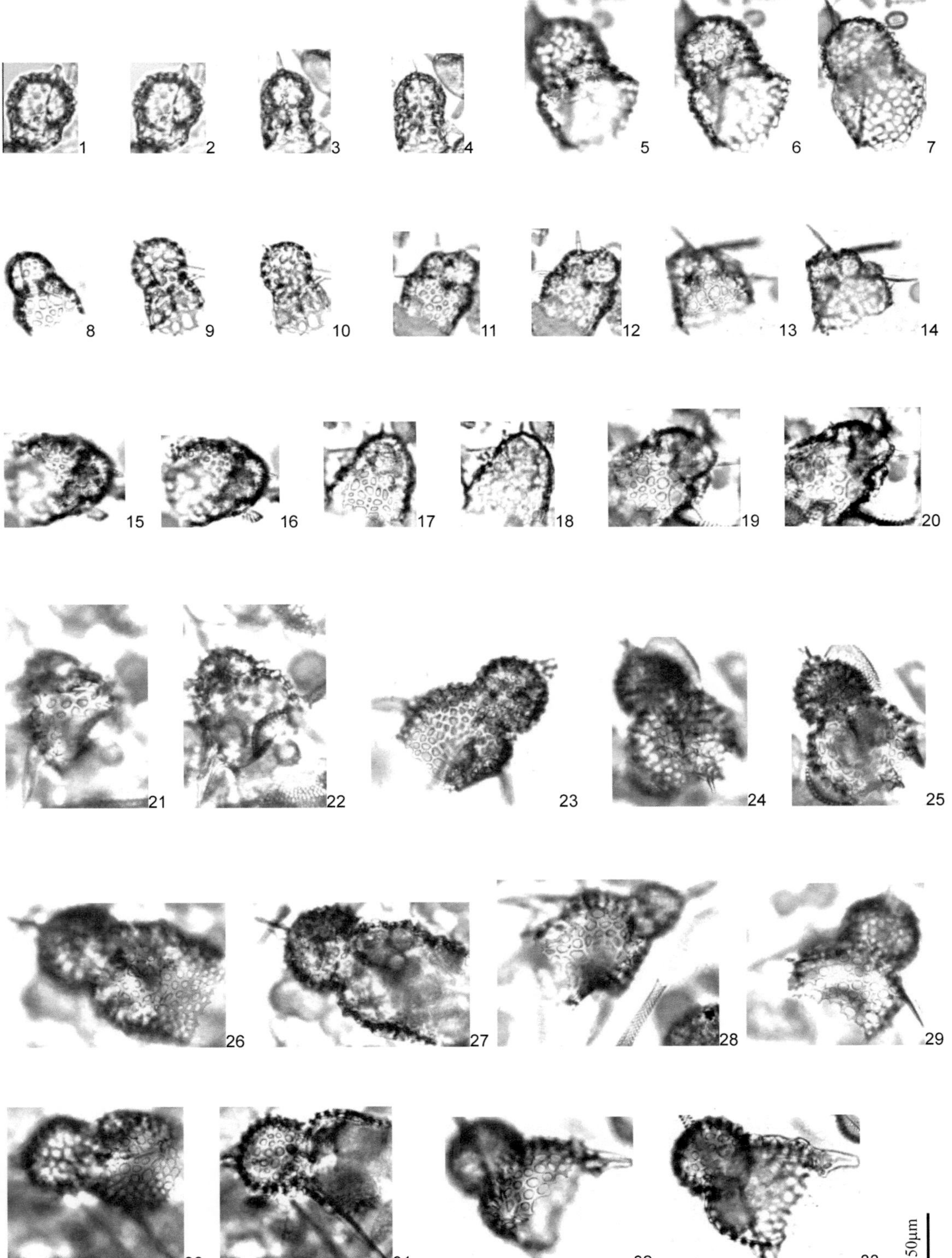

1–4. *Lithomelissa pachyoma* sp. nov.; 5–10. *Lithomelissa poljanskii* (Petrushevskaya); 11–20. *Lithomelissa preantarctica* (Pettrushevskaya); 21–33. *Lithomelissa robusta* Chen

图版 104

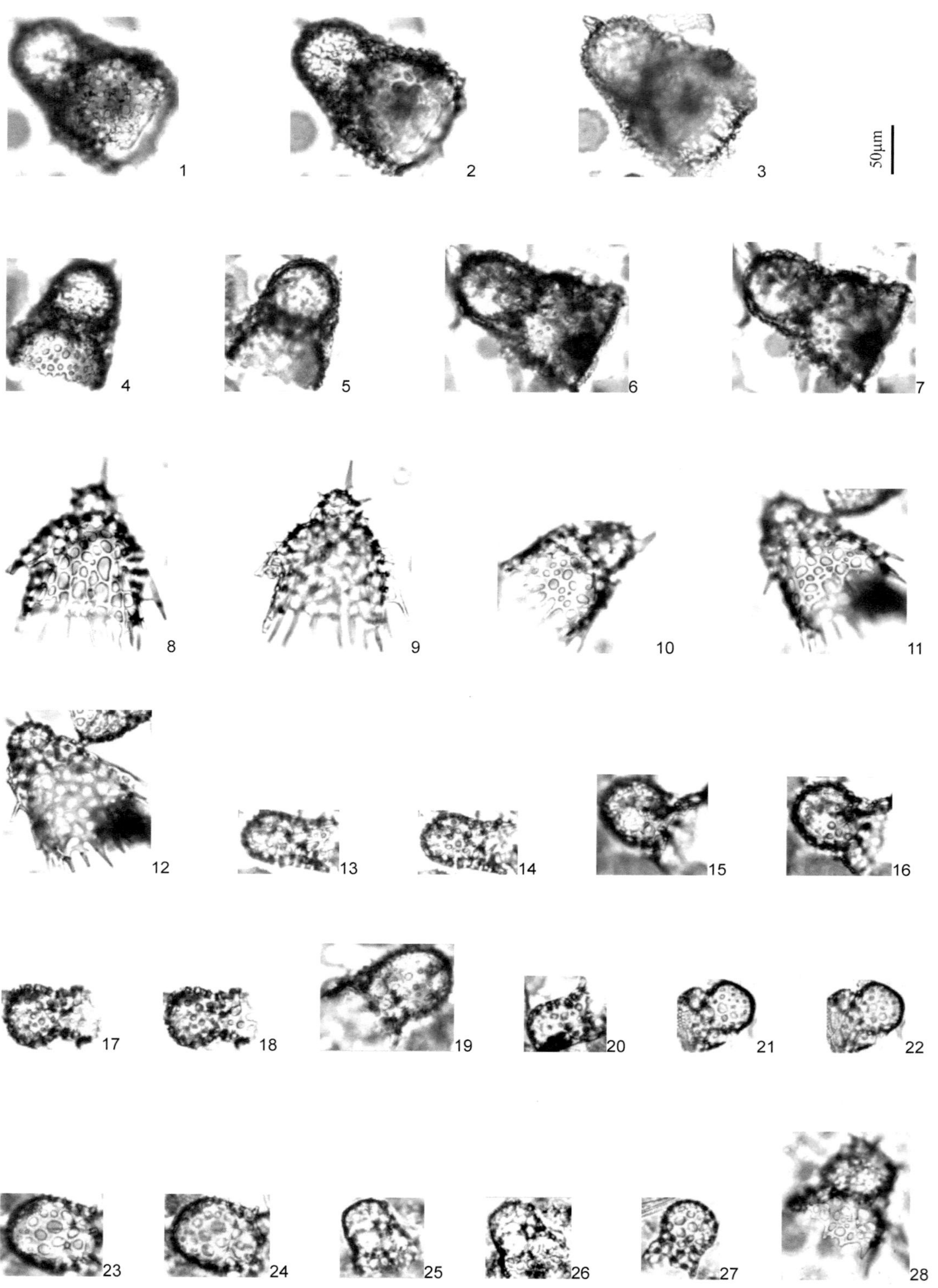

1–7. *Lithomelissa sphaerocephalis* Chen；8–12. *Lithomelissa stomaculeata* sp. nov.；13–28. *Lithomelissa thoracites* Haeckel

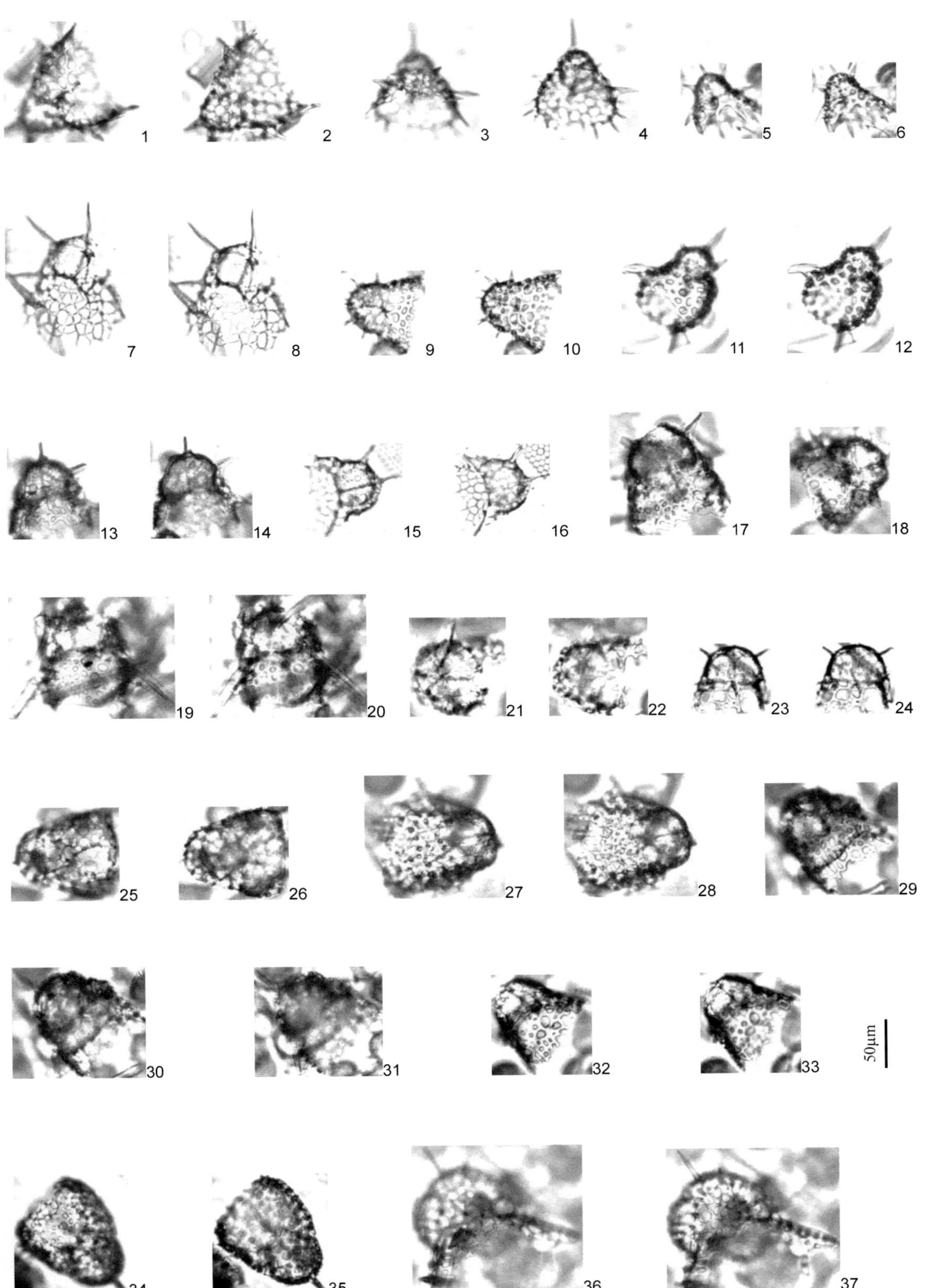

1–24. *Lithomelissa tricornis* Chen；25–33. *Lithomelissa tryphera* sp. nov.；34–35. *Lithomelissa* sp. 1；36–37. *Lithomelissa* sp. 2

图版 106

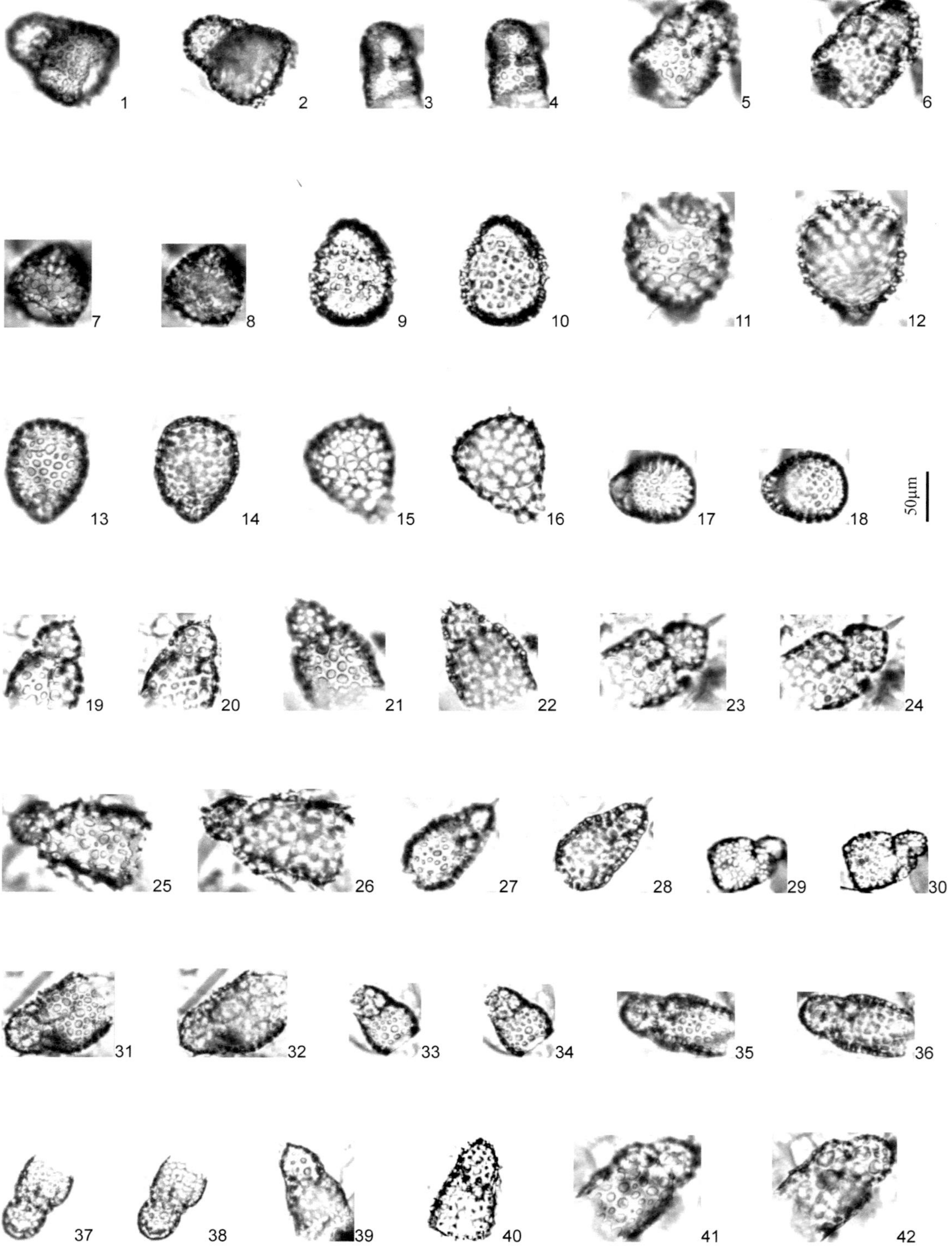

1–4. *Antarctissa antedenticulata* Chen；5–6. *Antarctissa cingocephalis* sp. nov.；7–18. *Antarctissa clausa* (Popofsky)；19–28. *Antarctissa conradae* Chen；29–42. *Antarctissa deflandrei* (Petrushevskaya)

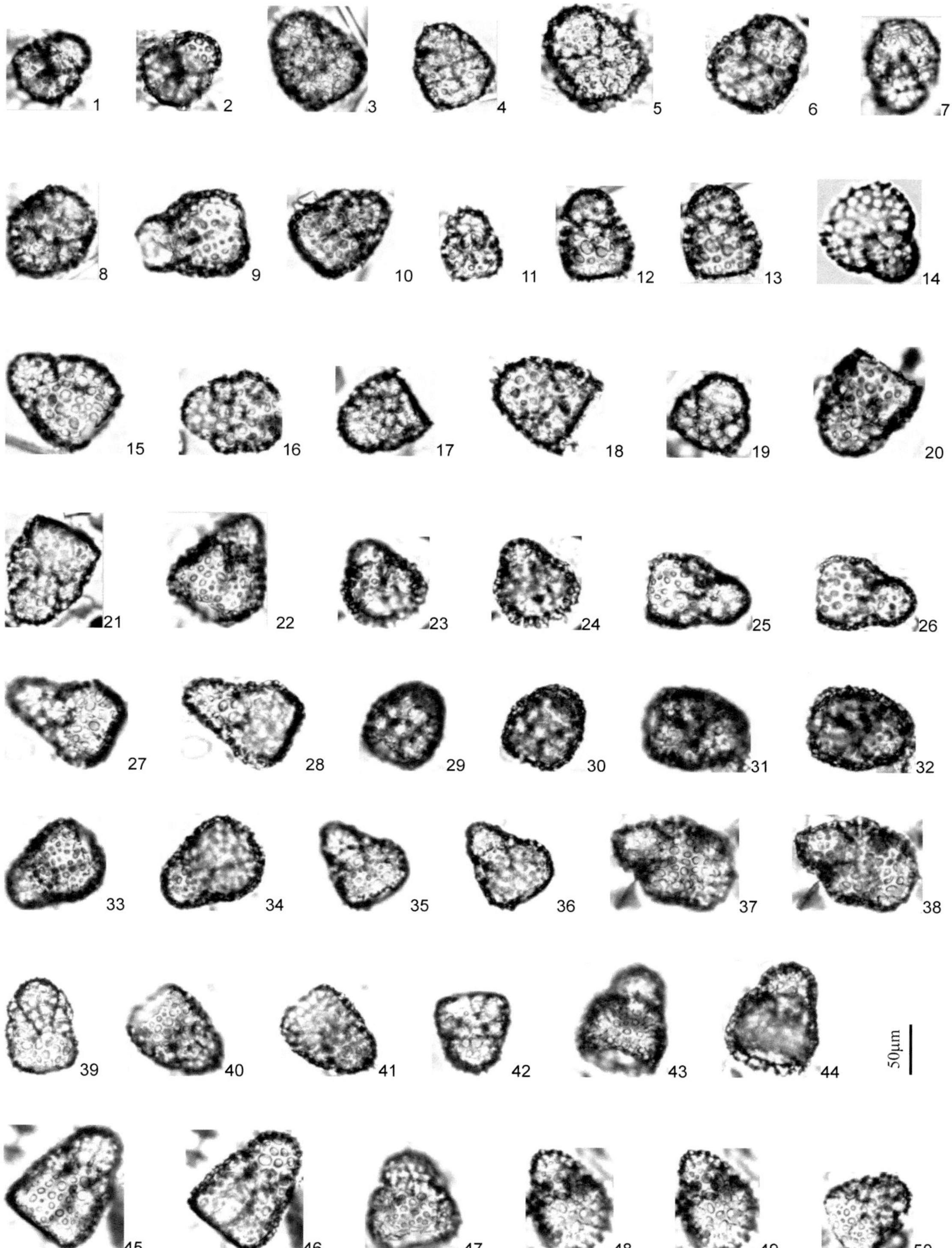

1–50. *Antarctissa denticulata* (Ehrenberg) group

图版 108

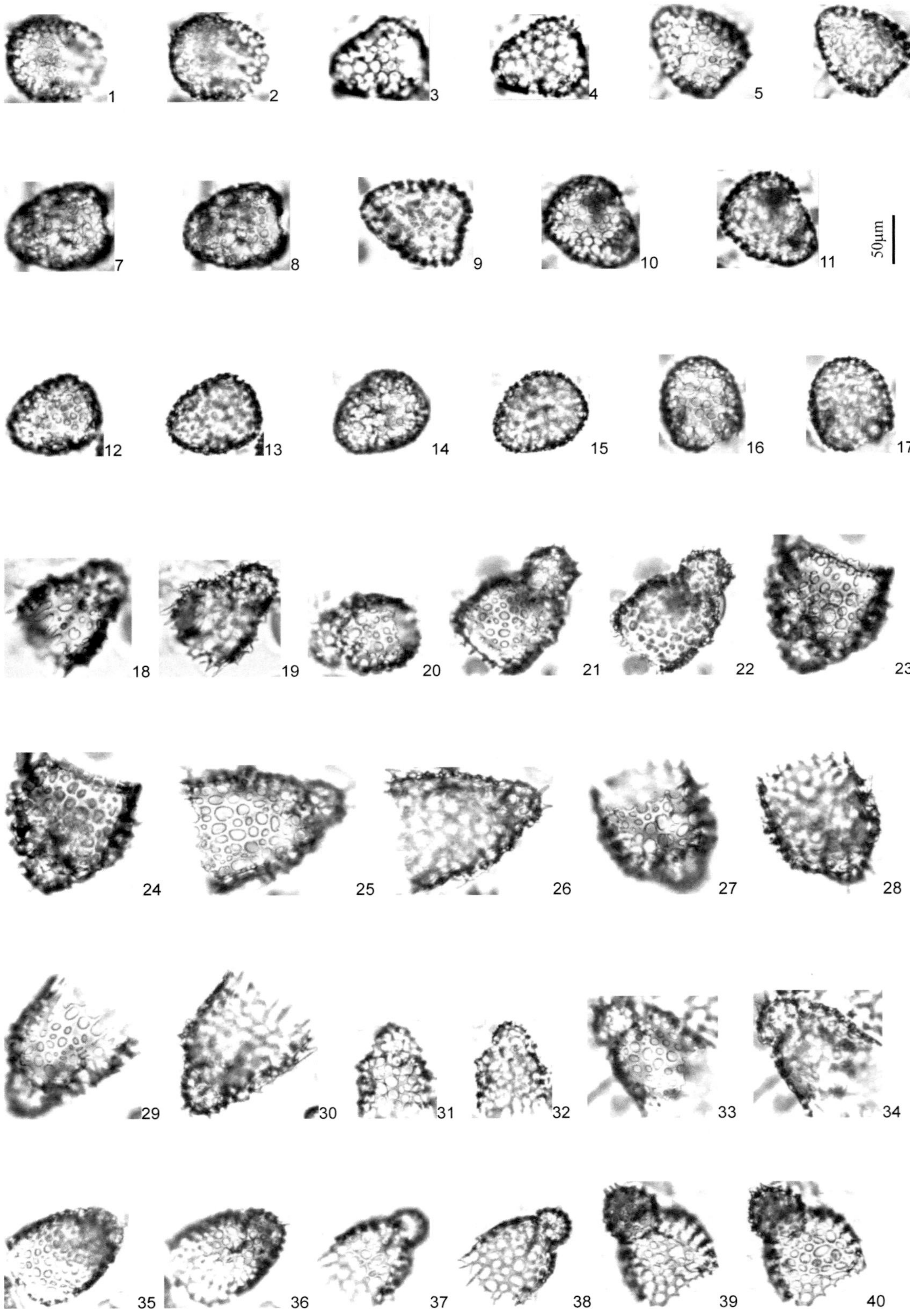

1–17. *Antarctissa denticulata clausa* Petrushevskaya；18–40. *Antarctissa longa* (Popofsky)

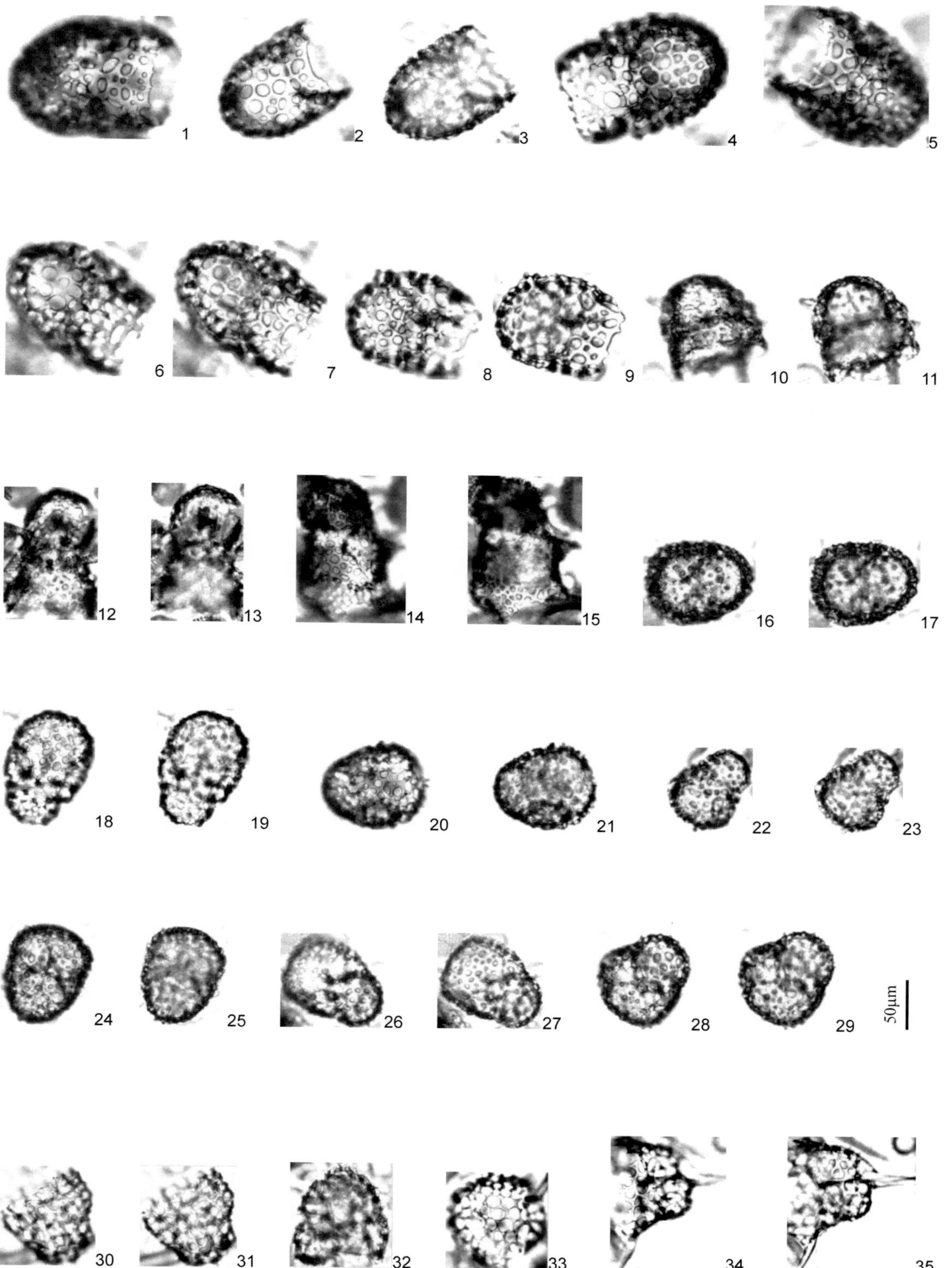

1–11. *Antarctissa pachyoma* sp. nov.; 12–15. *Antarctissa pterigyna* sp. nov.; 16–29. *Antarctissa strelkovi* Petrushevskaya; 30–33. *Antarctissa whitei* Bjørklund; 34–35. *Spongomelissa chaenothoraca* sp. nov.

图版 110

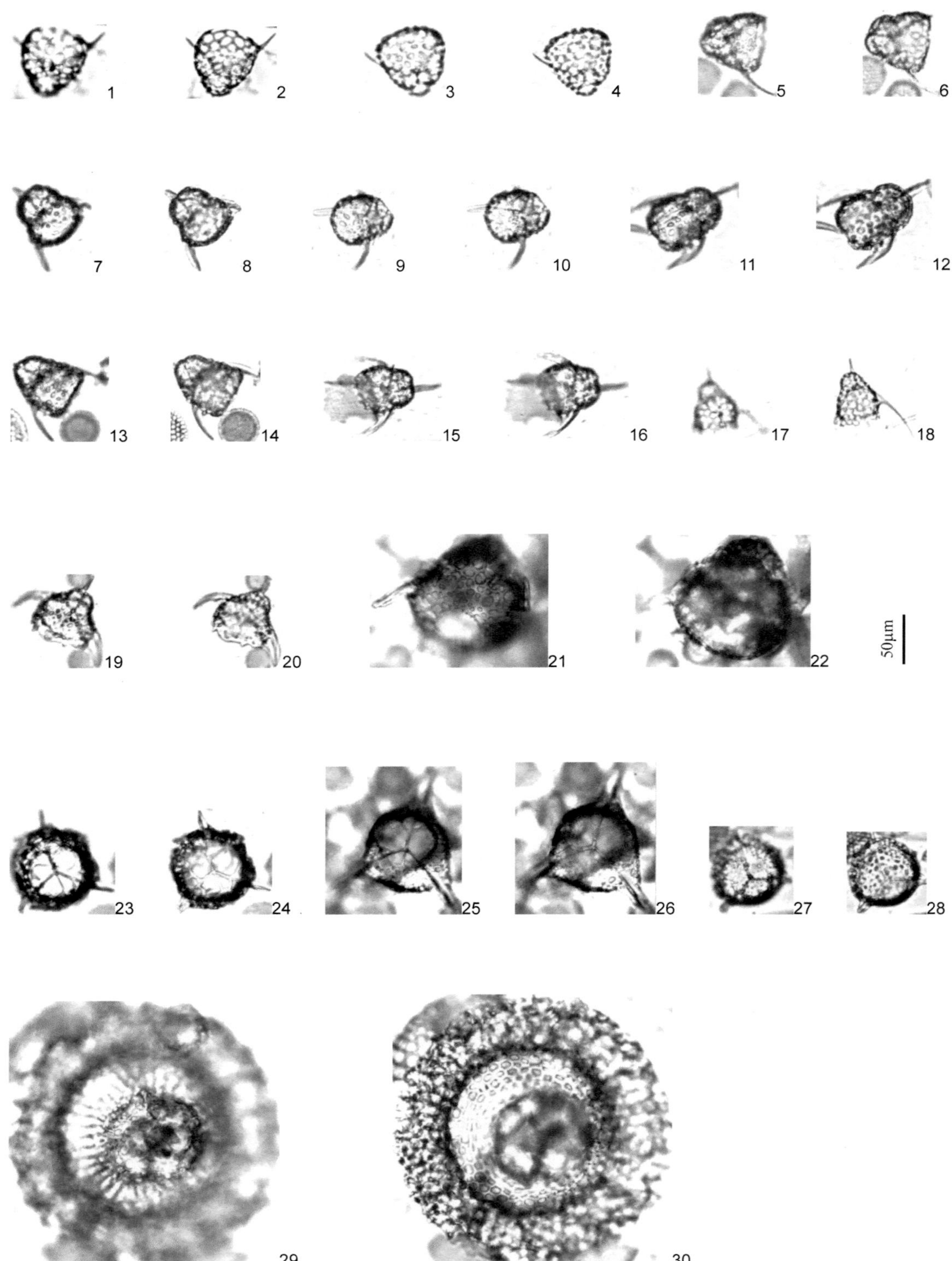

1–28. *Spongomelissa cucumella* Sanfilippo and Riedel；29–30. *Velicucullus altus* Abelmann

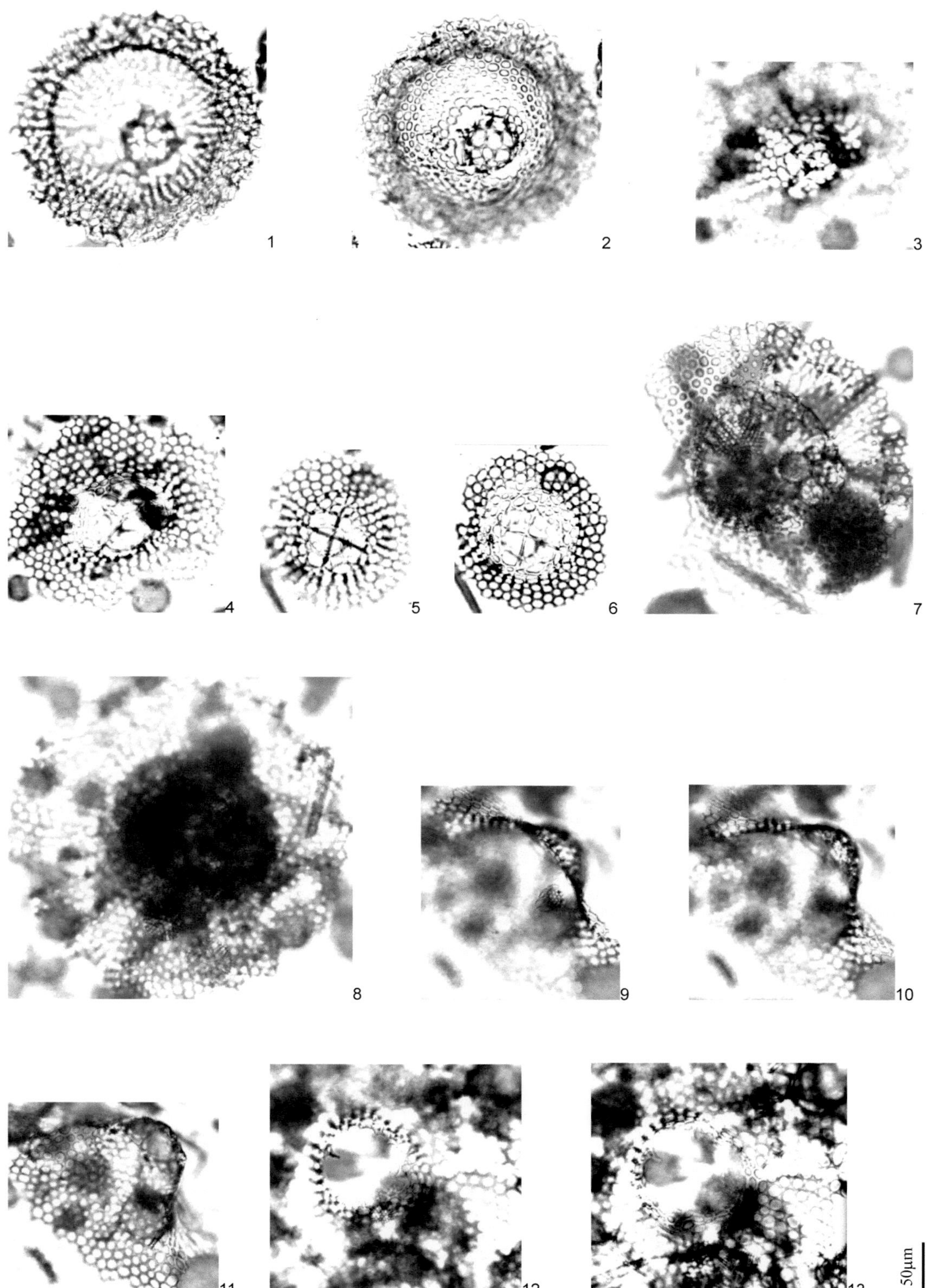

1–13. *Velicucullus oddgurneri* Bjørklund

图版 112

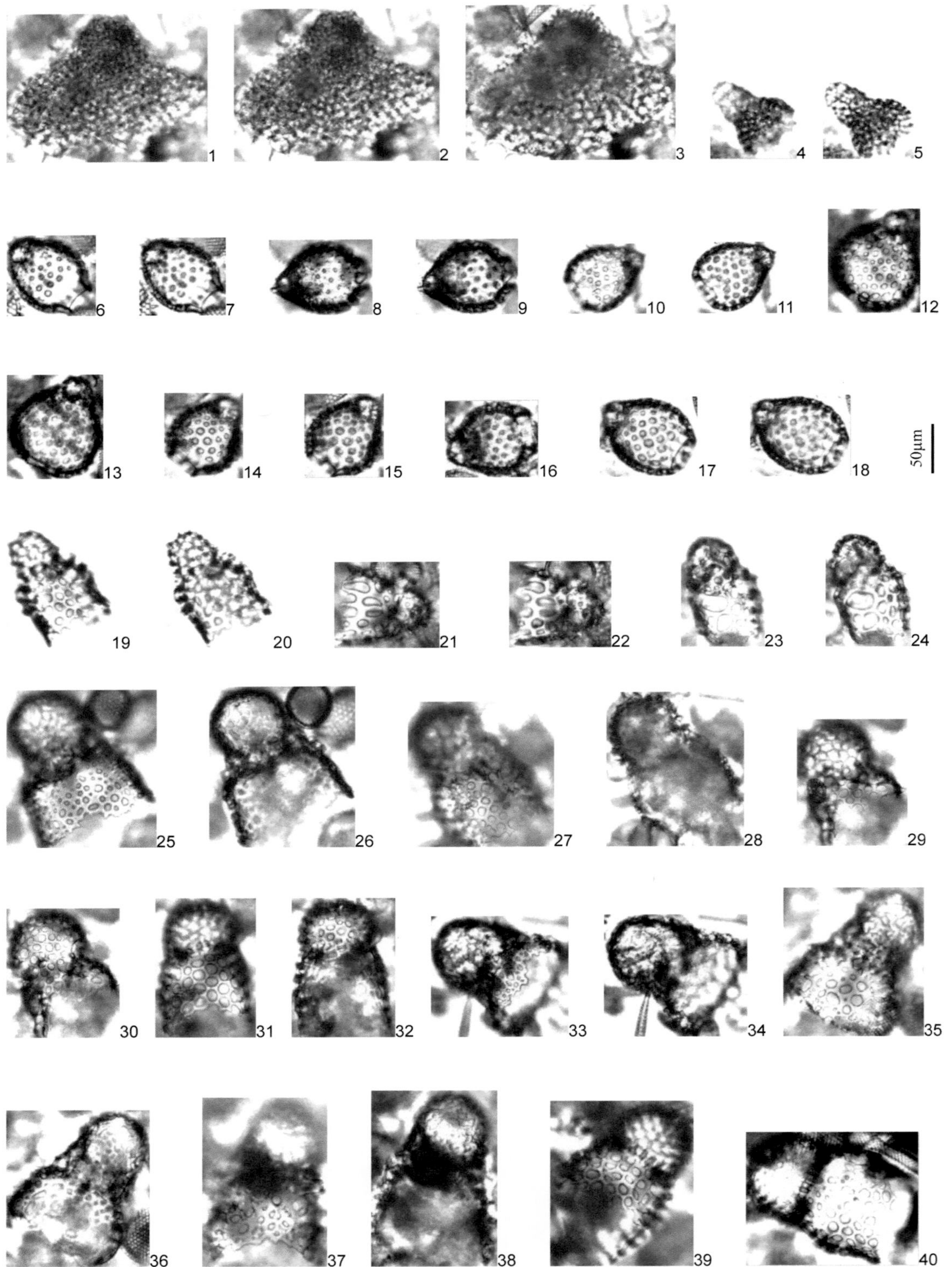

1–5. *Velicucullus spongiformus* sp. nov.; 6–18. *Sethocorys odysseus* Haeckel; 19–24. *Lophophaena cephalota* sp. nov.; 25–40. *Lophophaena cyclocapita* sp. nov.

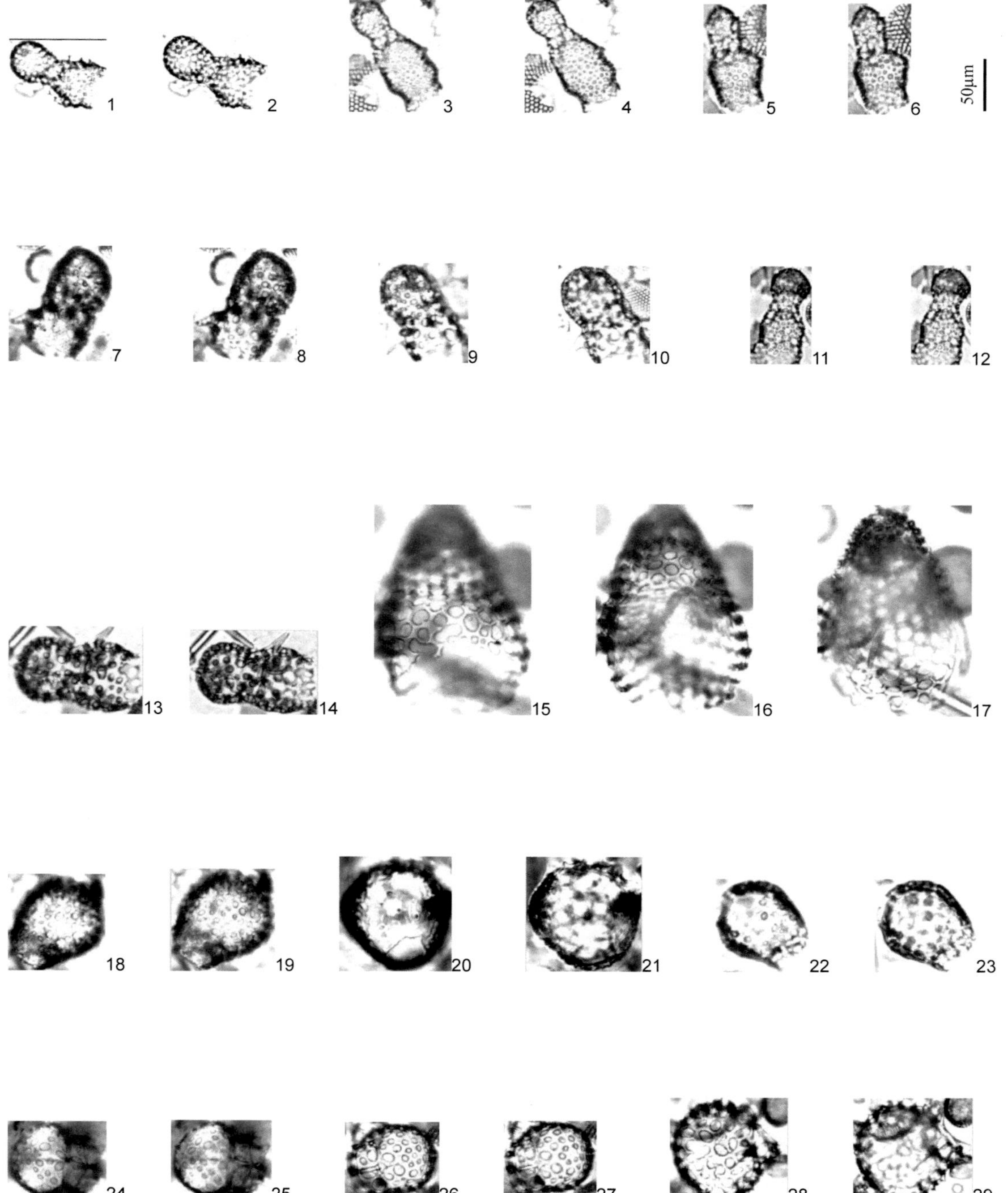

1–14. *Lophophaena cylindrica* (Cleve)； 15–17. *Lophophaena olivacea* sp. nov.； 18–29. *Lophophaena*? *orbicularis* sp. nov.

图版 114

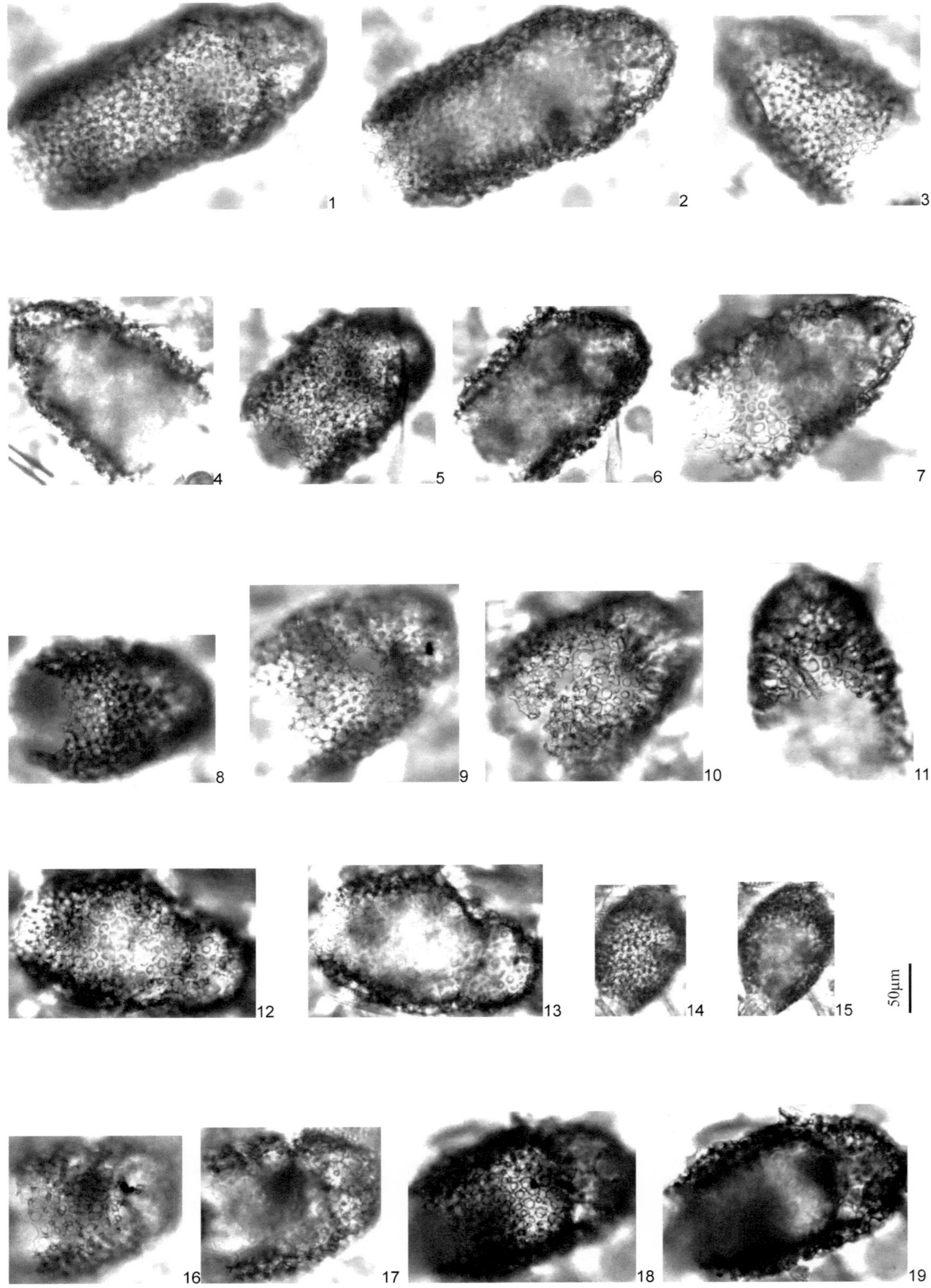

1–19. *Lophophaena spongiosa* (Petrushevskaya)

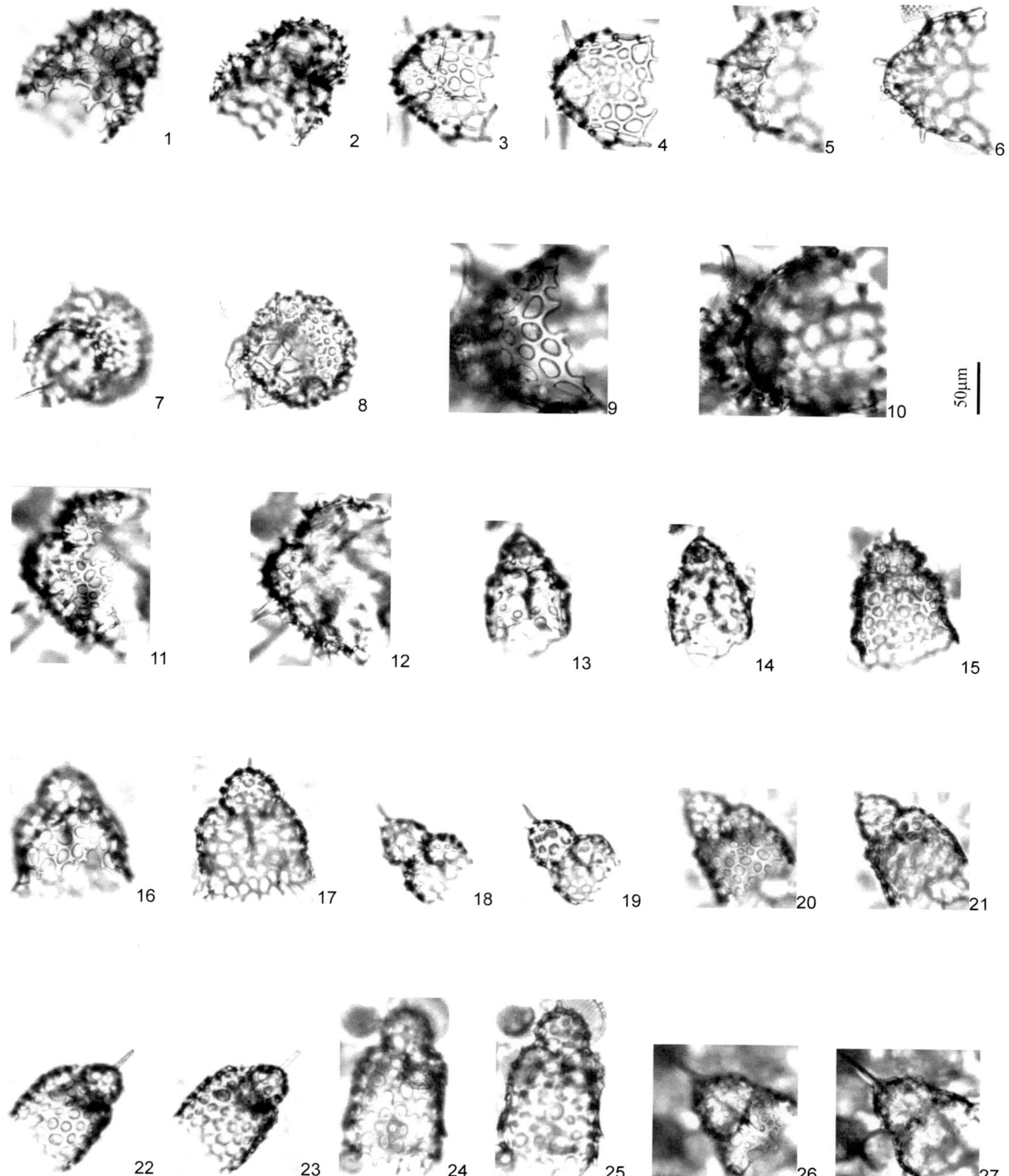

1–2. *Lophophaena* sp. 1； 3–6. *Helotholus ampliata* (Ehrenberg)； 7–8. *Helotholus bulbosus* sp. nov.；
9–12. *Helotholus histricosa* Jørgensen； 13–27. *Helotholus praevema* Weaver

图版 116

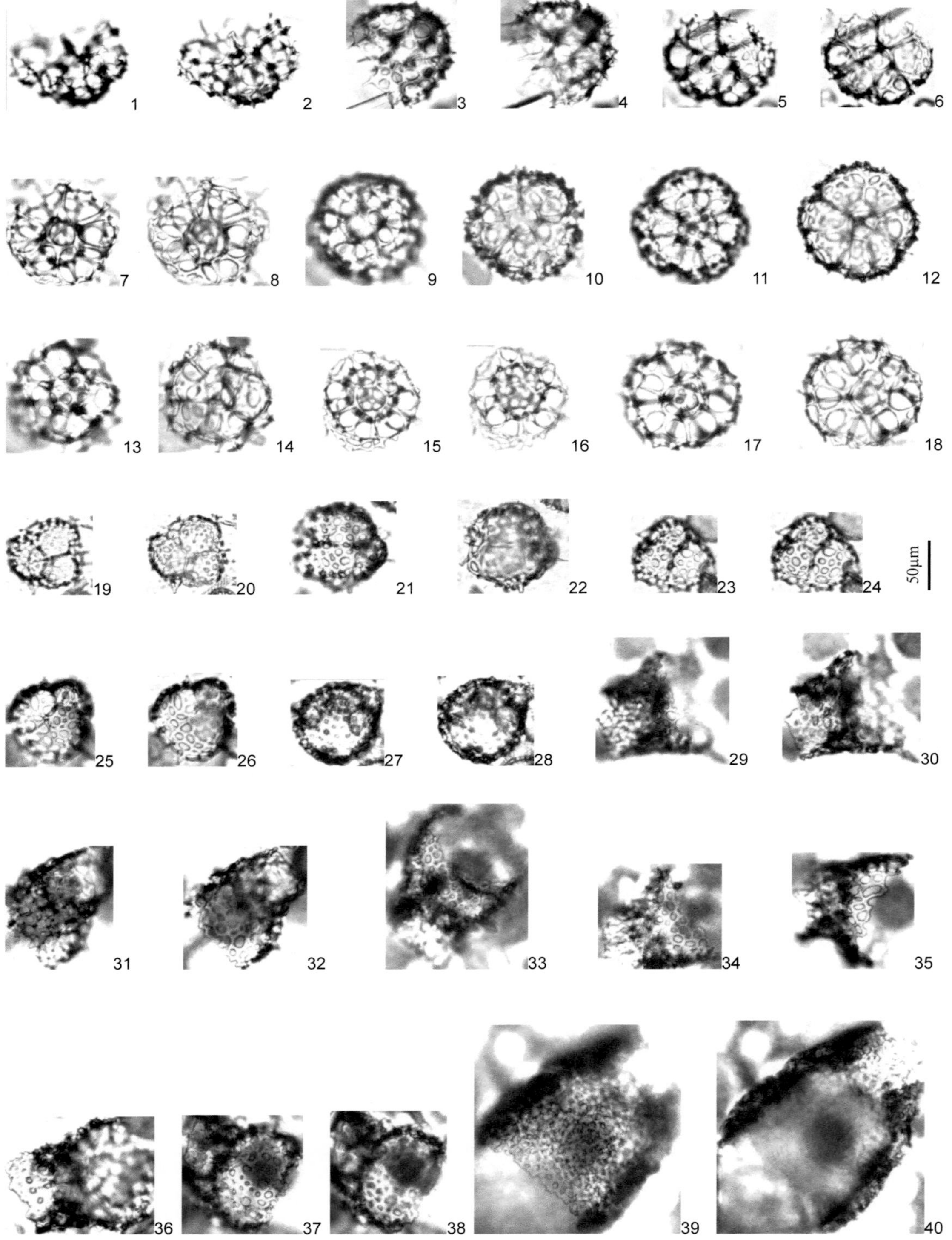

1–18. *Helotholus vema* Hays；19–26. *Helotholus vematella* sp. nov.；27–28. *Lithopera* sp. 1；29–38. *Amphiplecta siphona* sp. nov.；39–40. *Amphiplecta* sp. 1

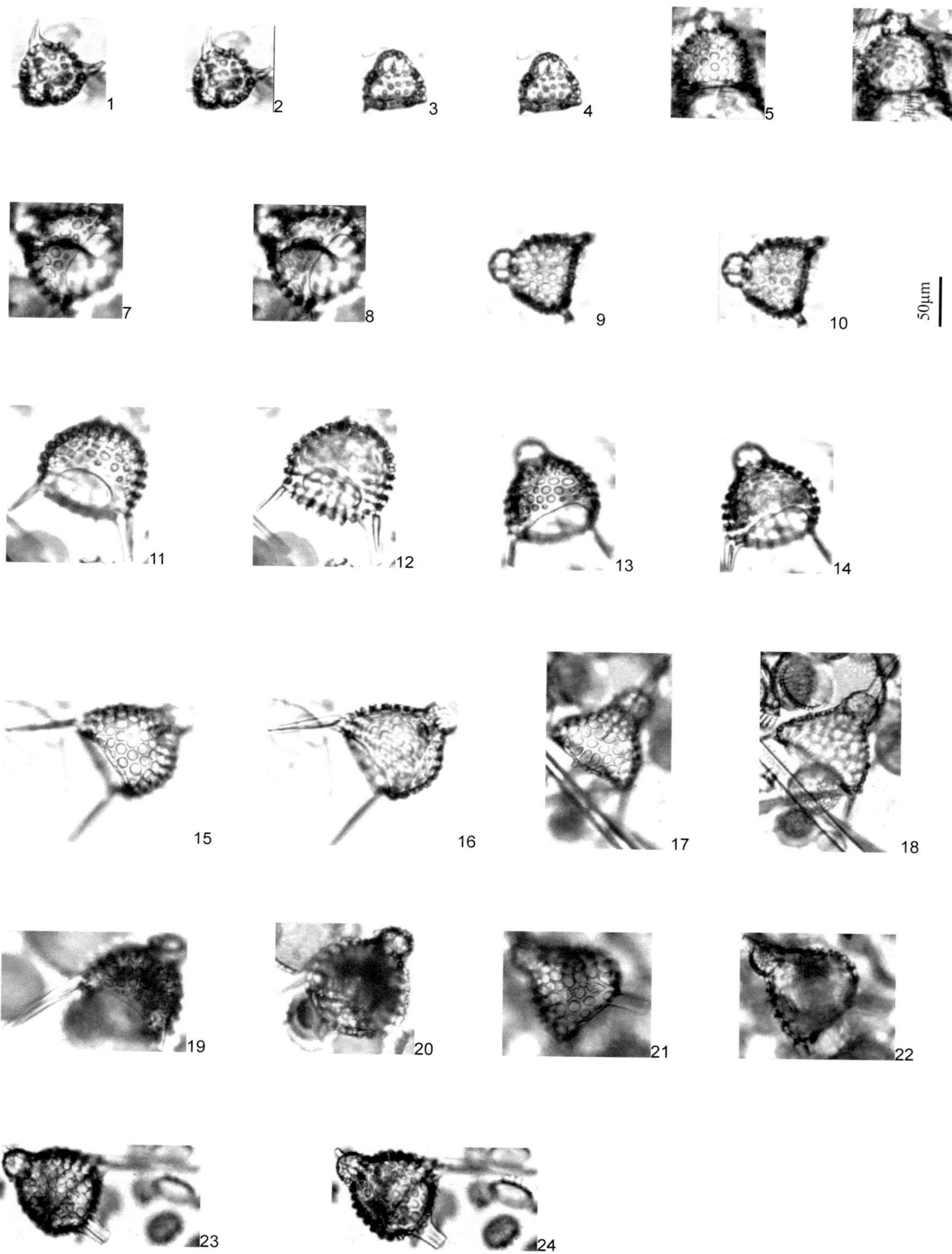

1–10. *Lychnocanoma bellum* (Clark and Campbell)；11–14. *Lychnocanoma conica* (Clark and Campbell)；
15–24. *Lychnocanoma eleganta* sp. nov.

图版 118

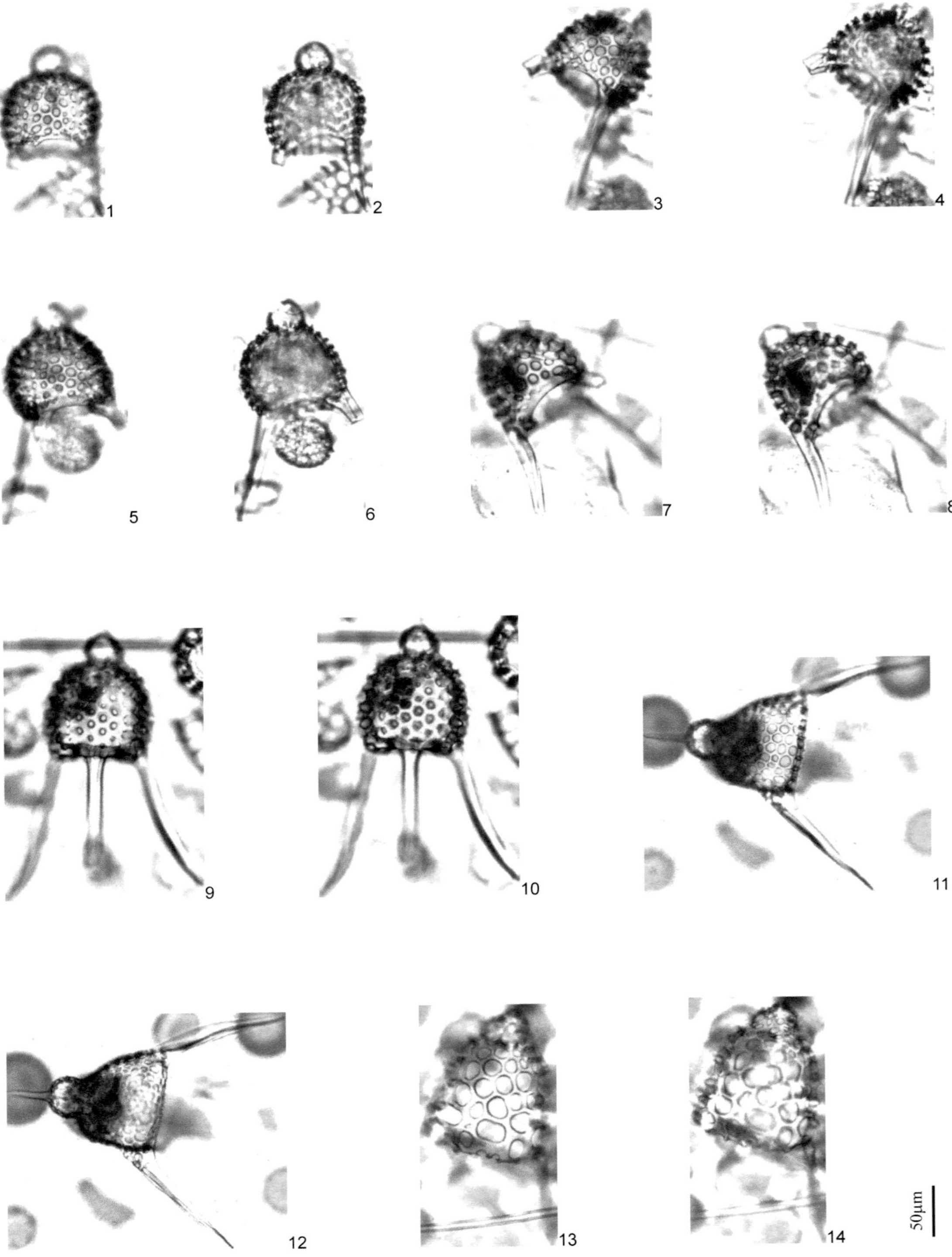

1–10. *Lychnocanoma grande* (Campbell and Clark) group；11–12. *Lychnocanoma grande rugosum* (Riedel)；
13–14. *Lychnocanoma macropora* sp. nov.

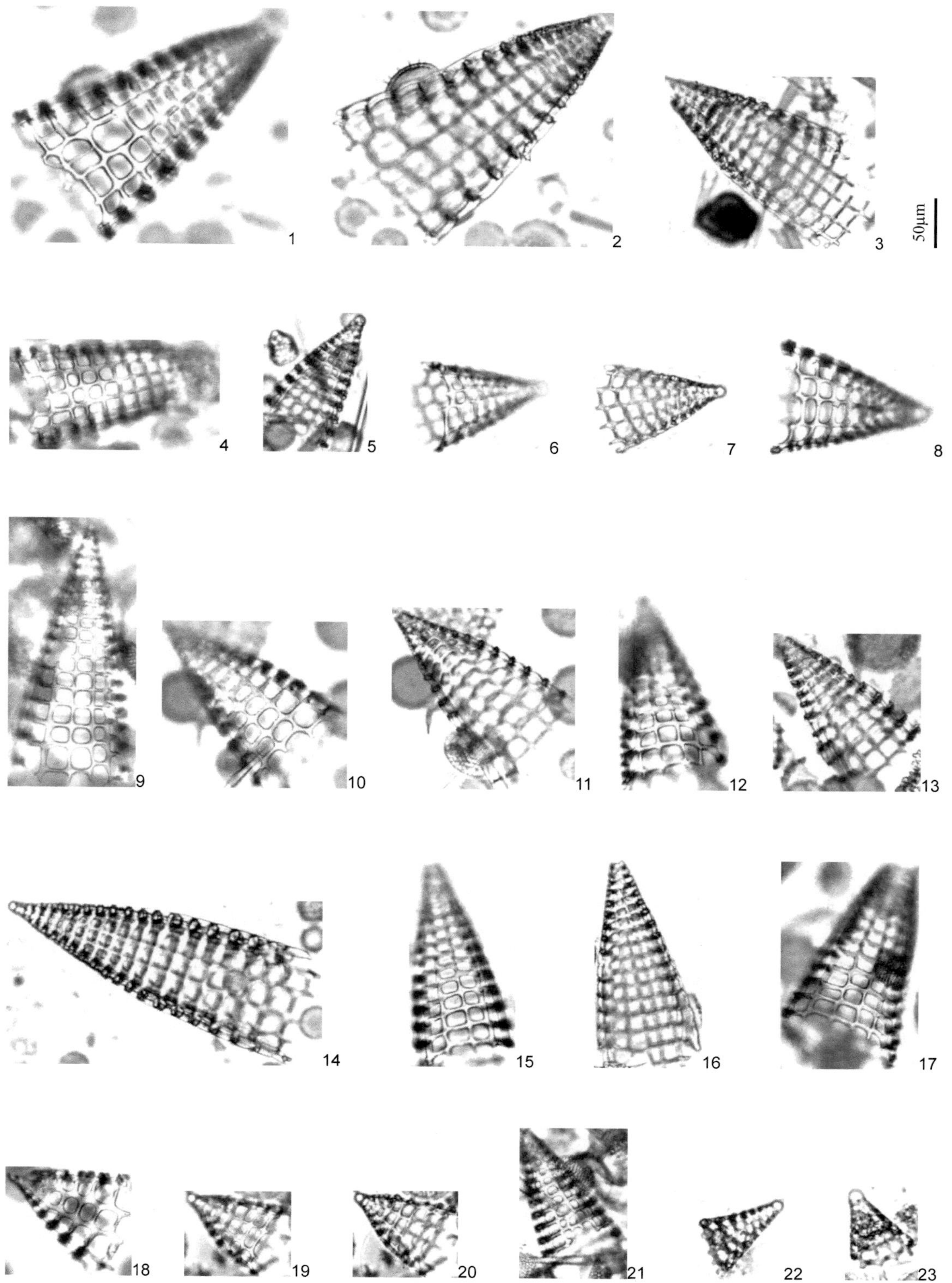

1–23. *Sethopyramis quadrata* Haeckel

图版 120

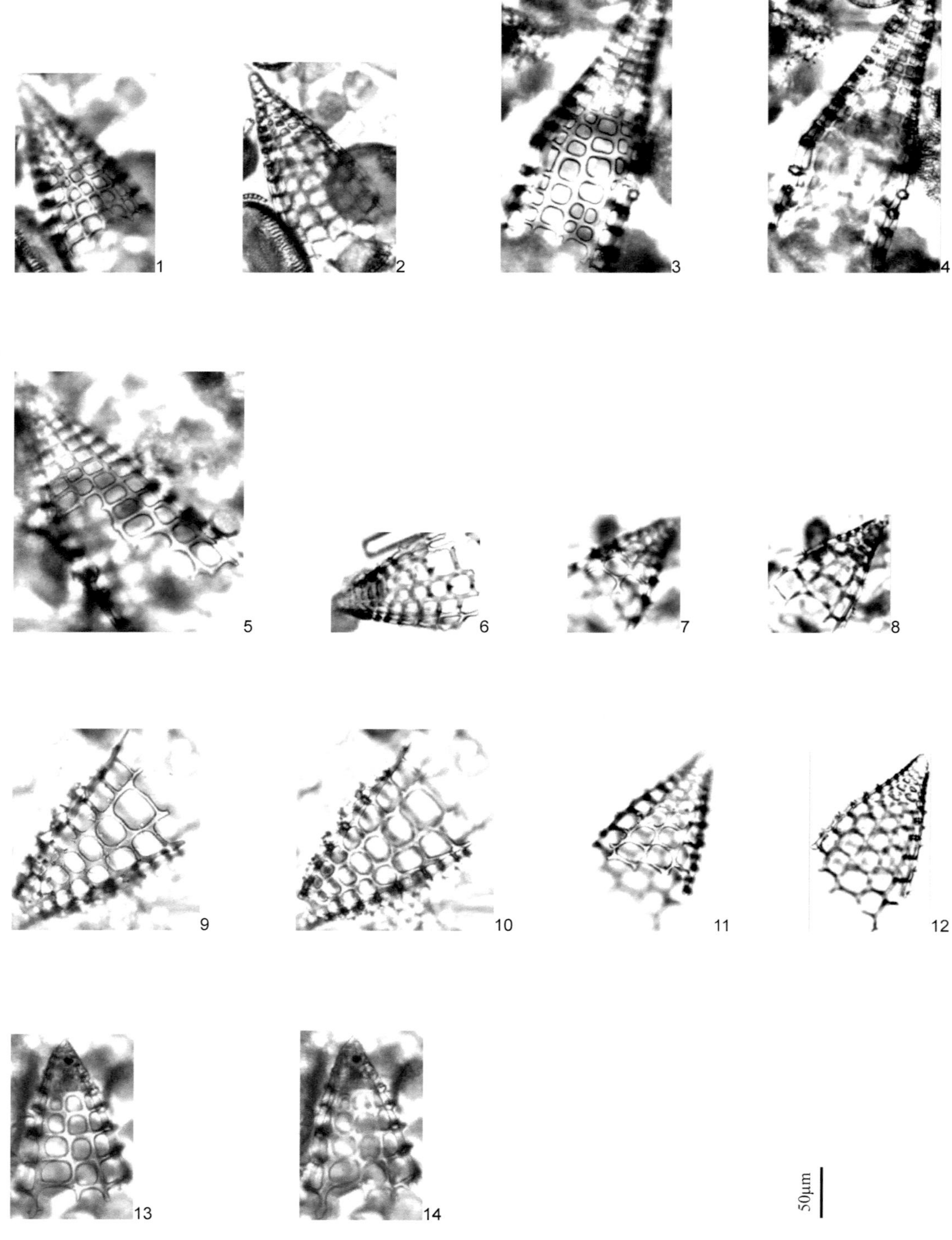

1–5. *Sethopyramis* sp. 1；6–14. *Peripyramis circumtexta* Haeckel

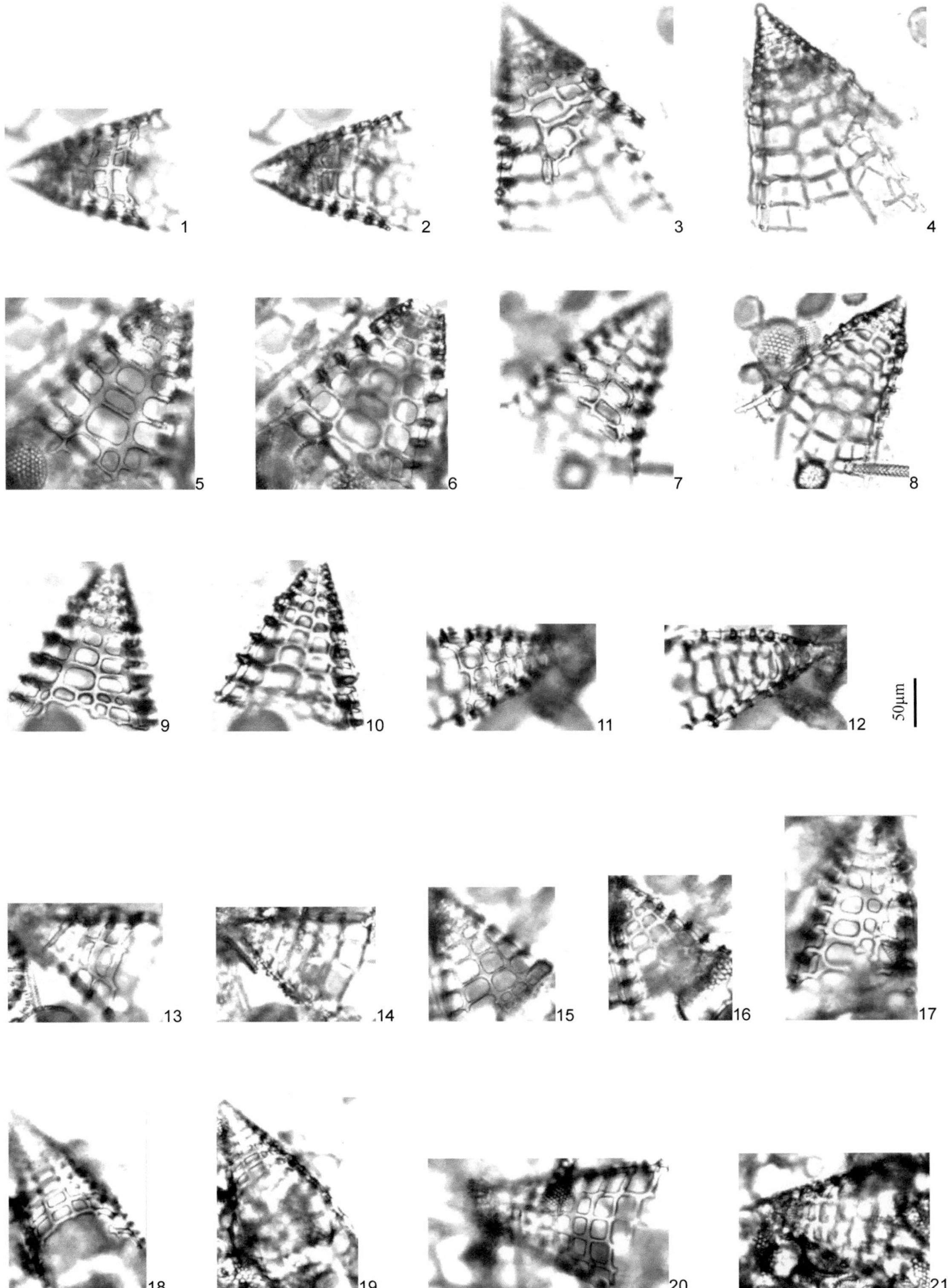

1–21. *Bathropyramis ramosa* Haeckel

图版 122

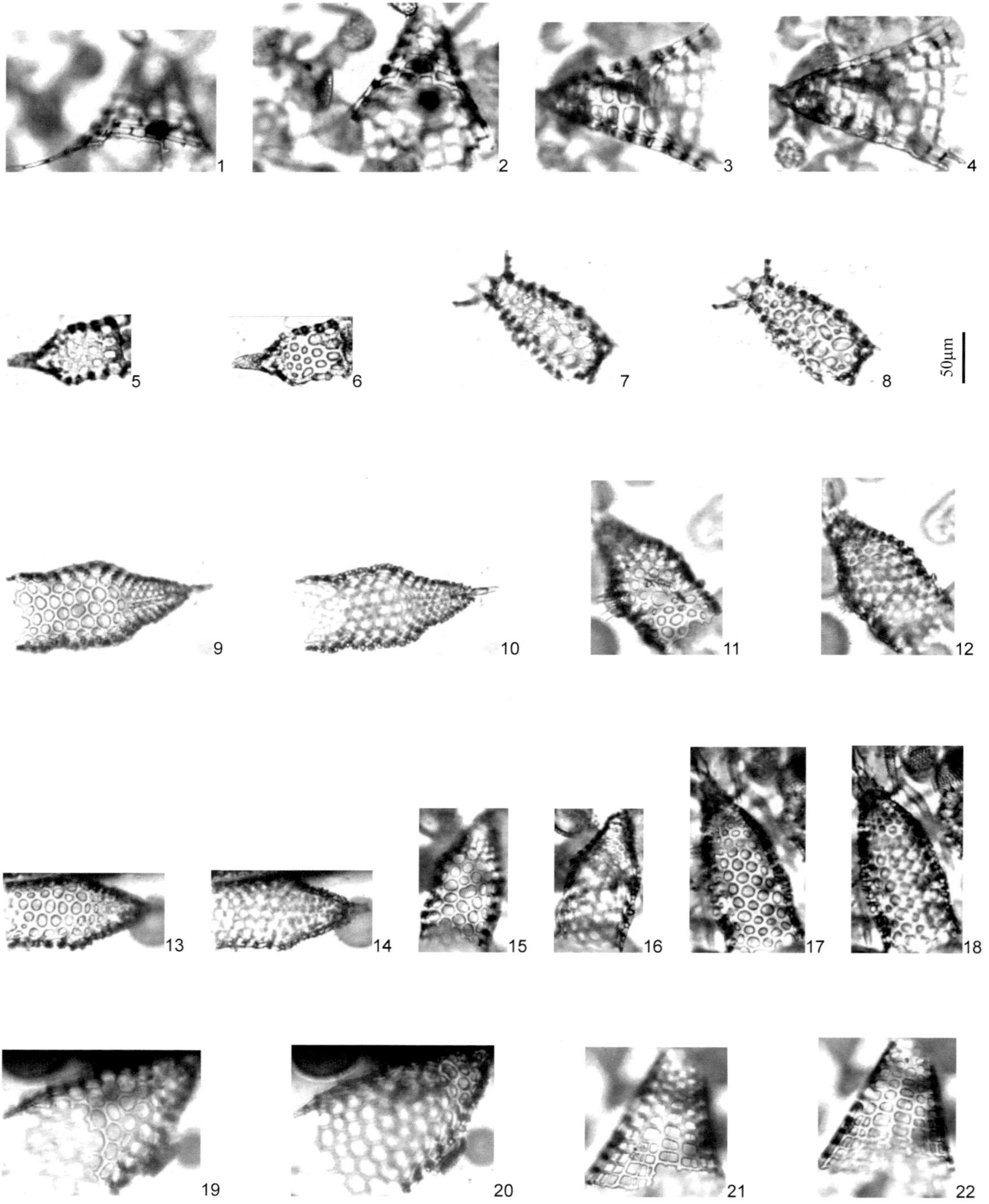

1–2. *Bathropyramis tenthorium* (Haeckel)；3–4. *Bathropyramis trapezoides* Haeckel；5–6. *Aspis ampullarus* sp. nov.；7–8. *Aspis cladocerus* sp. nov.；9–20. *Aspis murus* Nishimura；21–22. *Cinclopyramis* sp. 1

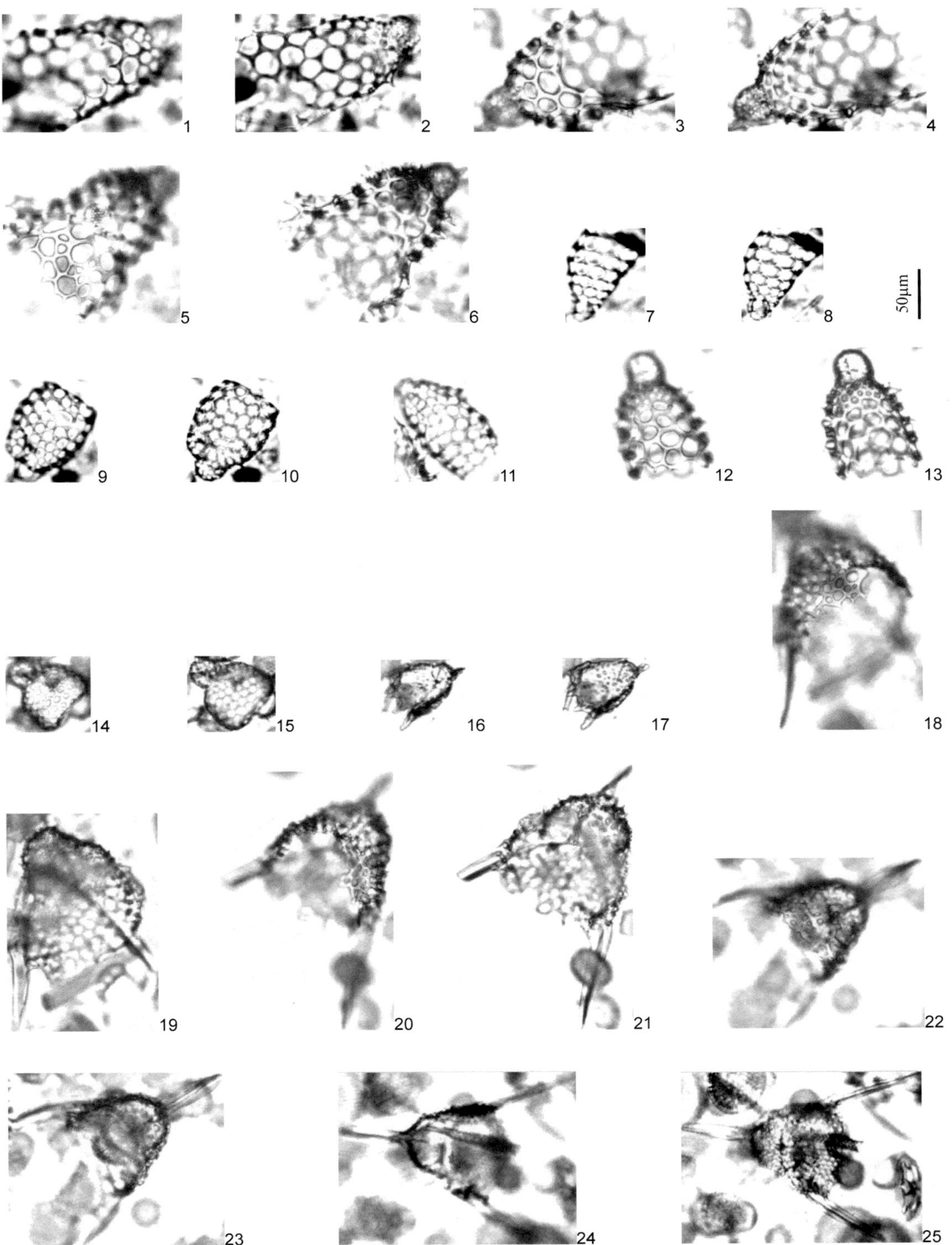

1–6. *Sethoconus*? *dogieli* Petrushevskaya；7–11. *Sethoconus tabulata* (Ehrenberg)；12–13. *Sethoconus* sp. 1；
14–15. *Anthocyrtium byronense* Clark；16–17. *Anthocyrtium quinquepedium* sp. nov.；
18–25. *Corythomelissa adunca* (Sanfilippo and Riedel)

图版 124

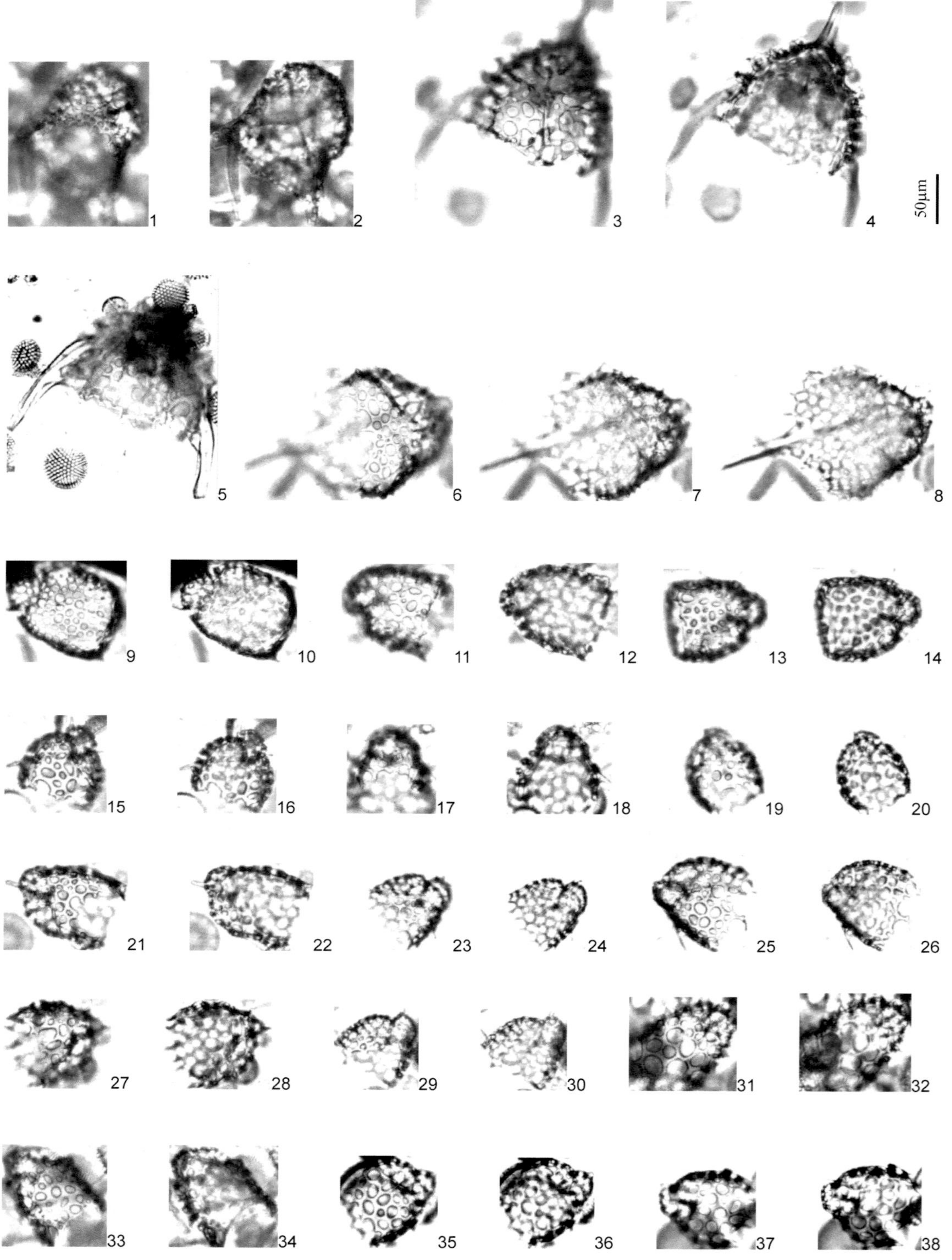

1–2. *Corythomelissa corysta* sp. nov.; 3–5. *Corythomelissa horrida* Petrushevskaya; 6–8. *Corythomelissa* sp. 1; 9–38. *Ceratocyrtis amplus* (Popofsky) group

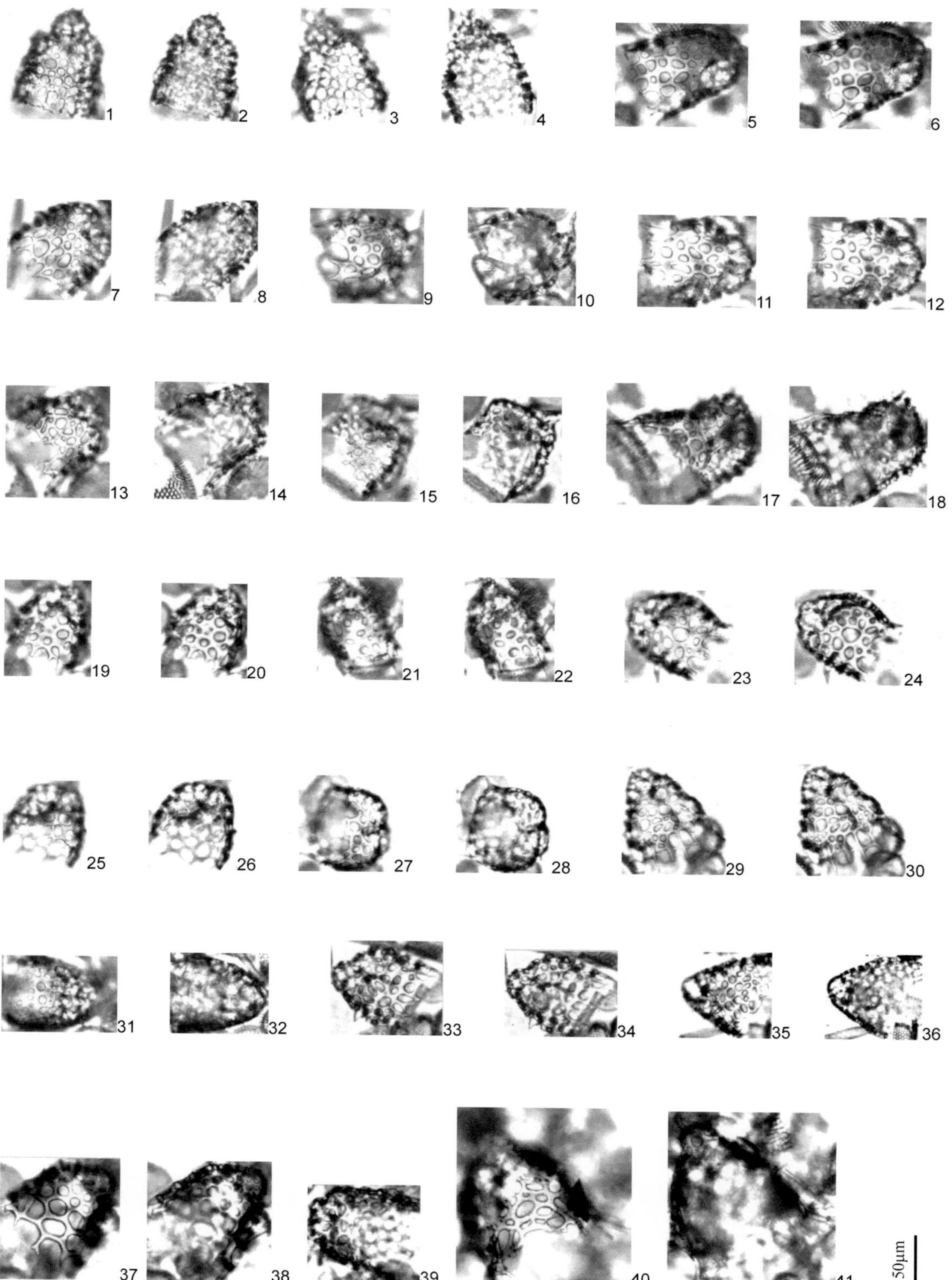

1–4. *Ceratocyrtis cylindris* sp. nov.; 5–39. *Ceratocyrtis irregularis* sp. nov.; 40–41. *Ceratocyrtis manumi* Goll and Bjørklund

图版 126

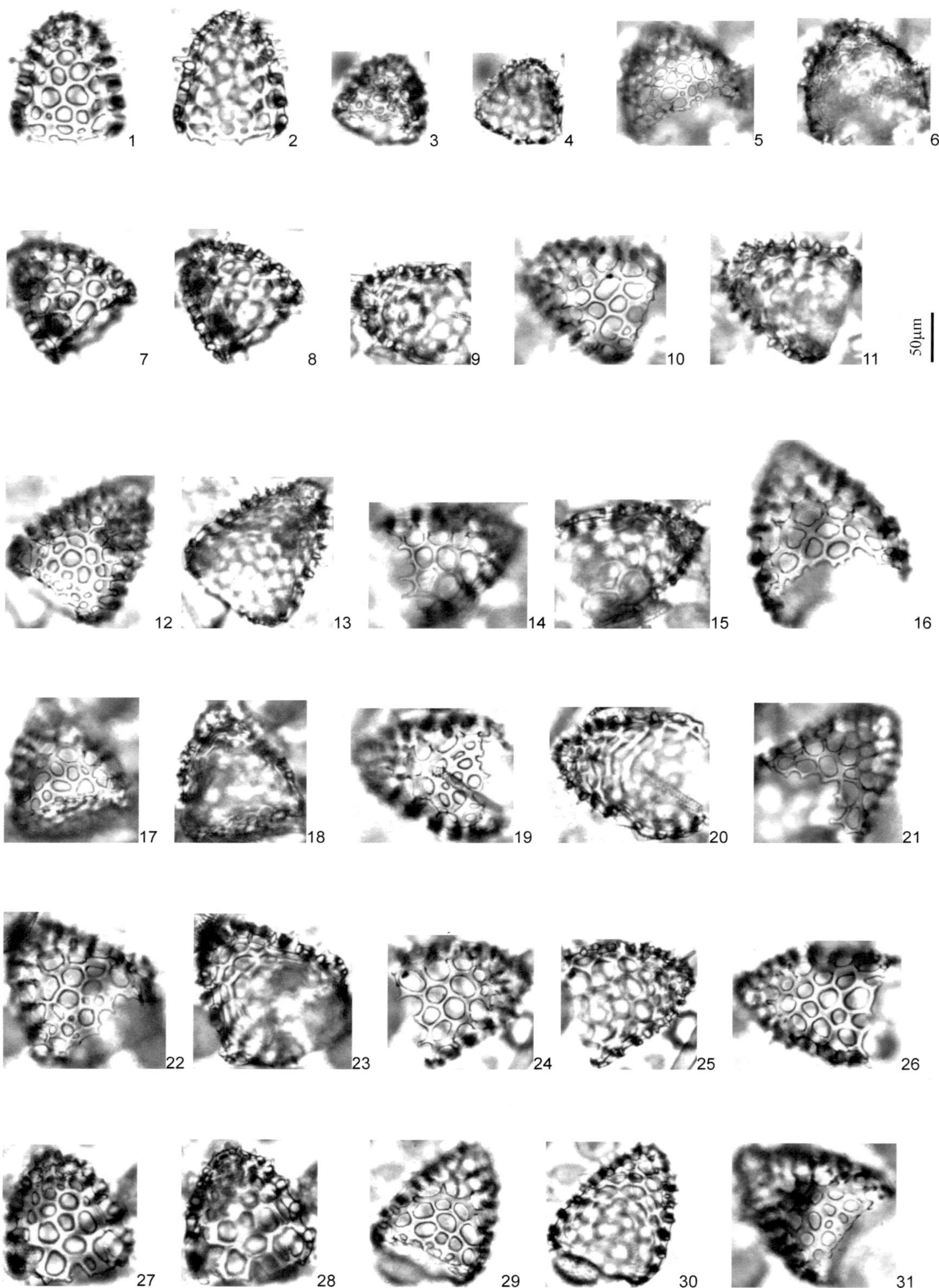

1–31. *Ceratocyrtis mashae* Bjørklund

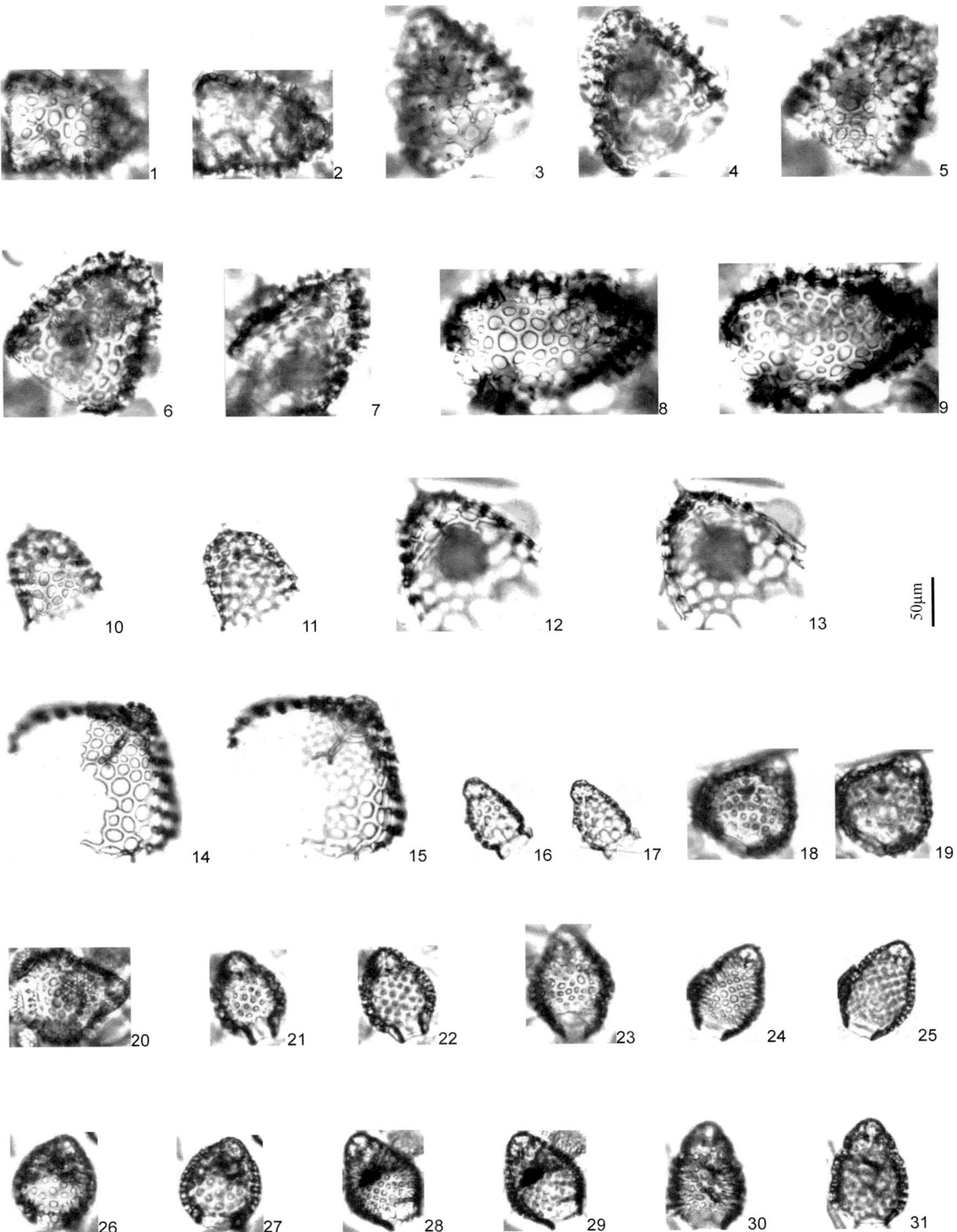

1–9. *Ceratocyrtis perimashae* sp. nov.; 10–13. *Ceratocyrtis robustus* Bjørklund; 14–15. *Ceratocyrtis* sp. 1;
16–17. *Dictyocephalus australis* Haeckel; 18–20. *Dictyocephalus crassus* Carnevale;
21–31. *Dictyocephalus papillosus* (Ehrenberg)

图版 128

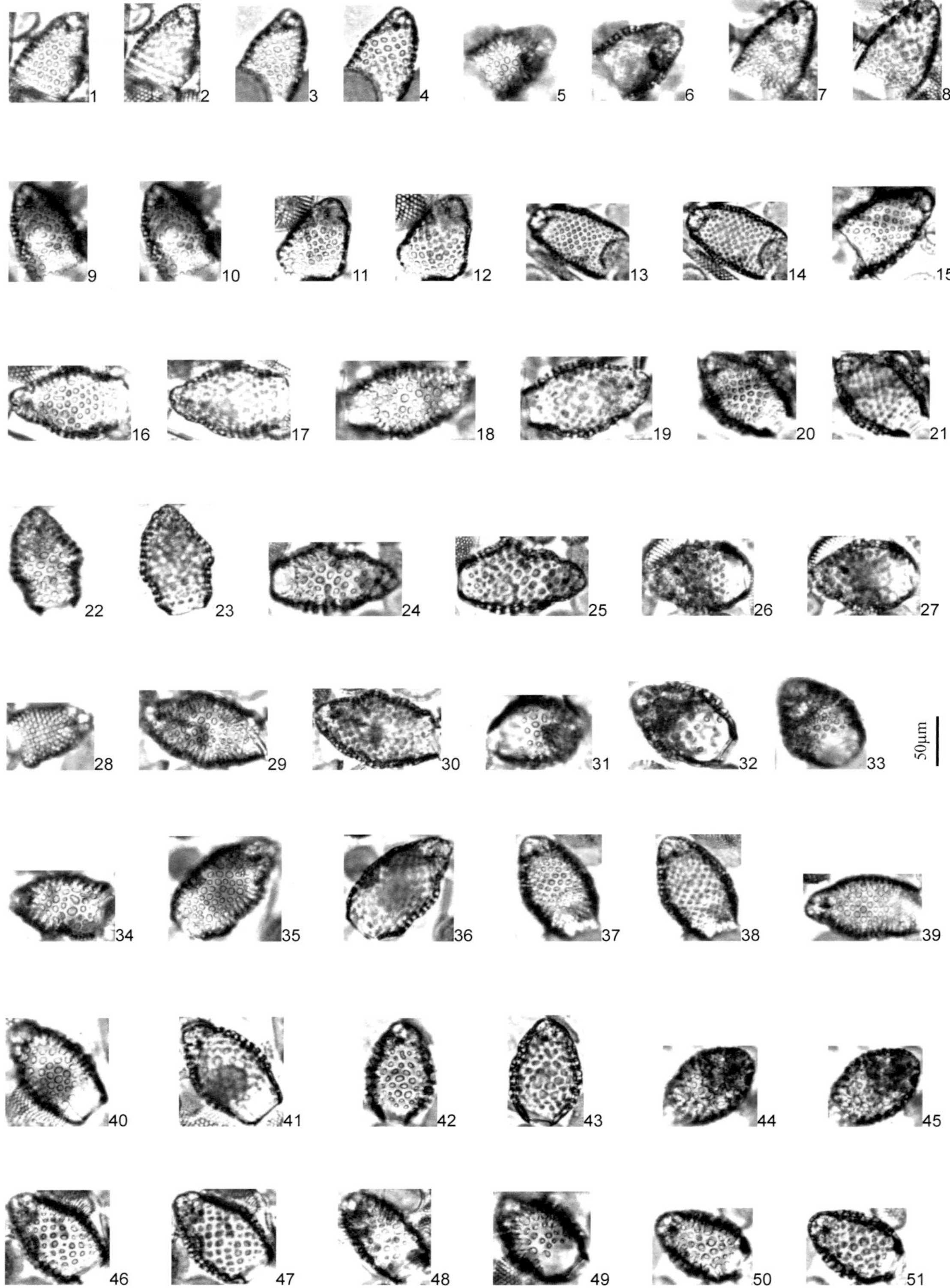

1–15. *Dictyocephalus turritus* sp. nov.; 16–51. *Dictyocephalus variabilus* sp. nov.

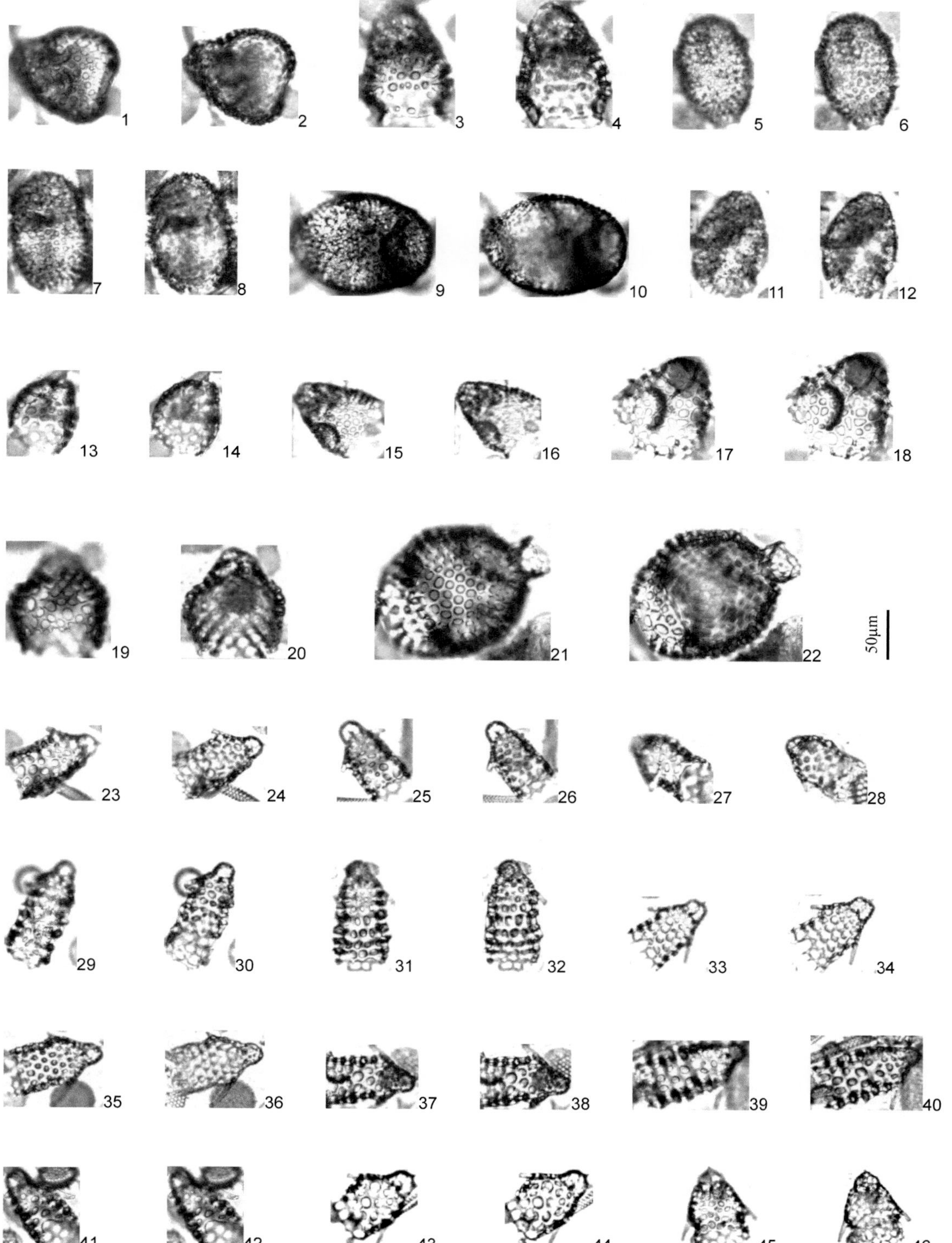

1–2. *Dictyocephalus* sp. 1； 3–4. *Dictyocephalus* sp. 2； 5–12. *Carpocanium spongiforma* sp. nov.； 13–20. *Carpocanium* sp. 1； 21–22. *Carpoglobatus tubulus* sp. nov.； 23–46. *Lipmanella archipilium* (Petrushevskaya)

图版 130

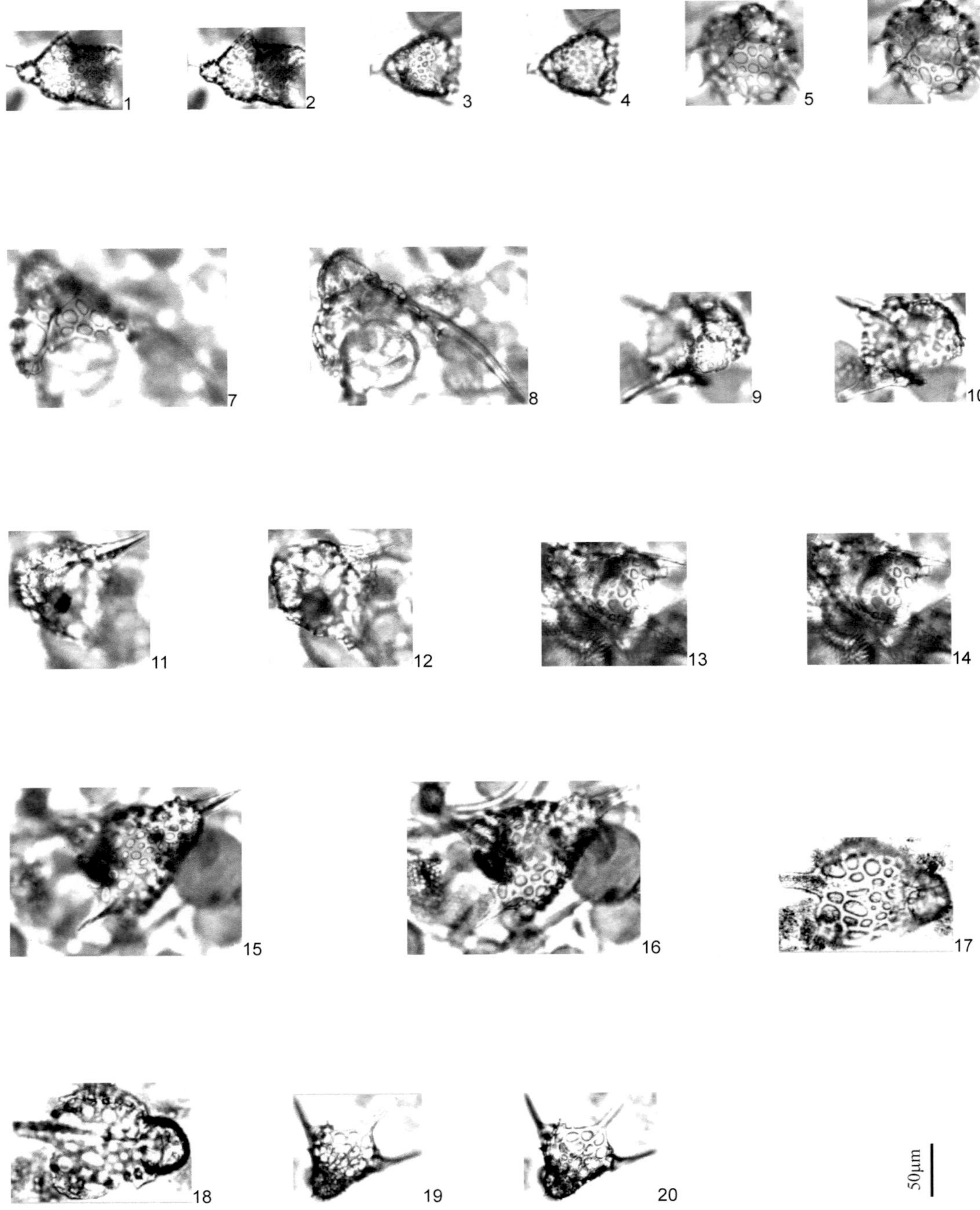

1–4. *Lipmanella dictyoceras* (Haeckel)；5–6. *Lipmanella* sp. 1；7–18. *Pterocanium polypylum* Popofsky；19–20. *Pterocanium* sp. 1

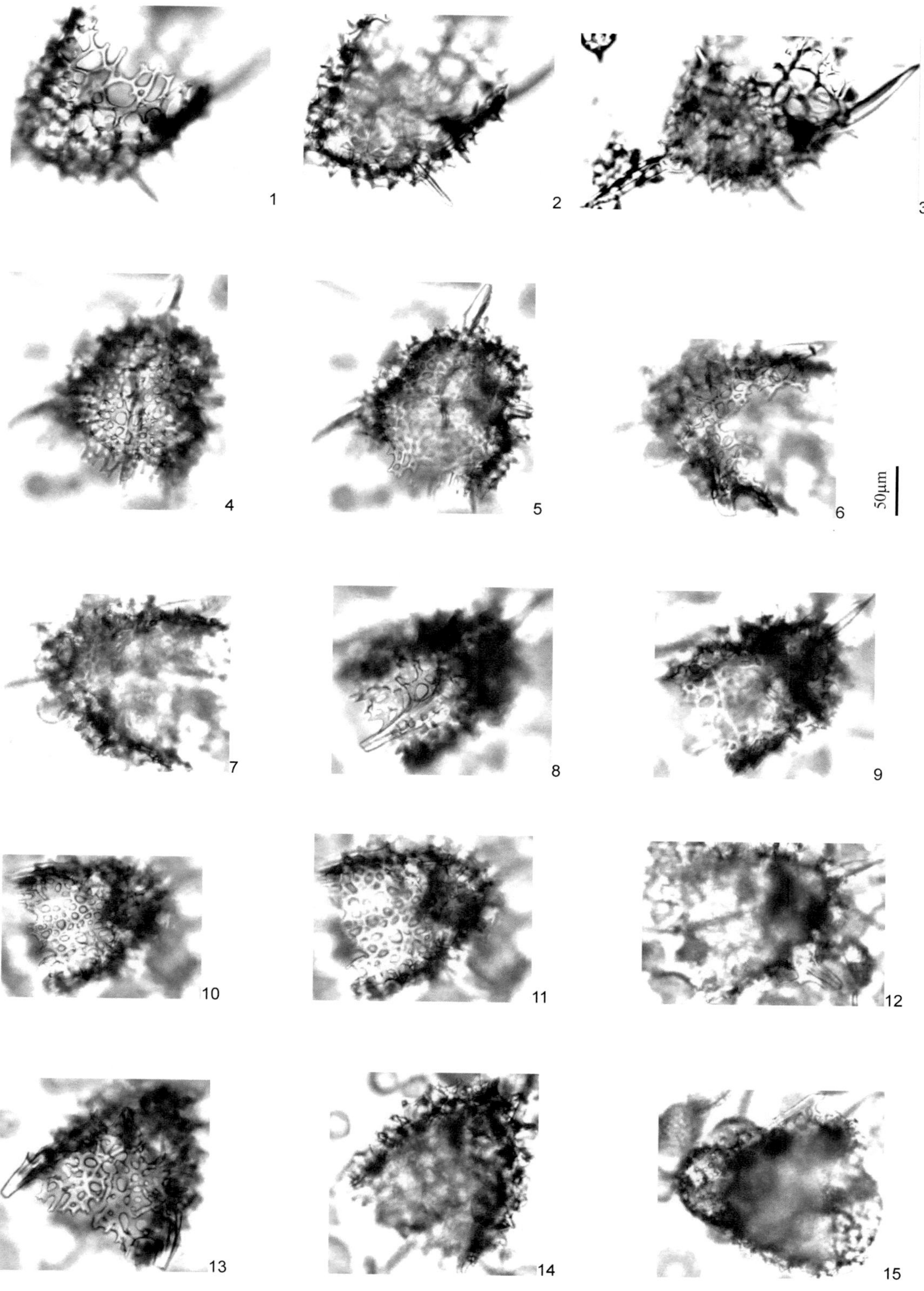

1–15. *Pseudodictyophimus galeatus* Caulet

图版 132

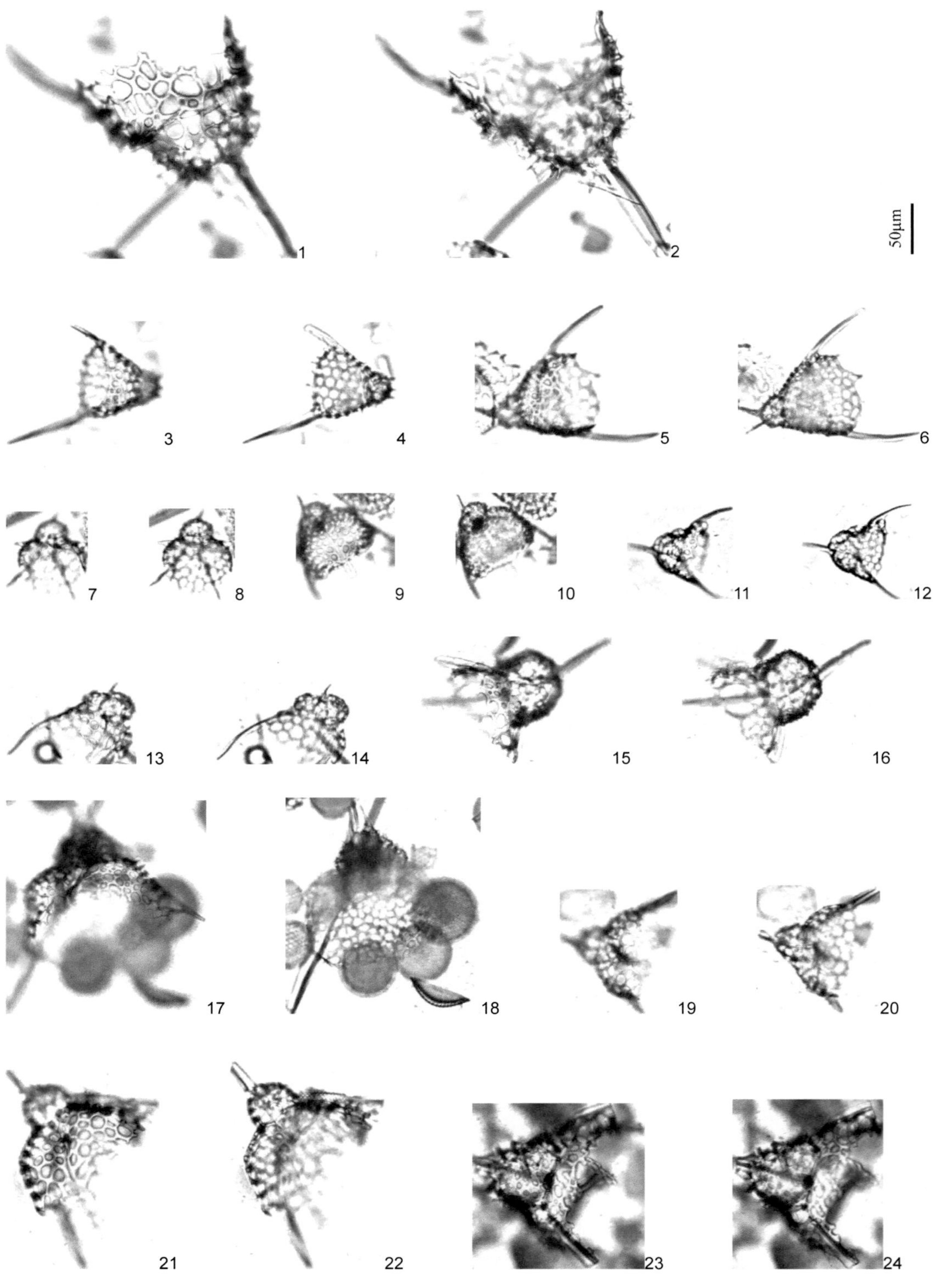

1–2. *Pseudodictyophimus gigantospinus* sp. nov.; 3–24. *Pseudodictyophimus gracilipes* (Bailey)

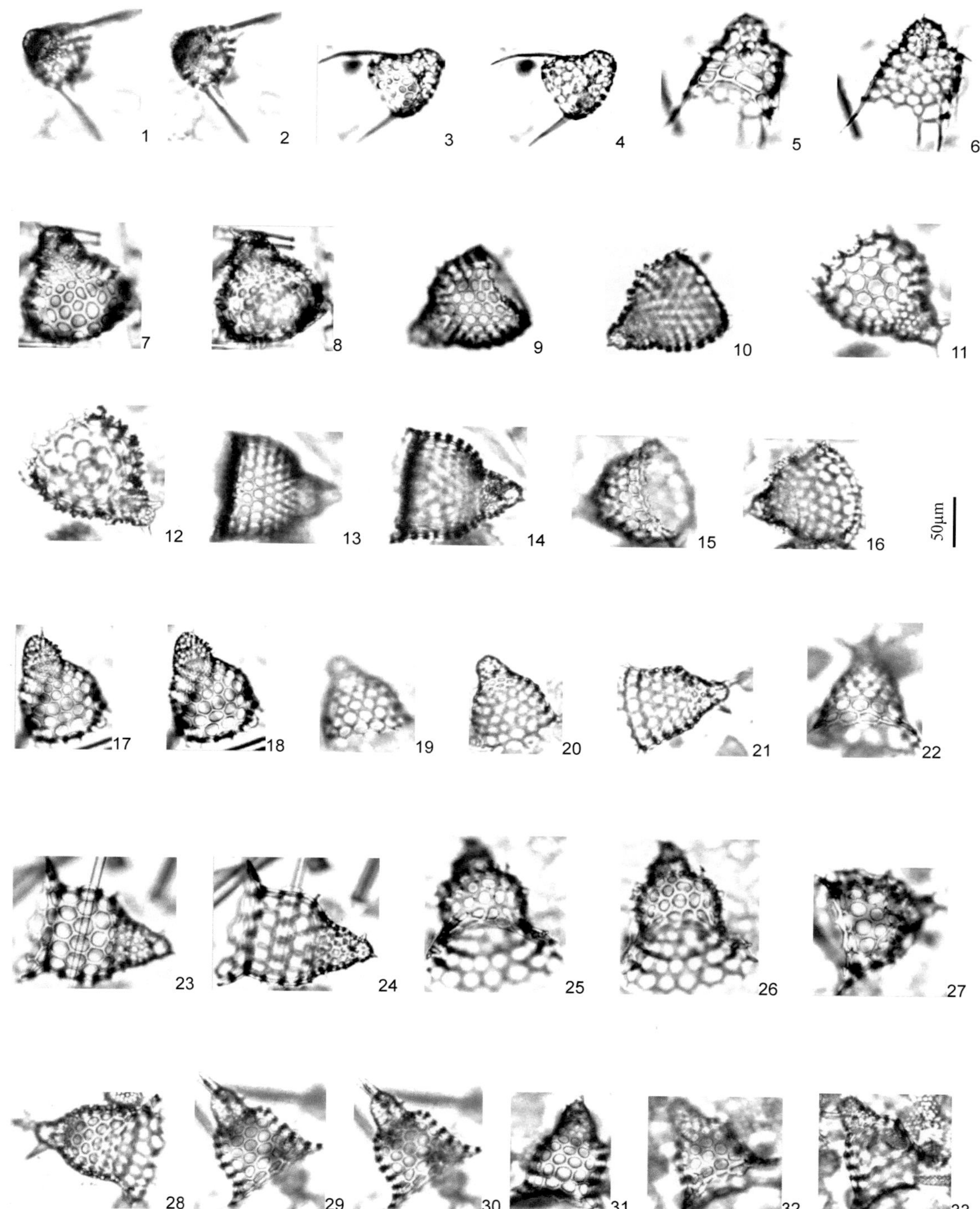

1–4. *Pseudodictyophimus* sp. aff. *P. gracilipes* (Bailey); 5–6. *Pseudodictyophimus tenellus* sp. nov.;
7–16. *Cycladophora bicornis* (Popofsky) group; 17–22. *Cycladophora conica* Lombari and Lazarus;
23–33. *Cycladophora davisiana* Ehrenberg group

图版 134

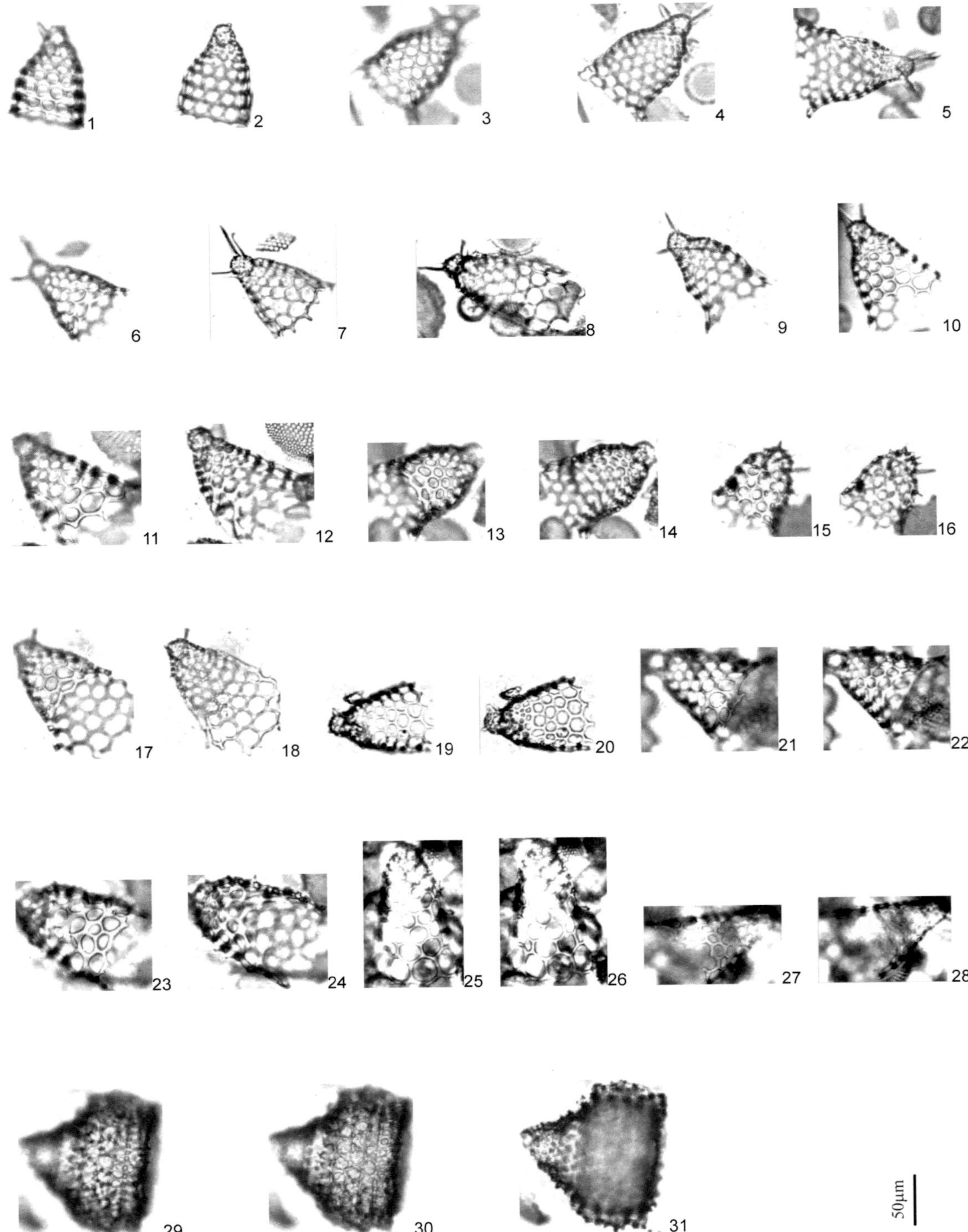

1–22. *Cycladophora davisiana cornutoides* (Petrushevskaya)；23–28. *Cycladophora pseudoadvena* (Kozlova)；29–31. *Cycladophora spongothorax* (Chen)

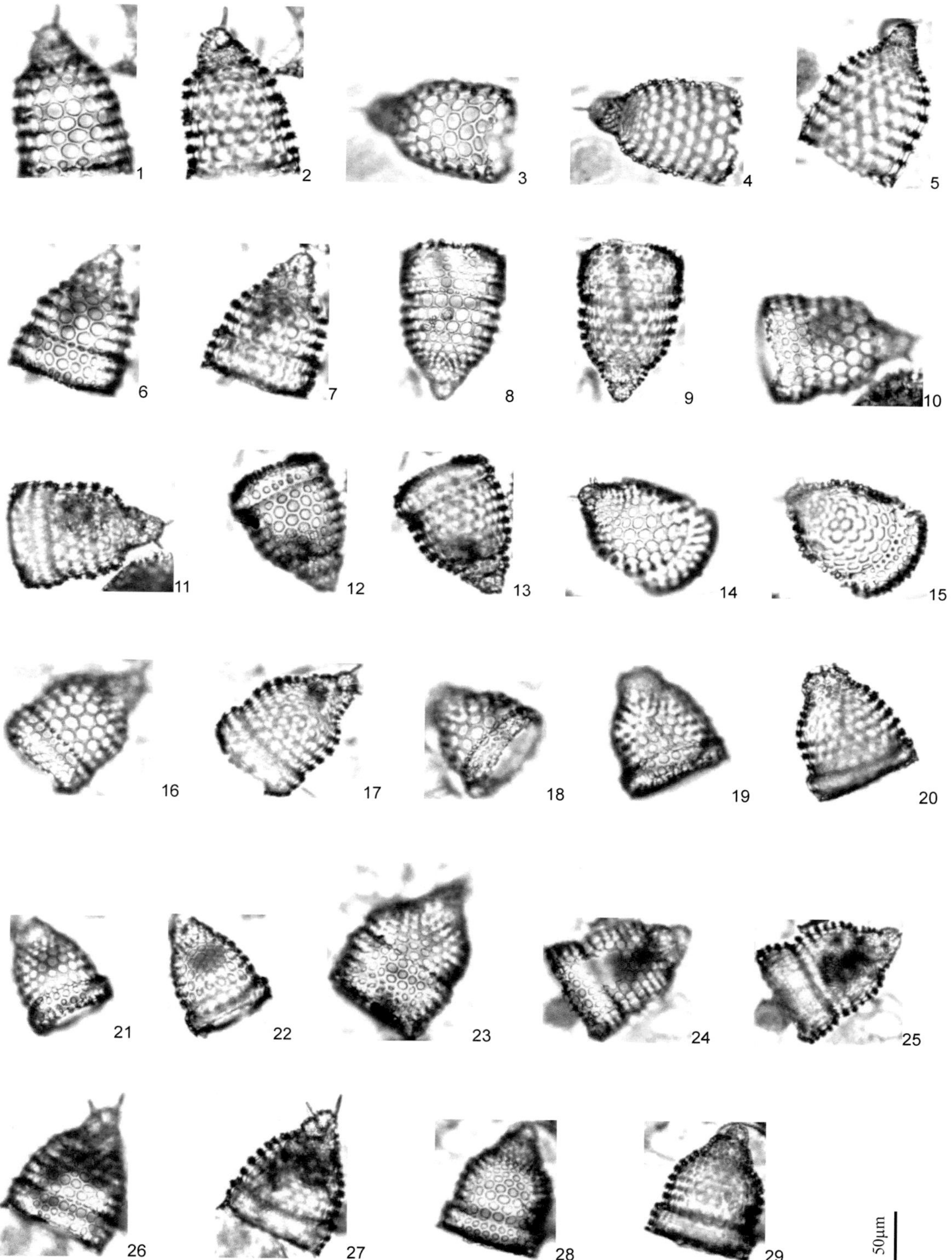

1–5. *Cycladophora tinocampanula* sp. nov.；6–29. *Lophocyrtis golli* Chen

图版 136

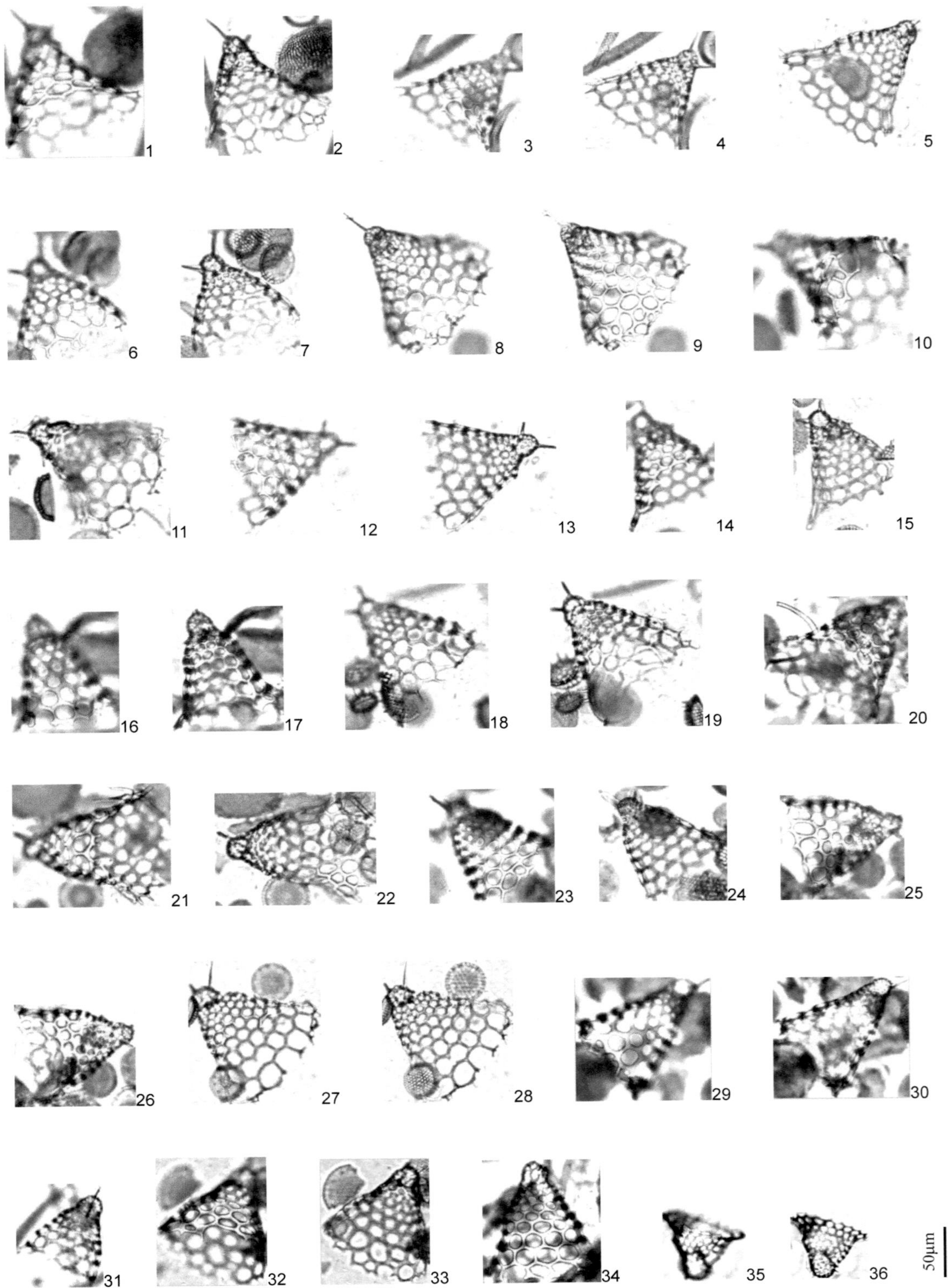

1–34. *Clathrocyclas alcmenae* Haeckel；35–36. *Clathrocyclas* cf. *C. danaes* Haeckel

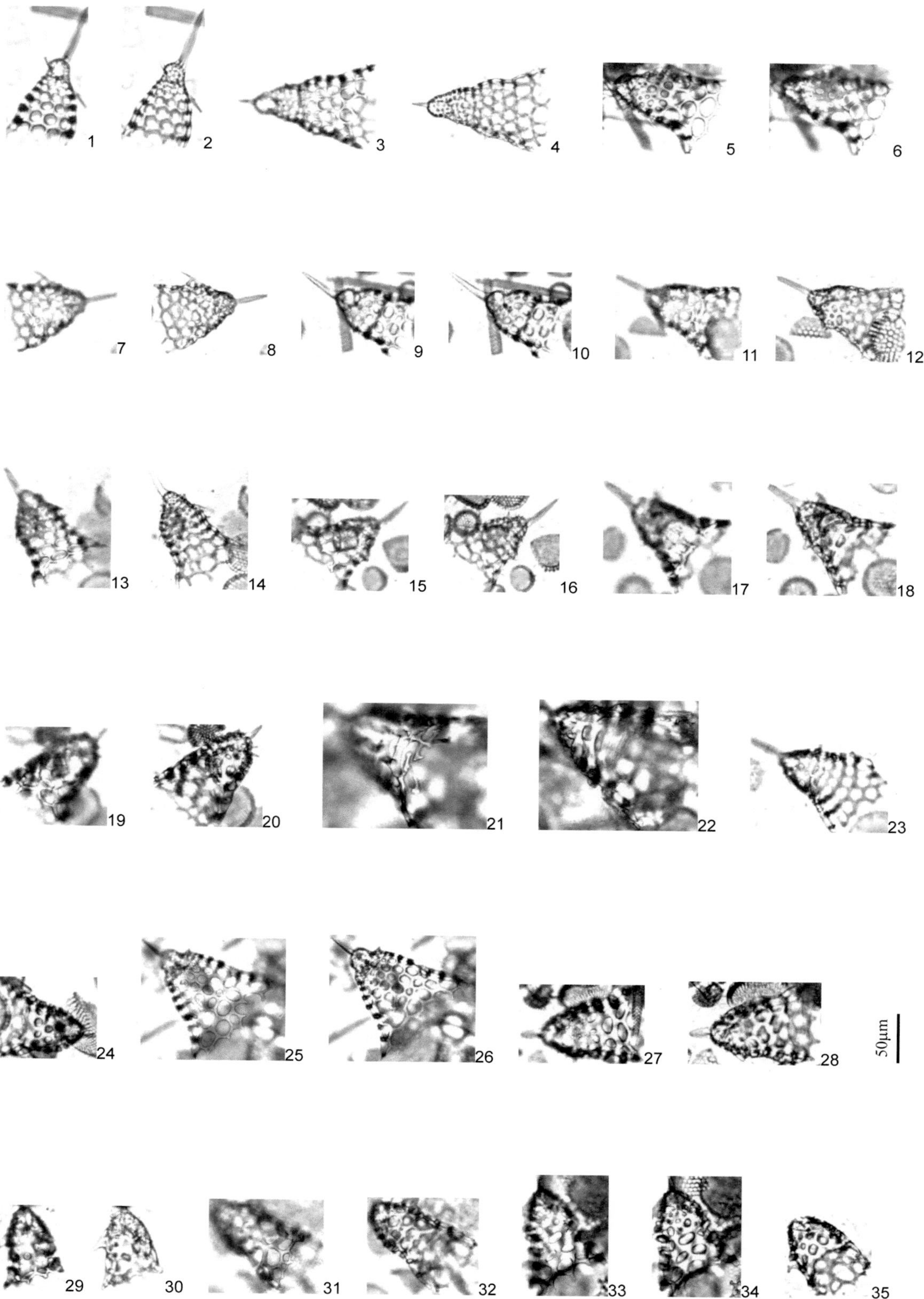

1–28. *Eurystomoskevos petrushevskaae* Caulet；29–35. *Eurystomoskevos* sp. 1

图版 138

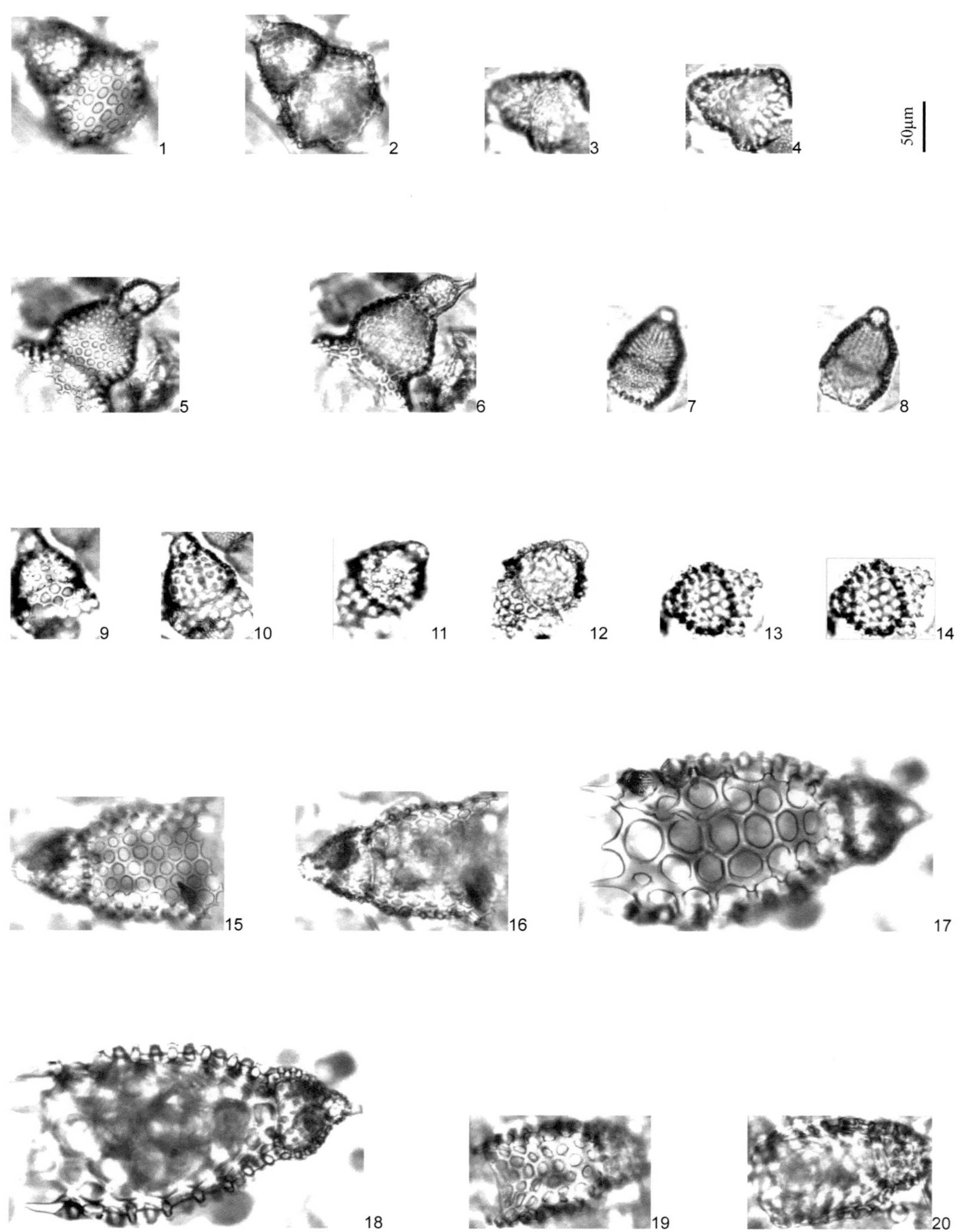

1–4. *Lamprocyrtis aegles* (Ehrenberg)；5–6. *Lamprocyrtis* sp. 1；7–8. *Gondwanaria deflandrei* Petrushevskaya；9–14. *Gondwanaria japonica* (Nakaseko) group；15–20. *Gondwanaria milowi* (Riedel and Sanfilippo) group

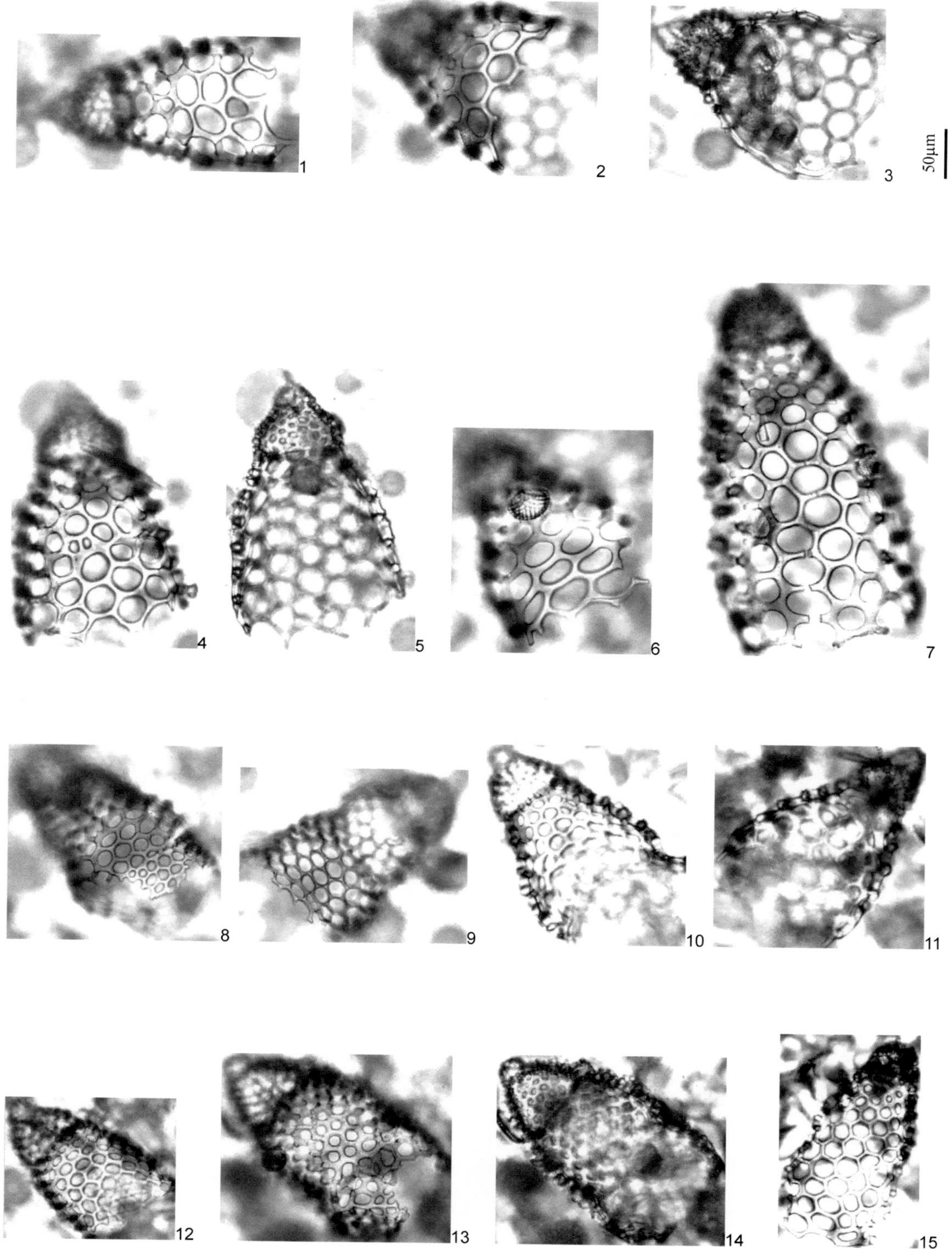

1–15. *Gondwanaria milowi* (Riedel and Sanfilippo) group

图版 140

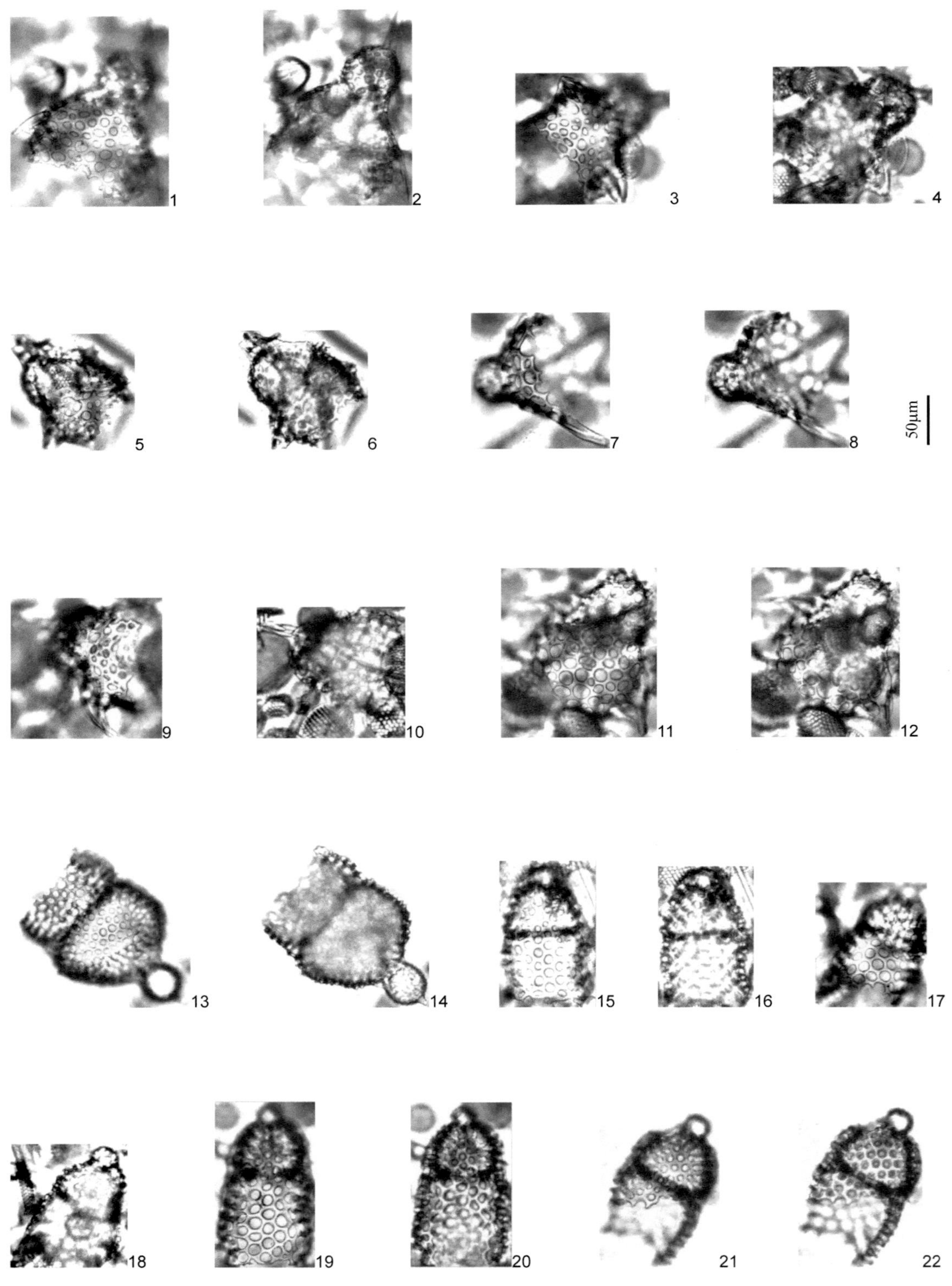

1–12. *Gondwanaria pteroforma* sp. nov.；13–14. *Gondwanaria redondoensis* (Campbell and Clark)；
15–22. *Gondwanaria robusta* (Abelmann)

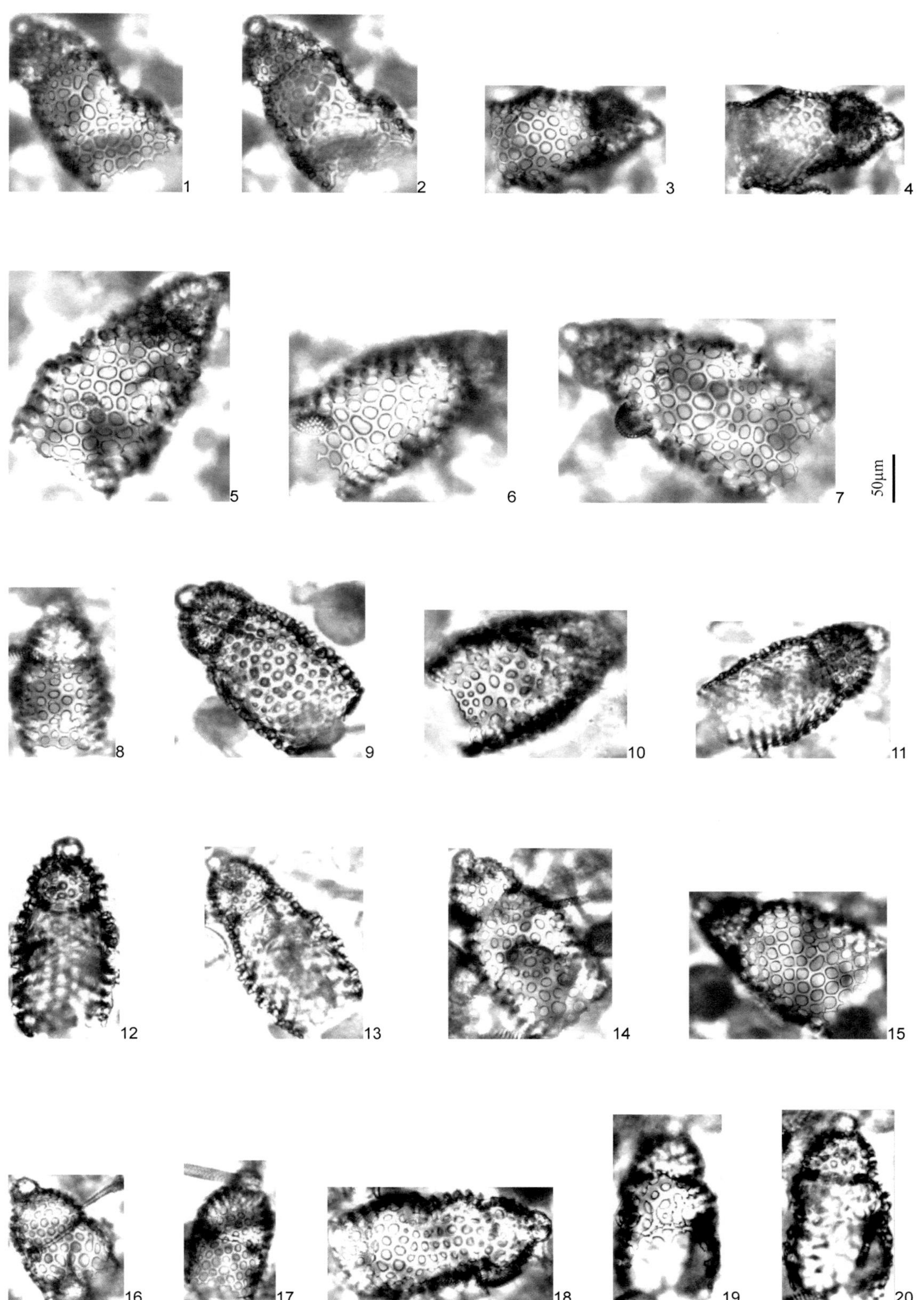

1–20. *Gondwanaria semipolita* (Clark and Campbell)

图版 142

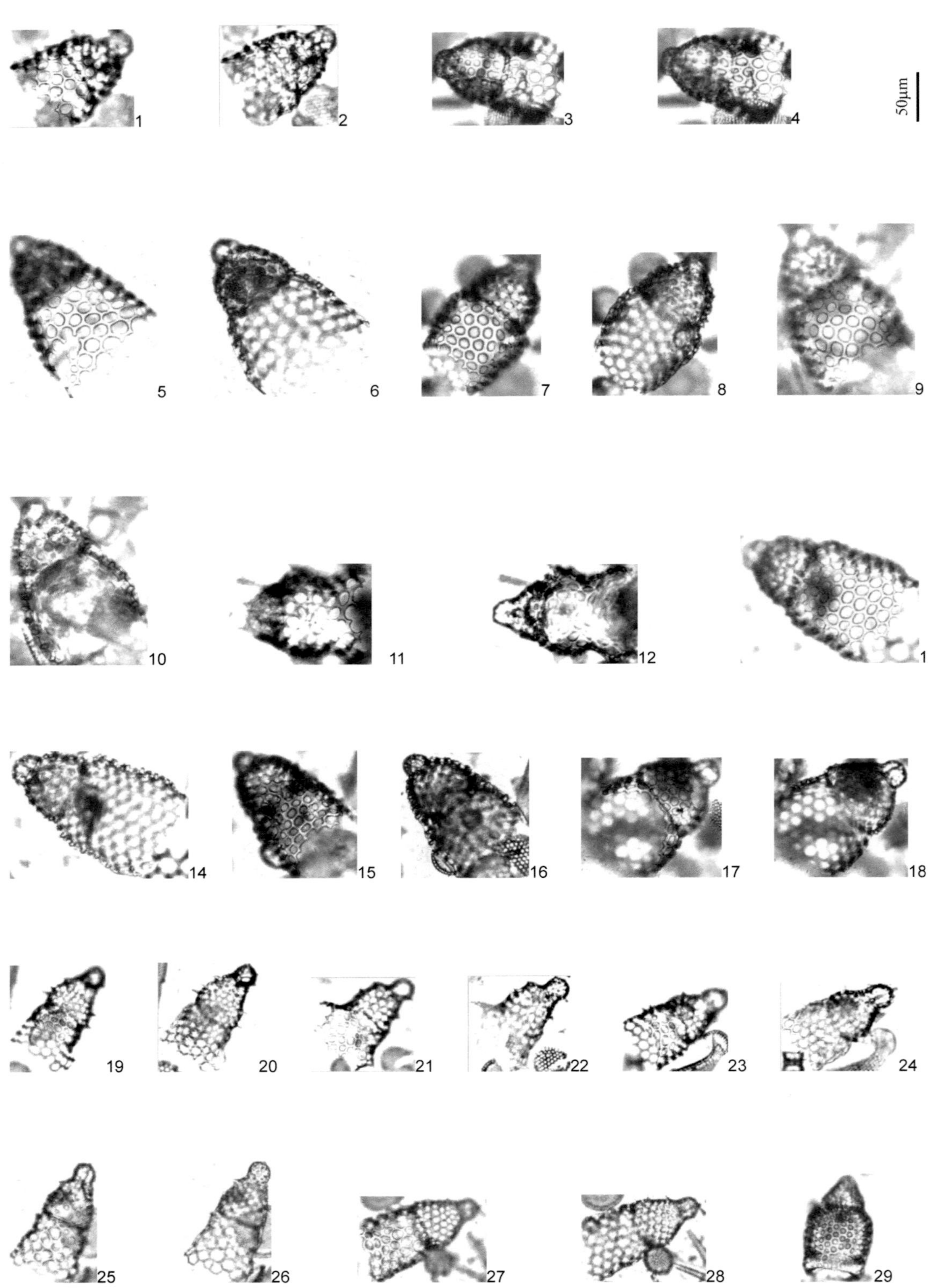

1–18. *Gondwanaria* cf. *semipolita* (Clark and Campbell)；19–28. *Gondwanaria tenuoria* sp. nov.；29. *Gondwanaria* sp. 1

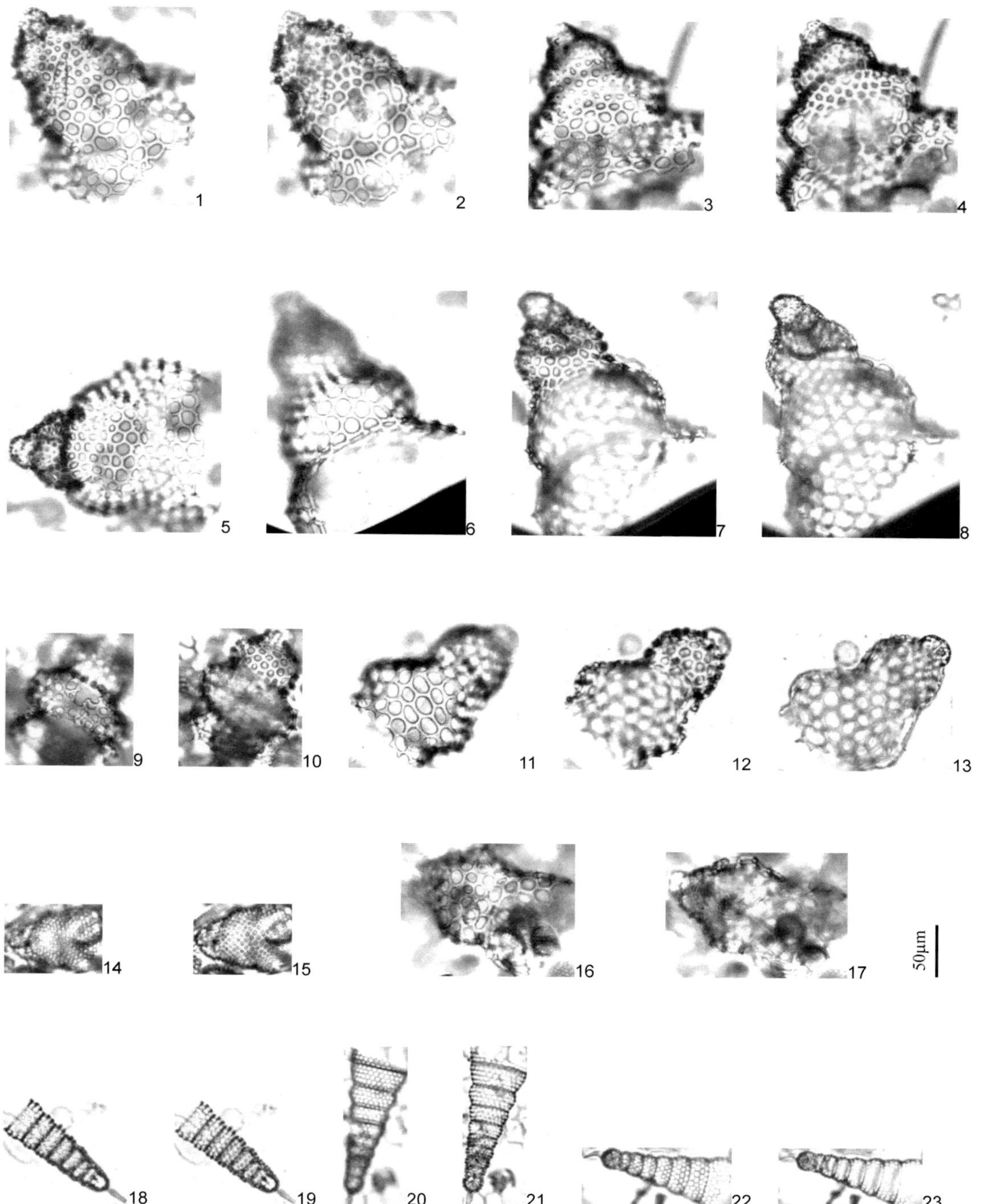

1–13. *Artophormis* sp. cf. *A. gracilis* Riedel；14–15. *Artophormis* sp. 1；16–17. *Artophormis* sp. 2；18–23. *Cyrtopera laguncula* Haeckel

图版 144

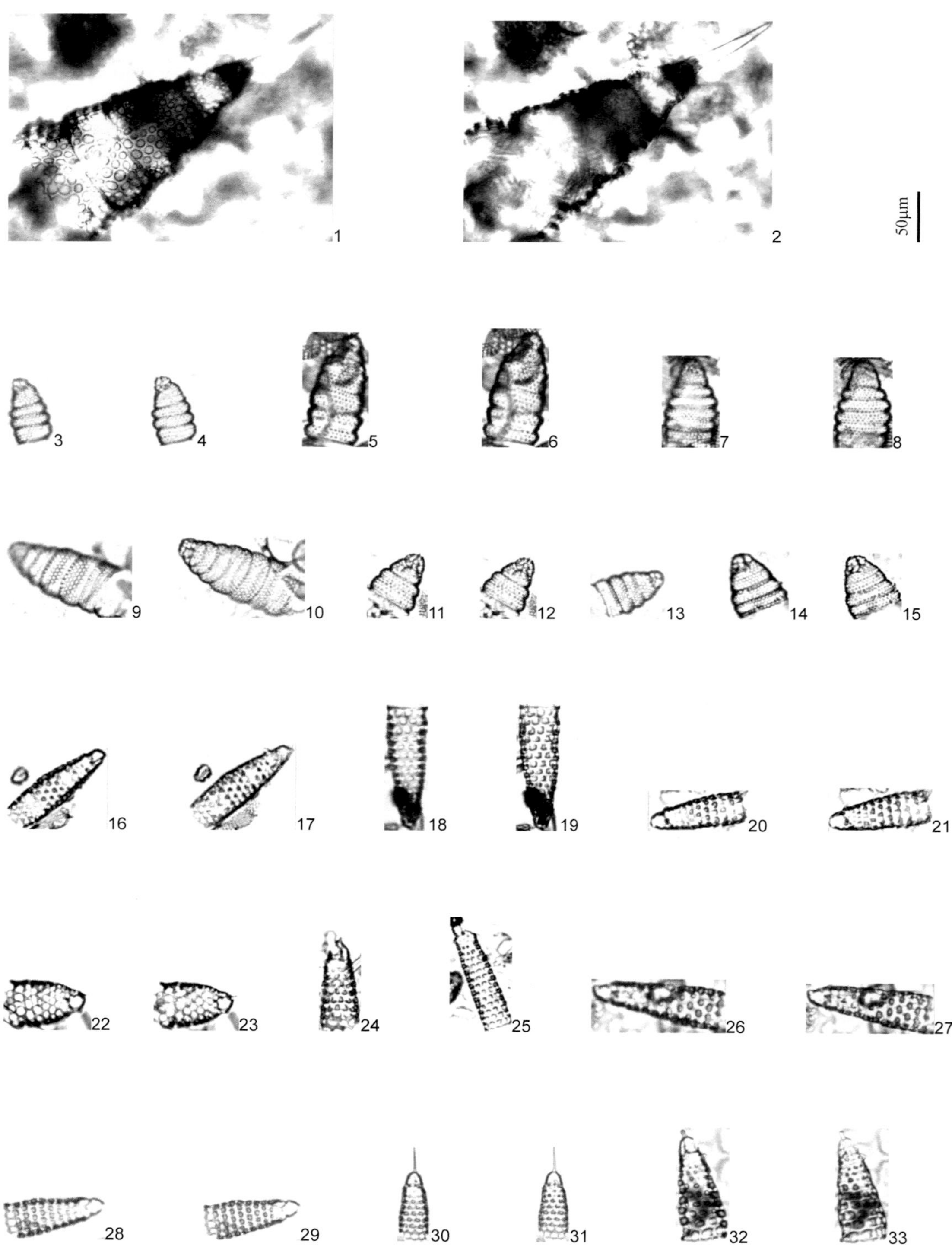

1–2. *Cyrtopera magnifica* sp. nov.；3–15. *Lithostrobus* cf. *longus* Grigorjeva；16–33. *Artostrobus annulatus* (Bailey)

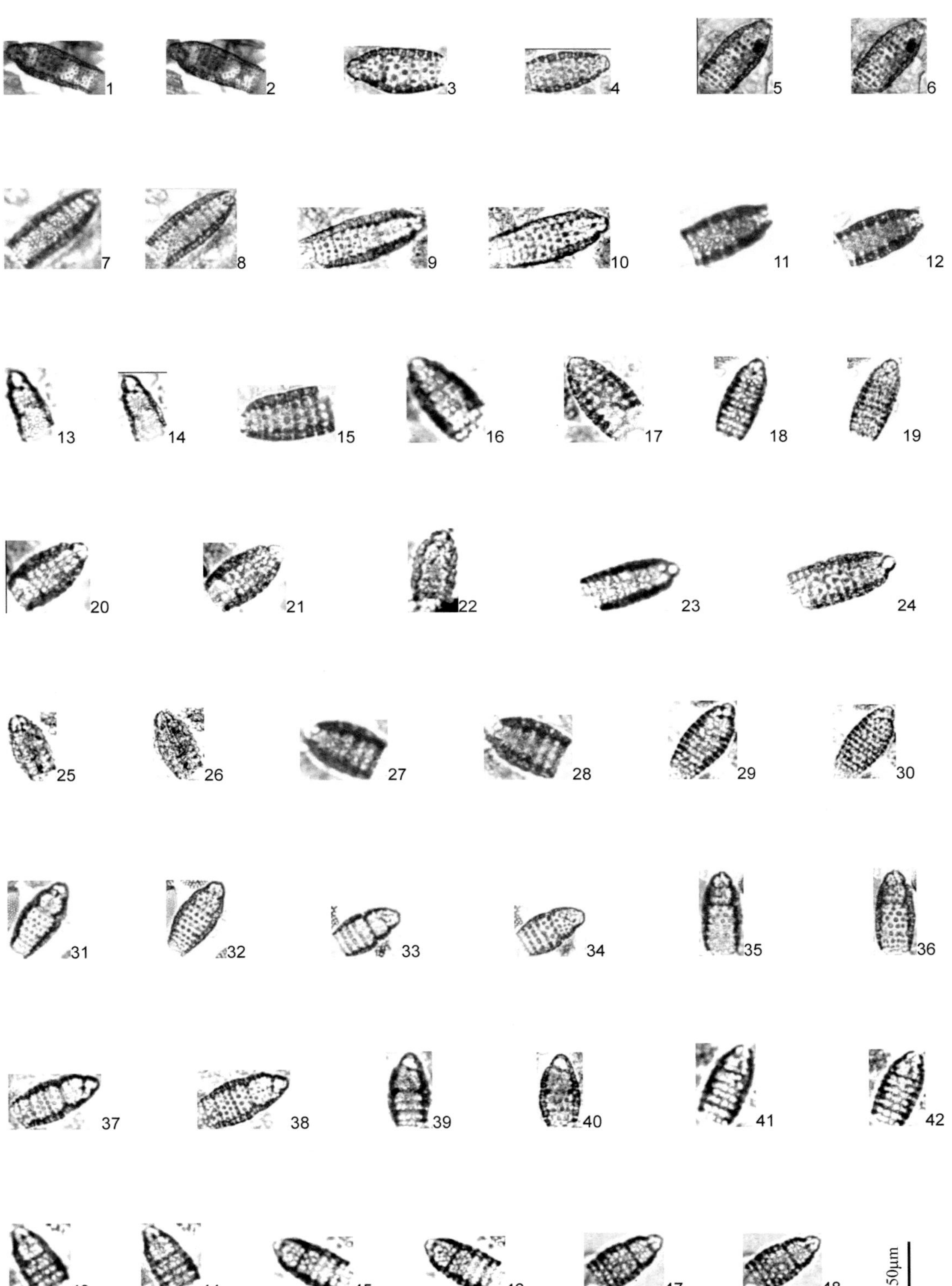

1–30. *Artostrobus missilis* (O'Connor)；31–48. *Artostrobus multiartus* sp. nov.

图版 146

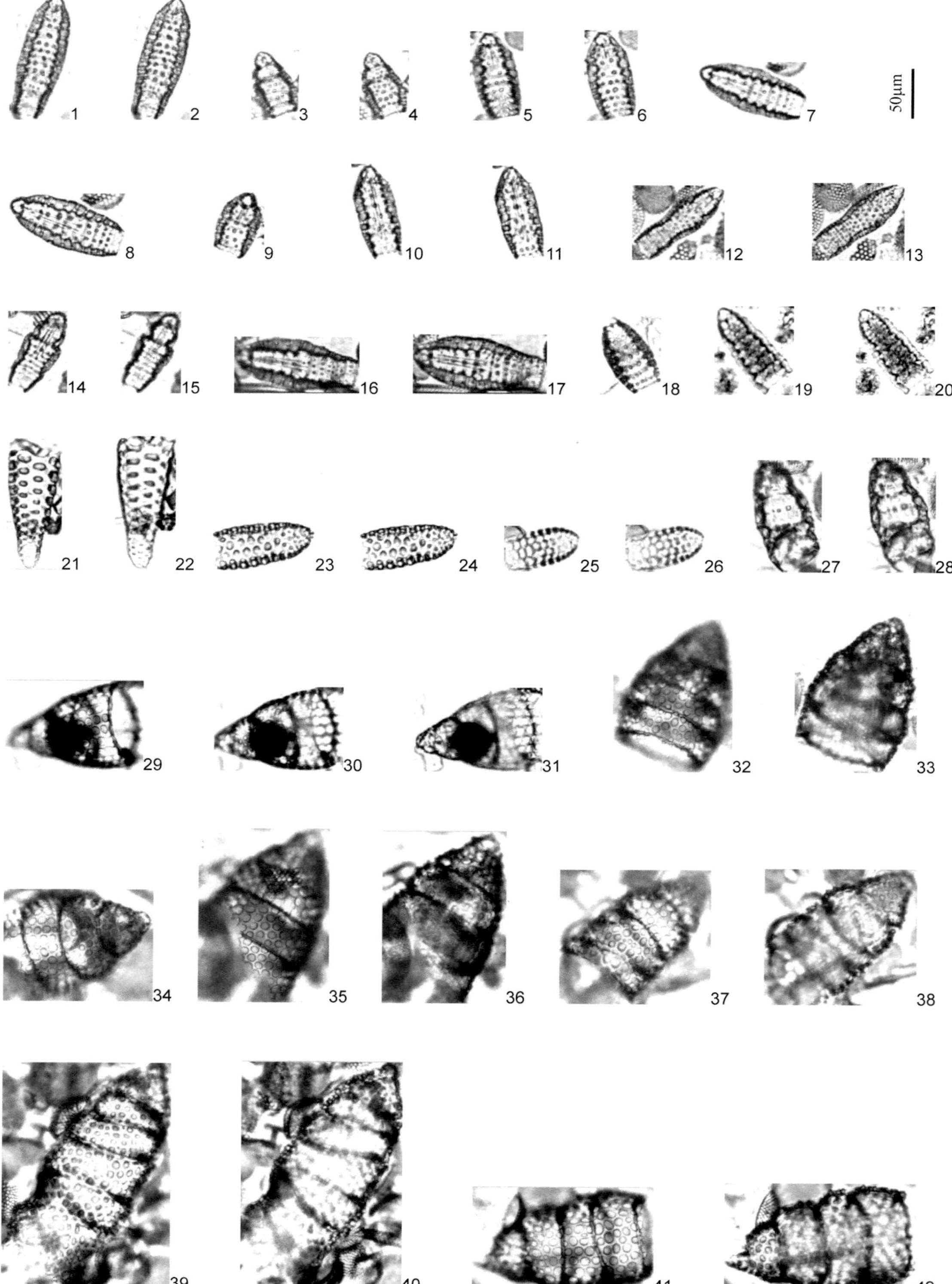

1–20. *Artostrobus pachyderma* (Ehrenberg) group；21–26. *Artostrobus*? *pretabulatus* Petrushevskaya；27–28. *Artostrobus* sp. 1；29–36. *Eucyrtidium annulatum* (Popofsky)；37–42. *Eucyrtidium anomurum* sp. nov.

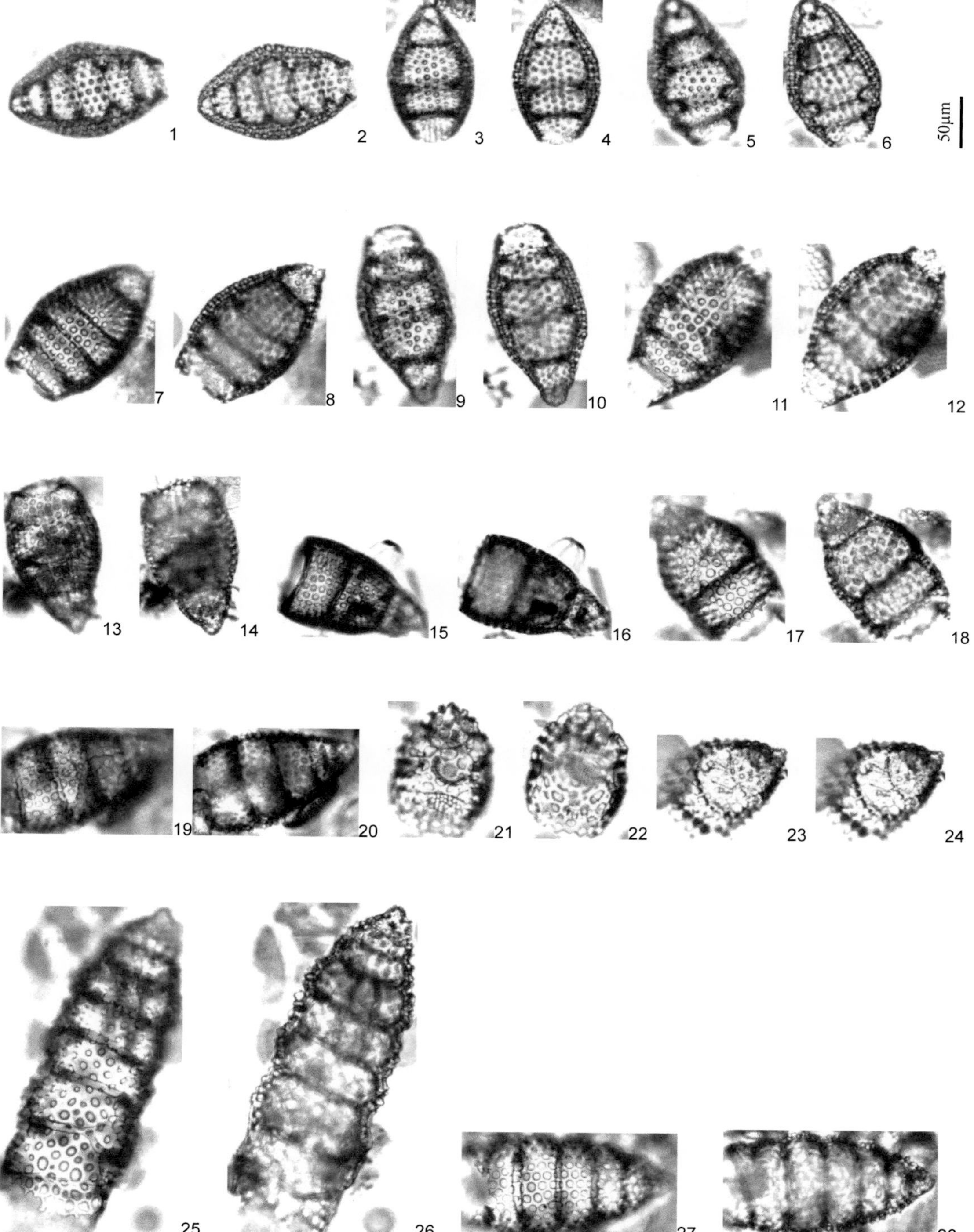

1–12. *Eucyrtidium calvertense* Martin；13–20. *Eucyrtidium cienkowskii* Haeckel group；21–24. *Eucyrtidium gelidium* sp. nov.；25–28. *Eucyrtidium granulatum* (Petrushevskaya) group

图版 148

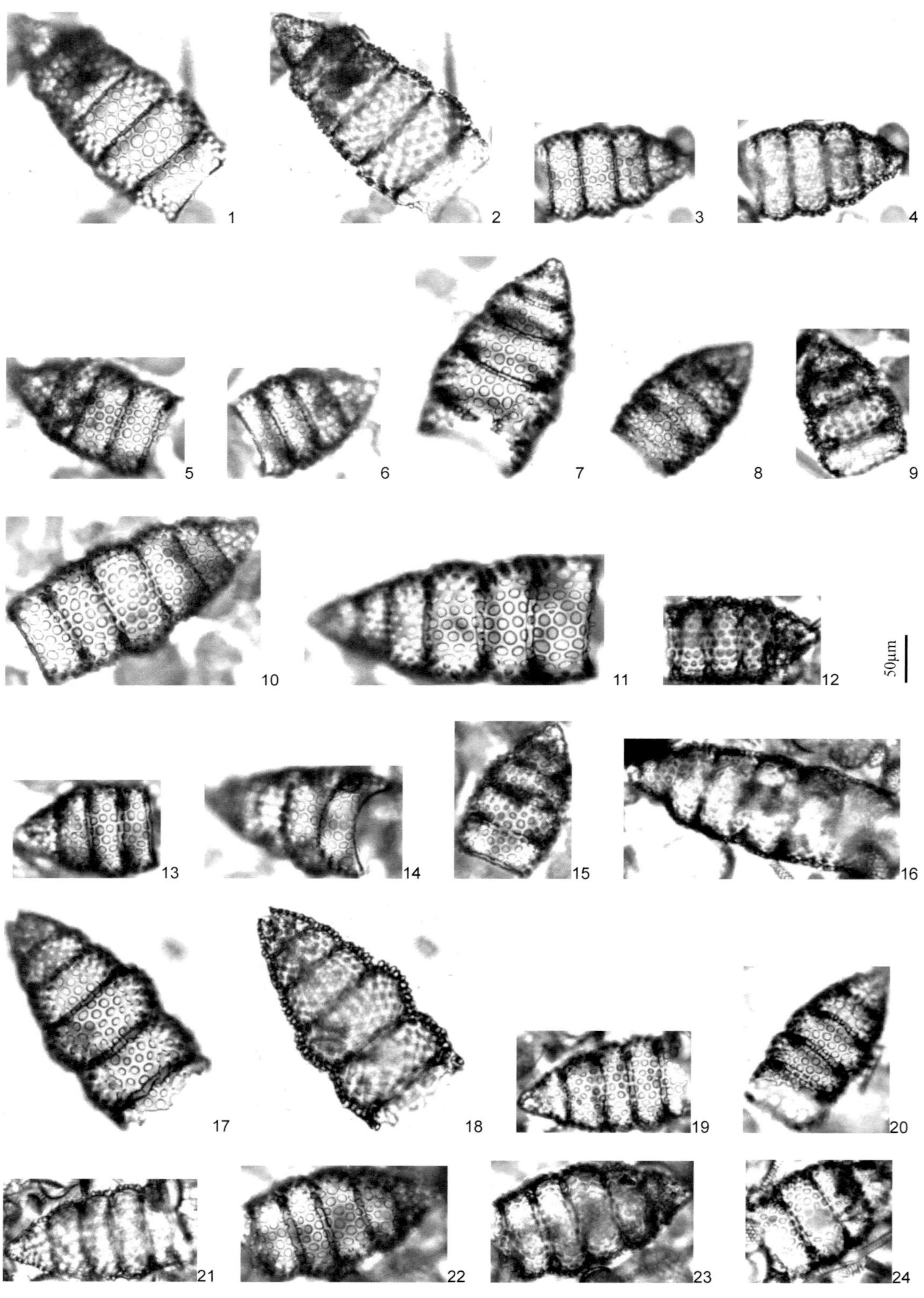

1–24. *Eucyrtidium granulatum* (Petrushevskaya) group

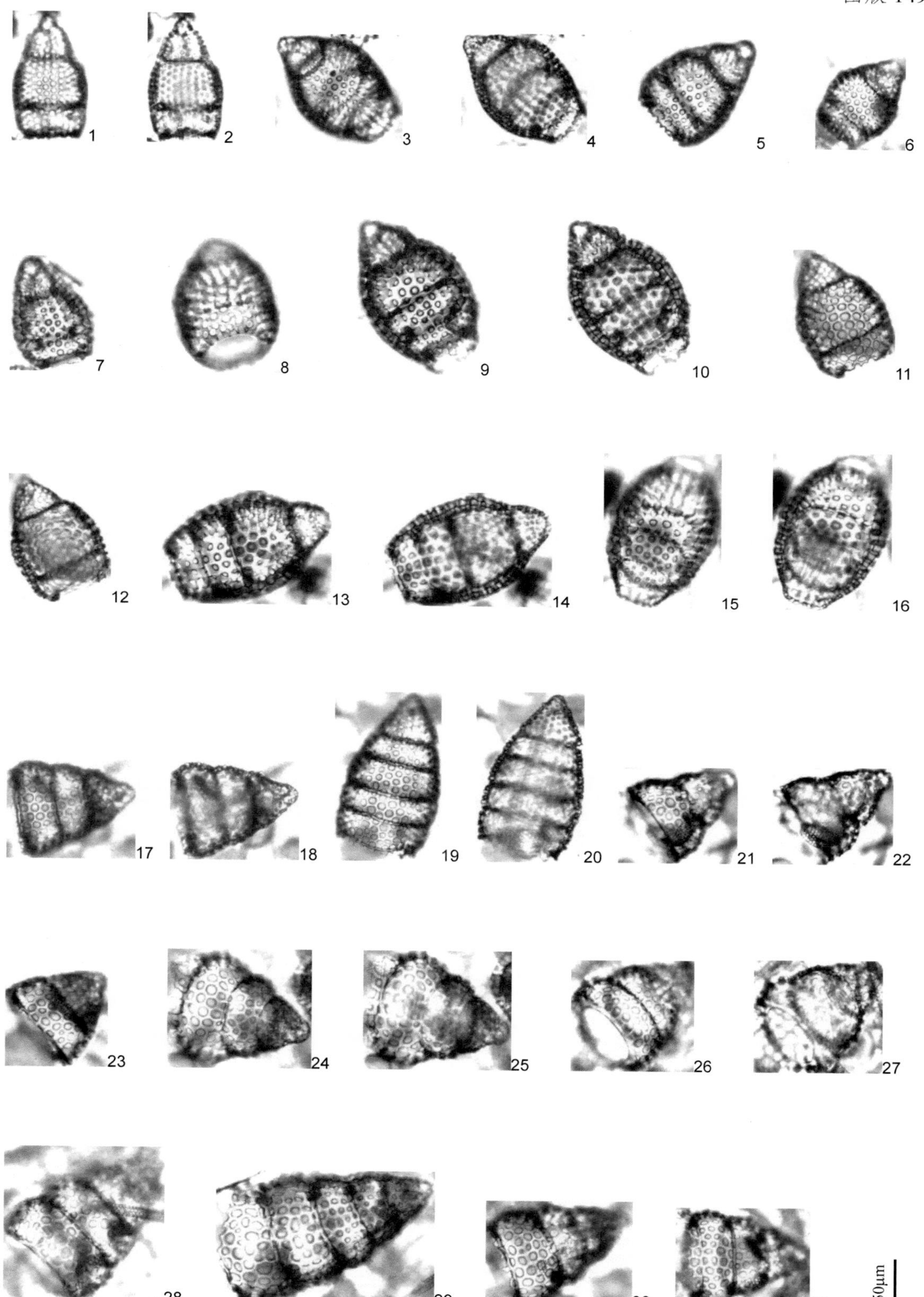

1–16. *Eucyrtidium punctatum* (Ehrenberg) group；17–31. *Eucyrtidium lepidosa* (Kozlova)

图版 150

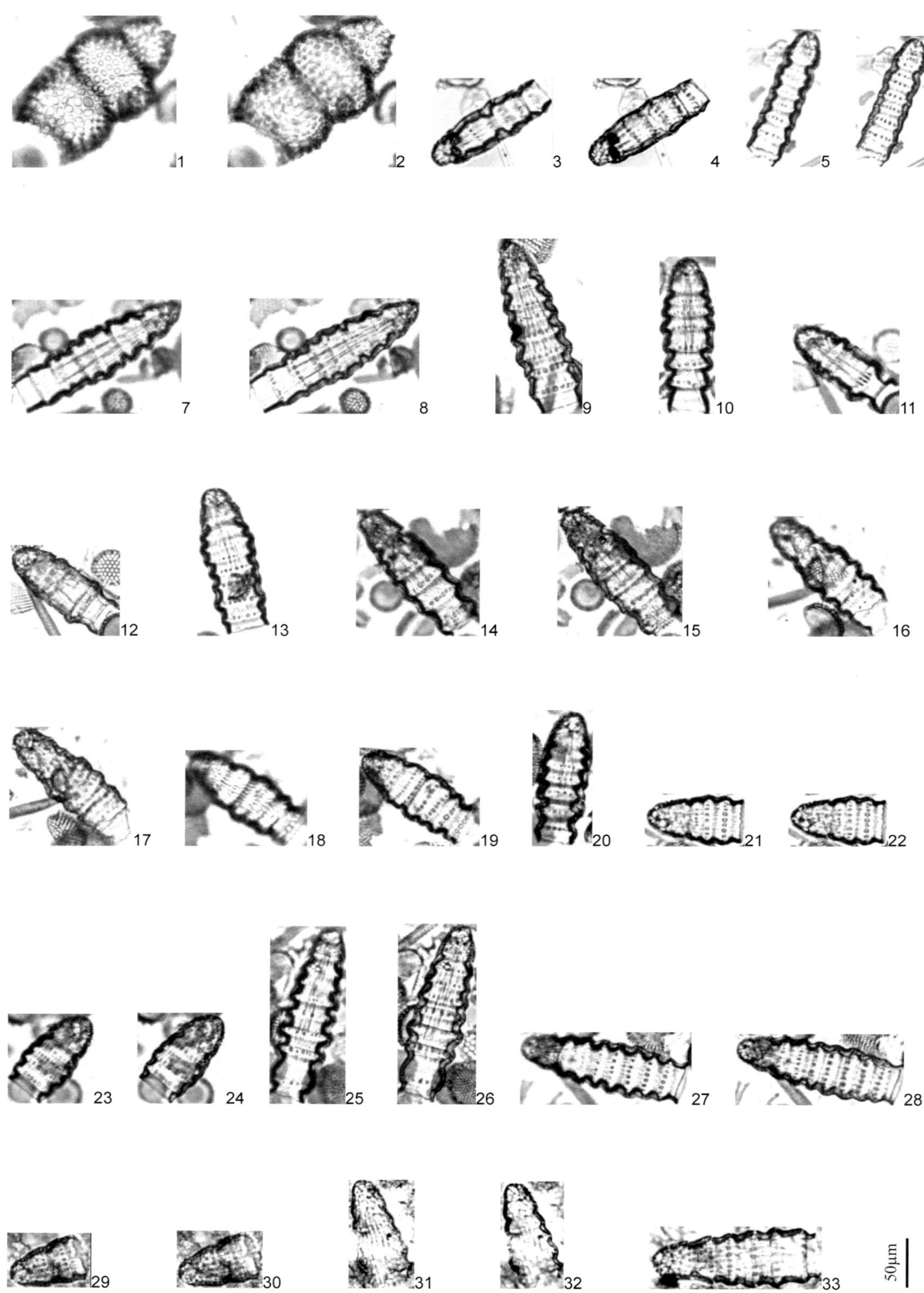

1–2. *Eucyrtidium* sp. 1；3–28. *Lithomitra lineata* (Ehrenberg) group；29–33. *Siphocampe altamiraensis* (Campbell and Clark)

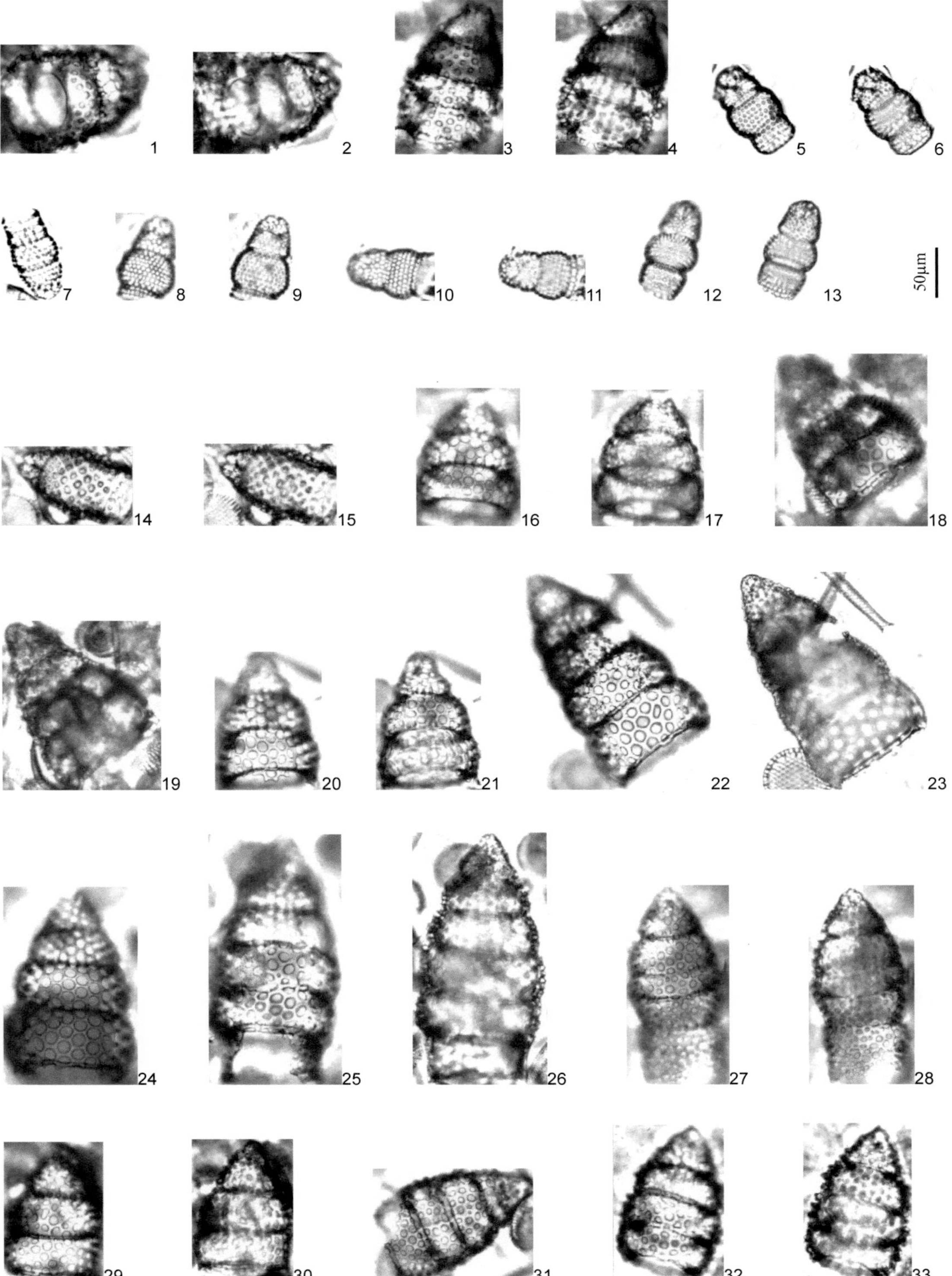

1–4. *Siphocampe elizabethae* (Clark and Campbell)； 5–13. *Siphocampe mesinflatis* sp. nov.；
14–15. *Siphocampe quadrata* (Petrushevskaya and Kozlova)； 16–33. *Lithocampe subligata* Stöhr group

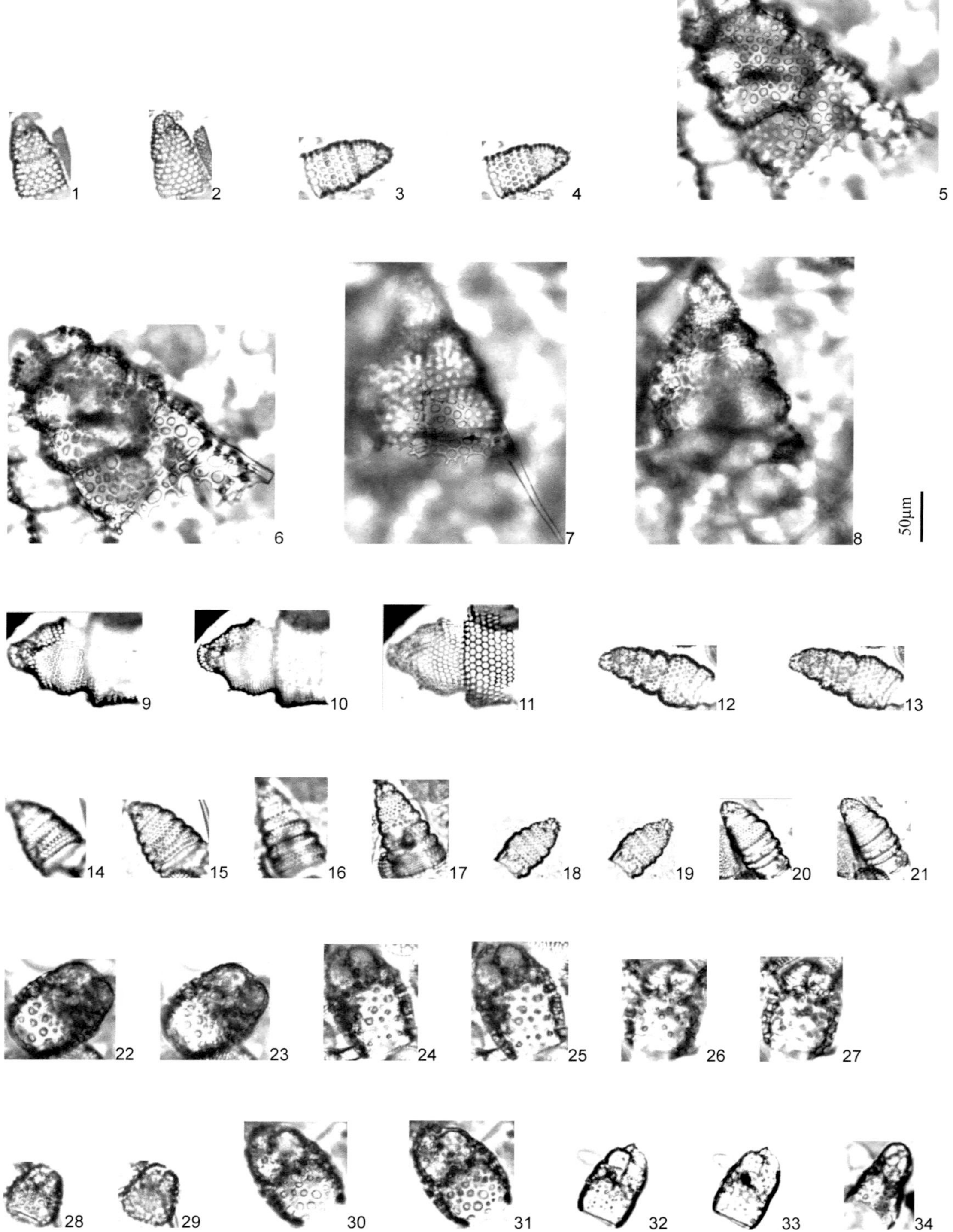

1–4. *Spirocyrtis bellulis* sp. nov.；5–8. *Spirocyrtis*? aff. *gyroscalaris* Nigrini；9–11. *Spirocyrtis rectangulis* sp. nov.；12–21. *Spirocyrtis scalaris* Haeckel；22–27. *Saccospyris antarctica* Haecker；28–29. *Saccospyris conithorax* Petrushevskaya；30–31. *Saccospyris dictyocephalus* (Haeckel) group；32–34. *Saccospyris* sp. 1

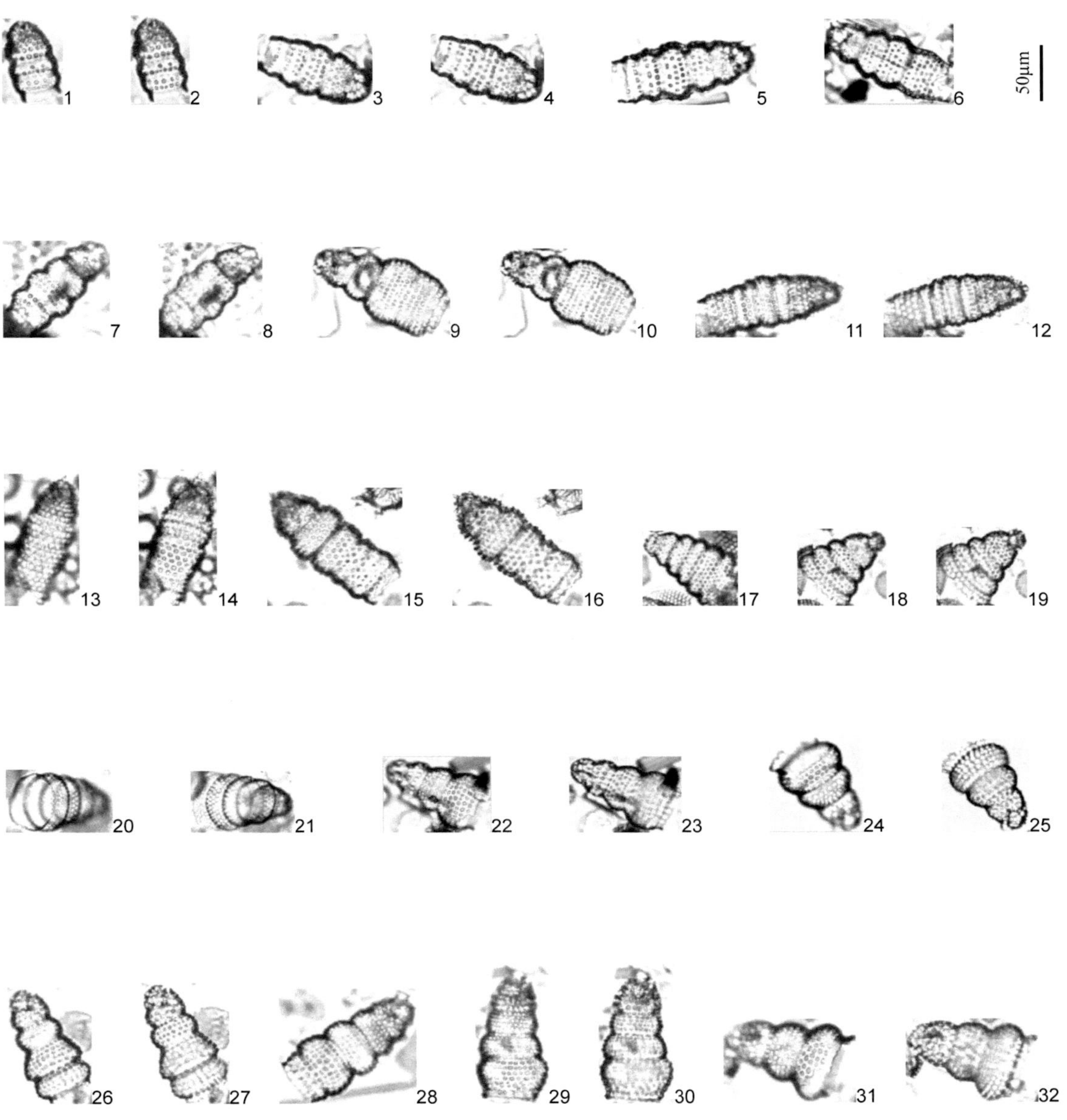

1–8. *Artostrobium auritum* (Ehrenberg) group；9–16. *Artostrobium rhinoceros* Sanfilippo and Riedel；
17–21. *Artostrobium undulatum* (Popofsky)；22–32. *Botryostrobus bramlettei* (Campbell and Clark)

图版 154

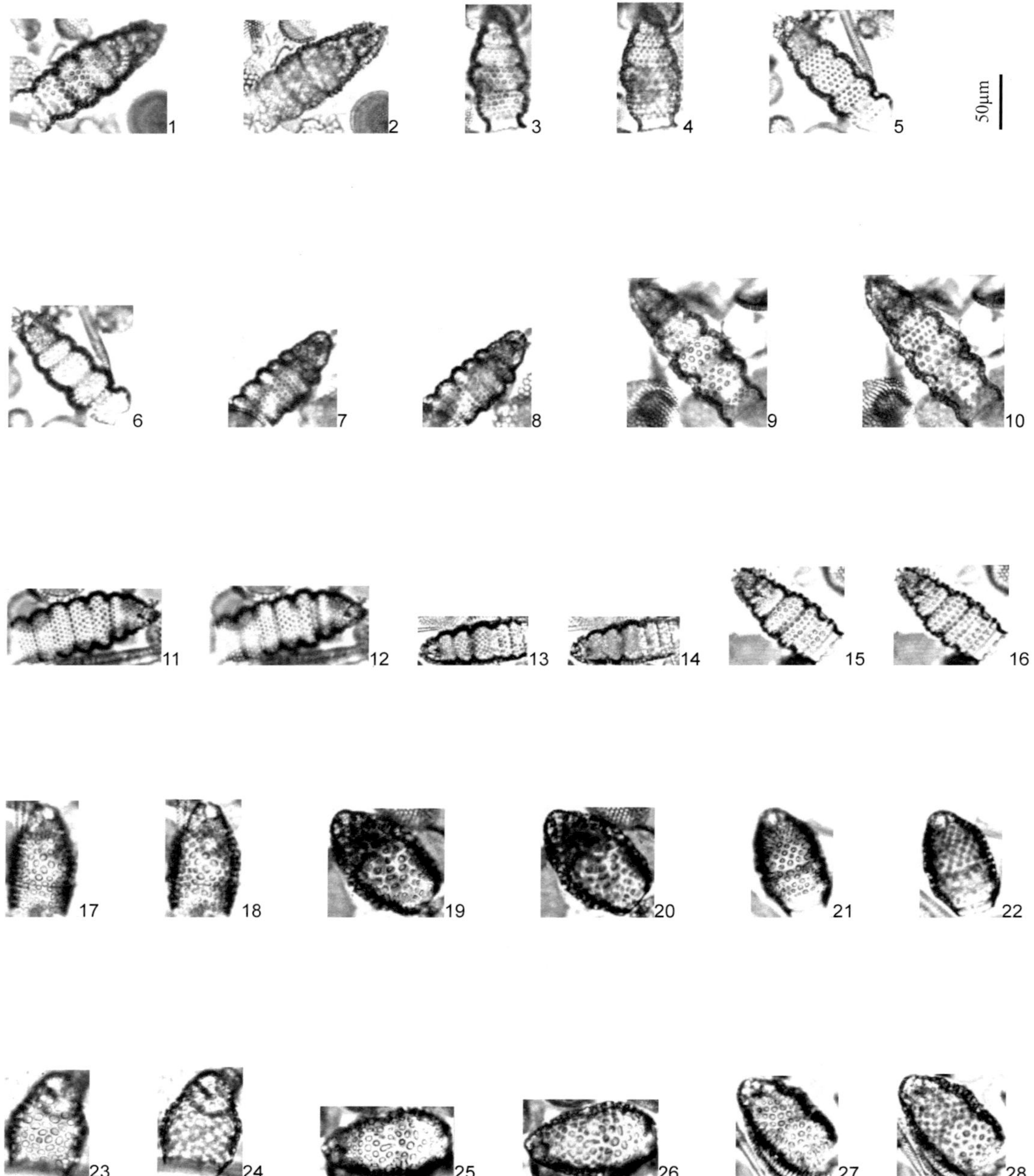

1–16. *Botryostrobus joides* Petrushevskaya；17–28. *Phormostichoartus fistula* Nigrini